TRANSLATIONS OF MATHEMATICAL MONOGRAPHS
VOLUME 52

Boundary Value Problems for Elliptic Pseudodifferential Equations

by G. I. ÈSKIN

American Mathematical Society · Providence · Rhode Island

Boundary Value Problems for Elliptic Pseudodifferential Equations

КРАЕВЫЕ ЗАДАЧИ ДЛЯ ЭЛЛИПТИЧЕСКИХ ПСЕВДОДИФФЕРЕНЦИАЛЬНЫХ УРАВНЕНИЙ

Г. П. ЭСКИН

ПЗДАТЕЛЬСТВО «НАУКА»
МОСКВА 1973

Translated from the Russian by S. Smith

1980 *Mathematics Subject Classification.* Primary 35S15;
Secondary 35S05, 47G05, 58G15.

Abstract. The present monograph is devoted to the class of equations known as pseudodifferential equations, which comprises differential and multidimensional singular integral equations as well as integral equations of the first kind and integrodifferential equations whose kernels have a weak singularity. Mixed boundary value problems for elliptic equations are investigated with the use of the Wiener-Hopf method, the asymptotic behavior of their solutions is obtained and various examples are analyzed. It is intended for mathematicians and theoreticians in continuum mechanics.

Library of Congress Cataloging in Publication Data

Ėskin, Grigoriĭ Il′ich.
Boundary value problems for elliptic pseudodifferential equations.

Translation of Kraevye zadachi dli͡a ėllipticheskikh psevdodifferent͡sial′nykh uravneniĭ.
Bibliography: p.
Includes index.
1. Boundary value problems. 2. Pseudodifferential operators. I. Title.
QA379.E8413 515.3′5 80-39789
ISBN 0-8218-4503-9

TABLE OF CONTENTS

PREFACE TO THE ENGLISH EDITION

The English edition differs from the Russian in that an Introduction and three new sections (§§25–27) have been added. Moreover, various corrections, improvements and remarks have been made throughout the book, especially in Chapter VI.

The author wishes to express his thanks to the American Mathematical Society, for its decision to publish an English edition of his book, and to Dr. S. F. Smith, whose competent translation brought to light a number of inaccuracies in the Russian edition.

PREFACE

The theory of boundary value problems for elliptic pseudodifferential equations has its roots in (i) the theory of multidimensional singular integral equations on a closed manifold (due to Giraud, Mihlin, Calderón and Zygmund, and others; see Mihlin's monograph [65] and the literature cited there), (ii) the theory of boundary value problems for elliptic differential equations with differential and pseudodifferential boundary conditions (due to Agmon, Douglis, Nirenberg, Browder, Lopatinskiĭ, Slobodeckiĭ, Peetre, Hörmander, Šapiro, Schechter, Agranovič, A. S. Dynin and others; see the survey by Agranovič [4]), and (iii) the theory of integral equations of convolution type on a halfline and singular integral equations on an open contour (see the works of Kreĭn, Gohberg and their students [50], [51], [60] and the monographs of Mushelišvili [67], Gahov [45] and Noble [70]).

In 1962 the author, starting from a paper [73] of Peetre and his own paper [35], considered mixed boundary value problems for multidimensional elliptic differential equations of chiefly second order. By reducing such a mixed boundary value problem to a pair of pseudodifferential equations and applying the Wiener-Hopf method, the author proved its solvability in a Sobolev-Slobodeckiĭ space determined by the factorization index of the corresponding symbol; in addition, he found the smoothness and asymptotic behavior of its solution.

These investigations were jointly continued by M. I. Višik and the author. Together they (i) constructed a theory of elliptic pseudodifferential equations and systems on a manifold with boundary that included a theory of multidimensional singular integral equations in domains with a smooth boundary, (ii) found a statement of boundary value problems for elliptic pseudodifferential equations and systems and proved their normal solvability in weighted spaces of functions of piecewise constant and variable order of smoothness, and (iii) distinguished a subclass of pseudodifferential equations whose solutions for smooth right sides are smooth up to the boundary.

The normal solvability in L_2 of a singular integral operator was independently investigated by Simonenko [85].

The theory of pseudodifferential operators in $\mathbf{R}^n$ and on a manifold without boundary is due to Kohn and Nirenberg [59] and Hörmander [55] (see also the works of Friedrichs, Seeley, Bokobza-Haggiag and Unterberger, M. I. Višik and the author, and others).

Boundary value problems for elliptic pseudodifferential equations in domains with a smooth boundary have also been studied by Shamir [79], Dynin [33], Boutet de Monvel [16], [17], Šubin [95], Dikanskiĭ [28], Šnirel′man, the author and others.

In the course of writing the book most of these investigations underwent an essential revision: sharper and more complete results have been obtained and new, simpler proofs have been found. The book also contains new results that have been obtained by the author and are being published for the first time.

The author wishes to take this opportunity to express his gratitude to his teacher G. E. Šilov for his help and support. The author also sincerely thanks V. B. Lidskiĭ for his constant interest in the work.

M. S. Agranovič, A. S. Dikanskiĭ and B. P. Panejah have read the manuscript and made many useful remarks, for which the author is deeply grateful.

To the memory of my teacher

Georgiĭ Evgen′evič Šilov

INTRODUCTION

In the present book we investigate two main types of problems:

1) mixed boundary value problems for elliptic differential equations; and

2) pseudodifferential equations in bounded domains of $\mathbf{R}^n$.

We are chiefly concerned with the determination of conditions (involving, in particular, the selection of function spaces) under which a given problem is normally solvable (i.e. the homogeneous equation has a finite number of linearly independent solutions and the inhomogeneous equation is solvable under the fulfillment of a finite number of "orthogonality" conditions on the right side).

Let G be a domain in $\mathbf{R}^n$ with a smooth boundary Γ and suppose Γ is divided into two parts Γ_1 and Γ_2 by a smooth $(n-2)$-dimensional surface γ. An example of a mixed boundary value problem is the problem of finding a function $u(x)$ satisfying Laplace's equation

$$\Delta u = 0 \tag{0.1}$$

in G and the boundary conditions

$$u|_{\Gamma_1} = g_1(x'), \tag{0.2}$$

$$\partial u/\partial n|_{\Gamma_2} = g_2(x'), \tag{0.3}$$

where $\partial/\partial n$ is the normal derivative on Γ. A more general mixed boundary value problem is obtained when (0.1) is replaced by an arbitrary elliptic differential equation or system of differential equations, and boundary conditions of a general form are posed on Γ_1 and Γ_2.

In addition to the question of normal solvability, in view of the fact that the solution of a mixed problem is not smooth on γ, we are concerned with determining the asymptotic behavior of the solution of a mixed problem in a neighborhood of γ.

An example of the second type of problem is a multidimensional elliptic singular integral equation (s.i.e.) in $G \subset \mathbf{R}^n$ of the form

$$a(x)u(x) + \text{p.v.} \int_G \frac{k(x, x - y)}{|x - y|^n} u(y)dy = f(x) \tag{0.4}$$

(for the definition of an s.i.e. see §3). A more general example is obtained when (0.4) is replaced by an arbitrary elliptic pseudodifferential equation or system of pseudodifferential equations (for the definition of a pseudodifferential equation (ψ d.e.) see §§3 and 18).

We note that the above two types of problems are interrelated inasmuch as a mixed problem for an elliptic differential equation can be reduced by means of potentials (or parametrices in the case of elliptic differential equations with variable coefficients) to a ψ d.e. or system of ψ d.e.'s on Γ_1 (or Γ_2). But it is usually simpler to investigate both types of problems independently.

Let us briefly outline the theory developed in the book for both types of problems, beginning with the second.

Whereas the ellipticity condition is the only necessary and sufficient condition for the normal solvability in L_2 of an s.i.e. on a manifold without boundary, additional normal solvability characteristics exist in the case of an s.i.e. on a manifold with boundary (or in a bounded domain of $\mathbf{R}^n$). In particular, every elliptic pseudodifferential operator in a bounded domain G has a normal solvability characteristic $\kappa(x')$ associated with it that is an infinitely differentiable function on $\Gamma = \partial G$ (see §§21 and 22). An elliptic equation of form (0.4), for example, is normally solvable in $L_2(G)$ if and only if $|\text{Re}\, \kappa(x')| < 1/2$ on Γ. In general, $\kappa(x')$ might turn out to be any smooth complex valued function on Γ. If the oscillation of Re $\kappa(x')$ on Γ is less than 1, then it is clearly possible to choose s so that $|s - \text{Re}\, \kappa(x')| < 1/2$, in which case equation (0.4) is normally solvable for $u(x) \in \mathring{H}_s(G)$ and $f(x) \in H_s(G)$ (for the definition of the spaces $\mathring{H}_s(G)$ and $H_s(G)$ see §§4 and 21). If, on the other hand, the oscillation of Re $\kappa(x')$ on Γ is greater than or equal to 1, then (0.4) (or a more general pseudodifferential equation) is normally solvable only for spaces of a piecewise constant or variable order of smoothness (see §25).

A slightly more detailed picture is available in the special case when $\kappa(x')$ is identically equal to a positive integer m. In this case (0.4), despite the fact that it is an integral equation, behaves like an elliptic differential equation of order $2m$ in the sense that it is normally solvable (i) for $u(x) \in \mathring{H}_m(G)$ (i.e. when $u(x)$ vanishes on Γ together with its normal derivatives up to order $m - 1$ inclusively) and (ii) in $L_2(G)$ only if m pseudodifferential boundary conditions are assigned on Γ (see §§7 and 8).

Suppose now $\kappa(x')$ is identically equal to a negative integer m. In this case the behavior of equation (0.4) does not have an analog in the class of differential equations but is similar to the behavior of, for example, the integral equation

$$\int_G |x - y|^{2-n} u(y)\, dy = f(x), \qquad n \geqslant 3, \tag{0.5}$$

for which $\kappa(x') \equiv -1$ and whose kernel $|x - y|^{2-n}$ is, to within a constant factor, a fundamental solution of Laplace's equation.

Let us consider (0.5) in more detail. Clearly it cannot have a solution $u(x) \in L_2(G)$ when $f(x)$ is a nontrivial smooth harmonic function in $\overline{G}$, since an application of the Laplacian Δ to (0.5) gives us $u = C_n \Delta f = 0$ (C_n being a constant), which is impossible when $f \not\equiv 0$. On the other hand, whenever $f(x)$ is a smooth function in $\overline{G}$, equation (0.5) has a solution of class $\mathring{H}_{-1}(G)$ of the form

$$u = v(x) + \rho(x')\delta(\Gamma), \tag{0.6}$$

where $v(x)$ is a smooth function in G, $\rho(x')$ is a smooth function on Γ and $\delta(\Gamma)$ is the delta function of the surface Γ in $\mathbf{R}^n$. By substituting (0.6) for u, we can convert (0.5) into an equation of the form

$$\int_G |x - y|^{2-n} v(y)\, dy + \int_\Gamma |x - y'|^{2-n} \rho(y')\, dS_{y'} = f(x), \tag{0.7}$$

where $dS_{y'}$ is the surface element of Γ (see §§7 and 21). (0.7) can be regarded as an equation for determining the two unknown functions $v(x)$, $x \in G$, and $\rho(x')$, $x' \in \Gamma$, from the known function $f(x)$. It can be proved that such an equation is normally solvable in L_2.

An operator of the form $\int_\Gamma |x - y'|^{2-n} \rho(y') dS_{y'}$ that takes a function on Γ into a function defined in G is called an *operator of potential type* or a *coboundary operator*. Analogously to (0.7), equation (0.4) for $\kappa(x') \equiv m < 0$ can be converted into a normally solvable equation for $u \in L_2(G)$ by adding to its left side $|m|$ operators of potential type with unknown densities $v_1(x'), \ldots, v_{|m|}(x')$ on Γ, the problem then being to determine $u(x)$ and $v_1(x'), \ldots, v_{|m|}(x')$ from a given function $f \in L_2(G)$ (see §§7 and 8).

The general normally solvable boundary value problem for equation (0.4) (or a general elliptic system of ψ d.e.'s in G) is constructed as follows: a certain number m_- of operators of potential type with unknown densities on Γ are added to (0.4), and a certain number m_+ of boundary conditions are given on Γ (see §§12 and 16). We note that, whereas a normally solvable boundary value problem can be constructed for a scalar elliptic ψ d.e. (through the choice of function spaces) without introducing potentials or boundary conditions, this is no longer so in the case of an elliptic system of ψ d.e.'s.

An interesting question that arises here is, When can a normally solvable boundary value problem be constructed for a given elliptic system of ψ d.e.'s? An answer to this question is given in topological terms in subsections 2 and 3 of §16 after the presentation of some necessary information from the theory of vector bundles and K-theory.

The investigation of boundary value problems for elliptic systems of ψ d.e.'s is based on a detailed study (in §§14, 15 and 17) of a system of ψ d.e.'s on a halfline, which is converted by the Fourier transform into a system of s.i.e.'s on a line whose symbol generally has a discontinuity at infinity. Also in §17 we construct the inverse operator of the general boundary value problem in a halfspace.

One of the important questions in the theory of ψ d.e.'s in a domain is that of the smoothness of a solution near a boundary. The asymptotic behavior near a boundary of a solution of an elliptic ψ d.e. is studied in §§9 and 26. In §10 (see also §23) we distinguish a subclass of ψ d.e.'s whose solutions are smooth right up to the boundary of a domain G when their right sides are smooth in $\overline{G}$. Such a pseudodifferential operator is said to be *smooth in a domain* or to satisfy the *transmission condition*. Boundary value problems for smooth pseudodifferential operators in a domain are considered in §§11 and 23.

Mixed boundary value problems for elliptic differential equations are considered in §§13 and 24. In the case of an elliptic equation of second order it is possible to select the function spaces (which, in general, are spaces with a weight factor vanishing on γ that ensure the smoothness of a solution everywhere off γ) so that a mixed problem of form (0.1)–(0.3) is normally solvable. Such a selection is not always possible, however, in the case of elliptic systems or elliptic equations of order greater than two. For this reason we consider in §24 a generalized mixed boundary value problem, which is constructed by adding operators of potential type with unknown densities on γ to the boundary conditions on Γ_1 and Γ_2 and assigning additional boundary conditions on γ.

To make the exposition selfcontained we have also presented, together with proofs and numerous examples, some of the fundamental theory on which the investigations of the book are based. In §§1–3 we give an account of the theory of slowly increasing generalized functions (essentially tempered distributions) and their Fourier transforms that, in the main, follows the monographs [49] of Gel′fand and Šilov. Here particular attention is paid to homogeneous generalized functions. The theory of Sobolev-Slobodeckiĭ and related spaces is set forth in §4. In §§5 and 6 we study integrals of Cauchy type and the factorization of symbols; these topics underlie the factorization (or Wiener-Hopf) method, which is used throughout the book. In §§18, 19 and 21 we present the theory of pseudodifferential operators with symbols having infinite smoothness in x and a finite smoothness in ξ.

In the final chapter (§§26 and 27) we consider applications of our theory to the solution of strongly elliptic ψ d.e.'s by the finite element method and to the construction of an asymptotic solution of an elliptic ψ d.e. with a small parameter. We note that the problems considered in §§26 and 27 are of interest in mechanics and, in particular, the theory of elasticity.

CHAPTER I

GENERALIZED FUNCTIONS AND THE FOURIER TRANSFORM

§1. Generalized functions

1. *The space S.* The space $S = S(\mathbf{R}^n)$ is defined as the totality of all infinitely differentiable functions $\varphi(x)$ in the n-dimensional space $\mathbf{R}^n$ that together with all of their derivatives decrease more rapidly than any negative power of $|x|$ as $|x| = \sqrt{x_1^2 + \ldots + x_n^2} \to \infty$. For an arbitrary function $\varphi(x) \in S$ we let

$$|[\varphi]|_m = \max_x (1 + |x|)^m \sum_{|p| \leqslant m} \left| \frac{\partial^p \varphi(x)}{\partial x^p} \right|, \qquad 0 \leqslant m < \infty, \tag{1.1}$$

where the multi-index $p = (p_1, \ldots, p_n)$ is an n-tuple of nonnegative integers, $|p| = p_1 + \ldots + p_n$ and

$$\frac{\partial^p \varphi(x)}{\partial x^p} = \frac{\partial^{p_1 + \ldots + p_n} \varphi(x)}{\partial x_1^{p_1} \ldots \partial x_n^{p_n}}.$$

The topology of S is given by the norms (1.1); in particular, a sequence $\varphi_n(x) \in S$ converges to $\varphi(x) \in S$ if $|[\varphi - \varphi_n]|_m \to 0$ for $m = 0, 1, \ldots$.

Let $C_0^\infty(\mathbf{R}^n)$ be the totality of all infinitely differentiable functions with compact support in $\mathbf{R}^n$. Clearly, $C_0^\infty(\mathbf{R}^n) \subset S$.

LEMMA 1.1. *The space C_0^∞ is dense in S in the topology of S.*

PROOF. Suppose $\chi(x) \in C_0^\infty$ and $\chi(x) \equiv 1$ for $|x| \leqslant 1$. For an arbitrary $\varphi(x) \in S$ we put $\varphi_\varepsilon(x) = \chi(\varepsilon x)\varphi(x)$. Then $\varphi_\varepsilon(x) \in C_0^\infty$ and, as is easily seen, $|[\varphi - \varphi_\varepsilon]|_m \to 0$ as $\varepsilon \to 0$ for $m = 0, 1, \ldots$. Lemma 1.1 is proved.

By the *convolution* $\varphi * \psi$ of a pair of functions $\varphi(x), \psi(x) \in S$ is meant the function

$$\varphi * \psi = \int_{-\infty}^{\infty} \varphi(x - y)\psi(y)dy. \tag{1.2}$$

Making the change of variables $x - y = z$, we get

$$\varphi * \psi = \int_{-\infty}^{\infty} \varphi(z)\psi(x - z)dz = \psi * \varphi. \tag{1.3}$$

Clearly,

$$\frac{\partial^p}{\partial x^p}(\varphi * \psi) = \frac{\partial^p \varphi}{\partial x^p} * \psi = \varphi * \frac{\partial^p \psi}{\partial x^p}. \tag{1.4}$$

The convolution operation can be defined on a significantly wider class of functions than S.

LEMMA 1.2. *Suppose $f(x)$ is a measurable function such that $|f(x)| \leqslant C(1 + |x|)^t$, and $\varphi(x) \in S$. Then the convolution $f * \varphi \in C^\infty$ and satisfies the estimate*

$$\left|\frac{\partial^p}{\partial x^p}(f * \varphi)\right| \leqslant C|[\varphi]|_m(1 + |x|)^t, \qquad 0 \leqslant |p| < \infty, \tag{1.5}$$

where $m = \max(|p|, |t| + n + 1)$ and C does not depend on φ.

PROOF. Clearly, properties (1.3) and (1.4) remain in force for $f * \varphi$, so that

$$\frac{\partial^p}{\partial x^p}(f * \varphi) = \int_{-\infty}^{\infty} \frac{\partial^p \varphi(x - y)}{\partial x^p} f(y)dy. \tag{1.6}$$

On the other hand, since $\varphi(x) \in S$, we have

$$|\partial^p \varphi(x)/\partial x^p| \leqslant |[\varphi]|_m(1 + |x|)^{-|t|-n-1},$$

where $m = \max(|p|, |t| + n + 1)$.

The following inequality holds for any real t:

$$(1 + |x - y|)^{-|t|} \leqslant (1 + |x|)^t / (1 + |y|)^t. \tag{1.7}$$

In fact, suppose for definiteness that $t > 0$. Then

$$1 + |y| \leqslant 1 + |y - x| + |x| \leqslant (1 + |x - y|)(1 + |x|). \tag{1.8}$$

Taking the tth power of each side of (1.8), we obtain (1.7). When $t < 0$, (1.7) is obtained by taking the $|t|$th power of each side of the inequality $1 + |x| \leqslant (1 + |x - y|)(1 + |y|)$.

Thus, from (1.6) and (1.7) we get

$$\left|\frac{\partial^p}{\partial x^p}(f * \varphi)\right| \leqslant C\int_{-\infty}^{\infty} \frac{(1 + |y|)^t |[\varphi]|_m \, dy}{(1 + |x - y|)^{|t|+n+1}} \leqslant C|[\varphi]|_m(1 + |x|)^t,$$
$$0 \leqslant |p| < \infty. \tag{1.9}$$

COROLLARY 1.1. *If $\varphi \in S$ and $\psi \in S$, then $\varphi * \psi \in S$.*

In fact, in this case it is possible to take $t = -1, -2, \ldots$ in (1.5). Consequently, $\varphi * \psi \in S$, with

$$\|[\psi * \varphi]\|_m \leqslant C_m \|[\varphi]\|_{m+n+1}, \qquad m = 0, 1, \ldots.$$

2. *Generalized functions.* A functional f on S is said to be *semilinear* if

$$1) \quad (f, \alpha_1\varphi_1 + \alpha_2\varphi_2) = \bar{\alpha}_1(f, \varphi_1) + \bar{\alpha}_2(f, \varphi_2) \tag{1.10}$$

for any $\varphi_1, \varphi_2 \in S$ and any complex numbers α_1 and α_2 (the value of a functional f at a function $\varphi \in S$ is denoted by (f, φ)) and *continuous* if for any convergent sequence $\varphi_n(x) \to \varphi(x)$ in S

$$(f, \varphi_n) \to (f, \varphi). \tag{1.11}$$

A continuous semilinear functional on S will be called a *generalized function*. The functions belonging to S will often be called *fundamental functions*.

EXAMPLE 1.1. Suppose $f(x)$ is a locally integrable function in the Lebesgue sense such that for some $N > 0$

$$\int_{-\infty}^{\infty} |f(x)|(1 + |x|)^{-N}\, dx < \infty. \tag{1.12}$$

Then a semilinear functional f on S can be associated with $f(x)$ by means of the formula

$$(f, \varphi) = \int_{-\infty}^{\infty} f(x)\, \overline{\varphi(x)}\, dx. \tag{1.13}$$

Clearly, $|(f, \varphi)| \leqslant C\|[\varphi]\|_N$, so that (1.13) is a continuous functional on S. A functional of form (1.13) will be called a *regular functional*. It is known (see [43]) that if $f_1(x) \neq f_2(x)$ almost everywhere, then there exists a $\varphi_0 \in S$ such that $(f_1, \varphi_0) \neq (f_2, \varphi_0)$, i.e. the functionals corresponding to $f_1(x)$ and $f_2(x)$ do not coincide. We will often not distinguish between a regular functional f and the function $f(x)$ corresponding to it, and we will denote a regular functional f by $f(x)$.

EXAMPLE 1.2. By the *delta function* δ is meant the generalized function defined by the formula $(\delta, \varphi) = \overline{\varphi(0)}$, $\forall \varphi \in S$, where $\forall \varphi \in S$ means "for every function belonging to S".

A linear combination $\alpha_1 f_1 + \alpha_2 f_2$ of generalized functions is defined in the same way as a linear combination of functionals:

$$(\alpha_1 f_1 + \alpha_2 f_2, \varphi) = \alpha_1(f_1, \varphi) + \alpha_2(f_2, \varphi), \qquad \forall \varphi \in S. \tag{1.14}$$

The space of generalized functions will be denoted by $S' = S'(\mathbf{R}^n)$.

REMARK 1.1. The class S' of generalized functions defined above is the class of so-called slowly increasing generalized functions. A wider class of generalized functions (viz. D' or K') is obtained if $C_0^\infty(\mathbf{R}^n)$ with the appropriate topology (see [21]) is taken instead of S as the space of test functions. However, the functionals on S will be sufficient for our purposes.

LEMMA 1.3. *Suppose* $f \in S'$. *Then there exists an integer* $m(f) \geqslant 0$ *such that*

$$|(f, \varphi)| \leqslant C_m \|[\varphi]\|_m \tag{1.15}$$

for any $\varphi \in S$.

PROOF. Suppose the contrary. Then there exists a sequence of functions $\varphi_m \in S$ such that $|(f, \varphi_m)| \geqslant m\|[\varphi_m]\|_m$. Let $\psi_m = \varphi_m / m\|[\varphi_m]\|$. Then $\|[\psi_m]\|_i = [\varphi_m]_i / m\|[\varphi_m]\|_m \leqslant 1/m$ for any $i \leqslant m$. Consequently, $\|[\psi_m]\|_i \to 0$ as $m \to \infty$ for $i = 0, 1, \ldots$. Thus $\psi_m \to 0$ in S. On the other hand, $|(f, \psi_m)| \geqslant 1$, which contradicts (1.11). Lemma 1.3 is proved.

Clearly, the converse assertion is also valid: *if a semilinear functional satisfies* (1.15), *then it is continuous.*

3. *Operations on generalized functions.* a) By the *derivative* $\partial^k f/\partial x^k$ of a generalized function $f \in S'$ is meant the generalized function satisfying the relation

$$(\partial^k f/\partial x^k, \varphi) = (-1)^{|k|}(f, \partial^k \varphi/\partial x^k), \qquad \forall \varphi \in S. \tag{1.16}$$

Clearly, $\partial^k f/\partial x^k$ is a semilinear functional that satisfies an estimate of form (1.15) and hence belongs to S'. Let $D^k = D_1^{k_1} \ldots D_n^{k_n}$, where $D_r^{k_r} = i^{k_r}\partial^{k_r}/\partial x_r^{k_r}$, $1 \leqslant r \leqslant n$. Then, by virtue of (1.10) and (1.16),

$$(D^k f, \varphi) = (f, D^k \varphi), \qquad \forall \varphi \in S, \qquad \forall k = (k_1, \ldots, k_n). \tag{1.16$'$}$$

We note that generalized functions have derivatives of all orders.

EXAMPLE 1.3. Suppose $\theta(t) = 1$ for $t > 0$ and $\theta(t) = 0$ for $t < 0$. Then $\theta(t)$ defines a regular functional on $S(\mathbf{R}^1)$. Let us find the functional $d\theta/dt$. We have, according to (1.16),

$$(d\theta/dt, \varphi) = -(\theta, d\varphi/dt) = -\int_0^\infty \overline{\frac{d\varphi(t)}{dt}}\,dt = \overline{\varphi(0)}, \quad \text{i.e. } d\theta/dt = \delta.$$

EXAMPLE 1.4. The kth derivative of the delta function δ is defined, according to (1.16′), by the formula $(D^k\delta, \varphi) = \overline{D^k\varphi(0)}$.

The definition of the derivative of a generalized function is natural in the sense that, when f is a regular functional corresponding to a slowly increasing $f(x)$ that is $|k|$ times continuously differentiable, $D^k f$ is also a regular functional corresponding to the function $D^k f(x)$. In an analogous sense, the definitions presented below are also natural.

b) Suppose $f \in S'$, $a(x) \in C^\infty$ and

$$|D^p a(x)| \leqslant C_p(1 + |x|)^{t_p}, \qquad 0 \leqslant |p| < \infty. \tag{1.17}$$

By the *product* $a(x)f$ is meant the generalized function satisfying the relation

$$(af, \varphi) = (f, \bar{a}\varphi), \qquad \forall \varphi \in S, \tag{1.18}$$

where $\bar{a}(x)$ is the complex conjugate of $a(x)$. If $\varphi \in S$ and $a(x)$ satisfies estimates (1.17), then $\bar{a}\varphi \in S$, with

$$\|[\bar{a}\varphi]\|_m \leqslant C_m \|[\varphi]\|_{N_m}, \qquad \text{where } N_m \leqslant \max_{0 \leqslant |p| \leqslant m} (m + |t_p|).$$

Consequently, (1.18) actually defines a continuous semilinear functional on S.

c) Let $a \in \mathbf{R}^n$ be an arbitrary vector. By a *translation* of a generalized function f is meant a generalized function f^a defined by

$$(f^a, \varphi) = (f, \varphi(x + a)). \tag{1.19}$$

When f is a regular functional corresponding to a function $f(x)$, f^a is also a regular functional corresponding to $f(x - a)$.

d) We will say that a sequence $f_n \in S'$ *converges* to $f \in S'$ if

$$(f_n, \varphi) \to (f, \varphi), \qquad \forall \varphi \in S. \tag{1.20}$$

We note that it follows from (1.20) and (1.16′) that for any k

$$(D^k f_n, \varphi) \to (D^k f, \varphi), \qquad \forall \varphi \in S. \tag{1.21}$$

EXAMPLE 1.5. Suppose $|f_n(x)| \leqslant f_0(x)$ for $n = 1, 2, \ldots,$ where $f_0(x)$ satisfies (1.12), and suppose $f_n(x) \to f(x)$ almost everywhere. Then

$$(f_n, \varphi) = \int_{-\infty}^{\infty} f_n(x)\overline{\varphi(x)}\, dx \to \int_{-\infty}^{\infty} f(x)\overline{\varphi(x)}\, dx = (f, \varphi) \tag{1.22}$$

by virtue of the Lebesgue convergence theorem.

EXAMPLE 1.6. The generalized functions $t_{\pm}^{\lambda}$, $\lambda \neq -1, -2, \ldots$. Let $t_+^{\lambda} = t^{\lambda} \equiv e^{\lambda \ln t}$ for $t > 0$ and $t_+^{\lambda} = 0$ for $t < 0$. When $\operatorname{Re} \lambda > -1$, the function t_+^{λ} defines a regular functional on $S(\mathbf{R}^1)$, which we also denote by t_+^{λ}:

$$(t_+^{\lambda}, \varphi) = \int_0^{\infty} t^{\lambda}\overline{\varphi(t)}\, dt.$$

We note that $t_+^0 = \theta(t)$ (see Example 1.3). Suppose now $\operatorname{Re}\lambda \leqslant -1$ and λ is not an integer. Then there exists an integer $k > 0$ such that $\lambda = \mu - k$, where $-1 < \operatorname{Re}\mu \leqslant 0$, and we can associate with t_+^{λ} a functional on $S(\mathbf{R}^1)$ by means of the formula

$$(t_+^{\lambda}, \varphi) = \frac{(-1)^k}{\mu(\mu - 1)\ldots(\mu - k + 1)} \int_0^{\infty} t^{\mu} \frac{d^k \overline{\varphi(t)}}{dt^k}\, dt. \tag{1.23}$$

Clearly, $t_+^\lambda \in S'$ since it is the kth derivative of the regular functional

$$\frac{t^\mu}{\mu(\mu-1)\ldots(\mu-k+1)}.$$

Let t_-^λ denote the function that is equal to 0 for $t > 0$ and to $|t|^\lambda = e^{\lambda \ln|t|}$ for $t < 0$. When $\operatorname{Re}\lambda > -1$, it defines a regular functional t_-^λ. When $\operatorname{Re}\lambda \leqslant -1$ and λ is not an integer, we put, analogously to (1.23),

$$(t_-^\lambda, \varphi) = \frac{1}{\mu(\mu-1)\ldots(\mu-k+1)} \int_{-\infty}^0 |t|^\mu \frac{d^k\varphi(t)}{dt^k}\, dt, \qquad (1.24)$$

i.e. $t_-^\lambda = \dfrac{1}{\mu(\mu-1)\ldots(\mu-k+1)} \dfrac{d^k}{dt^k} t_-^\mu.$

We note that if $\varphi(t) = 0$ in a neighborhood of $t = 0$, we get, upon integrating by parts in (1.23) and (1.24), that

$$(t_+^\lambda, \varphi) = \int_0^\infty t^\lambda \overline{\varphi(t)}\, dt, \qquad (t_-^\lambda, \varphi) = \int_{-\infty}^0 |t|^\lambda \overline{\varphi(t)}\, dt.$$

This fact reveals the naturalness of definitions (1.23) and (1.24).

EXAMPLE 1.7. **The generalized functions $(\sigma \pm i0)^\lambda$ and $(\sigma \pm i0)^\lambda \ln^p(\sigma \pm i0)$.** The function $\ln(\sigma + i\tau) = \ln|\sigma + i\tau| + i \arg(\sigma + i\tau)$ is analytic in the $\sigma + i\tau$ plane with a cut along the negative real axis. It is assumed that the branch of the logarithm has been chosen so that $\arg(\sigma + i\tau) = 0$ for $\sigma > 0$, $\tau = 0$. The function

$$(\sigma + i\tau)^\lambda \ln^p(\sigma + i\tau) = e^{\lambda \ln(\sigma + i\tau)} \ln^p(\sigma + i\tau)$$

is analytic for $\tau > 0$ (p is a nonnegative integer) and defines for each $\tau > 0$ a regular functional on $S(\mathbf{R}^1)$.

We introduce the notation

$$\ln(\sigma + i0) = \lim_{\tau \to +0} \ln(\sigma + i\tau) = \ln|\sigma| + i\pi\theta(-\sigma), \qquad (1.25)$$

$$(\sigma + i0)^\lambda = e^{\lambda \ln(\sigma + i0)} = |\sigma|^\lambda e^{i\lambda\pi\theta(-\sigma)} = \sigma_+^\lambda + e^{i\lambda\pi}\sigma_-^\lambda. \qquad (1.26)$$

When $\operatorname{Re}\lambda > -1$, the function $(\sigma + i0)^\lambda \ln^p(\sigma + i0)$ defines a regular functional on $S(\mathbf{R}^1)$, with

$$((\sigma + i\tau)^\lambda \ln^p(\sigma + i\tau),\ \varphi(\sigma)) \underset{\tau \to +0}{\to} ((\sigma + i0)^\lambda \ln^p(\sigma + i0),\ \varphi(\sigma)) \qquad (1.27)$$

for any function $\varphi(\sigma) \in S(\mathbf{R}^1)$ by virtue of the Lebesgue convergence theorem (see Example 1.5). When $\operatorname{Re}\lambda \leqslant -1$, the function $(\sigma + i\tau)^\lambda \ln^p(\sigma + i\tau)$ can be represented (clearly in many ways) as a linear combination of derivatives with respect to σ of functions of the form $(\sigma + i\tau)^\mu \ln^q(\sigma + i\tau)$, where $\operatorname{Re}\mu > -1$ and q is a nonnegative integer. Consequently, by virtue of (1.27) and property (1.21), the regular functionals $(\sigma + i\tau)^\lambda \ln^p(\sigma + i\tau)$ have a limit in $S'(\mathbf{R}^1)$ for $\tau \to +0$, which we denote by $(\sigma + i0)^\lambda \ln^p(\sigma + i0)$. Thus (1.27) is valid for any complex number λ and nonnegative integer p. When $p = 0$ and $\lambda \neq -1, -2, \ldots,$ the functional $(\sigma + i0)^\lambda$ is equal to the sum of functionals $\sigma_+^\lambda + e^{i\lambda\pi}\sigma_-^\lambda$, where σ_+^λ and σ_-^λ are defined in Example 1.6.

Analogously, the function $(\sigma + i\tau)^\lambda \ln^p(\sigma + i\tau)$, which is analytic in the halfplane $\tau < 0$, defines a regular functional on $S(\mathbf{R}^1)$ for each $\tau < 0$. Let

$$\ln(\sigma - i0) = \lim_{\tau \to -0} \ln(\sigma + i\tau) = \ln|\sigma| - i\pi\theta(-\sigma),$$

$$(\sigma - i0)^\lambda = e^{\lambda \ln(\sigma - i0)} = \sigma_+^\lambda + e^{-i\lambda\pi}\sigma_-^\lambda.$$

As above, it can be shown that for any complex λ and nonnegative integral p the regular functionals $(\sigma + i\tau)^\lambda \ln^p(\sigma + i\tau)$ have a limit in $S'(\mathbf{R}^1)$ for $\tau \to -0$, which we denote by $(\sigma - i0)^\lambda \ln^p(\sigma - i0)$.

4. *General form of a generalized function on $S(\mathbf{R}^n)$.*

THEOREM 1.1. *Any continuous semilinear functional f on S can be represented for certain p and N in the form*

$$f = \sum_{|k| \leq p} D^k f_k, \tag{1.28}$$

where the f_k are regular functionals corresponding to continuous functions $f_k(x)$ that are slowly increasing:

$$|f_k(x)| \leqslant C(1 + |x|)^N, \qquad 0 \leqslant |k| \leqslant p. \tag{1.29}$$

PROOF. By virtue of Lemma 1.3 there exists a nonnegative integer $m(f)$ such that

$$|(f, \varphi)| \leqslant C\|[\varphi]\|_m, \qquad \forall \varphi \in S. \tag{1.30}$$

We have

$$(1 + |x|)^m D^k\varphi(x) = \int_{-\infty}^{x_1} \cdots \int_{-\infty}^{x_n} \frac{\partial}{\partial y_1} \cdots \frac{\partial}{\partial y_n}\big((1 + |y|)^m D^k\varphi(y)\big)dy.$$

Consequently,

$$\|[\varphi]\|_m \leqslant C \sum_{|k| \leq m+n} \int_{-\infty}^{\infty} (1 + |y|)^m |D^k\varphi(y)|\,dy.$$

Applying the Cauchy-Schwarz-Bunjakovskiĭ inequality, we get

$$\begin{aligned}\|[\varphi]\|_m^2 &\leqslant C\left(\sum_{|k| \leq m+n} \int_{-\infty}^{\infty} (1 + |y|)^m |D^k\varphi(y)|\,dy\right)^2 \\ &\leqslant C\int_{-\infty}^{\infty} (1 + |y|)^{-n-1}\,dy \sum_{|k| \leq m+n} \int_{-\infty}^{\infty} (1 + |y|)^{2m+n+1} |D^k\varphi(y)|^2\,dy \\ &\leqslant C \sum_{|k| \leq m+n} \int_{-\infty}^{\infty} (1 + |y|)^{2m+n+1} |D^k\varphi(y)|^2\,dy. \end{aligned} \tag{1.31}$$

Let $L_2(\mathbf{R}^n)$ be the space of functions on $\mathbf{R}^n$ that are square integrable in the Lebesgue sense. We denote by $\vec{L}_2(\mathbf{R}^n) = \bigoplus_{|k|=0}^{m+n} L_2(\mathbf{R}^n)$ the direct sum of spaces $L_2(\mathbf{R}^n)$ with scalar product

$$\{\vec{u}, \vec{v}\} = \sum_{|k| \leqslant m+n} \int_{-\infty}^{\infty} u_k(x)\, \overline{v_k(x)}\, dx,$$

where $\vec{u} = \{u_k(x)\} \in \vec{L}_2(\mathbf{R}^n)$, $\vec{v} = \{v_k(x)\} \in \vec{L}_2(\mathbf{R}^n)$. With each function $\varphi(x) \in S$ we associate the vector

$$\vec{w} = \{(1 + |y|)^{m+(n+1)/2} D^k \varphi(y)\}, \qquad 0 \leqslant |k| \leqslant m + n,$$

which clearly belongs to $\vec{L}_2(\mathbf{R}^n)$. By virtue of (1.30) and (1.31), a functional $f \in S'(\mathbf{R}^n)$ is bounded on the linear manifold $W \subset \vec{L}_2(\mathbf{R}^n)$ of such vectors. Consequently, according to the Hahn-Banach theorem, it can be extended as a continuous semilinear functional onto all of $\vec{L}_2(\mathbf{R}^n)$. By virtue of Riesz's theorem there exists a vector $\{g_k(y)\}$, $|k| \leqslant m + n$, belonging to $\vec{L}_2(\mathbf{R}^n)$ such that

$$(f, \varphi) = \sum_{|k|=0}^{m+n} \int_{-\infty}^{\infty} g_k(y)(1 + |y|)^{m+(n+1)/2}\, \overline{D^k \varphi(y)}\, dy. \tag{1.32}$$

We put

$$f_k(x) = \int_0^{x_1} \cdots \int_0^{x_n} g_k(y)(1 + |y|)^{m+(n+1)/2}\, dy.$$

Then $f_k(x)$ is an absolutely continuous function and

$$(1 + |x|)^{m+(n+1)/2} g_k(x) = \frac{\partial}{\partial x_1} \cdots \frac{\partial}{\partial x_n} f_k(x). \tag{1.33}$$

By virtue of the Cauchy-Schwarz-Bunjakovskiĭ inequality,

$$|f_k(x)|^2 \leqslant \int_0^{x_1} \cdots \int_0^{x_n} (1 + |y|)^{2m+n+1}\, dy \int_0^{x_1} \cdots \int_0^{x_n} |g_k(y)|^2\, dy \leqslant C(1 + |x|)^{2m+2n+1}.$$

Substituting (1.33) in (1.32), we get

$$(f, \varphi) = \sum_{|k|=0}^{m+n} (-1)^n \int_{-\infty}^{\infty} f_k(y) \frac{\partial}{\partial y_1} \cdots \frac{\partial}{\partial y_n}\, \overline{D^k \varphi(y)}\, dy.$$

Theorem 1.1 is proved.

5. *Generalized functions in a domain. Support of a generalized function.* By the *support* of a continuous function $\varphi(x)$ is meant the closure of the set of points at which $\varphi(x) \neq 0$. Let U be an open domain in $\mathbf{R}^n$ that is generally unbounded and let $C_0^\infty(U)$ be the totality of infinitely differentiable functions with compact support in U. Let $S(U)$ denote the completion of $C_0^\infty(U)$ in the

topology of $S(\mathbf{R}^n)$, so that $S(U)$ is a closed subspace of $S(\mathbf{R}^n)$. We will say that a continuous semilinear functional on $S(U)$ is a *generalized function in the domain* U. The space of generalized functions in U will be denoted by $S'(U)$. Suppose $F \in S'(\mathbf{R}^n)$. A functional $f \in S'(U)$ is called the *restriction* of F to U if $(F, \varphi) = (f, \varphi)$ for any $\varphi \in S(U)$. The restriction operator will be denoted by p: $pF = f$. Since $S(U)$ is a closed subspace of $S(\mathbf{R}^n)$, by the Hahn-Banach theorem any continuous semilinear functional f on $S(U)$ can be extended (though, of course, not uniquely) to a continuous semilinear functional F on $S(\mathbf{R}^n)$. We will denote an extension of f by lf: $F = lf$.

REMARK 1.2. The fact that any generalized function in U can be extended onto all of $\mathbf{R}^n$ stems from the choice of $S(U)$ as the space of fundamental functions. Such an extension is not always possible for the spaces $D'(U)$ and $K'(U)$ (see [49]).

Suppose $F \in S'(\mathbf{R}^n)$. We will say that $F = 0$ in U if the restriction of F to U vanishes, i.e. is the null functional on $S(U)$: $(F, \varphi) = 0$ for all $\varphi \in S(U)$. Since $C_0^\infty(U)$ is dense in $S(U)$, $F = 0$ in U if and only if $(F, \varphi) = 0$ for any $\varphi \in C_0^\infty(U)$. Let $U_{\max}$ be the largest open set on which $F = 0$. Then the complement of $U_{\max}$ is denoted by supp F and called the *support* of the functional F. When F is a regular functional defined by a locally integrable function $F(x)$, supp F is the complement of the largest open set on which $F(x) = 0$ almost everywhere.

Clearly, supp D^pF is contained in supp F for any p.

EXAMPLE 1.8. The delta function δ vanishes everywhere except at the origin, so that supp $\delta = \{0\}$. Analogously, supp $\delta^{(k)} = \{0\}$.

EXAMPLE 1.9. The restriction of the generalized function t_+^λ (see Example 1.6) to the halfline $t > 0$ is a regular functional corresponding to the function t^λ:

$$(t_+^\lambda, \varphi) = \int_0^\infty t^\lambda \overline{\varphi(t)}\, dt, \tag{1.34}$$

where $\varphi(t) \in S(\mathbf{R}_+^1)$, i.e. $\varphi(t)$ has a zero of infinite order at $t = 0$. We note that t_+^λ for $\operatorname{Re}\lambda < -1$ and λ nonintegral is not a regular functional on the whole space $S(\mathbf{R}^1)$ and is defined at any $\varphi \in S(\mathbf{R}^1)$ by (1.23). When $\varphi \in S(\mathbf{R}_+^1)$, (1.23) can be transformed into (1.34) by integrating by parts.

6. *Direct product of generalized functions*. Suppose $x = (x', x_n)$, where $x' = (x_1, \ldots, x_{n-1}) \in \mathbf{R}^{n-1}$, $x_n \in \mathbf{R}^1$, $x \in \mathbf{R}^n = \mathbf{R}^{n-1} \times \mathbf{R}^1$, and suppose f_1 and f_2 are regular functionals on $S(\mathbf{R}^{n-1})$ and $S(\mathbf{R}^1)$ respectively. By the *direct product* $f_1 \times f_2$ is meant the functional defined on $S(\mathbf{R}^n)$ by

$$(f_1 \times f_2, \varphi) = \int_{-\infty}^{\infty} f_1(x')f_2(x_n)\,\overline{\varphi(x', x_n)}\, dx'\, dx_n, \qquad \forall \varphi \in S(\mathbf{R}^n). \quad (1.35)$$

Suppose now f_1 is an arbitrary continuous semilinear functional on $S(\mathbf{R}^{n-1})$ and f_2 is an arbitrary continuous semilinear functional on $S(\mathbf{R}^1)$. We will define their direct product $f_1 \times f_2$. Suppose $\varphi(x', x_n) \in S(\mathbf{R}^n)$. Then $\varphi(x', x_n) \in S(\mathbf{R}^1)$ for fixed $x' \in \mathbf{R}^{n-1}$. Consequently, the function $\varphi_1(x') = (f_2, \varphi(x', x_n))$ is defined.

Let us show that $\varphi_1(x') \in S(\mathbf{R}^{n-1})$. By Theorem 1.1,

$$\begin{aligned}\varphi_1(x') = (f_2, \varphi(x', x_n)) &= \sum_{k=0}^{m} (D_n^k f_{2k}, \varphi(x', x_n)) \\ &= \sum_{k=0}^{m} \int_{-\infty}^{\infty} f_{2k}(x_n)\, \overline{D_n^k \varphi(x', x_n)}\, dx_n, \end{aligned} \quad (1.36)$$

where the $f_{2k}(x_n)$ are continuous functions satisfying an estimate of form (1.29). Since $\varphi(x', x_n) \in S(\mathbf{R}^n)$, the integrals in (1.36) can be differentiated with respect to x' arbitrarily often and decrease more rapidly than any negative power of $|x'|$ as $|x'| \to \infty$. Thus $\varphi_1(x') = (f_2, \varphi(x', x_n)) \in S(\mathbf{R}^{n-1})$.

We now define $f_1 \times f_2 \in S'(\mathbf{R}^n)$ by means of the formula

$$(f_1 \times f_2, \varphi) = \left(f_1, \overline{(f_2, \varphi(x', x_n))}\right). \quad (1.37)$$

The functional (1.37) is clearly semilinear. Further, by virtue of (1.36), if $\varphi_m(x', x_n) \to \varphi(x', x_n)$ in $S(\mathbf{R}^n)$, then

$$\varphi_{1m}(x') = (f_2, \varphi_m(x', x_n)) \to \varphi_1(x') = (f_2, \varphi(x', x_n))$$

in $S(\mathbf{R}^{n-1})$. Consequently,

$$(f_1 \times f_2, \varphi_m) = \left(f_1, \overline{\varphi_{1m}}\right) \to \left(f_1, \overline{\varphi_1}\right) = (f_1 \times f_2, \varphi),$$

i.e. $f_1 \times f_2$ is a continuous functional on $S(\mathbf{R}^n)$.

EXAMPLE 1.10. Suppose $f_1 \in S'(\mathbf{R}^{n-1})$ and $\delta(x_n)$ is the delta function in $\mathbf{R}^1$. Then the functional $f_1 \times \delta(x_n) \in S'(\mathbf{R}^n)$ is defined by

$$(f_1 \times \delta, \varphi) = (f_1, \varphi(x', 0)). \quad (1.38)$$

If, in particular, f_1 is a regular functional, then

$$(f_1 \times \delta, \varphi) = \int_{-\infty}^{\infty} f_1(x')\,\overline{\varphi(x', 0)}\, dx'.$$

We note that $\operatorname{supp}(f_1 \times \delta) \in \mathbf{R}^{n-1}$.

We denote by 1 the regular functional in $\mathbf{R}^1$ corresponding to the function that is identically equal to 1. The direct product $f_1 \times 1$, where $f_1 \in S'(\mathbf{R}^{n-1})$, is defined by

$$(f_1 \times 1, \varphi) = \left(f_1, \int_{-\infty}^{\infty} \varphi(x', x_n) dx_n\right). \quad (1.39)$$

We note that $\int_{-\infty}^{\infty} \varphi(x', x_n) dx_n \in S(\mathbf{R}^{n-1})$.

We also note that the delta function in $\mathbf{R}^n$ is the direct product of n delta functions in $\mathbf{R}^1$: $\delta(x) = \delta(x_1) \times \ldots \times \delta(x_n)$.

EXAMPLE 1.11. Suppose $f(x', x_n)$ is a locally integrable function and

$$\left(f, \frac{\partial\varphi}{\partial x_n}\right) = \int_{-\infty}^{\infty} f(x)\overline{\frac{\partial\varphi(x)}{\partial x_n}}\, dx = 0, \qquad \forall\varphi(x) \in C_0^\infty(\mathbf{R}^n). \tag{1.40}$$

We will prove that $f(x', x_n)$ does not depend on x_n. Let $\varphi_0(x_n)$ be a fixed function in $C_0^\infty(\mathbf{R}^1)$ such that $\int_{-\infty}^{\infty} \varphi_0(x_n)dx_n = 1$, and suppose $\varphi(x) \in C_0^\infty(\mathbf{R}^n)$. We put

$$\varphi_1(x') = \int_{-\infty}^{\infty} \varphi(x', x_n)dx_n, \qquad \psi(x', x_n) = \varphi(x', x_n) - \varphi_1(x')\varphi_0(x_n). \tag{1.41}$$

Then, $\int_{-\infty}^{\infty} \psi(x', x_n)dx_n = 0$, and hence $\varphi_2(x', x_n) = \int_{-\infty}^{x_n} \psi(x', y_n)dy_n$ belongs to $C_0^\infty(\mathbf{R}^n)$, with $\partial\varphi_2(x)/\partial x_n = \psi(x)$. By (1.40),

$$\begin{aligned}(f, \varphi) &= \left(f, \frac{\partial\varphi_2}{\partial x_n} + \varphi_1(x')\varphi_0(x_n)\right) = (f, \varphi_1(x')\varphi_0(x_n)) \\ &= \int_{-\infty}^{\infty} f(x', x_n)\, \overline{\varphi_1(x')\varphi_0(x_n)}\; dx'\, dx_n. \end{aligned} \tag{1.42}$$

Let

$$f_0(x') = \int_{-\infty}^{\infty} f(x', x_n)\, \overline{\varphi_0(x_n)}\, dx_n.$$

Then (1.42) and (1.41) imply

$$\begin{aligned}\int_{-\infty}^{\infty}\int_{-\infty}^{\infty} f(x', x_n)\,\overline{\varphi(x', x_n)}\, dx'\, dx_n &= \int_{-\infty}^{\infty} f_0(x')\,\overline{\varphi_1(x')}\, dx' \\ &= \int_{-\infty}^{\infty}\int_{-\infty}^{\infty} f_0(x')\,\overline{\varphi(x', x_n)}\, dx'\, dx_n, \\ &\qquad \forall\varphi \in C_0^\infty(\mathbf{R}^n). \end{aligned} \tag{1.43}$$

Since $\varphi(x', x_n)$ is an arbitrary function in $C_0^\infty(\mathbf{R}^n)$, it follows by (1.43) that $f(x', x_n) = f_0(x')$ almost everywhere.

§2. Fourier transform

1. *Fourier transform of a fundamental function.* The Fourier transform of a function $\varphi(x) \in S(\mathbf{R}^n)$ is defined by

$$\tilde{\varphi}(\xi) = \int_{-\infty}^{\infty} \varphi(x)e^{i(x,\xi)}\, dx, \qquad (x, \xi) = x_1\xi_1 + \ldots + x_n\xi_n, \tag{2.1}$$

while the inverse Fourier transform has the form

$$\varphi(x) = \frac{1}{(2\pi)^n}\int_{-\infty}^{\infty} \tilde{\varphi}(\xi)e^{-i(x,\xi)}\, d\xi. \tag{2.2}$$

We will sometimes denote the Fourier transform operator by F, and the inverse Fourier transform operator by F^{-1}. We note the following properties of Fourier transforms of fundamental functions:

1)

$$F(D^k\varphi(x)) = \xi^k\tilde{\varphi}(\xi), \quad \text{where } \xi^k = \xi_1^{k_1}\cdots\xi_n^{k_n}, \qquad D^k = D_1^{k_1}\cdots D_n^{k_n},$$

$$D_r^{k_r} = \left(i\frac{\partial}{\partial x_r}\right)^{k_r}, \qquad 1 \leqslant r \leqslant n; \tag{2.3}$$

2)

$$F(\varphi * \psi) = \tilde{\varphi}(\xi)\tilde{\psi}(\xi); \tag{2.4}$$

3) the Parseval equality

$$\int_{-\infty}^{\infty} \varphi(x)\,\overline{\psi(x)}\,dx = \frac{1}{(2\pi)^n}\int_{-\infty}^{\infty} \tilde{\varphi}(\xi)\,\overline{\tilde{\psi}(\xi)}\,d\xi, \tag{2.5}$$

including, in particular, the equality

$$\int_{-\infty}^{\infty} |\varphi(x)|^2\,dx = \frac{1}{(2\pi)^n}\int_{-\infty}^{\infty} |\tilde{\varphi}(\xi)|^2\,d\xi; \tag{2.6}$$

4)

$$F(\varphi(x-a)) = e^{i(a,\xi)}\tilde{\varphi}(\xi), \qquad a = (a_1, \ldots, a_n) \in \mathbf{R}^n. \tag{2.7}$$

A proof of these properties can be found e.g. in [83]. Properties 1)–4) also hold for a wider class of functions than S. For example, the Parseval equality can be extended by means of the completion process to any $\varphi(x)$, $\psi(x) \in L_2(\mathbf{R}^n)$; it is also valid if $\psi(x)$ is an absolutely integrable function and $\varphi(x) \in S$.

LEMMA 2.1. *The Fourier transform operator is an injective continuous map of the space $S(\mathbf{R}^n)$ onto itself.*

PROOF. By virtue of (2.3),

$$\xi^p D_\xi^k\tilde{\varphi}(\xi) = \int_{-\infty}^{\infty} D_x^p[(-x)^k\varphi(x)]e^{i(x,\xi)}\,dx. \tag{2.8}$$

Consequently, according to (1.1),

$$|[\tilde{\varphi}]|_m \leqslant \sum_{|p|=0}^{m}\sum_{|k|=0}^{m}\int_{-\infty}^{\infty}|D_x^p(x^k\varphi(x))|\,dx \leqslant \int_{-\infty}^{\infty}\frac{C_m|[\varphi]|_{m+n+1}}{(1+|x|)^{n+1}}\,dx$$

$$= C_m^{(1)}|[\varphi]|_{m+n+1}, \qquad m = 0, 1, \ldots. \tag{2.9}$$

Thus $\tilde{\varphi}(\xi) \in S$ if $\varphi(x) \in S$, and, as a result of (2.9), $\tilde{\varphi}_i(\xi) \to \tilde{\varphi}(\xi)$ in S if $\varphi_i(x) \to \varphi(x)$ in S. Repeating the same arguments for F^{-1}, we get that F is an injective continuous map of S onto itself. Lemma 2.1 is proved.

Let $\mathbf{R}_+^n$ denote the halfspace $x_n > 0$ and let S^+ denote the subspace of S consisting of the functions $\varphi(x', x_n) \in S$ that vanish for $x_n < 0$, $x' = (x_1, \ldots, x_{n-1}) \in \mathbf{R}^{n-1}$.

LEMMA 2.2. *The Fourier transform $\tilde{S}^+ = FS^+$ of the subspace S^+ consists of those functions $\tilde{\varphi}(\xi', \zeta_n)$ that can be analytically continued with respect to $\zeta_n = \xi_n + i\tau$ into the halfplane $\tau > 0$, are infinitely differentiable for $(\xi', \xi_n) \in \mathbf{R}^n$, $\tau \geqslant 0$ and satisfy the estimates*

$$(1 + |\xi'| + |\xi_n| + \tau)^m \left| D_{\xi'}^{k'} D_{\zeta_n}^{k_n} \tilde{\varphi}(\xi', \zeta_n) \right| \leqslant C_{m,|k|}, \qquad 0 \leqslant m < \infty,$$
$$k = (k', k_n), \qquad 0 \leqslant |k| < \infty. \tag{2.10}$$

PROOF. Suppose $\varphi_+(x', x_n) \in S^+$. Then

$$\tilde{\varphi}_+(\xi', \xi_n) = \int_0^\infty \int_{-\infty}^\infty \varphi_+(x', x_n) e^{ix_n\xi_n + i(x', \xi')}\, dx'\, dx_n. \tag{2.11}$$

Since the integration with respect to x_n in (2.11) is taken from 0 to $+\infty$, the right side of (2.11) can be analytically continued with respect to $\zeta_n = \xi_n + i\tau$ into the halfplane $\tau > 0$:

$$\tilde{\varphi}_+(\xi', \zeta_n) = \int_0^\infty \int_{-\infty}^\infty \varphi_+(x', x_n) e^{ix_n\zeta_n + i(x', \xi')}\, dx'\, dx_n,$$
$$\zeta_n = \xi_n + i\tau, \quad \tau > 0. \tag{2.12}$$

Since $\varphi_+(x', x_n) \in S$, it follows that

$$(1 + |\xi'| + |\xi_n| + \tau)^m \left| D_{\xi'}^{k'} D_{\zeta_n}^{k_n} \tilde{\varphi}_+(\xi', \zeta_n) \right|$$
$$\leqslant C \int_0^\infty \int_{-\infty}^\infty \sum_{|p|=0}^m \left| D_x^p(x^k \varphi_+(x', x_n)) \right| dx \leqslant C_{m,|k|}.$$

Lemma 2.2 is proved.

Analogously, *if S^- denotes the subspace of $S(\mathbf{R}^n)$ consisting of the functions that vanish for $x_n > 0$, then $\tilde{S}^- = FS^-$ consists of those functions $\tilde{\varphi}(\xi', \zeta_n)$ that can be analytically continued with respect to $\zeta_n = \xi_n + i\tau$ into the halfplane $\tau < 0$ and satisfy estimates of form* (2.10) *with τ replaced by $|\tau|$.*

2. *Mellin transform.* By the *Mellin transform* of a function $f(t) \in C_0^\infty(\mathbf{R}_+^1)$ is meant the function

$$\hat{f}(s) = \int_0^\infty t^{s-1} f(t)\, dt. \tag{2.13}$$

Since $f(t) \in C_0^\infty(\mathbf{R}_+^1)$, the integral (2.13) converges for all complex s and is an entire function.

The change of variables $t = e^{t'}$ converts the Mellin transform of $f(t)$ into the Fourier transform of $f(e^{t'})$

$$\hat{f}(s) = \int_{-\infty}^\infty e^{(\sigma + i\tau)t'} f(e^{t'})\, dt', \qquad s = \sigma + i\tau. \tag{2.13'}$$

Consequently, the properties of the Mellin transform can be obtained from the corresponding properties of the Fourier transform. In particular, the inversion formula for the Mellin transform of $f(t) \in C_0^\infty(\mathbf{R}_+^1)$ has the form

$$f(t) = \frac{1}{2\pi}\int_{-\infty}^{\infty} \hat{f}(s)t^{-s}\,d\tau, \qquad s = \sigma + i\tau, \tag{2.14}$$

where σ is arbitrary. It follows from (2.13′) by virtue of (2.6) that

$$\int_{-\infty}^{\infty} \left|e^{\sigma t'}f(e^{t'})\right|^2 dt' = \frac{1}{2\pi}\int_{-\infty}^{\infty} \left|\hat{f}(s)\right|^2 d\tau. \tag{2.15}$$

Consequently, when we make the change of variables $t' = \ln t$ in (2.15), the Parseval equality takes the form

$$\int_0^{\infty} t^{2\sigma-1}|f(t)|^2\,dt = \frac{1}{2\pi}\int_{-\infty}^{\infty} \left|\hat{f}(s)\right|^2 d\tau, \qquad s = \sigma + i\tau. \tag{2.16}$$

In particular, when $\sigma = 1/2$ we get

$$\int_0^{\infty} |f(t)|^2\,dt = \frac{1}{2\pi}\int_{-\infty}^{\infty} \left|\hat{f}(s)\right|^2 d\tau, \qquad s = \frac{1}{2} + i\tau. \tag{2.16′}$$

Formulas (2.13), (2.14) and (2.16) also hold for a wider class of functions than $C_0^\infty(\mathbf{R}_+^1)$ (though generally not for arbitrary σ; see, for example, [96]). We note the following lemma.

LEMMA 2.3. *Suppose $f(t)$ is an infinitely differentiable function for $t > 0$ that satisfies the estimates*

$$t^k\left|\frac{d^k f(t)}{dt^k}\right| \leqslant \begin{cases} C_{\varepsilon,k}t^{-\delta-\varepsilon} & \text{for } 0 < t \leqslant 1, \\ C_{\varepsilon,k}t^{-1+\delta+\varepsilon} & \text{for } 1 \leqslant t < \infty, \quad \forall\varepsilon > 0, \forall k \geqslant 0, \end{cases} \tag{2.17}$$

where $0 \leqslant \delta < 1/2$. Then the Mellin transform $\hat{f}(s)$ of $f(t)$ is analytic in the strip $\delta < \operatorname{Re} s < 1 - \delta$, and

$$\left|s^m \hat{f}(s)\right| \leqslant C_{\varepsilon,m} \quad \text{for} \quad \delta + \varepsilon \leqslant \operatorname{Re} s \leqslant 1 - \delta - \varepsilon, \quad \forall\varepsilon > 0, \forall m \geqslant 0. \tag{2.18}$$

The converse assertion is also valid.

PROOF. Suppose $f(t)$ satisfies (2.17) for $k = 0$. Then the integral

$$\hat{f}(s) = \int_0^1 t^{s-1}f(t)dt + \int_1^{\infty} t^{s-1}f(t)dt$$

is absolutely convergent and an analytic function in the strip $\delta < \operatorname{Re} s < 1 - \delta$. In addition,

$$\left|\hat{f}(s)\right| \leqslant C_\varepsilon \quad \text{for} \quad \delta + \varepsilon \leqslant \operatorname{Re} s \leqslant 1 - \delta - \varepsilon, \quad \forall\varepsilon > 0. \tag{2.19}$$

Let $\hat{f}_k(s)$, $k \geqslant 1$, be the Mellin transform of the function $t^k \dfrac{d^k f(t)}{dt^k}$:

$$\hat{f}_k(s) = \int_0^\infty t^{s+k-1} \frac{d^k f(t)}{dt^k}\, dt, \qquad \delta < \operatorname{Re} s < 1 - \delta. \tag{2.20}$$

Analogously to (2.19),

$$|\hat{f}_k(s)| \leqslant C_{\varepsilon,k}, \qquad \delta + \varepsilon \leqslant \operatorname{Re} s \leqslant 1 - \delta - \varepsilon, \qquad \forall \varepsilon > 0. \tag{2.21}$$

Integrating by parts in (2.20), we get

$$\hat{f}_k(s) = (-1)^k (s + k - 1)(s + k - 2) \ldots s\hat{f}(s), \qquad \forall k \geqslant 1. \tag{2.22}$$

Since k is arbitrary, (2.18) follows from (2.19), (2.21) and (2.22).

Suppose now that an analytic function $\hat{f}(s)$ satisfying (2.18) is given in the strip $\delta < \operatorname{Re} s < 1 - \delta$, and let

$$f(t) = \frac{1}{2\pi} \int_{-\infty}^{\infty} \hat{f}(s) t^{-s}\, d\tau, \quad s = \sigma + i\tau, \quad \delta < \sigma < 1 - \delta. \tag{2.23}$$

By virtue of the fact that $\hat{f}(s)$ is analytic and rapidly decreasing for $\tau \to \pm\infty$, the integral (2.23) does not depend on the value of σ in the interval $(\delta, 1 - \delta)$. Let us show that $f(t)$ satisfies (2.17). From (2.23) we have

$$t^k \frac{d^k f(t)}{dt^k} = \frac{1}{2\pi} \int_{-\infty}^{\infty} (-1)^k s \ldots (s + k - 1) \hat{f}(s) t^{-s}\, d\tau,$$
$$\delta < \operatorname{Re} s < 1 - \delta, \qquad \forall k \geqslant 0. \tag{2.24}$$

It consequently follows by virtue of (2.18) that

$$t^k \left| \frac{d^k f(t)}{dt^k} \right| \leqslant C_{\varepsilon,k} t^{-\sigma}, \quad \delta + \varepsilon \leqslant \sigma \leqslant 1 - \delta - \varepsilon, \quad \forall \varepsilon > 0, \forall k \geqslant 0. \tag{2.25}$$

Setting $\sigma = \delta + \varepsilon$ ($\sigma = 1 - \delta - \varepsilon$) in (2.25) for $0 < t \leqslant 1$ ($t \geqslant 1$), we get (2.17). Lemma 2.3 is proved.

EXAMPLE 2.1. Let us calculate the Mellin transform

$$\hat{f}(s) = \int_0^\infty \frac{t^{\delta+s-1}\, dt}{t - a} \tag{2.26}$$

of the function $t^\delta/(t - a)$, where $|\delta| < 1/2$ and a is not a nonnegative real number. The integral (2.26) is absolutely convergent and an analytic function in the strip $-\delta < \operatorname{Re} s < 1 - \delta$. Let $\Gamma_{N,\varepsilon}$ be the contour consisting of the circles of radii ε and N and the segment εN, traversed twice, of the positive real axis. The function

$$\frac{z^{\delta+s-1}}{z - a} = \frac{e^{(\delta+s-1)\ln z}}{z - a}$$

is single valued in the domain bounded by the contour $\Gamma_{N,\varepsilon}$ and has a pole at $z = a$. Suppose we have chosen the branch of the logarithm with

$\arg z = 0$ on the upper edge of a cut along the positive real axis. Then, on the lower edge of this cut, $\arg z = 2\pi$ and hence the function $z^{\delta+s-1}/(z-a)$ is equal to

$$e^{2\pi i(\delta+s-1)}\frac{t^{\delta+s-1}}{t-a},$$

which implies, according to the residue theorem, that

$$\int_{\Gamma_{N,\varepsilon}} \frac{z^{\delta+s-1}}{z-a}\,dz = 2\pi i e^{(\delta+s-1)\ln a}.$$

Passing to the limit for $\varepsilon \to 0$, $N \to +\infty$ and taking into account the fact that the integrals over the circles tend to zero, we get

$$\hat{f}(s)(1 - e^{2\pi i(\delta+s-1)}) = 2\pi i e^{(\delta+s-1)\ln a}.$$

Consequently,

$$\hat{f}(s) = 2\pi i \frac{e^{(\delta+s-1)\ln a}}{1 - e^{2\pi i(\delta+s)}}. \tag{2.27}$$

3. *Fourier transform of a generalized function.* By the *Fourier transform* of a generalized function $f \in S'(\mathbf{R}^n)$ is meant the generalized function $\tilde{f} \in S'(\mathbf{R}^n)$ such that

$$(\tilde{f}, \tilde{\varphi}) = (2\pi)^n (f, \varphi), \qquad \forall \varphi \in S(\mathbf{R}^n), \tag{2.28}$$

where $\tilde{\varphi} = F\varphi \in S(\mathbf{R}^n)$. By virtue of Lemma 2.1, the functional $\tilde{f}$ is defined and continuous on $S(\mathbf{R}^n)$. When f is a regular functional corresponding to an absolutely integrable function $f(x)$, its Fourier transform $\tilde{f}$ is a regular functional corresponding to the Fourier transform $\tilde{f}(\xi)$ of $f(x)$, and (2.28) is a consequence of Parseval's equality. This fact reveals the naturalness of definition (2.28). The operator of the Fourier transform of a generalized function will also be denoted by F, and its inverse by F^{-1}.

Fourier transforms of generalized functions have properties analogous to (2.3) and (2.7):

$$F(D^k f) = \xi^k \tilde{f}, \tag{2.29}$$

$$F(f^a) = e^{i(a,\xi)}\tilde{f} \tag{2.30}$$

for any $f \in S'(\mathbf{R}^n)$ and any $a \in \mathbf{R}^n$. In fact, by virtue of (2.28), (1.16′), (2.3) and (1.18), we get

$$(F(D^k f), \tilde{\varphi}) = (2\pi)^n (D^k f, \varphi) = (2\pi)^n (f, D^k \varphi) = (\tilde{f}, \xi^k \tilde{\varphi}) = (\xi^k \tilde{f}, \tilde{\varphi})$$

for any $\varphi \in S$. Analogously, by virtue of (2.28), (1.19), (1.18) and (2.7),

$$(Ff^a, \tilde{\varphi}) = (2\pi)^n (f^a, \varphi) = (2\pi)^n (f, \varphi(x+a)) = (\tilde{f}, e^{-i(a,\xi)}\tilde{\varphi}) = (e^{i(a,\xi)}\tilde{f}, \tilde{\varphi}).$$

The operator F is continuous in $S'(\mathbf{R}^n)$, i.e. $\tilde{f}_n \to \tilde{f}$ in $S'(\mathbf{R}^n)$ if $f_n \to f$ in $S'(\mathbf{R}^n)$. In fact, by virtue of (2.28),

$$(\tilde{f}_n, \tilde{\varphi}) = (2\pi)^n(f_n, \varphi) \to (2\pi)^n(f, \varphi) = (\tilde{f}, \tilde{\varphi}), \qquad \forall\varphi \in S. \tag{2.31}$$

EXAMPLE 2.2. **$F(\delta^{(k)}) = \xi^k$. In fact,**

$$\begin{aligned}(F(D^k\delta), \tilde{\varphi}) &= (2\pi)^n(D^k\delta, \varphi) = (2\pi)^n\, \overline{\varphi^{(k)}}(0)\\ &= \int_{-\infty}^{\infty} \xi^k\, \overline{\tilde{\varphi}(\xi)}\, d\xi = (\xi^k, \tilde{\varphi}(\xi)).\end{aligned} \tag{2.32}$$

EXAMPLE 2.3. **Calculation of $F^{-1}(\sigma + i\tau)^\lambda \ln^p(\sigma + i\tau)$ and**

$$F^{-1}(\sigma + i0)^\lambda \ln^p(\sigma + i0).$$

Let us find the Fourier transform of the function $t_+^\lambda e^{-\tau t}$ for $\operatorname{Re}\lambda > -1$. By definition,

$$F(t_+^\lambda e^{-\tau t}) = \int_0^\infty t^\lambda e^{it(\sigma + i\tau)}\, dt.$$

We make the change of variables $z = -it(\sigma + i\tau)$. When t ranges over the positive real axis, z ranges over a ray lying in the right half of the complex $z = x + iy$ plane, which we denote by $C_{\sigma + i\tau}$. We have

$$\int_0^\infty t^\lambda e^{it(\sigma + i\tau)}\, dt = e^{(i(\lambda+1)\pi)/2}(\sigma + i\tau)^{-\lambda-1}\int_{C_{\sigma+i\tau}} z^\lambda e^{-z}\, dz. \tag{2.33}$$

Applying Cauchy's theorem to the sector bounded by an arc of radius N, a segment of the ray $C_{\sigma+i\tau}$ and a segment of the positive real axis, and then letting N tend to ∞, we get

$$\int_{C_{\sigma+i\tau}} z^\lambda e^{-z}\, dz = \int_0^\infty t^\lambda e^{-t}\, dt = \Gamma(\lambda + 1),$$

where $\Gamma(\lambda + 1)$ is the gamma function. Thus, for $\operatorname{Re}\lambda > -1$

$$F(t_+^\lambda e^{-\tau t}) = e^{(i\pi(\lambda+1))/2}\Gamma(\lambda + 1)(\sigma + i\tau)^{-\lambda-1}. \tag{2.34}$$

Replacing $\lambda + 1$ by $-\mu$ in (2.34), we get that for $\operatorname{Re}\mu < 0$

$$(\sigma + i\tau)^\mu = \frac{e^{i(\pi)/2\mu}}{\Gamma(-\mu)}\int_0^\infty t^{-\mu-1}e^{-\tau t}e^{it\sigma}\, dt. \tag{2.35}$$

Let us find $F^{-1}((\sigma + i\tau)^\mu \ln^p(\sigma + i\tau))$ for $\operatorname{Re}\mu < 0$. We denote by $\ln t_+$ the function that is equal to $\ln t$ for $t > 0$ and vanishes for $t < 0$. Differentiating (2.35) p times with respect to the parameter μ, we get

$$(\sigma + i\tau)^\mu \ln^p(\sigma + i\tau) = \sum_{k=0}^{p} C_{pk}(\mu)F(t_+^{-\mu-1}e^{-\tau t}\ln^k t_+), \quad \operatorname{Re}\mu < 0, \tag{2.36}$$

where

$$C_{pk}(\mu)=\frac{(-1)^k p!}{k!\,(p-k)!}\,\frac{d^{p-k}}{d\mu^{p-k}}\,\frac{e^{i(\pi)/2\mu}}{\Gamma(-\mu)},\qquad 0\leqslant k\leqslant p. \tag{2.37}$$

Let us now calculate $F^{-1}((\sigma+i\tau)^\lambda\ln^p(\sigma+i\tau))$ for $\operatorname{Re}\lambda\geqslant 0$. In this case there exists a nonnegative integer m such that $\lambda=m+\mu$ and $-1\leqslant\operatorname{Re}\mu<0$. Multiplying (2.36) by $(\sigma+i\tau)^m$, we get by virtue of (2.29) that

$$\begin{aligned}(\sigma+i\tau)^\lambda\ln^p(\sigma+i\tau)&=\sum_{k=0}^{p}C_{pk}(\mu)F\left[\left(i\frac{d}{dt}+i\tau\right)^m\left(t_+^{-\mu-1}e^{-\tau t}\ln^k t_+\right)\right]\\&=\sum_{k=0}^{p}C_{pk}(\mu)F\left[\left(\left(i\frac{d}{dt}\right)^m\left(t_+^{-\mu-1}\ln^k t_+\right)\right)e^{-\tau t}\right],\\&\qquad\lambda=m+\mu,\quad -1\leqslant\operatorname{Re}\mu<0.\end{aligned}\tag{2.38}$$

In (2.38) the derivatives of the regular functionals $t_+^{-\mu-1}e^{-\tau t}\ln^k t_+$ are taken in the sense of definition (1.16) and the Fourier transform is understood in the sense of (2.28).

Let us pass to the limit in (2.36) and (2.38) for $\tau\to+0$. When $\operatorname{Re}\mu<0$, the regular functionals $t_+^{-\mu-1}e^{-\tau t}\ln^k t_+$ converge to the regular functional $t_+^{-\mu-1}\ln^k t_+$ (see Example 1.5). Consequently, by virtue of (1.21), (1.27) and (2.31),

$$(\sigma+i0)^\mu\ln^p(\sigma+i0)=\sum_{k=0}^{p}C_{pk}(\mu)F(t_+^{-\mu-1}\ln^k t_+),\quad \operatorname{Re}\mu<0, \tag{2.39}$$

$$\begin{aligned}(\sigma+i0)^\lambda\ln^p(\sigma+i0)&=\sum_{k=0}^{p}C_{pk}(\mu)F\left[\left(i\frac{d}{dt}\right)^m\left(t_+^{-\mu-1}\ln^k t_+\right)\right],\\ \lambda=m+\mu,\qquad -1&\leqslant\operatorname{Re}\mu<0,\qquad m>0\text{–integer}.\end{aligned}\tag{2.40}$$

We note some special cases of (2.36) and (2.40). Suppose $\mu=-1$ and $p=1$ in (2.36). Since

$$C_{11}(-1)=i,\ C_{10}(-1)=\left.\frac{d}{d\mu}\frac{e^{i(\pi/2)\mu}}{\Gamma(-\mu)}\right|_{\mu=-1}=\frac{\pi}{2}-i\Gamma'(1),\ F\left(t_+^0e^{-\tau t}\right)=\frac{i}{\sigma+i\tau},$$

we get from (2.36) that

$$F\left(\ln t_+e^{-\tau t}\right)=-i\,\frac{\ln(\sigma+i\tau)}{\sigma+i\tau}-\frac{\pi/2-i\Gamma'(1)}{\sigma+i\tau}. \tag{2.41}$$

Suppose now $p=0$ and $\operatorname{Re}\lambda\geqslant 0,\ \lambda\neq 0,+1,+2,\dots$ in (2.40). Then

$$(\sigma+i0)^\lambda=\frac{e^{i(\pi/2)\mu}i^m}{\Gamma(-\mu)}F\left(\frac{d^m}{dt^m}t_+^{-\mu-1}\right),\quad \lambda=m+\mu,\ -1<\operatorname{Re}\mu<0.$$

Since

$$i^m = e^{i(\pi/2)m}, \Gamma(-\mu) = (-1)^m \mu(\mu + 1) \ldots (\mu + m - 1)\Gamma(-\lambda),$$

we get

$$(\sigma + i0)^\lambda = \frac{e^{i(\pi/2)\lambda}}{\Gamma(-\lambda)} F\left(\frac{(-1)^m}{\mu \ldots (\mu + m - 1)} \frac{d^m}{dt^m} t_+^{-\mu-1} \right) = \frac{e^{i(\pi/2)\lambda}}{\Gamma(-\lambda)} F(t_+^{-\lambda-1}), \tag{2.41'}$$

where $t_+^{-\lambda-1}$ is defined for Re $\lambda \geqslant 0, \lambda \neq 0, +1, +2, \ldots$ as in Example 1.6.

4. *Fourier transform of a functional with compact support.* A functional $f \in S'(\mathbf{R}^n)$ is said to have *compact support* if its support is contained in a ball $|x| < R$, i.e. $(f, \varphi) = 0$ for any $\varphi(x) \in C_0^\infty(\mathbf{R}^n)$ that vanishes for $|x| \leqslant R$.

LEMMA 2.4. *Any functional $g \in S'(\mathbf{R}^n)$ with compact support can be represented in the form*

$$g = \sum_{|k|=0}^{m} D^k g_k, \tag{2.42}$$

where the g_k are regular functionals defined by continuous functions $g_k(x)$ with compact support.

PROOF. Suppose $\chi(x) \in C_0^\infty(\mathbf{R}^n)$ and $\chi(x) = 1$ for $|x| \leqslant R$. Then $(g, (1 - \chi)\varphi) = 0$ for any $\varphi \in S(\mathbf{R}^n)$ and, consequently, $(g, \varphi) = (g, \chi\varphi)$. It follows from Theorem 1.1 that

$$(g, \varphi) = (g, \chi\varphi) = \sum_{|k|=0}^{m} (D^k h_k, \chi\varphi) = \sum_{|k|=0}^{m} (h_k, D^k(\chi\varphi)), \tag{2.43}$$

where the $h_k(x)$ are continuous functions, and hence

$$(g, \varphi) = \sum_{|k|=0}^{m} (h_k, D^k(\chi\varphi)) = \sum_{|p|=0}^{m} (g_p, D^p\varphi),$$

where the $g_p(x)$ are continuous functions with compact support.

THEOREM 2.1. *The Fourier transform of a functional $g \in S'(\mathbf{R}^n)$ with compact support is an entire function $\tilde{g}(\xi + i\eta)$ satisfying*

$$\left| D_\zeta^k \tilde{g}(\xi + i\eta) \right| \leqslant C_k (1 + |\xi| + |\eta|)^m e^{b|\eta|}, \qquad \zeta = \xi + i\eta, \quad k \geqslant 0. \tag{2.44}$$

PROOF. If $g_p(x)$ is a continuous function that vanishes for $|x| \geqslant b$, then its Fourier transform

$$\tilde{g}_p(\xi + i\eta) = \int_{|x|<b} g_p(x) e^{i(x, \xi + i\eta)} \, dx$$

is an entire function of $\zeta = \xi + i\eta$ that satisfies

$$\left|D_\zeta^k \tilde{g}_p(\xi + i\eta)\right| \leqslant \int_{|x|<b} |x|^{|k|} |g_p(x)| e^{-(x,\eta)}\, dx \leqslant C_k e^{b|\eta|}. \qquad (2.44')$$

By Lemma 2.4, a functional $g \in S'(\mathbf{R}^n)$ with compact support can be represented in the form (2.42). Consequently, $\tilde{g} = Fg$ is a regular functional corresponding to the entire function

$$\tilde{g}(\xi + i\eta) = \sum_{|p|=0}^{m} (\xi + i\eta)^p \tilde{g}_p(\xi + i\eta),$$

which satisfies (2.44) by virtue of (2.44′).

5. *Convolution of a generalized function and a fundamental function.* By the *convolution* of a generalized function $f \in S'$ and a fundamental function $\varphi \in S$ is meant the function

$$f * \varphi = \left(f(y), \overline{\varphi(x-y)}\right), \qquad (2.45)$$

where $\overline{\varphi(x-y)}$ is regarded as a function of y (x is fixed) and the notation $f(y)$ means that the functional f acts on the fundamental function with respect to the variable y. The function $f * \varphi$ is defined for any $x \in \mathbf{R}^n$.

LEMMA 2.5. *The function $f * \varphi$ is infinitely differentiable, and*

$$|D^p(f * \varphi)| \leqslant C_p(1 + |x|)^N, \qquad 0 \leqslant |p| < \infty. \qquad (2.46)$$

PROOF. By (2.45) and Theorem 1.1,

$$\begin{aligned} f * \varphi &= \sum_{|k|\leqslant m} D^k f_k * \varphi = \sum_{|k|\leqslant m} \left(D_y^k f_k(y), \overline{\varphi(x-y)}\right) \\ &= \sum_{|k|\leqslant m} \left(f_k(y), D_y^k \overline{\varphi(x-y)}\right) = \sum_{|k|\leqslant m} \left(f_k(y), \overline{D_x^k \varphi(x-y)}\right), \end{aligned} \qquad (2.47)$$

where $|f_k(y)| \leqslant C(1 + |y|)^N$. By Lemma 1.2 the function

$$\left(f_k(y), \overline{D_x^k \varphi(x-y)}\right) = \int_{-\infty}^{\infty} f_k(y) D_x^k \varphi(x-y)\, dy$$

belongs to the class C^∞ and satisfies (2.46). Consequently, $f * \varphi$ also belongs to C^∞ and satisfies (2.46).

EXAMPLE 2.4. Suppose $\varphi \in S$. Then

$$\delta * \varphi = \left(\delta(y), \overline{\varphi(x-y)}\right) = \varphi(x), \qquad \text{i.e. } \delta * \varphi = \varphi, \quad \forall \varphi \in S(\mathbf{R}^n).$$

Analogously, if $P(D) = \sum_{|k|\leqslant m} a_k D^k$ is a differential operator with constant coefficients, then

$$P(D)\delta * \varphi = \sum_{|k|\leqslant m} a_k \left(D_y^k \delta(y), \overline{\varphi(x-y)}\right) = \sum_{|k|\leqslant m} a_k D^k \varphi(x),$$

i.e. $P(D)\delta * \varphi = P(D)\varphi$ for any $\varphi(x) \in S(\mathbf{R}^n)$.

LEMMA 2.6. *Suppose* $f \in S'$ *and* $\varphi, \psi \in S$. *Then*

$$(f * \varphi, \psi) = (f, \varphi_1 * \psi), \tag{2.48}$$

where $\varphi_1(x) = \overline{\varphi(-x)}$.

PROOF. By (2.47),

$$f * \varphi = \sum_{|k| \leq m} \int_{-\infty}^{\infty} f_k(y) D_x^k \varphi(x - y) dy,$$

where $|f_k(y)| \leq C(1 + |y|)^N$. Consequently,

$$(f * \varphi, \psi) = \sum_{|k| \leq m} \int_{-\infty}^{\infty} \int_{-\infty}^{\infty} f_k(y) D_x^k \varphi(x - y) \overline{\psi(x)} \, dx \, dy. \tag{2.49}$$

Further, by (1.28),

$$(f, \varphi_1 * \psi) = \sum_{|k| \leq m} \left(D_y^k f_k(y), \varphi_1 * \psi\right) = \sum_{|k| \leq m} \int_{-\infty}^{\infty} f_k(y) \overline{D_y^k(\varphi_1 * \psi)} \, dy. \tag{2.50}$$

On the other hand, by definition of the convolution of fundamental functions, we have

$$D_y^k(\varphi_1 * \psi) = \int_{-\infty}^{\infty} D_y^k \varphi_1(y - x) \psi(x) dx = \int_{-\infty}^{\infty} \overline{D_x^k \varphi(x - y)} \, \psi(x) dx. \tag{2.51}$$

Therefore, substituting (2.51) and (2.50), we get

$$(f, \varphi_1 * \psi) = \sum_{|k| \leq m} \int_{-\infty}^{\infty} \left(\int_{-\infty}^{\infty} f_k(y) D_x^k \varphi(x - y) \overline{\psi(x)} \, dx \right) dy, \tag{2.52}$$

and the lemma follows upon comparing (2.49) with (2.52).

6. *Fourier transform of a convolution.*

THEOREM 2.2. *Suppose* $f \in S'$ *and* $\varphi \in S$. *Then*

$$F(f * \varphi) = \tilde{\varphi}(\xi) \tilde{f}. \tag{2.53}$$

PROOF. By Lemma 2.6,

$$\left(F(f * \varphi), \tilde{\psi}\right) = (2\pi)^n (f * \varphi, \psi) = (2\pi)^n (f, \varphi_1 * \psi), \tag{2.54}$$

where $\varphi_1(x) = \overline{\varphi(-x)}$. We note that

$$F\varphi_1 = \int_{-\infty}^{\infty} \varphi_1(x) e^{i(x,\xi)} \, dx = \int_{-\infty}^{\infty} \overline{\varphi(-x)} \, e^{i(x,\xi)} \, dx = \overline{\tilde{\varphi}(\xi)} \,. \tag{2.55}$$

Consequently, by (2.4),

$$(2\pi)^n (f, \varphi_1 * \psi) = \left(\tilde{f}, F(\varphi_1 * \psi)\right) = \left(\tilde{f}, \overline{\tilde{\varphi}(\xi)} \, \tilde{\psi}(\xi)\right) = \left(\tilde{\varphi}(\xi) \tilde{f}, \tilde{\psi}\right). \tag{2.56}$$

Thus $(F(f * \varphi), \tilde{\psi}) = (\tilde{\varphi}(\xi)\tilde{f}, \tilde{\psi})$ for $\forall \psi \in S$, i.e. $F(f * \varphi) = \tilde{\varphi}(\xi)\tilde{f}$. Theorem 2.2 is proved.

7. *Convolution of a generalized function and a generalized function with compact support.*

LEMMA 2.7. *Suppose $g \in S'$ is a generalized function with compact support and $\varphi \in S$. Then $g * \varphi \in S$, and the operator of convolution with g is a continuous operator in S.*

PROOF. By Lemma 2.4,

$$g * \varphi = \sum_{|k| \leq m} D^k g_k * \varphi = \sum_{|k| \leq m} \int_{-\infty}^{\infty} g_k(y) D_x^k \varphi(x - y) dy, \tag{2.57}$$

where the $g_k(y)$ are continuous functions with compact support. Consequently, applying Lemma 1.2 to (2.57), we get that $g * \varphi \in S$ and

$$\|[g * \varphi]\|_p \leq C_p \|[\varphi]\|_{p+m+n+1}, \qquad p = 0, 1, \ldots. \tag{2.58}$$

Lemma 2.7 is proved.

We now proceed to define the convolution of an arbitrary generalized function f and a generalized function g with compact support. Let g_1 denote the functional defined by

$$(g_1, \varphi) = \overline{(g, \overline{\varphi(-x)})}, \qquad \forall \varphi \in S. \tag{2.59}$$

Clearly, g_1 is also a generalized function with compact support. When g is a regular functional defined by a function $g(x)$, the functional g_1 is also regular and corresponds to the function $g_1(x) = \overline{g(-x)}$. In fact, making the change of variables $x = -y$, we get

$$(g_1, \varphi) = \overline{(g, \overline{\varphi(-x)})} = \overline{\int_{-\infty}^{\infty} g(x) \overline{\varphi(-x)}\, dx} = \int_{-\infty}^{\infty} \overline{g(-x)}\, \varphi(x)\, dx.$$

By Lemma 2.7, $g_1 * \varphi \in S$ and hence the following continuous functional on S is defined:

$$(h, \varphi) = (f, g_1 * \varphi). \tag{2.60}$$

The functional h is called the *convolution* of a generalized function f and a generalized function g with compact support.

When $g \in C_0^\infty(\mathbf{R}^n)$, it follows from Lemma 2.6 that (2.60) agrees with the definition (2.45) of the convolution of a generalized function and a fundamental function.

THEOREM 2.3. *Suppose $f \in S'$ and g is a generalized function with compact support. Then*

$$F(f * g) = \tilde{g}(\xi) \tilde{f}. \tag{2.61}$$

PROOF. By Theorem 2.2, $F(g_1 * \varphi) = \tilde{\varphi}(\xi)\tilde{g}_1$. Further, by (2.59) and (2.55),

$$(\tilde{g}_1, \tilde{\varphi}) = (2\pi)^n \overline{(g, \overline{\varphi(-x)})} = \overline{(\tilde{g}, \overline{\tilde{\varphi}(\xi)})} = \int_{-\infty}^{\infty} \overline{\tilde{g}(\xi)}\ \overline{\tilde{\varphi}(\xi)}\, d\xi, \quad (2.62)$$

i.e. $\tilde{g}_1(\xi) = \overline{\tilde{g}(\xi)}$. Consequently (see (2.60) and (2.62)),

$$(F(f * g), \tilde{\varphi}) = (\tilde{f}, F(g_1 * \varphi)) = (\tilde{f}, \overline{\tilde{g}(\xi)}\,\tilde{\varphi}(\xi)) = (\tilde{g}(\xi)\tilde{f}, \varphi(\xi)),$$

i.e. $F(f * g) = \tilde{g}(\xi)\tilde{f}$. Theorem 2.3 is proved.

8. *Fourier transform of a direct product of generalized functions.*

THEOREM 2.4. *Suppose $f_1 \in S'(\mathbf{R}^{n-1})$, $f_2 \in S'(\mathbf{R}^1)$, and $f = f_1 \times f_2 \in S'(\mathbf{R}^n)$ is the direct product of f_1 and f_2. Then*

$$F(f_1 \times f_2) = \tilde{f}_1 \times \tilde{f}_2, \quad (2.63)$$

where $\tilde{f}_1 \in S'(\mathbf{R}^{n-1})$ is the Fourier transform of f_1 and $\tilde{f}_2 \in S'(\mathbf{R}^1)$ is the Fourier transform of f_2.

PROOF. Let $\varphi_1(x') = (f_2, \varphi(x', x_n))$. By (1.36),

$$\tilde{\overline{\varphi}}_1(\xi') = \int_{-\infty}^{\infty} \overline{\varphi_1(x')}\, e^{i(x',\xi')}\, dx' = \sum_{k=0}^{m} \int_{-\infty}^{\infty} \int_{-\infty}^{\infty} \overline{f_{2k}(x_n)}\, D_n^k \varphi(x', x_n)$$

$$\times e^{i(x',\xi')}\, dx_n\, dx' = \sum_{k=0}^{m} \int_{-\infty}^{\infty} \overline{f_{2k}(x_n)}\, D_n^k \tilde{\psi}(\xi', x_n) dx_n = \overline{(f_2, \overline{\tilde{\psi}(\xi', x_n)})}, \quad (2.64)$$

where

$$\tilde{\psi}(\xi', x_n) = F_{x'}\varphi(x', x_n) = \int_{-\infty}^{\infty} \varphi(x', x_n) e^{i(x',\xi')}\, dx'.$$

The order of integration can be changed in (2.64) since $\varphi(x', x_n) \in S(\mathbf{R}^n)$. According to (2.28),

$$\tilde{\overline{\varphi}}_1(\xi') = \overline{(f_2, \overline{\tilde{\psi}(\xi', x_n)})} = \frac{1}{2\pi} \overline{(\tilde{f}_2, \overline{\tilde{\varphi}(\xi', \xi_n)})}, \quad (2.65)$$

where $\tilde{\varphi}(\xi', \xi_n) = F_{x_n}\tilde{\psi}(\xi', x_n) = F\varphi$. Consequently, by (2.28), (2.64) and (2.65),

$$(F(f_1 \times f_2), \tilde{\varphi}) = (2\pi)^n (f_1 \times f_2, \varphi) = (2\pi)^n (f_1, \overline{\varphi_1(x')}) = 2\pi(\tilde{f}_1, \tilde{\overline{\varphi}}_1)$$
$$= (\tilde{f}_1, \overline{(\tilde{f}_2, \overline{\tilde{\varphi}(\xi', \xi_n)})}) = (\tilde{f}_1 \times \tilde{f}_2, \tilde{\varphi}), \qquad \forall \varphi \in S(\mathbf{R}^n). \quad (2.66)$$

Theorem 2.4 is proved.

EXAMPLE 2.5. Suppose $f_1 \in S'(\mathbf{R}^{n-1})$ and $f = f_1 \times \delta(x_n)$. Then $F(f_1 \times \delta(x_n)) = \tilde{f}_1 \times F\delta = \tilde{f}_1 \times 1$. In particular, if $\tilde{f}_1$ is the regular functional on $S(\mathbf{R}^{n-1})$ corresponding to $\tilde{f}_1(\xi')$, then $\tilde{f}_1 \times 1$ is the regular functional on $S(\mathbf{R}^n)$ corresponding to the same function $\tilde{f}_1(\xi')$:

$$(\tilde{f}_1 \times 1, \tilde{\varphi}) = \int_{-\infty}^{\infty} \tilde{f}_1(\xi')\, \overline{\tilde{\varphi}(\xi', \xi_n)}\, d\xi'\, d\xi_n, \qquad \forall \tilde{\varphi} \in S(\mathbf{R}^n).$$

9. *Fourier transform of a homogeneous function.*

EXAMPLE 2.6. Suppose $A_0(\xi) \in C^\infty$ for $\xi = (\xi_1, \dots, \xi_n) \neq 0$ and is a homogeneous function of degree $\alpha + i\beta$, i.e.

$$A_0(t\xi) = t^{\alpha+i\beta}A_0(\xi), \qquad \forall t > 0. \tag{2.67}$$

The class of such functions will be denoted by $O^\infty_{\alpha+i\beta}$. Suppose $-n < \alpha < 0$. Then $A_0(\xi)$ defines a regular functional on $S(\mathbf{R}^n)$.

We will find the inverse Fourier transform $a_0 = F^{-1}A_0(\xi)$. Let

$$a_0(x, \varepsilon) = F^{-1}A_0(\xi)e^{-\varepsilon|\xi|} = \frac{1}{(2\pi)^n}\int_{-\infty}^{\infty} A_0(\xi)e^{-\varepsilon|\xi| - i(x,\xi)}\, d\xi, \qquad \varepsilon > 0. \tag{2.68}$$

Going over to spherical coordinates $\rho = |\xi|$, $\eta = \xi/|\xi|$ in (2.68), we get

$$a_0(x, \varepsilon) = \frac{1}{(2\pi)^n}\int_{S^{n-1}}\left(\int_0^\infty \rho^{\alpha+i\beta+n-1}e^{-\rho(\varepsilon+i(x,\eta))}\, d\rho\right)A_0(\eta)\,ds_\eta,$$

where S^{n-1} is the sphere of unit radius in $\mathbf{R}^n$ and ds_η is the surface element of S^{n-1}. Analogously to (2.34), we have

$$\int_0^\infty \rho^{\alpha+i\beta+n-1}e^{-\rho(\varepsilon+i(x,\eta))}\, d\rho = \frac{e^{-i(\pi/2)(\alpha+i\beta+n)}\Gamma(\alpha+i\beta+n)}{((x,\eta) - i\varepsilon)^{\alpha+i\beta+n}}. \tag{2.69}$$

Consequently,

$$\begin{aligned} a_0(x, \varepsilon) &= \frac{1}{(2\pi)^n}e^{-i(\pi/2)(\alpha+i\beta+n)}\Gamma(\alpha+i\beta+n)\int_{S^{n-1}}\frac{A_0(\eta)\,ds_\eta}{((x,\eta) - i\varepsilon)^{\alpha+i\beta+n}} \\ &= \frac{1}{(2\pi)^n}\,\frac{e^{-i(\pi/2)(\alpha+i\beta+n)}\Gamma(\alpha+i\beta+n)}{|x|^{\alpha+i\beta+n}}\int_{S^{n-1}}\frac{A_0(\eta)\,ds_\eta}{((y,\eta) - i\varepsilon_1)^{\alpha+i\beta+n}}, \end{aligned} \tag{2.70}$$

where $y = x/|x|$ and $\varepsilon_1 = \varepsilon/|x|$. Clearly, $a_0(x, \varepsilon) \in C^\infty$ for $\varepsilon > 0$. As $\varepsilon \to 0$,

$$(A_0(\xi)e^{-\varepsilon|\xi|}, \tilde{\varphi}(\xi)) \to (A_0(\xi), \tilde{\varphi}(\xi))$$

for any $\tilde{\varphi} \in S(\mathbf{R}^n)$ (see Example 1.5). It consequently follows from (2.31) that the regular functional $a_0(x, \varepsilon)$ converges in S' to the functional $a_0 = F^{-1}A_0(\xi)$ as $\varepsilon \to 0$.

Let us show that a_0 is a regular functional corresponding to a function $a_0(x) \in O^\infty_{-\alpha-i\beta-n}$. Suppose $\varphi(t) \in C_0^\infty(\mathbf{R}^1)$, $\varphi(t) = 1$ for $|t| < 1/4$ and $\varphi(t) = 0$ for $|t| > 1/2$. We have

$$a_0(x, \varepsilon) = a_1(x, \varepsilon) + a_2(x, \varepsilon), \tag{2.71}$$

where

$$a_1(x, \varepsilon) = \frac{C_{\alpha+i\beta}}{|x|^{\alpha+i\beta+n}} \int_{S^{n-1}} \frac{A_0(\eta)\varphi((y, \eta))ds_\eta}{((y, \eta) - i\varepsilon_1)^{\alpha+i\beta+n}}, \tag{2.72}$$

$$a_2(x, \varepsilon) = \frac{C_{\alpha+i\beta}}{|x|^{\alpha+i\beta+n}} \int_{S^{n-1}} \frac{A_0(\eta)(1 - \varphi((y, \eta)))}{((y, \eta) - i\varepsilon_1)^{\alpha+i\beta+n}} ds_\eta,$$

$$C_{\alpha+i\beta} = \frac{e^{-i(\pi/2)(\alpha+i\beta+n)}\Gamma(\alpha + i\beta + n)}{(2\pi)^n}. \tag{2.72'}$$

Clearly, $a_2(x, \varepsilon) \to a_2(x, 0) \in O^{\infty}_{-\alpha-i\beta-n}$ uniformly in $y \in S^{n-1}$ as $\varepsilon \to 0$.

We introduce in a neighborhood of the set $\operatorname{supp} \varphi((y, \eta)) \subset S^{n-1}$ new coordinates (t, η'), where $t = (y, \eta)$ and η' ranges over the unit sphere S^{n-2} in $\mathbf{R}^{n-1}$. Then

$$a_1(x, \varepsilon) = \frac{C_{\alpha+i\beta}}{|x|^{\alpha+i\beta+n}} \int_{-\infty}^{\infty} \int_{S^{n-1}} \frac{\varphi(t)A_1(y, t, \eta')}{(t - i\varepsilon_1)^{\alpha+i\beta+n}} ds_{\eta'}\, dt, \tag{2.73}$$

where $A_1(y, t, \eta') \in C^\infty$ with respect to the totality of variables. Let m be a positive integer such that $m + 1 > \alpha + n$. Then, integrating by parts m times with respect to t in (2.73), we get, as in Example 1.7, that $a_1(x, \varepsilon) \to a_1(x) \in O^{\infty}_{-\alpha-i\beta-n}$ uniformly in $y \in S^{n-1}$ as $\varepsilon \to +0$. Thus $a_0(x, \varepsilon) \to a_0(x) \in O^{\infty}_{-\alpha-i\beta-n}$ uniformly in $y = x/|x|$ as $\varepsilon \to 0$. Consequently, the regular functional $a_0(x, \varepsilon)$ converges in S' to a regular functional $a_0(x) = F^{-1}A_0(\xi)$.

Let us prove that

$$\int_{S^{n-1}} a_0(y)ds_y = \frac{2^{\alpha+i\beta}\Gamma((\alpha + i\beta + n)/2)}{\pi^{n/2}\Gamma(-(\alpha + i\beta)/2)} \int_{S^{n-1}} A_0(\eta)ds_\eta. \tag{2.74}$$

By (2.28), $(A_0(\xi), \tilde{\varphi}(\xi)) = (2\pi)^n(a_0(x), \varphi(x))$ for any $\varphi(x) \in S(\mathbf{R}^n)$. We take $\varphi(x) = e^{-|x|^2/2}$. Then

$$\tilde{\varphi}(\xi) = (2\pi)^{n/2}e^{-|\xi|^2/2}.$$

Consequently,

$$(2\pi)^{n/2}\int_{-\infty}^{\infty} A_0(\xi)e^{-|\xi|^2/2}\, d\xi = (2\pi)^n \int_{-\infty}^{\infty} a_0(x)e^{-|x|^2/2}\, dx. \tag{2.75}$$

Going over to spherical coordinates $\rho = |\xi|$, $\eta = \xi/|\xi|$ and $r = |x|$, $y = x/|x|$ in (2.75), we get

$$\int_0^\infty \rho^{\alpha+i\beta+n-1}e^{-\rho^2/2}\, d\rho \int_{S^{n-1}} A_0(\eta)ds_\eta$$
$$= (2\pi)^{n/2}\int_0^\infty r^{-\alpha-i\beta-1}e^{-r^2/2}\, dr \int_{S^{n-1}} a_0(y)ds_y.$$

Thus (2.74) holds, since

$$\int_0^\infty \rho^\lambda e^{-\rho^2/2}\,d\rho = 2^{(\lambda-1)/2}\int_0^\infty t^{(\lambda+1)/2-1}e^{-t}\,dt = 2^{(\lambda-1)/2}\Gamma\left(\frac{\lambda+1}{2}\right). \tag{2.76}$$

EXAMPLE 2.7. Suppose $A_0(\xi) \in O^\infty_{\alpha+i\beta}$, $\alpha \leqslant -n$ and $\alpha+i\beta+n \neq 0, -1, \ldots$. Then the function $A_0(\xi)$ does not define a regular functional on $S(\mathbf{R}^n)$.

Let m be a positive integer such that $-n < \alpha + m < 0$ and let $\mathcal{C}_0$ denote the functional acting on the fundamental functions $\tilde{\varphi}(\xi) \in S(\mathbf{R}^n)$ according to the formula (cf. (1.23))

$$(\mathcal{C}_0, \tilde{\varphi}) = (-1)^m \int_0^\infty \frac{r^{\alpha+i\beta+m+n-1}}{(\alpha+i\beta+m+n-1)\ldots(\alpha+i\beta+n)}\,\frac{d^m\psi(r)}{dr^m}\,dr, \tag{2.77}$$

where

$$\psi(r) = \int_{S^{n-1}} A_0(\eta)\,\overline{\tilde{\varphi}(r\eta)}\,ds_\eta. \tag{2.78}$$

Since

$$\frac{\partial\varphi(r\eta)}{\partial r} = \sum_{k=1}^n \frac{\partial\varphi(r\eta)}{\partial\xi_k}\eta_k, \qquad \xi_k = r\eta_k,$$

it follows that $\psi(r) \in C^\infty$ for $r \geqslant 0$ and that $d^p\psi(r)/dr^p$, $p \geqslant 0$, decreases more rapidly than any negative power of r as $r \to \infty$. If

$$N > \max(\alpha + m + n + 1, m),$$

then

$$|d^m\psi(r)/dr^m| \leqslant C_N(1+r)^{-N}\|[\tilde{\varphi}]\|_N$$

and hence $|(\mathcal{C}_0, \tilde{\varphi})| \leqslant C_N^{(1)}\|[\tilde{\varphi}]\|_N$. Thus $\mathcal{C}_0$ is a continuous functional on $S(\mathbf{R}^n)$.

Let us calculate the inverse Fourier transform $a_0 = F^{-1}\mathcal{C}_0$. By definition (2.28),

$$(a_0, \varphi) = \frac{1}{(2\pi)^n}(\mathcal{C}_0, \tilde{\varphi}), \qquad \forall\varphi \in S(\mathbf{R}^n),$$

where $\tilde{\varphi} = F\varphi$. Substituting $\tilde{\varphi}(r\eta) = \int_{-\infty}^\infty \varphi(x)e^{i(x,r\eta)}\,dx$ in (2.78), we get

$$\psi(r) = \int_{-\infty}^\infty \left(\int_{S^{n-1}} A_0(\eta)e^{-i(x,r\eta)}\,ds_\eta\right)\overline{\varphi(x)}\,dx. \tag{2.79}$$

Consequently,

$$(a_0, \varphi) = \frac{1}{(2\pi)^n}(\mathcal{a}_0, \tilde{\varphi}) = \frac{i^m}{(2\pi)^n}\int_0^\infty \frac{r^{\alpha+i\beta+m+n-1}}{(\alpha+i\beta+m+n-1)\dots(\alpha+i\beta+n)}$$

$$\times\left(\int_{-\infty}^{\infty}\left(\int_{S^{n-1}} A_0(\eta)(x,\eta)^m e^{-i(x,r\eta)}\,ds_\eta\right)\overline{\varphi(x)}\,dx\right)dr. \tag{2.80}$$

The order of integration in (2.80) can be changed if we replace $e^{-i(x,r\eta)}$ by $e^{-i(x,r\eta)-\varepsilon r}$, and then let ε tend to zero. In this way, calculating the integral with respect to r by a formula of form (2.69), we get

$$(a_0, \varphi)$$

$$= \lim_{\varepsilon\to 0}\int_{-\infty}^{\infty}\left(\int_{S^{n-1}}\left(\int_0^\infty \frac{i^m r^{\alpha+i\beta+m+n-1}e^{-r(\varepsilon+i(x,\eta))}}{(2\pi)^n(\alpha+i\beta+m+n-1)\dots(\alpha+i\beta+n)}\,dr\right)\right.$$

$$\left.\times\,(x,\eta)^m A_0(\eta)\,ds_\eta\right)\tilde{\varphi}(x)dx = \int_{-\infty}^{\infty} a_0(x)\,\overline{\varphi(x)}\,dx, \tag{2.81}$$

where

$$a_0(x) = \lim_{\varepsilon\to 0}\frac{i^m e^{-i(\pi/2)(\alpha+i\beta+m+n)}\Gamma(\alpha+i\beta+m+n)}{(2\pi)^n(\alpha+i\beta+m+n-1)\dots(\alpha+i\beta+n)}$$

$$\times\int_{S^{n-1}}\frac{(x,\eta)^m A_0(\eta)\,ds_\eta}{((x,\eta)-i\varepsilon)^{\alpha+i\beta+m+n}}$$

$$= \frac{e^{-i(\pi/2)(\alpha+i\beta+n)}\Gamma(\alpha+i\beta+n)}{(2\pi)^n}\int_{S^{n-1}}\frac{A_0(\eta)\,ds_\eta}{((x,\eta)-i0)^{\alpha+i\beta+n}}. \tag{2.82}$$

Here we have made use of the fact that

$$\Gamma(\alpha+i\beta+m+n)$$
$$= (\alpha+i\beta+n)\dots(\alpha+i\beta+m+n-1)\Gamma(\alpha+i\beta+n). \tag{2.83}$$

The limit for $\varepsilon\to 0$ in (2.82) exists uniformly with respect to $y = x/|x|$ since $\alpha + n \leqslant 0$. As in the case $\alpha + n > 0$, it can be shown that $a_0(x) \in O^\infty_{-\alpha-i\beta-n}$. Thus $a_0(x) = F^{-1}\mathcal{a}_0$ is a regular funtional on $S(\mathbf{R}^n)$.

EXAMPLE 2.8. Suppose $A_0(\xi) \in O_\alpha^\infty$, where $\alpha \leqslant -n$ is an integer. We define a functional $\mathcal{a}_0$ by

$$(\mathcal{a}_0, \tilde{\varphi}) = -\int_0^\infty \frac{\ln r}{(-\alpha-n)!}\,\frac{d^{|\alpha|-n+1}}{dr^{|\alpha|-n+1}}\left(\int_{S^{n-1}} A_0(\eta)\,\overline{\tilde{\varphi}(r\eta)}\,ds_\eta\right)dr,$$

$$\forall\tilde{\varphi}\in S(\mathbf{R}^n). \tag{2.84}$$

It can be shown in the same way as in Example 2.7 that $\mathcal{Q}_0$ is a continuous functional on $S(\mathbf{R}^n)$.

Let $a_0 = F^{-1}\mathcal{Q}_0$. Analogously to (2.80),

$$(a_0, \varphi) = \frac{1}{(2\pi)^n}(\mathcal{Q}_0, \tilde{\varphi}) = \frac{-(-i)^{|\alpha|-n+1}}{(2\pi)^n(-\alpha-n)!} \times \int_0^\infty \ln r\left(\int_{-\infty}^{\infty}\left(\int_{S^{n-1}}(x,\eta)^{|\alpha|-n+1}A_0(\eta)e^{-i(x,\eta)r}\,ds_\eta\right)\overline{\varphi(x)}\,dx\right)dr. \tag{2.85}$$

By (2.41),

$$\int_0^\infty \ln r e^{-r(\varepsilon+i(x,\eta))}\,dr = -\frac{i\ln(-(x,\eta)+i\varepsilon)}{-(x,\eta)+i\varepsilon} - \frac{\pi/2 - i\Gamma'(1)}{-(x,\eta)+i\varepsilon}. \tag{2.86}$$

Calculating the integral with respect to r in (2.85) by (2.86), we get, analogously to (2.81),

$$(a_0, \varphi) = \lim_{\varepsilon\to 0}\frac{-(-i)^{|\alpha|-n+1}}{(2\pi)^n(-\alpha-n)!}\int_0^\infty \ln r\left(\int_{-\infty}^{\infty}\left(\int_{S^{n-1}}(x,\eta)^{|\alpha|-n+1}A_0(\eta) \times e^{-r(\varepsilon+i(x,\eta))}\,ds_\eta\right)\overline{\varphi}(x)dx\right)dr = \int_{-\infty}^{\infty} a_0(x)\overline{\varphi}(x)dx, \tag{2.87}$$

where

$$a_0(x) = \lim_{\varepsilon\to 0}\frac{-(-i)^{|\alpha|-n+1}}{(2\pi)^n(-\alpha-n)!}\int_{S^{n-1}}(x,\eta)^{|\alpha|-n+1}A_0(\eta) \times\left[\frac{i\ln(-(x,\eta)+i\varepsilon)}{(x,\eta)-i\varepsilon} + \frac{\pi/2 - i\Gamma'(1)}{(x,\eta)-i\varepsilon}\right]ds_\eta = \frac{-(-i)^{|\alpha|-n+1}}{(2\pi)^n(|\alpha|-n)!} \times\int_{S^{n-1}} A_0(\eta)(x,\eta)^{|\alpha|-n}[i\ln(-(x,\eta)+i0) + \pi/2 - i\Gamma'(1)]\,ds_\eta. \tag{2.88}$$

The limit for $\varepsilon \to 0$ exists since $|\alpha| \geqslant n$. As in the case when $-n < \alpha < 0$, it can be proved that $a_0(x) \in C^\infty$ for $x \neq 0$.

REMARK 2.1. Suppose $\tilde{\varphi}(\xi) = 0$ for $|\xi| < \varepsilon$ in (2.77) and (2.84). Then $\psi(r) = 0$ for $r < \varepsilon$. Consequently, integrating by parts in (2.77) and (2.84), we get

$$(\mathcal{Q}_0, \tilde{\varphi}) = \int_{-\infty}^{\infty} A_0(\xi)\,\overline{\tilde{\varphi}(\xi)}\,d\xi. \tag{2.89}$$

The functionals (2.77) and (2.84) are not the only functionals satisfying (2.89) at those functions $\tilde{\varphi}(\xi) \in S(\mathbf{R}^n)$ that vanish for $|\xi| < \varepsilon$. The functionals

$$\mathcal{Q}_0' = \mathcal{Q}_0 + \sum_{|k| < N} c_k D^k \delta(\xi)$$

clearly also have this property for any coefficients c_k. Among all such functionals, the functional (2.77) is characterized by the fact that $a_0(x) = F^{-1}\mathcal{Q}_0$ is a homogeneous function of class $O^{\infty}_{-\alpha-i\beta-n}$. We note that $a_0'(x) = F^{-1}\mathcal{Q}_0'$ differs from $a_0(x) = F^{-1}\mathcal{Q}_0$ by the polynomial $\sum_{|k|<N} c_k^{(1)} x^k$.

REMARK 2.2. Let us determine when the function $a_0(x)$ defined by (2.88) is homogeneous. Since

$$\ln(-(x,\eta) + i0) = \ln|x| + \ln\left(-\left(\frac{x}{|x|}, \eta\right) + i0\right),$$

we have

$$a_0(x) = a_0^{(1)}(x)\ln|x| + a_0^{(2)}(x), \tag{2.90}$$

where

$$a_0^{(1)}(x) = -\frac{(-i)^{|\alpha|-n}}{(2\pi)^n(|\alpha|-n)!}\int_{S^{n-1}} A_0(\eta)(x,\eta)^{|\alpha|-n}\, ds_\eta, \tag{2.91}$$

$$a_0^{(2)}(x) = -\frac{(-i)^{|\alpha|-n+1}}{(2\pi)^n(|\alpha|-n)!}\int_{S^{n-1}} A_0(\eta)(x,\eta)^{|\alpha|-n} \times\left[i\ln(-(x/|x|,\eta) + i0) + \frac{\pi}{2} - i\Gamma'(1)\right] ds_\eta. \tag{2.92}$$

Clearly, $a_0^{(1)}(x) \in O^{\infty}_{|\alpha|-n}$ and $a_0^{(2)}(x) \in O^{\infty}_{|\alpha|-n}$, with $a_0^{(1)}(x)$ being a homogeneous polynomial in x. The function $a_0(x)$ will be homogeneous if and only if $a_0^{(1)}(x) \equiv 0$ or, equivalently,

$$\int_{S^{n-1}} \eta^k A_0(\eta)\, ds_\eta = 0, \qquad \forall |k| = |\alpha| - n, \tag{2.93}$$

where $k = (k_1, \ldots, k_n)$, $|k| = k_1 + \ldots + k_n$, in which case $a_0(x) = a_0^{(2)}(x) \in O^{\infty}_{|\alpha|-n}$.

§3. Pseudodifferential operators

1. *The class* S_α^0. Let $A(\xi)$ be a measurable function satisfying

$$|A(\xi)| \leqslant C(1 + |\xi|)^\alpha. \tag{3.1}$$

The class of such functions will be denoted by S_α^0. By a *pseudodifferential operator* (p.o.) is meant an operator A defined on $S(\mathbf{R}^n)$ by a formula of the form

$$Au = \frac{1}{(2\pi)^n}\int_{-\infty}^{\infty} A(\xi)\tilde{u}(\xi)e^{-i(x,\xi)}\, d\xi, \tag{3.2}$$

where $\tilde{u}(\xi)$ is the Fourier transform of $u(x) \in S(\mathbf{R}^n)$. The function $A(\xi)$ is called the *symbol* of A. If $A(\xi)$ is a polynomial in ξ, i.e. $A(\xi) = \sum_{|k|\leqslant m} a_k\xi^k$, then, by (3.2),

$$Au = \frac{1}{(2\pi)^n}\int_{-\infty}^{\infty} \sum_{|k|\leq m} a_k\xi^k\tilde{u}(\xi)e^{-i(x,\xi)}\,d\xi = \sum_{|k|\leq m} a_k D^k u(x), \tag{3.3}$$

i.e. A is a differential operator. Thus the class of pseudodifferential operators contains the class of differential operators. We will sometimes denote a pseudodifferential operator with symbol $A(\xi)$ by $A(D)$ in analogy with differential operators.

Suppose $\alpha < -n$ in (3.1). Then $A(\xi)$ is an absolutely integrable function and hence $a(x) = F^{-1}A(\xi)$ is a bounded continuous function. By virtue of the formula for the Fourier transform of a convolution,

$$Au = F^{-1}(A(\xi)\tilde{u}(\xi)) = \int_{-\infty}^{\infty} a(x-y)u(y)dy, \tag{3.4}$$

i.e. when $A(\xi) \in S_\alpha^0$ and $\alpha < -n$, the p.o. A is an integral operator of convolution type.

When $\alpha \geq -n$, there exists a positive integer m such that $\alpha + n < 2m$. Let

$$A_1(\xi) = \frac{A(\xi)}{(1+|\xi|^2)^m}.$$

Then $|A_1(\xi)| \leq C(1+|\xi|)^{\alpha-2m}$, and hence $A_1(\xi)$ is absolutely integrable. Let $a_1(x) = F^{-1}A_1(\xi)$. Then

$$Au = (-\Delta+1)^m\int_{-\infty}^{\infty} a_1(x-y)u(y)dy = \int_{-\infty}^{\infty} a_1(x-y)(-\Delta+1)^m u(y)dy, \tag{3.5}$$

where $\Delta = \sum_{k=1}^n \frac{\partial^2}{\partial x_k^2}$ is the Laplace operator. Thus the p.o. A can be represented in the form of an integrodifferential operator.

We note that $A(\xi)\tilde{u}(\xi)$ satisfies the estimate

$$|A(\xi)\tilde{u}(\xi)| \leq C_N(1+|\xi|)^{-N}, \tag{3.6}$$

for any N, since $\tilde{u}(\xi) \in S(\mathbf{R}^n)$. Consequently, $Au = F^{-1}A(\xi)\tilde{u}(\xi)$ is a bounded infinitely differentiable function.

2. *P.o.'s with a homogeneous symbol.* It is possible to define pseudodifferential operators for a wider class of symbols than S_α^0.

Let $\mathcal{A} \in S'(\mathbf{R}^n)$ be an arbitrary generalized function. Then the product $\mathcal{A}\tilde{u}(\xi) \in S'(\mathbf{R}^n)$ is defined for any $\tilde{u}(\xi) \in S(\mathbf{R}^n)$ by (1.18). We now define a pseudodifferential operator A acting from $S(\mathbf{R}^n)$ into $S'(\mathbf{R}^n)$ by means of the formula

$$Au = F^{-1}(\mathcal{Q}\tilde{u}), \qquad \tilde{u} = Fu, \qquad u \in S(\mathbf{R}^n), \tag{3.7}$$

where F^{-1} is defined by (2.28).

If $u_n(x) \to u(x)$ in $S(\mathbf{R}^n)$, then $\tilde{u}_n(\xi) \to \tilde{u}(\xi)$ in $S(\mathbf{R}^n)$ by virtue of Lemma 2.1, $\mathcal{Q}\tilde{u}_n \to \mathcal{Q}\tilde{u}$ in $S'(\mathbf{R}^n)$ by virtue of (1.18), and $F^{-1}\mathcal{Q}\tilde{u}_n \to F^{-1}\mathcal{Q}\tilde{u}$ by virtue of (2.31). Thus the p.o. A continuously maps $S(\mathbf{R}^n)$ into $S'(\mathbf{R}^n)$. Let $a = F^{-1}\mathcal{Q} \in S'(\mathbf{R}^n)$. By Theorem 2.2,

$$Au = a * u, \qquad \forall u \in S(\mathbf{R}^n), \tag{3.8}$$

while Lemma 2.5 implies that Au is an infinitely differentiable function satisfying estimates of form (2.46).

In the applications one often encounters pseudodifferential operators with symbol $\mathcal{Q}_0 \in S'(\mathbf{R}^n)$ induced by a homogeneous function $A_0(\xi)$. Suppose $A_0(\xi) \in O^\infty_{\alpha+i\beta}$, i.e. $A_0(\xi)$ is a homogeneous function of degree $\alpha + i\beta$ that is infinitely differentiable for $\xi = (\xi_1, \ldots, \xi_n) \neq 0$, $n \geqslant 2$.

Suppose $\alpha > -n$. Then the function $A_0(\xi) \in O^\infty_{\alpha+i\beta}$ defines a regular functional, which we also denote by $A_0(\xi)$, and the p.o. A_0 with symbol $A_0(\xi)$ has the form

$$A_0u = \frac{1}{(2\pi)^n}\int_{-\infty}^{\infty} A_0(\xi)\tilde{u}(\xi)e^{-i(x,\xi)}\,d\xi, \qquad \forall u \in S(\mathbf{R}^n), \quad \tilde{u} = Fu, \tag{3.9}$$

which is analogous to (3.2).

If $-n < \alpha < 0$, then $a_0(x) = F^{-1}A_0(\xi) \in O^\infty_{-\alpha-i\beta-n}$ is a regular functional by virtue of Example 2.6. According to (3.8),

$$A_0u = \int_{-\infty}^{\infty} a_0(x-y)u(y)dy, \tag{3.10}$$

i.e. when $A_0(\xi) \in O^\infty_{\alpha+i\beta}$ and $-n < \alpha < 0$, the p.o. A_0 is an integral operator with a weak singularity.

Suppose $\alpha \geqslant 0$. Then there exists a positive integer m such that $-1 \leqslant \gamma = \alpha - m < 0$. Consequently, $A_0(\xi) = |\xi|^m B_0(\xi)$, where $B_0(\xi) \in O^\infty_{\gamma+i\beta}$. If m is even, $m = 2m_1$, then

$$A_0(\xi) = |\xi|^{2m_1}B_0(\xi) = \left(\sum_{k=1}^{m}\xi_k^2\right)^{m_1} B_0(\xi).$$

If m is odd, $m = 2m_1 + 1$, then

$$A_0(\xi) = |\xi|^{2m_1}|\xi|B_0(\xi) = |\xi|^{2m_1}\left(\sum_{k=1}^{n}\xi_k\frac{\xi_k}{|\xi|}\right)B_0(\xi) = \sum_{k=1}^{n}|\xi|^{2m_1}\xi_k B_k(\xi),$$

where $B_k(\xi) \in O^\infty_{\gamma+i\beta}$. Thus, for any $\alpha \geqslant 0$ the function $A_0(\xi)$ can be represented in the form

$$A_0(\xi) = \sum_{k=1}^{n} P_k(\xi)B_k(\xi), \tag{3.11}$$

where $B_k(\xi) \in O^{\infty}_{\gamma+i\beta}$, $-1 \leqslant \gamma < 0$, and the $P_k(\xi)$ are homogeneous polynomials in ξ of degree $m = \alpha - \gamma$. Consequently, the p.o. A_0 with symbol $A_0(\xi)$ can be represented in the form of an integrodifferential operator:

$$\begin{aligned} A_0 u &= \sum_{k=1}^{n} P_k(D_x)\int_{-\infty}^{\infty} b_k(x-y)u(y)dy \\ &= \sum_{k=1}^{n} \int_{-\infty}^{\infty} b_k(x-y)P_k(D_y)u(y)dy, \qquad \forall u \in S(\mathbf{R}^n), \end{aligned} \tag{3.12}$$

where $b_k(x) = F^{-1}B_k(\xi) \in O^{\infty}_{-\gamma-i\beta-n}$, $1 \leqslant k \leqslant n$. Clearly, there are many other ways of representing the p.o. A_0 in the form of an integrodifferential operator.

EXAMPLE 3.1. Let $\Lambda^{\alpha+i\beta}$ denote the p.o. with symbol $|\xi|^{\alpha+i\beta}$, $-n < \alpha < 0$. By virtue of Example 2.6, $\lambda_{\alpha+i\beta}(x) = F^{-1}|\xi|^{\alpha+i\beta} \in O^{\infty}_{-\alpha-i\beta-n}$. Let us show that $\lambda_{\alpha+i\beta}(x)$ does not depend on $y = x/|x|$. By (2.70),

$$\lambda_{\alpha+i\beta}(x, \varepsilon) = \frac{e^{-i(\pi/2)(\alpha+i\beta+n)}\Gamma(\alpha+i\beta+n)}{(2\pi)^n|x|^{\alpha+i\beta+n}} \int_{S^{n-1}} \frac{ds_\eta}{((y,\eta)-i\varepsilon_1)^{\alpha+i\beta+n}}, \tag{3.13}$$

where $y = x/|x|$ and $\varepsilon_1 = \varepsilon/|x|$. Making an orthogonal transformation in (3.13) by directing the η_1 axis along the vector y, we get

$$\begin{aligned} \lambda_{\alpha+i\beta}(x, \varepsilon) &= \frac{e^{-i(\pi/2)(\alpha+i\beta+n)}\Gamma(\alpha+i\beta+n)}{(2\pi)^n|x|^{\alpha+i\beta+n}} \int_{S^{n-1}} \frac{ds_\eta}{(\eta_1-i\varepsilon_1)^{\alpha+i\beta+n}} \\ &= \frac{C^{(1)}_{\alpha+i\beta}(\varepsilon_1)}{|x|^{\alpha+i\beta+n}}. \end{aligned}$$

As in (2.73), it can be proved that there exists a constant $C^{(1)}_{\alpha+i\beta}$ such that

$$\lim_{\varepsilon_1\to 0} C^{(1)}_{\alpha+i\beta}(\varepsilon_1) = C^{(1)}_{\alpha+i\beta}$$

and hence $\lambda_{\alpha+i\beta}(x) = C^{(1)}_{\alpha+i\beta}|x|^{-\alpha-i\beta-n}$. We determine this constant by setting $a_0(y) = C^{(1)}_{\alpha+i\beta}$ and $A_0(\eta) = 1$ in (2.74). Thus

$$C^{(1)}_{\alpha+i\beta} = \frac{2^{\alpha+i\beta}\Gamma((\alpha+i\beta+n)/2)}{\pi^{n/2}\Gamma(-(\alpha+i\beta)/2)},$$

and, consequently,

$$\lambda_{\alpha+i\beta}(x) = \frac{2^{\alpha+i\beta}\Gamma((\alpha+i\beta+n)/2)}{\pi^{n/2}\Gamma(-(\alpha+i\beta)/2)}|x|^{-\alpha-i\beta-n}, \qquad -n<\alpha<0. \tag{3.14}$$

By (3.14) and (3.10), the p.o. $\Lambda^{\alpha+i\beta}$ has the form

$$\Lambda^{\alpha+i\beta}u = \frac{2^{\alpha+i\beta}\Gamma((\alpha+i\beta+n)/2)}{\pi^{n/2}\Gamma(-(\alpha+i\beta)/2)}\int_{-\infty}^{\infty}\frac{u(y)dy}{|x-y|^{\alpha+i\beta+n}}, \qquad -n<\alpha<0. \tag{3.15}$$

In particular, when $\alpha = -1$ and $\beta = 0$, we get

$$\Lambda^{-1}u = \frac{\Gamma((n-1)/2)}{2\pi^{(n+1)/2}} \int_{-\infty}^{\infty} \frac{u(y)dy}{|x-y|^{n-1}}, \qquad n > 2. \tag{3.15'}$$

Here we have made use of the fact that $\Gamma(1/2) = \sqrt{\pi}$. If $\alpha = -2$, $\beta = 0$ and $n > 3$, then

$$\Lambda^{-2}u = \frac{\Gamma((n-2)/2)}{4\pi^{n/2}} \int_{-\infty}^{\infty} \frac{u(y)dy}{|x-y|^{n-2}}, \qquad n > 3. \tag{3.15''}$$

The function (3.15″) is called a *Newtonian potential.*

EXAMPLE 3.2. With the use of (3.15″), we can find a representation of the form (3.10) for the p.o. P_n with symbol $\xi_n/|\xi|^2$ when $n > 3$. In view of the fact that

$$F^{-1}\frac{\xi_n}{|\xi|^2} = i\frac{\partial}{\partial x_n}F^{-1}\frac{1}{|\xi|^2},$$

we have

$$\begin{aligned} P_n u &= -\frac{i(n-2)\Gamma((n-2)/2)}{4\pi^{n/2}} \int_{-\infty}^{\infty} \frac{(x_n - y_n)u(y)dy}{|x-y|^n} \\ &= \frac{-i\Gamma(n/2)}{2\pi^{n/2}} \int_{-\infty}^{\infty} \frac{(x_n - y_n)u(y)}{|x-y|^n}\,dy, \qquad n > 3, \end{aligned} \tag{3.16}$$

since

$$\frac{n-2}{2}\Gamma\left(\frac{n-2}{2}\right) = \Gamma\left(\frac{n}{2}\right).$$

EXAMPLE 3.3. Let k be a positive integer. Then $\Lambda^{2k}(\xi) = |\xi|^{2k}$ and hence $\Lambda^{2k}u = (-\Delta)^k u$, where Δ is the Laplace operator. If $\alpha > 0$ is not even, then there exists an even integer $2m > 0$ such that $\alpha = 2m + \gamma$, where $-2 < \gamma < 0$. Consequently,

$$\Lambda^{\alpha}u = \Lambda^{2m+\gamma}u = (-\Delta)^m \frac{2^{\gamma}\Gamma((n+\gamma)/2)}{\pi^{n/2}\Gamma(-\gamma/2)} \int_{-\infty}^{\infty} \frac{u(y)dy}{|x-y|^{n+\gamma}},$$
$$\alpha = 2m + \gamma, \qquad -2 < \gamma < 0. \tag{3.17}$$

In particular,

$$\begin{aligned} \Lambda u &= -\Delta\Lambda^{-1}u = -\Delta\frac{\Gamma((n-1)/2)}{2\pi^{(n+1)/2}} \int_{-\infty}^{\infty} \frac{u(y)dy}{|x-y|^{n-1}} \\ &= -\frac{\Gamma((n-1)/2)}{2\pi^{(n+1)/2}} \int_{-\infty}^{\infty} \frac{\Delta_y u(y)}{|x-y|^{n-1}}\,dy. \end{aligned} \tag{3.17'}$$

Let us return to a consideration of the p.o. (3.12) under the assumption that $-1 < \gamma < 0$. The functional

$$a_0 = \sum_{k=1}^{n} P_k(D)b_k = F^{-1}A_0(\xi)$$

is not regular for $\alpha \geqslant 0$. Since $b_k(x) \in O^{\infty}_{-\gamma-i\beta-n}$ and $P_k(D)b_k(x) \in O^{\infty}_{-m-\gamma-i\beta-n}$, the restriction of the generalized function a_0 to the open domain $\mathbf{R}^n \setminus \{0\}$ is a regular functional corresponding to the infinitely differentiable function $a_0(x) = \sum_{k=1}^n P_k(D)b_k(x) \in O^\infty_{-\alpha-i\beta-n}$. We have

$$A_0 u = \lim_{\varepsilon\to 0} \sum_{k=1}^{n} \int_{1/\varepsilon>|x-y|>\varepsilon} b_k(x-y)P_k(D_y)u(y)dy. \tag{3.18}$$

Since $P_k(D_y)$ is a homogeneous polynomial in D_y of degree m,

$$P_k(D_y)u(y) = P_k(D_y)\left(u(y) - \sum_{|p|=0}^{m-1} \frac{1}{p!}\frac{\partial^p u(x)}{\partial x^p}(y-x)^p\right), \tag{3.19}$$

where $p! = p_1!p_2!\ldots p_n!$. Substituting (3.19) in (3.18), integrating by parts m times and then passing to the limit for $\varepsilon \to 0$, we get

$$A_0 u = \int_{-\infty}^{\infty} a_0(x-y)\left(u(y) - \sum_{|p|=0}^{m-1} \frac{1}{p!}\frac{\partial^p u(x)}{\partial x^p}(y-x)^p\right)dy. \tag{3.20}$$

The integral (3.20) converges, since $|a_0(x-y)| \leqslant \dfrac{C}{|x-y|^{m+n+\gamma}}$, $-1<\gamma<0$, and

$$\left|u(y) - \sum_{|p|=0}^{m-1} \frac{1}{p!}\frac{\partial^p u(x)}{\partial x^p}(y-x)^p\right| \leqslant \begin{cases} C|x-y|^m & \text{for } |x-y| \leqslant 1, \\ C|x-y|^{m-1} & \text{for } |x-y| \geqslant 1. \end{cases}$$

We note that a fundamental role in the study of pseudodifferential operators will be played in the sequel by representation (3.9) but not representations (3.10), (3.12) or (3.20).

3. *Singular integral operator*. Suppose $A_0(\xi) \in O_0^\infty$ and $A_k(\xi) = \xi_k|\xi|^{-2}A_0(\xi)$. Then $A_k(\xi) \in O^\infty_{-1}$, $A_0(\xi) = \sum_1^n \xi_k A_k(\xi)$, and hence, analogously to (3.12), the p.o. A_0 with symbol $A_0(\xi)$ can be represented in the form

$$A_0 u = \sum_{k=1}^{n} \int_{-\infty}^{\infty} a_k(x-y) i\frac{\partial u(y)}{\partial y_k}\, dy, \tag{3.20$'$}$$

where $a_k(x) = F^{-1}A_k \in O^\infty_{-n+1}$. We will show that A_0 can also be represented in the form of a singular integral operator. Suppose $b(x) \in O^\infty_{-n}$ and $u(x) \in S(\mathbf{R}^n)$. If the limit

$$\lim_{\varepsilon\to 0} \int_{|x-y|>\varepsilon} b(x-y)u(y)dy \tag{3.21}$$

exists, we will say that *the singular integral* (3.21) *converges to its Cauchy principal value*, and write

$$\lim_{\varepsilon\to 0} \int_{|x-y|>\varepsilon} b(x-y)u(y)dy = \text{p.v.} \int_{-\infty}^{\infty} b(x-y)u(y)dy.$$

LEMMA 3.1. *Suppose* $b(x) \in O^{\infty}_{-n}$. *The singular integral*

$$\text{p.v.} \int_{-\infty}^{\infty} b(x-y)u(y)dy,$$

exists for any $u(x) \in S(\mathbf{R}^n)$ *if and only if*

$$\int_{S^{n-1}} b(\omega)ds_\omega = 0. \tag{3.22}$$

PROOF. We have

$$\int_{|x-y|>\varepsilon} b(x-y)u(y)dy = \int_{\varepsilon<|x-y|<1} b(x-y)(u(y)-u(x))dy$$
$$+ \int_{\varepsilon<|x-y|<1} b(x-y)u(x)dy + \int_{|x-y|>1} b(x-y)u(y)dy. \tag{3.23}$$

The last integral in (3.23) clearly converges since $u(y) \in S(\mathbf{R}^n)$. Further, the limit

$$\lim_{\varepsilon\to 0} \int_{\varepsilon<|x-y|<1} b(x-y)(u(y)-u(x))dy$$

exists, since $|u(y)-u(x)| \leqslant c|x-y|$. The second integral in (3.23) is calculated by going over to the spherical coordinates

$$|x-y| = r, \quad \frac{y-x}{|y-x|} = \omega$$

as follows:

$$\int_{\varepsilon<|x-y|<1} b(x-y)u(x)dy = u(x)\int_{\varepsilon}^{1}\frac{dr}{r}\int_{S^{n-1}} b(\omega)ds_\omega$$
$$= u(x)\ln\frac{1}{\varepsilon}\int_{S^{n-1}} b(\omega)ds_\omega. \tag{3.24}$$

Thus the limit

$$\lim_{\varepsilon\to 0} \int_{|x-y|>\varepsilon} b(x-y)u(y)dy$$

exists for any $u(x) \in S(\mathbf{R}^n)$ if and only if (3.22) is satisfied. We note that (3.23) and (3.22) imply

$$\text{p.v.} \int_{-\infty}^{\infty} b(x-y)u(y)dy = \int_{|x-y|<1} b(x-y)(u(y)-u(x))dy$$
$$+ \int_{|x-y|>1} b(x-y)u(y)dy. \tag{3.25}$$

THEOREM 3.1. *A pseudodifferential operator A_0 with symbol $A_0(\xi) \in O_0^\infty$ can be represented in the form of a singular integral operator*

$$A_0 u = c_0 u(x) + \text{p.v.} \int_{-\infty}^{\infty} a_0(x-y)u(y)dy, \tag{3.26}$$

where

$$c_0 = \frac{\Gamma(n/2)}{2\pi^{n/2}} \int_{S^{n-1}} A_0(\eta)ds_\eta, \tag{3.27}$$

$$a_0(x) \in O_{-n}^\infty \quad \textit{and} \quad \int_{S^{n-1}} a_0(\omega)ds_\omega = 0. \tag{3.28}$$

PROOF. Since the integrals in (3.20′) converge,

$$A_0 u = \lim_{\varepsilon \to 0} \sum_{k=1}^{n} i \int_{|x-y|>\varepsilon} a_k(x-y)\frac{\partial u(y)}{\partial y_k}\, dy. \tag{3.29}$$

Integrating by parts in (3.29), we get

$$A_0 u = \lim_{\varepsilon\to 0}\left(\int_{|x-y|>\varepsilon} \sum_{k=1}^{n} i\frac{\partial a_k(x-y)}{\partial x_k} u(y)dy \right.$$

$$\left. + \sum_{k=1}^{n} i \int_{|x-y|=\varepsilon} a_k(x-y)\frac{x_k - y_k}{|x-y|} u(y)ds^\varepsilon \right), \tag{3.30}$$

where ds^ε is the surface element of the sphere $|x-y| = \varepsilon$. Introducing the spherical coordinates

$$|x-y| = r, \quad \frac{y-x}{|y-x|} = \omega,$$

we get

$$\sum_{k=1}^{n} i \int_{|x-y|=\varepsilon} a_k(x-y)\frac{x_k-y_k}{|x-y|}u(y)ds^\varepsilon = \sum_{k=1}^{n} i \int_{S^{n-1}} \frac{a_k(\omega)}{\varepsilon^{n-1}}\omega_k u(x+\varepsilon\omega)\varepsilon^{n-1}\, ds_\omega$$

$$\to \sum_{k=1}^{n} iu(x)\int_{S^{n-1}} a_k(\omega)\omega_k\, ds_\omega \quad \text{as} \quad \varepsilon \to 0. \tag{3.31}$$

Let

$$a_0(x) = \sum_{k=1}^{n} i\frac{\partial a_k(x)}{\partial x_k}.$$

Since $a_k(x) \in O_{-n+1}^\infty$, we have $a_0(x) \in O_{-n}^\infty$. We note that the functional

$$a_0 = F^{-1}A_0(\xi) = \sum_{k=1}^{n} i\frac{\partial}{\partial x_k}F^{-1}A_k(\xi)$$

is not regular, but that its restriction to $\mathbf{R}^n \setminus \{0\}$ is the regular functional corresponding to $a_0(x)$.

By (3.29), the limit (3.30) exists. Since the second integral in (3.30) has a limiting value for $\varepsilon \to 0$, the limit

$$\lim_{\varepsilon\to 0} \int_{|x-y|>\varepsilon} a_0(x-y)u(y)dy = \text{p.v.} \int_{-\infty}^{\infty} a_0(x-y)u(y)dy$$

exists. It therefore follows from Lemma 3.1 that $\int_{S^{n-1}} a_0(\omega)ds_\omega = 0$. Thus

$$A_0 u = c_0 u(x) + \text{p.v.} \int_{-\infty}^{\infty} a_0(x-y)u(y)dy,$$

where $c_0 = \sum_{k=1}^n i\int_{S^{n-1}} a_k(\omega)\omega_k ds_\omega$. It remains to show that

$$\sum_{k=1}^n i\int_{S^{n-1}} a_k(\omega)\omega_k\, ds_\omega = \frac{\Gamma(n/2)}{2\pi^{n/2}} \int_{S^{n-1}} A_0(\eta)ds_\eta.$$

Since $a_k = F^{-1}A_k$, we have

$$i\sum_{k=1}^n x_k a_k = \sum_{k=1}^n F^{-1}\frac{\partial A_k}{\partial \xi_k}.$$

Consequently, by (2.28) and (1.16),

$$\left(i\sum_{k=1}^n x_k a_k(x), \varphi(x)\right) = \frac{1}{(2\pi)^n}\left(\sum_{k=1}^n \frac{\partial A_k}{\partial \xi_k}, \tilde{\varphi}(\xi)\right) = -\frac{1}{(2\pi)^n}\sum_{k=1}^n \left(A_k(\xi), \frac{\partial \tilde{\varphi}}{\partial \xi_k}\right). \tag{3.32}$$

Analogously to (2.75), we put $\varphi(x) = e^{-x^2/2}$ in (3.32). Going over to spherical coordinates, we get

$$i\int_{S^{n-1}}\left(\sum_{k=1}^n y_k a_k(y)\right)ds_y \int_0^\infty \frac{1}{r^{n-2}} e^{-r^2/2} r^{n-1} dr$$
$$= \frac{1}{(2\pi)^{n/2}}\left(\sum_{k=1}^n \int_{S^{n-1}} A_k(\eta)\eta_k\, ds_\eta\right)\int_0^\infty e^{-\rho^2/2}\rho^{n-1}\, d\rho. \tag{3.33}$$

Since $\sum_1^n A_k(\eta)\eta_k = A_0(\eta)$, we have

$$i\int_{S^{n-1}} \sum_{k=1}^n y_k a_k(y)ds_y = \frac{2^{(n-2)/2}\Gamma(n/2)}{2^{n/2}\pi^{n/2}} \int_{S^{n-1}} A_0(\eta)ds_\eta$$
$$= \frac{\Gamma(n/2)}{2\pi^{n/2}} \int_{S^{n-1}} A_0(\eta)ds_\eta.$$

Theorem 3.1 is proved.

We note that $\Omega_n = 2\pi^{n/2}/\Gamma(n/2)$ is equal to the area of S^{n-1} (see [49]), so that c_0 is the average value of $A_0(\eta)$ on S^{n-1}.

EXAMPLE 3.4. Let Y_k, $1 < k < n$, denote the p.o. with symbol $\eta_k = \xi_k/|\xi|$. We note that $\int_{S^{n-1}} \eta_k \, ds_\eta = 0$ by virtue of the oddness of η_k. Since

$$F^{-1}\frac{\xi_k}{|\xi|} = i\frac{\partial}{\partial x_k}F^{-1}\frac{1}{|\xi|},$$

it follows by virtue of (3.14) and Theorem 3.1 that

$$\begin{aligned} Y_k u &= \text{p.v.} \frac{-i\Gamma((n-1)/2)\cdot(n-1)}{2\pi^{(n+1)/2}} \int_{-\infty}^{\infty} \frac{(x_k - y_k)}{|x-y|^{n+1}} u(y)dy \\ &= -i\frac{\Gamma((n+1)/2)}{\pi^{(n+1)/2}} \text{ p.v.} \int_{-\infty}^{\infty} \frac{x_k - y_k}{|x-y|^{n+1}} u(y)dy. \end{aligned} \tag{3.34}$$

Here we have made use of the fact that

$$\frac{n-1}{2}\Gamma\left(\frac{n-1}{2}\right) = \Gamma\left(\frac{n+1}{2}\right).$$

The singular integral operators Y_k are called *Riesz operators*.

EXAMPLE 3.5. Let Λ be the p.o. with symbol $|\xi|$. Since

$$|\xi| = \sum_{k=1}^{n} \xi_k \frac{\xi_k}{|\xi|},$$

we can represent Λ in the form

$$\Lambda u = \sum_{k=1}^{n} Y_k D_k u = \sum_{k=1}^{n} \frac{\Gamma((n+1)/2)}{\pi^{(n+1)/2}} \text{ p.v.} \int_{-\infty}^{\infty} \frac{x_k - y_k}{|x-y|^{n+1}} \frac{\partial u(y)}{\partial y_k} dy. \tag{3.35}$$

The operator Λ is called the *Calderón-Zygmund integrodifferential operator*. Formula (3.17′) provides another representation for Λ.

We note that any p.o. with symbol $A_0(\xi) \in O_m^\infty$, where m is a positive integer, can be represented in the form

$$A_0 u = \sum_{k=1}^{n} Q_k(D) B_k u, \tag{3.36}$$

where the $Q_k(D)$ are homogeneous differential operators of order m and the B_k are singular integral operators. In fact, the symbol $A_0(\xi)$ can be represented in the form $A_0(\xi) = \sum_{k=1}^{n} Q_k(\xi) B_k(\xi)$, where $B_k(\xi) \in O_0^\infty$, from which representation (3.36) follows.

4. *P.o.'s with a homogeneous symbol (continuation)*. Suppose $A_0(\xi) \in O_{\alpha+i\beta}^\infty$ and $\alpha \leqslant -n$. We can associate a functional $\mathcal{A}_0 \in S'$ with $A_0(\xi)$ by (2.77) for $\alpha + i\beta + n \neq 0, -1, \ldots$ and by (2.84) for $\beta = 0$, $\alpha + n = 0, -1, \ldots$. The p.o. A_0 with symbol $\mathcal{A}_0$ has the form

$$A_0 u = \int_{-\infty}^{\infty} a_0(x-y)u(y)dy, \qquad \forall u \in S(\mathbf{R}^n), \tag{3.37}$$

where $a_0(x) = F^{-1}\mathcal{A}_0$ is defined by (2.82) for $\alpha + i\beta + n \neq 0, -1, \ldots$ and by (2.88) for $\alpha + i\beta + n = 0, -1, \ldots$.

EXAMPLE 3.6. Let Λ^{-n} denote the functional defined by (2.84) for $A_0(\xi) = |\xi|^{-n}$, and let $\lambda_{-n}(x) = F^{-1}\Lambda^{-n}$. Then (2.90) takes the form

$$\lambda_{-n}(x) = -\frac{\Omega_n \ln|x|}{(2\pi)^n} - c_n, \tag{3.38}$$

where $\Omega_n = 2\pi^{n/2}/\Gamma(n/2)$ is the area of S^{n-1} and

$$c_n = \frac{i}{(2\pi)^n}\int_{S^{n-1}}\left[i\ln\left(-\left(\frac{x}{|x|},\eta\right)+i0\right)+\frac{\pi}{2}-i\Gamma'(1)\right]ds_\eta. \tag{3.39}$$

If, analogously to (3.13), we make an orthogonal transformation by directing the η_1 axis along the vector $x/|x|$, we get that

$$c_n = \frac{i}{(2\pi)^n}\int_{S^{n-1}}\left[i\ln(-\eta_1+i0)+\frac{\pi}{2}-i\Gamma'(1)\right]ds_\eta \tag{3.40}$$

and hence c_n does not depend on $x/|x|$. The p.o. $\Lambda^{-n}(D)$ with symbol Λ^{-n} has the form

$$\Lambda^{-n}(D)u = -\frac{1}{2^{n-1}\pi^{n/2}\Gamma(n/2)}\int_{-\infty}^{\infty}\ln|x-y|u(y)dy - c_n\int_{-\infty}^{\infty}u(y)dy. \tag{3.41}$$

EXAMPLE 3.7. With the use of (3.38), we can find the p.o. P_2 with symbol $\xi_n/|\xi|^2$ for $n = 2$. Since

$$F^{-1}\frac{\xi_2}{|\xi|^2} = i\frac{\partial}{\partial x_2}F^{-1}\Lambda^{-2} = i\frac{\partial}{\partial x_2}\lambda_{-2}(x),$$

we have

$$P_2 u = -\frac{i}{2\pi}\int_{-\infty}^{\infty}\frac{x_2-y_2}{|x-y|^2}u(y)dy, \qquad n = 2. \tag{3.42}$$

We note that (3.42) can be formally obtained from (3.16) for $n = 2$. We also note the following formulas for $n = 2$, which readily follow from (3.42) and (3.15′):

$$F^{-1}\left(\frac{1}{\xi_1+i\xi_2}\right) = F^{-1}\left(\frac{\xi_1-i\xi_2}{|\xi|^2}\right) = \frac{1}{2\pi}\frac{-ix_1-x_2}{|x|^2} = \frac{1}{2\pi i}\frac{1}{x_1+ix_2}, \tag{3.43}$$

$$F^{-1}\left(\frac{\xi_1-i\xi_2}{\xi_1+i\xi_2}\right) = \left(i\frac{\partial}{\partial x_1}+\frac{\partial}{\partial x_2}\right)\frac{1}{2\pi i}\frac{1}{x_1+ix_2} = -\frac{1}{\pi(x_1+ix_2)^2}, \qquad x \neq 0, \tag{3.44}$$

$$F^{-1}\left(\frac{\xi_1-i\xi_2}{|\xi|}\right) = \left(i\frac{\partial}{\partial x_1}+\frac{\partial}{\partial x_2}\right)\frac{1}{2\pi|x|} = \frac{1}{2\pi i(x_1+ix_2)|x|}, \qquad x \neq 0. \tag{3.45}$$

5. Suppose $A_0(\xi) \in O^{\infty}_{\alpha+i\beta}$ and $\alpha < 0$, i.e. the p.o. A_0 with symbol $\mathcal{A}_0$ ($\mathcal{A}_0 = A_0(\xi)$ for $-n < \alpha < 0$) is an integral operator, and let $\chi(\xi)$ denote a function of class $C_0^\infty(\mathbf{R}^n)$ that is equal to 1 for $|\xi| \leqslant 1$. Then the function $A_1(\xi) = (1-\chi(\xi))A_0(\xi)$ belongs to S^0_α and is infinitely differentiable, with

$$|D_\xi^k A_1(\xi)| \leqslant C_k(1+|\xi|)^{\alpha-|k|}, \qquad 0 \leqslant |k| < \infty. \tag{3.46}$$

Let us compare the p.o. A_1 with symbol $A_1(\xi)$ with the p.o. A_0. Let $a_1 = F^{-1}A_1(\xi)$ and $a_0 = F^{-1}\mathcal{Q}_0$. Then $a_0 - a_1 = F^{-1}\chi\mathcal{Q}_0$. Since $\chi(\xi) \in C_0^\infty(\mathbf{R}^n)$, it follows that $\chi\mathcal{Q}_0$ is a functional with compact support, and hence, according to Theorem 2.1, $a_2(x) = F^{-1}\chi\mathcal{Q}_0$ is an entire function. Consequently, when $\alpha < 0$, the function $a_1(x) = a_0(x) - a_2(x) \in C^\infty$ for $x \neq 0$, and A_1 is an integral operator with kernel $a_1(x - y)$. Let $B_N(\xi) = \Delta_\xi^N A_1(\xi)$. By (3.46), $|B_N(\xi)| \leqslant C_N(1 + |\xi|)^{\alpha - 2N}$. When $N > n/2$, the function $b_N(x) = F^{-1}B_N(\xi)$ is bounded. Since

$$b_N(x) = F^{-1}\Delta_\xi^N A_1(\xi) = (-1)^N|x|^{2N}a_1(x),$$

we have $|a_1(x)| \leqslant \dfrac{C_{N1}}{|x|^{2N}}$, $\forall N$. Thus the integral operators A_1 and A_0 differ by an integral operator with an infinitely differentiable kernel $a_2(x - y)$. The multiplication of the symbol $A_0(\xi)$ by $(1 - \chi(\xi))$ leads to the fact that the kernel $a_1(x - y)$ of A_1 decreases more rapidly than any power as $|x - y| \to \infty$, whereas $a_0(x - y) = O(|x - y|^{|\alpha| - n})$ for $\alpha + i\beta + n \neq 0, -1, \ldots$ and $a_0(x - y) = O(|x - y|^{|\alpha| - n}\ln|x - y|)$ for $\alpha + i\beta + n = 0, -1, \ldots$.

§4. Sobolev-Slobodeckiĭ spaces

1. *The space $H_s(\mathbf{R}^n)$.* Let s be an arbitrary real number. By the *Sobolev-Slobodeckiĭ space $H_s(\mathbf{R}^n)$* is meant the space of those generalized functions u whose Fourier transform $\tilde{u}(\xi)$ is locally integrable in the Lebesgue sense and such that

$$\|u\|_s^2 = \int_{-\infty}^{\infty} |\tilde{u}(\xi)|^2(1 + |\xi|)^{2s}\, d\xi < \infty. \tag{4.1}$$

The Fourier transform of the space $H_s = H_s(\mathbf{R}^n)$ will be denoted by $\tilde{H}_s = \tilde{H}_s(\mathbf{R}^n)$. Formula (4.1) defines the norm in H_s and $\tilde{H}_s$. We note that $\tilde{H}_s$, and hence H_s, is a Hilbert space relative to the scalar product

$$\langle u, v\rangle_s = \int_{-\infty}^{\infty} \tilde{u}(\xi)\,\overline{\tilde{v}(\xi)}\,(1 + |\xi|)^{2s}\, d\xi. \tag{4.2}$$

It follows, in particular, that H_s and $\tilde{H}_s$ are complete. For $u \in H_s$ and $\phi \in S$, (2.28) acquires the form

$$(u, \varphi) = \frac{1}{(2\pi)^n}\int_{-\infty}^{\infty} \tilde{u}(\xi)\,\overline{\tilde{\varphi}(\xi)}\, d\xi. \tag{4.3}$$

Applying the Cauchy-Schwarz-Bunjakovskiĭ inequality to (4.3), we get

$$|(u, \varphi)| \leqslant \frac{1}{(2\pi)^n}\int_{-\infty}^{\infty} |\tilde{u}(\xi)|\,|\tilde{\varphi}(\xi)|\frac{(1 + |\xi|)^s}{(1 + |\xi|)^s}\, d\xi \leqslant \frac{1}{(2\pi)^n}\|u\|_s\,\|\varphi\|_{-s}. \tag{4.4}$$

Clearly, the topology of S is stronger than the norm topology of H_s, i.e. $\varphi_n \to \varphi$ in S implies $\|\varphi_n - \varphi\|_s \to 0$. On the other hand, by (4.4), if $\|u_n - u\|_s \to 0$, then $(u_n, \varphi) \to (u, \varphi)$ for any $\varphi \in S$, i.e. $u_n \to u$ in S'.

Suppose $s = 0$. Then $\tilde{H}_s$ is the space $L_2(\mathbf{R}^n)$ of square integrable functions in the Lebesgue sense. By virtue of Plancherel's theorem, $H_0 = F^{-1}\tilde{H}_0 = L_2(\mathbf{R}^n)$.

Suppose $s = m$, where m is a positive integer. Then $\xi^k\tilde{u}(\xi) \in L_2(\mathbf{R}^n)$ for $|k| \leqslant m$, and hence $D^k u = F^{-1}(\xi^k\tilde{u}(\xi)) \in L_2(\mathbf{R}^n)$ by virtue of (2.29) and Plancherel's theorem. Thus $H_m(\mathbf{R}^n)$ consists of those square integrable functions $u(x)$ whose generalized derivatives $D^k u(x)$ are also square integrable functions for $1 \leqslant |k| \leqslant m$. The norm (4.1) for $s = m$ is equivalent to the norm

$$\|u\|_m'^2 = \sum_{|k|\leqslant m} \int_{-\infty}^{\infty} |D^k u(x)|^2\, dx = \sum_{|k|\leqslant m} \frac{1}{(2\pi)^n} \int_{-\infty}^{\infty} |\xi^k \tilde{u}(\xi)|^2\, d\xi. \tag{4.5}$$

Suppose now $s = -m$, where m is a positive integer, and let $\tilde{v}(\xi) = (1 + |\xi|)^{-m}\tilde{u}(\xi)$, where $\tilde{u}(\xi) \in \tilde{H}_{-m}$. Then $\tilde{v}(\xi) \in \tilde{H}_0 = L_2(\mathbf{R}^n)$. Using the expansion

$$|\xi| = \sum_{k=1}^{n} \xi_k \frac{\xi_k}{|\xi|},$$

we can represent $\tilde{u}(\xi)$, analogously to (3.11), in the form

$$\tilde{u}(\xi) = (1 + |\xi|)^m \tilde{v}(\xi) = \sum_{|k|\leqslant m} \xi^k \tilde{v}_k(\xi), \tag{4.6}$$

where $\tilde{v}_k(\xi) \in L_2(\mathbf{R}^n)$. Taking the inverse Fourier transform of both sides of (4.6), we get

$$u = \sum_{|k|\leqslant m} D^k v_k, \tag{4.7}$$

where $v_k(x) \in L_2(\mathbf{R}^n)$. Thus the generalized functions in H_{-m} are derivatives of order at most m of the functions in $L_2(\mathbf{R}^n)$.

EXAMPLE 4.1. The function $(1 + |\sigma|)^r$ clearly belongs to $\tilde{H}_{-r-1/2-\varepsilon}(\mathbf{R}^1)$ for any $\varepsilon > 0$, but not to $\tilde{H}_{-r-1/2}(\mathbf{R}^1)$; the function $\delta^{(r)}(t) = (i\frac{d}{dt})^r\delta(t)$ belongs to $H_s(\mathbf{R}^1)$ for $s < -r - 1/2$ but not to $H_{-r-1/2}(\mathbf{R}^1)$, since $F(\delta^{(r)}) = \sigma^r$ by virtue of Example 2.2.

THEOREM 4.1. *The functions of class $C_0^\infty(\mathbf{R}^n)$ are dense in $H_s(\mathbf{R}^n)$ in the norm* (4.1).

PROOF. Let $\alpha_\varepsilon(x) = \varepsilon^{-n}\alpha(x/\varepsilon)$, where $\alpha(x)$ is a nonnegative function belonging to $C_0^\infty(\mathbf{R}^n)$ that vanishes for $|x| \geqslant 1$ and satisfies the condition $\int_{-\infty}^{\infty} \alpha(x)dx = 1$. The function $\alpha_\varepsilon(x)$ is usually called an *averaging kernel.*

Let $u_\varepsilon = u * \alpha_\varepsilon$, where $u \in H_s$. By Lemma 2.5 and Theorem 2.2, $u_\varepsilon(x) \in C^\infty$ and $\tilde{u}_\varepsilon = \tilde{\alpha}_\varepsilon \tilde{u}$. Making the change of variables $x = \varepsilon y$, we get

$$\tilde{\alpha}_\varepsilon(\xi) = \int_{-\infty}^{\infty} \alpha\left(\frac{x}{\varepsilon}\right) e^{i(x,\xi)} \frac{dx}{\varepsilon^n} = \int_{-\infty}^{\infty} \alpha(y) e^{i(y,\varepsilon\xi)}\, dy = \tilde{\alpha}(\varepsilon\xi), \tag{4.8}$$

with

$$|\tilde{\alpha}_\varepsilon(\xi)| \leqslant \int_{-\infty}^{\infty} \alpha(y)dy = \tilde{\alpha}(0) = 1. \tag{4.9}$$

Consequently, $\tilde{u}_\varepsilon \in \tilde{H}_s$ and

$$\|u - u_\varepsilon\|_s^2 \leqslant \int_{-\infty}^{\infty} |\tilde{u}(\xi)|^2\, |1 - \tilde{\alpha}(\varepsilon\xi)|^2 (1 + |\xi|)^{2s}\, d\xi \to 0 \tag{4.10}$$

as $\varepsilon \to 0$. In fact, $1 - \tilde{\alpha}(\varepsilon\xi) \to 0$ as $\varepsilon \to 0$ for any fixed ξ, and $|1 - \tilde{\alpha}(\varepsilon\xi)| \leqslant 2$, so that (4.10) follows, for example, from the Lebesgue convergence theorem (see Example 1.5). Thus for any $\delta > 0$ there exists an $\varepsilon_1 > 0$ such that

$$\|u - u_{\varepsilon_1}\| < \delta/2. \tag{4.11}$$

Since $\tilde{\alpha}(\varepsilon_1\xi)$ belongs to $S(\mathbf{R}^n)$ and hence decreases more rapidly than any negative power of $|\xi|$ as $|\xi| \to \infty$, we have $\tilde{u}_{\varepsilon_1} = \tilde{\alpha}(\varepsilon_1\xi)\tilde{u}(\xi) \in \tilde{H}_N$ for any N. Let $v_\varepsilon(x) = \chi(\varepsilon x) u_{\varepsilon_1}(x)$, where $\chi(x)$ is a function belonging to $C_0^\infty(\mathbf{R}^n)$ that is equal to 1 for $|x| \leqslant 1$. Then $v_\varepsilon(x) \in C_0^\infty(\mathbf{R}^n)$, since $u_{\varepsilon_1}(x) \in C^\infty(\mathbf{R}^n)$. Let us show that $\|u_{\varepsilon_1} - v_\varepsilon\|_N \to 0$ as $\varepsilon \to 0$ for any N. Making use of the norm (4.5), which is equivalent to (4.1), we get that

$$\|u_{\varepsilon_1}(x) - v_\varepsilon(x)\|_N^2 = \sum_{|k| \leqslant N} \int_{|x| > 1/\varepsilon} \left|D^k[(1 - \chi(\varepsilon x)) u_{\varepsilon_1}(x)]\right|^2 dx \to 0$$

as $\varepsilon \to 0$. Consequently, if $N \geqslant s$, there exists an ε_2 such that

$$\|v_{\varepsilon_2}(x) - u_{\varepsilon_1}(x)\|_s \leqslant \|v_{\varepsilon_2} - u_{\varepsilon_1}\|_N < \delta/2. \tag{4.12}$$

It follows from (4.11) and (4.12) that there exists a function $v_{\varepsilon_2} \in C_0^\infty(\mathbf{R}^n)$ such that $\|u - v_{\varepsilon_2}\|_s < \delta$. Theorem 4.1 is proved.

By virtue of Theorem 4.1, the space H_s could have been defined as the completion of $C_0^\infty(\mathbf{R}^n)$ in the norm (4.1). When s is positive and nonintegral, an expression for the norm (4.1) can also be given in the x space.

LEMMA 4.1. *Suppose* $0 < \lambda < 1$. *The norm* (4.1) *is equivalent to the norm*

$$\|u\|_\lambda'^2 = \int_{-\infty}^{\infty} \int_{-\infty}^{\infty} \frac{|u(x+y) - u(x)|^2}{|y|^{n+2\lambda}}\, dx\, dy + \int_{-\infty}^{\infty} |u(x)|^2\, dx. \tag{4.13}$$

PROOF. Suppose $u(x) \in C_0^\infty(\mathbf{R}^n)$. We have

$$u(x+y) - u(x) = \frac{1}{(2\pi)^n} \int_{-\infty}^{\infty} \tilde{u}(\xi)(e^{-i(y,\xi)} - 1)e^{-i(x,\xi)}\, d\xi. \qquad (4.14)$$

Applying Parseval's equality to (4.14) for fixed y, we get

$$\int_{-\infty}^{\infty} |u(x+y) - u(x)|^2\, dx = \frac{1}{(2\pi)^n} \int_{-\infty}^{\infty} |\tilde{u}(\xi)|^2 \left|e^{-i(y,\xi)} - 1\right|^2 d\xi.$$

Consequently,

$$\|u\|_\lambda'^2 = \frac{1}{(2\pi)^n} \int_{-\infty}^{\infty} \int_{-\infty}^{\infty} |\tilde{u}(\xi)|^2 \frac{|e^{-i(y,\xi)} - 1|^2}{|y|^{n+2\lambda}}\, dy\, d\xi + \frac{1}{(2\pi)^n} \int_{-\infty}^{\infty} |\tilde{u}(\xi)|^2\, d\xi. \qquad (4.15)$$

Let us show that

$$\int_{-\infty}^{\infty} \frac{|e^{-i(y,\xi)} - 1|^2}{|y|^{n+2\lambda}}\, dy = C_\lambda |\xi|^{2\lambda}. \qquad (4.16)$$

We make an orthogonal transformation of coordinates $y = Az$ in (4.16) such that the z_1 axis is directed along the vector ξ. Then $(y, \xi) = z_1|\xi|$, $|y| = |z|$, and, after replacing z by $z/|\xi|$, we get

$$\int_{-\infty}^{\infty} \frac{|e^{-i(y,\xi)} - 1|^2}{|y|^{n+2\lambda}}\, dy = \int_{-\infty}^{\infty} \frac{|e^{-iz_1|\xi|} - 1|^2}{|z|^{n+2\lambda}}\, dz$$

$$= |\xi|^{2\lambda} \int_{-\infty}^{\infty} \frac{|e^{-iz_1} - 1|^2}{|z|^{n+2\lambda}}\, dz = C_\lambda |\xi|^{2\lambda}. \qquad (4.17)$$

We note that the integral (4.17) converges, since $|e^{-iz_1} - 1|^2 \leqslant C|z_1|^2$ in a neighborhood of zero and $0 < \lambda < 1$.

Substituting (4.16) in (4.15), we get that $C^1\|u\|_\lambda \leqslant \|u\|_\lambda' \leqslant C^2\|u\|_\lambda$, $\forall u \in C_0^\infty(\mathbf{R}^n)$. By Theorem 4.1, the norm (4.13) remains finite for any $u \in H_\lambda$ and $\|u\|_\lambda' \approx \|u\|_\lambda$. Lemma 4.1 is proved.

If $s = m + \lambda$, where m is a nonnegative integer and $0 < \lambda < 1$, then the norm $\|u\|_{m+\lambda}$ is equivalent by virtue of (4.5) and Lemma 4.1 to the norm

$$\|u\|_\lambda'^2 = \sum_{|k|=m} \int_{-\infty}^{\infty} \int_{-\infty}^{\infty} \frac{|D_x^k u(x+y) - D_x^k u(x)|^2}{|y|^{n+2\lambda}}\, dx\, dy + \int_{-\infty}^{\infty} |u(x)|^2\, dx. \qquad (4.18)$$

Thus $H_{m+\lambda}(\mathbf{R}^n)$ could have been defined as the completion of $C_0^\infty(\mathbf{R}^n)$ in the norm (4.18).

2. *Restriction to a hyperplane.* Let $x' = (x_1, \ldots, x_{n-1}) \in \mathbf{R}^{n-1}$, and let $[v]_s$ denote the norm in the space $H'_s = H_s(\mathbf{R}^{n-1})$:

$$[v]_s^2 = \int_{-\infty}^{\infty} (1 + |\xi'|)^{2s} |\tilde{v}(\xi')|^2 \, d\xi'. \tag{4.19}$$

THEOREM 4.2. *Suppose $s > 1/2$. Then any function $u(x', x_n) \in H_s(\mathbf{R}^n)$ is a continuous function of $x_n \in \mathbf{R}^1$ with values in $H'_{s-1/2} = H_{s-1/2}(\mathbf{R}^{n-1})$, and the following estimate holds*:

$$\max_{x_n \in \mathbf{R}^1} [u(x', x_n)]_{s-1/2} \leqslant C\|u\|_s, \qquad \forall u \in H_s(\mathbf{R}^n), \tag{4.20}$$

so that the operator of restriction to the hyperplane x_n = const *is a bounded operator from $H_s(\mathbf{R}^n)$ into $H_{s-1/2}(\mathbf{R}^{n-1})$.*

PROOF. Suppose first that $u(x', x_n) \in C_0^\infty(\mathbf{R}^n)$, and let $\tilde{w}(\xi', x_n)$ denote the Fourier transform of $u(x', x_n)$ with respect to x': $\tilde{w}(\xi', x_n) = F_{x'}u(x', x_n)$. By (2.2),

$$\tilde{w}(\xi', x_n) = \frac{1}{2\pi} \int_{-\infty}^{\infty} \tilde{u}(\xi', \xi_n) e^{-ix_n\xi_n} \, d\xi_n. \tag{4.21}$$

Using the Cauchy-Schwarz-Bunjakovskiĭ inequality, we get

$$|\tilde{w}(\xi', x_n)|^2 \leqslant \frac{1}{(2\pi)^2} \int_{-\infty}^{\infty} |\tilde{u}(\xi', \xi_n)|^2 (1 + |\xi'| + |\xi_n|)^{2s} \, d\xi_n \times \int_{-\infty}^{\infty} (1 + |\xi'| + |\xi_n|)^{-2s} \, d\xi_n. \tag{4.22}$$

We make the change of variables $\xi_n = (1 + |\xi'|)\eta_n$. Then

$$\int_{-\infty}^{\infty} (1 + |\xi'| + |\xi_n|)^{-2s} \, d\xi_n = (1 + |\xi'|)^{-2s+1} \int_{-\infty}^{\infty} (1 + |\eta_n|)^{-2s} \, d\eta_n = C_1 (1 + |\xi'|)^{-2s+1}. \tag{4.23}$$

Multiplying (4.22) by $(1 + |\xi'|)^{2s-1}$ and integrating with respect to ξ', we get

$$[u(x', x_n)]_{s-1/2}^2 \leqslant C_0 \|u\|_s^2, \qquad \forall u \in C_0^\infty(\mathbf{R}^n), \tag{4.24}$$

where C_0 does not depend on x_n or $u(x', x_n)$.

Let us now show that the functions of class $C_0^\infty(\mathbf{R}^n)$ are continuous functions of x_n with values in $H_{s-1/2}(\mathbf{R}^{n-1})$. By (2.2),

$$\tilde{w}(\xi', x_n + y_n) - \tilde{w}(\xi', x_n) = \frac{1}{2\pi} \int_{-\infty}^{\infty} \tilde{u}(\xi', \xi_n)(e^{-iy_n\xi_n} - 1) e^{-ix_n\xi_n} \, d\xi_n. \tag{4.25}$$

Estimating (4.25) analogously to (4.22), we get

$$|\tilde{w}(\xi', x_n + y_n) - \tilde{w}(\xi', x_n)|^2 \leqslant a(\xi', y_n) \times \int_{-\infty}^{\infty} |\tilde{u}(\xi', \xi_n)|^2 (1 + |\xi'| + |\xi_n|)^{2s} \, d\xi_n, \tag{4.26}$$

where

$$a(\xi', y_n) = \frac{1}{(2\pi)^2} \int_{-\infty}^{\infty} |e^{-iy_n\xi_n} - 1|^2 (1 + |\xi'| + |\xi_n|)^{-2s} \, d\xi_n. \tag{4.27}$$

We note that

$$|a(\xi', y_n)| \leqslant \frac{1}{\pi^2} \int_{-\infty}^{\infty} (1 + |\xi'| + |\xi_n|)^{-2s} \, d\xi_n = c(1 + |\xi'|)^{-2s+1},$$

with $a(\xi', y_n) \to 0$ as $y_n \to 0$ for fixed ξ' (cf. (4.10)). Multiplying (4.26) by $(1 + |\xi'|)^{2s-1}$ and integrating with respect to ξ', we get

$$[u(x', x_n + y_n) - u(x', x_n)]_{s-1/2}^2 \leqslant \int_{-\infty}^{\infty} |\tilde{u}(\xi', \xi_n)|^2 (1 + |\xi'| + |\xi_n|)^{2s} a(\xi', y_n)(1 + |\xi'|)^{2s-1} \, d\xi' \, d\xi_n. \tag{4.28}$$

Since $a(\xi', y_n) \to 0$ as $y_n \to 0$, $\forall \xi'$ and $a(\xi', y_n) \leqslant c(1 + |\xi'|)^{-2s+1}$, it follows from (4.28), analogously to (4.10), that

$$[u(x', x_n + y_n) - u(x', x_n)]_{s-1/2}^2 \to 0 \tag{4.29}$$

as $y_n \to 0$ for a fixed function $u(x', x_n) \in C_0^\infty(\mathbf{R}^n)$.

Suppose now $u_0(x', x_n) \in H_s(\mathbf{R}^n)$. Theorem 4.1 implies the existence of a sequence $u_m(x', x_n) \in C_0^\infty(\mathbf{R}^n)$ such that $\|u_0 - u_m\|_s \to 0$. By (4.29), the functions $u_m(x', x_n)$ are continuous functions of x_n with values in $H_{s-1/2}(\mathbf{R}^{n-1})$. It follows from (4.24) that the sequence $u_m(x', x_n)$ is a Cauchy sequence in the norm $\max_{x_n \in \mathbf{R}^1} [u(x', x_n)]_{s-1/2}$. Since the space $H_{s-1/2}(\mathbf{R}^{n-1})$ is complete, the limit $\lim_{m\to\infty} u_m(x', x_n) = v(x', x_n)$ exists in $H_{s-1/2}(\mathbf{R}^{n-1})$ for each fixed $x_n \in \mathbf{R}^1$. The function $v(x', x_n)$, as the limit of a uniformly convergent sequence of continuous functions:

$$\max_{x_n \in \mathbf{R}^1} [v(x', x_n) - u_m(x', x_n)]_{s-1/2} \to 0 \tag{4.30}$$

is a continuous function of x_n with values in $H_{s-1/2}(\mathbf{R}^{n-1})$. It clearly follows from (4.30) that $(u_m, \varphi(x', x_n)) \to (v, \varphi(x', x_n))$ for any $\varphi \in S(\mathbf{R}^n)$, i.e. $u_m \to v$ in S'. In addition, $u_m \to u_0$ in S'. Consequently, $u_0 = v$. Thus $u_0(x', x_n)$ is a continuous function of x_n with values in $H_{s-1/2}(\mathbf{R}^{n-1})$, while (4.24) implies (4.20) for any $u_0 \in H_s(\mathbf{R}^n)$. Theorem 4.2 is proved.

Let p' denote the operator of restriction to the hyperplane $x_n = 0$:

$$p'u(x', x_n) = u(x', 0) \tag{4.31}$$

and let Π' denote the operator defined by

$$\Pi'\tilde{u}(\xi', \xi_n) = \frac{1}{2\pi}\int_{-\infty}^{\infty} \tilde{u}(\xi', \xi_n)d\xi_n. \tag{4.32}$$

By (4.22),

$$p'\tilde{w}(\xi', x_n) = \Pi'\tilde{u}(\xi', \xi_n), \tag{4.32'}$$

where $\tilde{w}(\xi', x_n) = F_{x'}u(x', x_n)$ and $\tilde{u} = F_x u(x)$.

EXAMPLE 4.2. Suppose

$$\tilde{w}(\xi', x_n) = \tilde{v}(\xi')e^{-|x_n|(|\xi'|+1)}.$$

We will show that $\|u\|_1^2 = c[v]_{1/2}^2$, which implies that (4.20) cannot be improved. The following equivalence between norms clearly holds:

$$\|u\|_1^2 \approx \sum_{|k|<1} \int |D^k u(x)|^2\, dx \approx \int_{-\infty}^{\infty} \left(\left| \frac{\partial \tilde{w}(\xi, x_n)}{\partial x_n} \right|^2 \right.$$
$$\left. + (1 + |\xi'|)^2 |\tilde{w}(\xi', x_n)|^2 \right) d\xi'\, dx_n. \tag{4.33}$$

Substituting $\tilde{w}(\xi', x_n) = \tilde{v}(\xi')e^{-|x_n|(|\xi'|+1)}$ in (4.33), we get

$$\|u\|_1^2 \approx 2\int_{-\infty}^{\infty} (1 + |\xi'|)^2 e^{-2|x_n|(|\xi'|+1)} |\tilde{v}(\xi')|^2\, d\xi'\, dx_n.$$

Since

$$\int_{-\infty}^{\infty} e^{-|x_n|(|\xi'|+1)}\, dx_n = 2\int_0^{\infty} e^{-x_n(|\xi'|+1)}\, dx_n = \frac{2}{|\xi'| + 1},$$

it follows that

$$\|u\|_1^2 \approx \int_{-\infty}^{\infty} (1 + |\xi'|)|\tilde{v}(\xi')|^2\, d\xi' = [v]_{1/2}^2.$$

It is not difficult to also construct examples showing the sharpness of estimate (4.20) for any $s > 1/2$ (see, for example, [112]).

Let r be a nonnegative integer. We denote by $C_0^r(\mathbf{R}^n)$ the space of functions that together with their derivatives up to order r inclusively are continuous and tend to zero as $|x| \to \infty$. The norm in $C_0^r(\mathbf{R}^n)$ is

$$\sum_{|p|=0}^{r} \max_{x\in\mathbf{R}^n} |D^p u(x)|.$$

THEOREM 4.3 (Sobolev's imbedding theorem). *Suppose* $0 \leqslant r < s - n/2$. *Then*

$$\sum_{|p|=0}^{r} \max_{x \in \mathbf{R}^n} |D^p u(x)| \leqslant C\|u\|_s, \qquad \forall u(x) \in H_s(\mathbf{R}^n), \tag{4.34}$$

and hence $H_s(\mathbf{R}^n) \subset C_0^r(\mathbf{R}^n)$.

PROOF. Suppose $u(x) \in C_0^\infty(\mathbf{R}^n)$ and $\tilde{u}(\xi)$ is the Fourier transform of $u(x)$. By (2.2),

$$D^p u(x) = \frac{1}{(2\pi)^n} \int_{-\infty}^{\infty} \xi^p \tilde{u}(\xi) e^{-i(x,\xi)}\, d\xi.$$

Using the Cauchy-Schwarz-Bunjakovskiĭ inequality, we get

$$|D^p u(x)|^2 \leqslant \frac{1}{(2\pi)^{2n}} \left(\int_{-\infty}^{\infty} |\xi^p|\, |\tilde{u}(\xi)|\, d\xi \right)^2$$
$$\leqslant \frac{1}{(2\pi)^{2n}} \int_{-\infty}^{\infty} \frac{|\xi|^{2|p|}}{(1+|\xi|)^{2s}}\, d\xi \int_{-\infty}^{\infty} (1+|\xi|)^{2s} |\tilde{u}(\xi)|^2\, d\xi.$$

Suppose $|p| \leqslant r < s - n/2$. Then

$$\int_{-\infty}^{\infty} \frac{|\xi|^{2|p|}}{(1+|\xi|)^{2s}}\, d\xi \leqslant \int_{-\infty}^{\infty} \frac{d\xi}{(1+|\xi|)^{2(s-r)}} < \infty.$$

Consequently, $\max_{x \in \mathbf{R}^n} |D^p u(x)| \leqslant C\|u\|_s$, which implies (4.34) for $u(x) \in C_0^\infty(\mathbf{R}^n)$. By Theorem 4.1, any function $u(x) \in H_s(\mathbf{R}^n)$ is the limit in the norm of $H_s(\mathbf{R}^n)$ of a sequence $u_m(x) \in C_0^\infty(\mathbf{R}^n)$. But then (4.34) implies that $u_m(x)$ is a Cauchy sequence in $C_0^r(\mathbf{R}^n)$. Since $C_0^r(\mathbf{R}^n)$ is clearly complete, there exists a function $v(x', x_n) \in C_0^r(\mathbf{R}^n)$ such that

$$\sum_{|p|=0}^{r} \max_{x \in \mathbf{R}^n} |D^p v(x) - D^p u_m(x)| \to 0.$$

Hence, in particular, $u_m \to v$ in S' (see Example 1.5). On the other hand, the fact that $\|u_m - u\|_s \to 0$ implies that $u_m \to u$ in S'. Consequently, $u = v$ and hence $H_s(\mathbf{R}^n) \subset C_0^r(\mathbf{R}^n)$. Inequality (4.34) remains in force for any $u \in H_s(\mathbf{R}^n)$ after passing to the limit for $m \to \infty$. Theorem 4.3 is proved.

It follows from Theorem 4.3 that the functions in $H_s(\mathbf{R}^n)$ are sufficiently smooth if s is sufficiently large. In particular, $u(x) \in C^\infty(\mathbf{R}^n)$ if $u(x) \in H_s(\mathbf{R}^n)$ for all s.

3. *The spaces* H_s^+ *and* H_s^-. The space H_s^+ consists of those functions $u(x) \in H_s$ with support in the closed halfspace $\overline{\mathbf{R}_+^n}$. Analogously, H_s^- consists of those functions in H_s with support in $\overline{\mathbf{R}_-^n}$. The norm in $H_s^\pm$ is also given by (4.1). For $s \geqslant 0$ the membership of $u(x)$ in H_s^+ means that $u(x) = 0$

almost everywhere for $x_n < 0$. For arbitrary, in particular negative, s the inclusion $u \in H_s^+$ means that $u \in H_s$ and, by definition of the support of a generalized function, that $(u, \varphi) = 0$ for any $\varphi(x) \in C_0^\infty(\mathbf{R}^n_-)$.

LEMMA 4.2. *H_s^+ is a closed subspace of H_s.*

PROOF. Suppose $\{u_n\} \subset H_s^+$ and $\|u - u_n\|_s \to 0$. We will show that $u \in H_s^+$. We have $(u_n, \varphi) = 0$ for any $\varphi \in C_0^\infty(\mathbf{R}^n_-)$. By (4.4), $(u_n, \varphi) \to (u, \varphi)$. Consequently, $(u, \varphi) = 0$ for any $\varphi \in C_0^\infty(\mathbf{R}^n_-)$, i.e. $u \in H_s^+$. Lemma 4.2 is proved.

It follows from Lemma 4.2 that the space H_s^+ is complete.

LEMMA 4.3. *The functions of class $C_0^\infty(\mathbf{R}^n_+)$ are dense in H_s^+ in the norm* (4.1).

PROOF. Suppose $u \in H_s^+$, $\vec{\varepsilon} = (0, \ldots, 0, \varepsilon) \in \mathbf{R}^n$, $\varepsilon > 0$, and let u^ε denote the translation of u by $\vec{\varepsilon}$ (see (1.19)). By (2.30), $\tilde{u}^\varepsilon = e^{i\varepsilon\xi_n}\tilde{u}(\xi', \xi_n)$; hence $u^\varepsilon \in H_s$ and

$$\|u^\varepsilon - u\|_s^2 = \int_{-\infty}^{\infty} |\tilde{u}(\xi)|^2 \, |e^{i\varepsilon\xi_n} - 1|^2 (1 + |\xi|)^{2s} \, d\xi \to 0$$

as $\varepsilon \to 0$ (cf. (4.10)). Inasmuch as $\operatorname{supp} u \subset \overline{\mathbf{R}^n_+}$, it follows according to (1.19) that

$$(u^\varepsilon, \varphi) = (u, \varphi(x', x_n + \varepsilon)) = 0, \tag{4.35}$$

if $\varphi(x', x_n) = 0$ for $x_n > \varepsilon/2$, since in this case $\varphi(x', x_n + \varepsilon) \in C_0^\infty(\mathbf{R}^n_-)$. Thus $u^\varepsilon \in H_s^+$. For an arbitrary $\delta > 0$ there exists an $\varepsilon_1 > 0$ such that

$$\|u - u^{\varepsilon_1}\|_s < \delta/2. \tag{4.36}$$

We will approximate u^{ε_1} by functions of class $C_0^\infty(\mathbf{R}^n)$ in the same way as in the proof of Theorem 4.1. Let $\alpha_{\varepsilon_2}(x)$ and $\chi(\varepsilon_3 x)$ be the same as in Theorem 4.1. Then

$$v(x) = (u^{\varepsilon_1} * \alpha_{\varepsilon_2})\chi(\varepsilon_3 x) \in C_0^\infty(\mathbf{R}^n)$$

and, as was shown in the proof of Theorem 4.1,

$$\|u^{\varepsilon_1} - v\|_s < \delta/2, \tag{4.37}$$

if ε_2 and ε_3 are sufficiently small. Let us show that $v(x) \in C_0^\infty(\mathbf{R}^n_+)$ for $\varepsilon_2 < \varepsilon_1/4$. It suffices to establish that $\operatorname{supp}(u^{\varepsilon_1} * \alpha_{\varepsilon_2}) \subset \mathbf{R}^n_+$. By (2.45), we have

$$u^{\varepsilon_1} * \alpha_{\varepsilon_2} = \left(u^{\varepsilon_1}(y), \frac{1}{\varepsilon_2^n}\alpha\left(\frac{x' - y'}{\varepsilon_2}, \frac{x_n - y_n}{\varepsilon_2}\right)\right).$$

But

$$\alpha\left(\frac{x' - y'}{\varepsilon_2}, \frac{x_n - y_n}{\varepsilon_2}\right) = 0$$

for $x_n < \varepsilon_1/4$ and $y_n > \varepsilon_1/2$, since

$$\operatorname{supp} \alpha\left(\frac{x'-y'}{\varepsilon_2}, \frac{x_n-y_n}{\varepsilon_2}\right)$$

is contained in the halfspace $y_n - x_n < \varepsilon_2$. Consequently, by (4.35), $u^{\varepsilon_1} * \alpha_{\varepsilon_2} = 0$ for $x_n < \varepsilon_1/4$ and $\varepsilon_2 < \varepsilon_1/4$, i.e. $u^{\varepsilon_1} * \alpha_{\varepsilon_2} \in C_0^\infty(\mathbf{R}^n_+)$. It follows from (4.36) and (4.37) that $\|u - v\|_s < \delta$. Lemma 4.3 is proved.

It is proved analogously that H_s^- is a closed subspace of H_s and that the functions from $C_0^\infty(\mathbf{R}^n_-)$ are dense in H_s^-.

REMARK 4.1. Suppose $s = 0$. Then H_s^+ is the space of functions in $L_2(\mathbf{R}^n)$ that vanish for $x_n < 0$. Suppose now $s = m + \delta$, where m is a positive integer and $|\delta| < 1/2$. If $u(x', x_n) \in H_s^+$, Theorem 4.2 implies the existence of the restriction of $D_{x_n}^{k_n} u(x', x_n)$ to the hyperplane $x_n = 0$ for $0 \leqslant k_n \leqslant m - 1$, where $D_{x_n}^{k_n} = i^{k_n}\dfrac{\partial^{k_n}}{\partial x_n^{k_n}}$. Let us show that $D_{x_n}^{k_n}u(x', 0) = 0$ for $0 \leqslant k_n \leqslant m - 1$. By Theorem 4.2, $D_{x_n}^{k_n}u(x', x_n)$ is a continuous function of x_n with values in $H_{s-k_n-1/2}(\mathbf{R}^{n-1})$. Since $D_{x_n}^{k_n}u(x', x_n) = 0$ for $x_n < 0$ by virtue of the definition of H_s^+, it follows by continuity that $D_{x_n}^{k_n}u(x', 0) = 0$ for $0 \leqslant k_n \leqslant m - 1$.

EXAMPLE 4.3. The generalized function $\delta^{(k)}(x_n) = i^k \dfrac{\partial^{k_\delta}}{\partial x_n^k}$ belongs to both $H^+_{-k-1/2-\varepsilon}(\mathbf{R}^1)$, and $H^-_{-k-1/2-\varepsilon}(\mathbf{R}^1)$ for any $\varepsilon > 0$.

Suppose $f_k \in H_{-\varepsilon}(\mathbf{R}^{n-1})$. Then the direct product $f_k \times \delta^{(k)}(x_n)$ belongs to

$$H^+_{-k-1/2-\varepsilon}(\mathbf{R}^n) \cap H^-_{-k-1/2-\varepsilon}(\mathbf{R}^n).$$

In fact, $F(f_k \times \delta^{(k)}(x_n)) = \tilde{f}_k(\xi')\xi_n^k$, and, analogously to (4.24),

$$\int_{-\infty}^{\infty} \left|\tilde{f}_k(\xi')\right|^2 \xi_n^{2k}(1 + |\xi'| + |\xi_n|)^{-2k-1-2\varepsilon}\, d\xi'\, d\xi_n$$
$$= c\int_{-\infty}^{\infty} \left|\tilde{f}_k(\xi')\right|^2 (1 + |\xi'|)^{-2\varepsilon}\, d\xi' < \infty.$$

Thus

$$f_k \times \delta^{(k)}(x_n) \in H_{-k-1/2-\varepsilon}(\mathbf{R}^n).$$

In addition, $\operatorname{supp}(f_k \times \delta^{(k)}(x_n)) \subset \mathbf{R}^{n-1}$, and hence

$$f_k \times \delta^{(k)}(x_n) \in H^+_{-k-1/2-\varepsilon} \cap H^-_{-k-1/2-\varepsilon}.$$

Suppose $A(\xi) \in S_\alpha^0$, i.e. $A(\xi)$ is a measurable function satisfying

$$|A(\xi)| \leqslant C(1 + |\xi|)^\alpha.$$

The pseudodifferential operator with symbol $A(\xi)$ is defined on $S(\mathbf{R}^n)$ by the formula (see (3.2))

$$Au = \frac{1}{(2\pi)^n}\int_{-\infty}^{\infty} A(\xi)\tilde{u}(\xi)e^{-i(x,\xi)}\, d\xi \equiv F^{-1}A(\xi)Fu, \qquad \forall u \in S(\mathbf{R}^n),$$

where $\tilde{u} = Fu$ is the Fourier transform of $u(x)$. Thus the p.o. A is the inverse Fourier transform of the operator of multiplication by $A(\xi)$.

The following simple lemma holds.

LEMMA 4.4. *Suppose $A(\xi) \in S_\alpha^0$. Then the pseudodifferential operator A with symbol $A(\xi)$ satisfies*

$$\|Au\|_{s-\alpha} \leqslant C_s\|u\|_s, \qquad \forall u \in S(\mathbf{R}^n) \tag{4.38}$$

for any s and hence is extendable by continuity to a bounded operator from $H_s(\mathbf{R}^n)$ into $H_{s-\alpha}(\mathbf{R}^n)$.

In fact,

$$\begin{aligned}\|Au\|_{s-\alpha}^2 &= \int_{-\infty}^{\infty} (1+|\xi|)^{2(s-\alpha)}|A(\xi)\tilde{u}(\xi)|^2\, d\xi \\ &\leqslant c\int_{-\infty}^{\infty} (1+|\xi|)^{2s}|\tilde{u}(\xi)|^2\, d\xi = c\|u\|_s^2,\end{aligned}$$

which proves Lemma 4.4.

We note that if $A(\xi)$ is an unbounded function in a neighborhood of some point, e.g. in a neighborhood of $\xi = 0$, then (4.38) does not hold and hence A is an unbounded operator from $H_s(\mathbf{R}^n)$ into $H_{s-\alpha}(\mathbf{R}^n)$.

Let $\tilde{H}_s^{\pm}$ denote the Fourier transforms of the spaces $H_s^{\pm}$.

THEOREM 4.4. *Suppose the function $A_+(\xi', \xi_n + i\tau)$ $(A_-(\xi', \xi_n + i\tau))$ is continuous with respect to the totality of its variables for $\xi' \neq 0$, $\tau \geqslant 0$ $(\tau \leqslant 0)$, is analytic with respect to $\zeta_n = \xi_n + i\tau$ for $\tau > 0$ $(\tau < 0)$ and satisfies*

$$|A_\pm(\xi', \xi_n + i\tau)| \leqslant C(1+|\xi'|+|\xi_n|+|\tau|)^\alpha. \tag{4.39}$$

Then the operator of multiplication by $A_\pm(\xi', \xi_n)$ is a bounded operator from $\tilde{H}_s^{\pm}$ into $\tilde{H}_{s-\alpha}^{\pm}$.

PROOF. The operator of multiplication by $A_+(\xi', \xi_n)$ is a bounded operator from $\tilde{H}_s^+$ into $\tilde{H}_{s-\alpha}$ by virtue of (4.39) (see Lemma 4.4).

It remains to show that $A_+\tilde{u}_+ \in \tilde{H}_{s-\alpha}^+$ for any $\tilde{u}_+ \in \tilde{H}_s^+$. Suppose first that $u_+(x) \in C_0^\infty(\mathbf{R}_+^n)$. Then, by Lemma 2.2, $\tilde{u}_+(\xi', \xi_n + i\tau)$ is analytic with respect to $\zeta_n = \xi_n + i\tau$ for $\tau > 0$ and satisfies

$$(1+|\xi'|+|\xi_n|+\tau)^N|\tilde{u}_+(\xi', \xi_n + i\tau)| \leqslant C_N, \qquad \forall N. \tag{4.40}$$

Let $\tilde{v}(\xi', \xi_n) = A_+(\xi', \xi_n)\tilde{u}_+(\xi', \xi_n)$. We have

$$v(x', x_n) = \frac{1}{(2\pi)^n}\int_{-\infty}^{\infty} A_+(\xi', \xi_n)\tilde{u}_+(\xi', \xi_n)e^{-(x',\xi')-ix_n\xi_n}\, d\xi'\, d\xi_n, \tag{4.41}$$

where $v(x', x_n) \in C^\infty(\mathbf{R}^n)$ by virtue of (4.39) and (4.40). Using the analyticity of the function $A_+(\xi', \xi_n + i\tau)\tilde{u}_+(\xi', \xi_n + i\tau)$ with respect to $\xi_n + i\tau$ in the halfplane $\tau > 0$ and its continuity in the closed halfplane $\tau \geqslant 0$ for $\xi' \neq 0$, we get by Cauchy's theorem that

$$v(x', x_n) = \frac{1}{(2\pi)^n} \int_{-\infty}^{\infty} A_+(\xi', \xi_n + i\tau)\tilde{u}_+(\xi', \xi_n + i\tau)e^{-i(x',\xi') - ix_n(\xi_n + i\tau)}\, d\xi'\, d\xi_n, \tag{4.42}$$

where τ is arbitrary. It follows by virtue of (4.39) and (4.40) that $|v(x', x_n)| \leqslant Ce^{x_n\tau}$, where C does not depend on τ. When $x_n < 0$, we get $v(x', x_n) = 0$, i.e. $\operatorname{supp} v \subset \overline{\mathbf{R}^n_+}$, by letting τ tend to infinity. Consequently, $v \in H^+_{s-\alpha}$ and $\tilde{v}(\xi) \in \tilde{H}^+_{s-\alpha}$. If $u_+ \in H_s^+$, then Lemma 4.3 implies the existence of a sequence $\varphi_n \in C_0^\infty(\mathbf{R}^n_+)$ such that $\|u_+ - \varphi_n\|_s \to 0$. As was proved above, $A_+(\xi)\tilde{\varphi}_n(\xi) \in \tilde{H}^+_{s-\alpha}$. Since $A_+(\xi)\tilde{\varphi}_n(\xi) \to A_+(\xi)\tilde{u}_+(\xi)$ in $\tilde{H}_{s-\alpha}$, it follows by virtue of Lemma 4.2 that $A_+(\xi)\tilde{u}_+(\xi) \in \tilde{H}^+_{s-\alpha}$. Theorem 4.4 is proved for $A_+(\xi', \xi_n)$.

The boundedness from $\tilde{H}_s^-$ into $\tilde{H}^-_{s-\alpha}$ of the operator of multiplication by $A_-(\xi)$ is proved analogously.

We note that the assertion of Theorem 4.4 is equivalent to the fact that the p.o.'s $A_\pm$ with symbols $A_\pm(\xi)$ satisfying the conditions of Theorem 4.4 are bounded operators from $H_s^\pm$ into $H^\pm_{s-\alpha}$.

EXAMPLE 4.4. Let t be an arbitrary real number, and let

$$\Lambda^t_+(\xi', \xi_n + i\tau) = (\xi_n + i\tau + i|\xi'|)^t = e^{t\ln(\xi_n + i\tau + |\xi'|)},$$

where $\tau \geqslant 0$ and, as in Example 1.7, the branch of the logarithm has been chosen so that

$$\operatorname{Im}\ln(\xi_n + i\tau + i|\xi'|) = \arg(\xi_n + i|\xi'| + i\tau) \to 0 \qquad \text{as } \xi_n \to +\infty.$$

Clearly, $\Lambda^t_+(\xi', \xi_n + i\tau)$ is analytic with respect to $\zeta_n = \xi_n + i\tau$ for $\tau > 0$ and any t. Analogously, let $\Lambda^t_-(\xi', \xi_n + i\tau)$ denote the function

$$(\xi_n + i\tau - i|\xi'|)^t = e^{t(\ln|\xi_n + i\tau - i|\xi'|| + i\arg(\xi_n + i\tau - i|\xi'|))},$$

where $\tau \leqslant 0$ and $\arg(\xi_n + i\tau - i|\xi'|) \to 0$ as $\xi_n \to +\infty$. The function $\Lambda^t_-(\xi', \xi_n + i\tau)$ is analytic with respect to $\zeta_n = \xi_n + i\tau$ for $\tau < 0$ and any t.

Also, let $\hat{\Lambda}^t_+(\xi) = \Lambda^t_+(\xi', \xi_n + i)$ and $\hat{\Lambda}^t_-(\xi) = \Lambda^t_-(\xi', \xi_n - i)$. By Theorem 4.4, the operator of multiplication by $\hat{\Lambda}^t_\pm(\xi)$ is a continuous operator from $\tilde{H}_s^\pm$ into $\tilde{H}^\pm_{s-t}$. Since the operators of multiplication by $\hat{\Lambda}^t_\pm(\xi)$ and $\hat{\Lambda}^{-t}_\pm(\xi)$ are the inverses of each other, the operator of multiplication by $\hat{\Lambda}^t_\pm(\xi)$ effects an isomorphism between $\tilde{H}_s^\pm$ and $\tilde{H}^+_{s-t}$: $\hat{\Lambda}^t_\pm(\xi)\tilde{H}_s^\pm = \tilde{H}^\pm_{s-t}$. Consequently, the p.o. $\hat{\Lambda}^t_\pm$ with symbol $\hat{\Lambda}^t_\pm(\xi)$ effects an isomorphism between $H_s^\pm$ and $H^\pm_{s-t}$.

Suppose $e(x_n\tau) \in C^\infty(\mathbf{R}^1)$, $e(x_n\tau) = e^{-x_n\tau}$ for $x_n\tau > 0$ and $e(x_n\tau) = 0$ for $x_n\tau \leqslant -1$. If $f \in S'(\mathbf{R}^n)$ and $\operatorname{supp} f \subset \overline{\mathbf{R}^n_+}$, then the product $e(x_n\tau)f$ does not depend on the values of $e(x_n\tau)$ for $x_n\tau < 0$. In particular, if f is a regular functional corresponding to a locally summable function $f(x', x_n)$, then

$e(x_n\tau)f = e^{-x_n\tau}f(x', x_n)$. In the sequel we will always denote the product $e(x_n\tau)f$ by $e^{-x_n\tau}f$. Suppose $f \in H_s^+$. Then, by Example 4.4, $g = \hat{\Lambda}_+^s f \in H_0^+$, where $\hat{\Lambda}_+^s$ is the p.o. with symbol

$$\hat{\Lambda}_+^s(\xi', \xi_n) = (\xi_n + i|\xi'| + i)^s.$$

Let us prove that

$$F(ge^{-x_n\tau}) = \hat{\Lambda}_+^s(\xi', \xi_n + i\tau)F(fe^{-x_n\tau}), \qquad \tau \geqslant 0, \tag{4.43}$$

where F is the Fourier transform operator with respect to (x', x_n). Suppose first $f \in C_0^\infty(\mathbf{R}_+^n)$. Then for $\tau > 0$ the function $\tilde{f}(\xi', \xi_n + i\tau) = F(fe^{-x_n\tau})$ is analytic with respect to $\xi_n + i\tau$ and satisfies (2.10). By Cauchy's theorem,

$$F^{-1}\big(\hat{\Lambda}_+^s(\xi', \xi_n + i\tau)\tilde{f}(\xi', \xi_n + i\tau)\big) = e^{-x_n\tau}g,$$

which is equivalent to (4.43). If $f \in H_s^+$, then there exists a sequence $f_m \in C_0^\infty(\mathbf{R}_+^n)$ such that $\|f - f_m\|_s \to 0$ as $m \to \infty$, in which case $g_m = \hat{\Lambda}_+^s f_m \to g$ in H_0^+, $g_m e^{-x_n\tau} \to ge^{-x_n\tau}$ in H_0^+ and hence

$$\hat{\Lambda}_+^{-s}(\xi', \xi_n + i\tau)F(g_m e^{-x_n\tau}) \to \Lambda_+^{-s}(\xi', \xi_n + i\tau)F(ge^{-x_n\tau})$$

in $\tilde{H}_s^+(\mathbf{R}^n)$ for any τ. On the other hand, since $f_m \to f$ in H_s^+, we obviously have $f_m e^{-x_n\tau} \to fe^{-x_n\tau}$ in $S'(\mathbf{R}^n)$, and hence $F(f_m e^{-x_n\tau}) \to F(fe^{-x_n\tau})$ in $S'(\mathbf{R}^n)$. Thus, passing to the limit in S' for $m \to \infty$ in the equality

$$\Lambda_+^{-s}(\xi', \xi_n + i\tau)F(g_m e^{-x_n\tau}) = F(f_m e^{-x_n\tau}),$$

we get (4.43).

THEOREM 4.5 (Paley-Wiener theorem). *Suppose $f \in H_s^+$. Then $\tilde{f}(\xi', \xi_n + i\tau) = F(fe^{-x_n\tau})$ is a continuous function of τ for $\tau \geqslant 0$ with values in $\tilde{H}_s(\mathbf{R}^n)$ that is analytic with respect to $\xi_n + i\tau$ in the halfplane $\tau > 0$ for almost all $\xi' \in \mathbf{R}^{n-1}$ and satisfies*

$$\int_{-\infty}^{\infty} |\tilde{f}(\xi', \xi_n + i\tau)|^2(1 + |\xi'| + |\xi_n| + \tau)^{2s}\, d\xi'\, d\xi_n \leqslant C, \qquad \tau \geqslant 0, \tag{4.44}$$

where C does not depend on τ.

Conversely, suppose for $\tau > 0$ we are given a function $\tilde{f}(\xi', \xi_n + i\tau)$ that is locally integrable in $\mathbf{R}^n \times (0, +\infty)$, is analytic with respect to $\xi_n + i\tau$ for almost all $\xi' \in \mathbf{R}^{n-1}$, and satisfies (4.44). Then there exists a function $f \in H_s^+$ such that $\tilde{f}(\xi', \xi_n + i\tau) = F(fe^{-x_n\tau})$.

PROOF. Suppose $f \in H_s^+$, $g = \hat{\Lambda}_+^s f \in H_0^+$, and let $\tilde{g}(\xi', \xi_n + i\tau) = F(ge^{-x_n\tau})$. Since

$$\int_0^\infty \int_{-\infty}^\infty |g(x', x_n)e^{-x_n\tau'} - g(x', x_n)e^{-x_n\tau''}|^2\, dx'\, dx_n \to 0$$

as $\tau' - \tau'' \to 0$, $\tau' \geqslant 0$, $\tau'' \geqslant 0$ (cf. (4.10)), it follows that $g(x', x_n)e^{-x_n\tau}$ is a continuous function of $\tau \in [0, +\infty)$ with values in $H_0^+(\mathbf{R}^n)$. Consequently, Parseval's equality implies $\tilde{g}(\xi', \xi_n + i\tau)$ is a continuous function of $\tau \in [0, +\infty)$ with values in $\tilde{H}_0(\mathbf{R}^n)$ such that

$$\int_{-\infty}^{\infty} |\tilde{g}(\xi', \xi_n + i\tau)|^2 \, d\xi' \, d\xi_n = (2\pi)^n \int_{-\infty}^{\infty} \int_0^{\infty} |g(x', x_n)e^{-x_n\tau}|^2 \, dx' \, dx_n$$
$$\leqslant (2\pi)^n \|g\|_0^2. \tag{4.45}$$

Let $h(\xi', x_n) = F_{x'} g(x', x_n)$. We have $h(\xi', x_n) \in H_0^+(\mathbf{R}^n)$ by virtue of Plancherel's theorem, and

$$\int_0^{\infty} |h(\xi', x_n)|^2 \, dx_n < \infty$$

for almost all $\xi' \in \mathbf{R}^{n-1}$. For any such ξ' the function

$$\tilde{g}(\xi', \xi_n + i\tau) = F_{x_n}(h(\xi', x_n)e^{-x_n\tau}) = \int_0^{\infty} h(\xi', x_n) e^{ix_n(\xi_n + i\tau)} \, dx_n$$

is analytic with respect to $\xi_n + i\tau$ for $\tau > 0$. By (4.43),

$$\tilde{f}(\xi', \xi_n + i\tau) = [\xi_n + i(|\xi'| + \tau + 1)]^{-s} \tilde{g}(\xi', \xi_n + i\tau).$$

Consequently, $\tilde{f}(\xi', \xi_n + i\tau)$ has all of the properties indicated in Theorem 4.5.

Suppose now for $\tau > 0$ we are given a function $\tilde{f}(\xi', \xi_n + i\tau)$ that is analytic with respect to $\xi_n + i\tau$ in the halfplane $\tau > 0$ for almost all $\xi' \in \mathbf{R}^{n-1}$. Suppose, in addition, that (4.44) is satisfied for $\tau > 0$, and let

$$\tilde{g}(\xi', \xi_n + i\tau) = \hat{\Lambda}_+^s(\xi', \xi_n + i\tau)\tilde{f}(\xi', \xi_n + i\tau).$$

As a result of the analyticity of $\tilde{g}(\xi', \xi_n + i\tau)$ for almost all $\xi' \in \mathbf{R}^{n-1}$, we have

$$\frac{\partial \tilde{g}(\xi', \xi_n + i\tau)}{\partial \tau} = i \frac{\partial \tilde{g}(\xi', \xi_n + i\tau)}{\partial \xi_n}. \tag{4.46}$$

Consider $\tilde{g}(\xi', \xi_n + i\tau)$ as a regular functional on $S(\mathbf{R}^n \times (0, +\infty))$. (4.46) can be regarded as an equality between derivatives of the functional $\tilde{g}(\xi', \xi_n + i\tau)$. Let $g(x', x_n, \tau)$ denote the inverse Fourier transform of $\tilde{g}(\xi', \xi_n + i\tau)$ with respect to (ξ', ξ_n):

$$(g, \varphi(x', x_n, \tau)) = \frac{1}{(2\pi)^n}(\tilde{g}(\xi', \xi_n + i\tau), F_x\varphi(x, \tau)),$$
$$\forall \varphi \in C_0^{\infty}(\mathbf{R}^n \times (0, +\infty)). \tag{4.46'}$$

From (4.44), (4.46′) and Parseval's equality we have

$$|(g, \varphi(x, \tau))| \leqslant C \int_0^{\infty} \|\tilde{g}\|_0 \|\tilde{\varphi}\|_0 \, d\tau \leqslant C_1 \int_0^{\infty} \|\varphi\|_0 \, d\tau, \tag{4.47}$$

where $\|\varphi\|_0$ is the norm in $H_0(\mathbf{R}^n) = L_2(\mathbf{R}^n)$. Let $L_1(H_0)$ denote the Banach space of functions $f(x, \tau)$ that are locally integrable in $\mathbf{R}^n \times (0, +\infty)$ and have the finite norm $\int_0^\infty \|f\|_0\, d\tau$. By virtue of (4.47) and the fact that $C_0^\infty(\mathbf{R}^n \times (0, +\infty))$ is dense in $L_1(H_0)$ (cf. Theorem 4.1), g is a continuous semilinear functional on $L_1(H_0)$. Consequently, the theorem on the general form of a functional on $L_1(H_0)$ implies that $g(x', x_n, \tau)$ is locally integrable in $\mathbf{R}^n \times (0, +\infty)$ and satisfies

$$\int_{-\infty}^{\infty} |g(x', x_n, \tau)|^2\, dx'\, dx_n \leqslant C, \tag{4.47'}$$

where C does not depend on τ. From (2.29) and (4.46) it follows that

$$\partial g(x', x_n, \tau)/\partial\tau = -x_n g(x', x_n, \tau),$$

which, according to (1.16), means that

$$\left(g(x', x_n, \tau), \frac{\partial\varphi(x', x_n, \tau)}{\partial\tau} - x_n\varphi(x', x_n, \tau)\right) = 0,$$

$$\forall\varphi \in C_0^\infty(\mathbf{R}^n \times (0, +\infty)). \tag{4.48}$$

Substituting $\varphi = e^{x_n\tau}\psi(x', x_n, \tau)$, where $\psi \in C_0^\infty(\mathbf{R}^n \times (0, +\infty))$, in (4.48), we get

$$\left(e^{x_n\tau}g(x', x_n, \tau), \frac{\partial\psi(x', x_n, \tau)}{\partial\tau}\right) = 0, \qquad \forall\psi \in C_0^\infty(\mathbf{R}^n \times (0, +\infty)). \tag{4.49}$$

This implies by virtue of Example 1.11 that

$$g(x', x_n, \tau) = e^{-x_n\tau}g(x', x_n),$$

where $g(x', x_n)$ is a locally integrable function. By (4.47′),

$$\int_{-\infty}^{\infty} |g(x', x_n)|^2 e^{-2x_n\tau}\, dx'\, dx_n \leqslant C, \tag{4.50}$$

where $\tau > 0$ and C does not depend on τ. Consequently, letting τ tend to $+\infty$, we get $g(x', x_n) = 0$ for $x_n < 0$. If we now let τ tend to zero, we get $g(x', x_n) \in H_0^+(\mathbf{R}^n)$. Thus

$$\tilde{g}(\xi', \xi_n + i\tau) = F(ge^{-x_n\tau}).$$

Let $f = \hat{\Lambda}_+^{-s}g$. Then, by Example 4.4, $f \in H_s^+$ and, by (4.43),

$$\tilde{f}(\xi', \xi_n + i\tau) = F(fe^{-x_n\tau}).$$

Theorem 4.5 is completely proved.

4. *The space* $H_s(\mathbf{R}_+^n)$. Let $H_s(\mathbf{R}_+^n)$ denote the space of generalized functions $f \in S'(\mathbf{R}_+^n)$ (see §1.5) admitting an extension lf onto $\mathbf{R}^n$ that belongs to $H_s(\mathbf{R}^n)$. The norm in $H_s(\mathbf{R}_+^n)$ is defined by

$$\|f\|_s^+ = \inf_l \|lf\|_s, \tag{4.51}$$

where the infimum is taken over all extensions of f that belong to $H_s(\mathbf{R}^n)$. We note that if $l_0 f$ is an extension of f, then $l_0 f + f_-$, where $f_- \in H_s^-$, is also an extension of f since $pf_- = 0$, where p is the operator of restriction to $\mathbf{R}^n_+$. Consequently, the space $H_s(\mathbf{R}^n_+)$ is isomorphic to the quotient space $H_s(\mathbf{R}^n)/H_s^-$.

Consider the case $s = 0$. Here $H_0(\mathbf{R}^n_+) = L_2(\mathbf{R}^n_+)$ and the infimum of the norms of the extensions of $f \in H_0(\mathbf{R}^n_+)$ is achieved at the function f_+ that equals f for $x_n > 0$ and vanishes for $x_n < 0$. Thus

$$(\|f\|_0^+)^2 = \int_{\mathbf{R}^n_+} |f(x)|^2\, dx. \tag{4.52}$$

Suppose $s = m$, where m is a positive integer. We introduce the norm

$$(\|f\|_m'^+)^2 = \sum_{|k| \leq m} \int_{\mathbf{R}^n_+} |D^k f(x)|^2\, dx. \tag{4.53}$$

LEMMA 4.5. *The norms* (4.51) *and* (4.53) *are equivalent in* $H_m(\mathbf{R}^n_+)$.

PROOF. Since $\|lf\|_m$ is equivalent to the norm $\|lf\|'_m$ defined by (4.5), and $\|lf\|'_m \geq \|f\|_m'^+$ for any extension lf, it follows that

$$\|f\|_m'^+ \leq C\|f\|_m^+. \tag{4.54}$$

Let us now prove the opposite inequality. We construct a special operator of extension of a function from $\mathbf{R}^n_+$ onto $\mathbf{R}^n$. Let $f(x', x_n)$ denote a function belonging to $C^\infty(\overline{\mathbf{R}^n_+})$ that vanishes for sufficiently large $|x|$, and let L_N denote the operator such that

$$L_N f = \begin{cases} f(x', x_n) & \text{for } x_n > 0, \\ \sum_{p=1}^{N} \lambda_p f(x', -px_n) & \text{for } x_n < 0. \end{cases} \tag{4.55}$$

The numbers λ_p are chosen so that $L_N f \in C^{N-1}(\mathbf{R}^n)$, i.e. so that $L_N f$ together with all of its derivatives up to order $N - 1$ is continuous at $x_n = 0$. It is accordingly necessary that the numbers λ_p satisfy the system

$$\sum_{p=1}^{N} (-p)^k \lambda_p = 1, \qquad 0 \leq k \leq N - 1. \tag{4.56}$$

Since the determinant of this system (a Vandermonde determinant) does not vanish, the numbers λ_p $(p = 1, \ldots, N)$ are uniquely defined.

For $N \geq m$ we have

$$\|L_N f\|'_m \leq C\|f\|_m'^+, \tag{4.57}$$

where the constant C does not depend on f. Since the functions of class $C_0^\infty(\mathbf{R}^n)$ are dense in $H_s(\mathbf{R}^n)$, their restrictions to $\mathbf{R}^n_+$ are dense in $H_s(\mathbf{R}^n_+)$.

Consequently, taking the closure, we get (4.57) for all $f \in H_m(\mathbf{R}^n_+)$. Thus $\|f\|^+_m \leqslant \|L_N f\|'_m \leqslant C\|f\|'^+_m$. Q.E.D.

By virtue of Lemma 4.5, $H_m(\mathbf{R}^n_+)$ could have been defined as the space of all functions on $\mathbf{R}^n_+$ with finite norm (4.53).

REMARK 4.2. Suppose $s = m + \lambda$, where m is a nonnegative integer and $0 < \lambda < 1$. Using the extension operator L_N with $N > m + 1$, we can show analogously to the proof of Lemma 4.5 (see [36]) that the norm $\|f\|^+_{m+\lambda}$ is equivalent to the norm

$$\sum_{|k|=m} \int_{\mathbf{R}^n_+} \int_{\mathbf{R}^n_+} \frac{|D^k f(x) - D^k f(y)|^2}{|x-y|^{n+2\lambda}}\, dx\, dy + \int_{\mathbf{R}^n_+} |f(x)|^2\, dx. \quad (4.58)$$

It will be more convenient in the sequel to use definition (4.51) of the norm in $H_s(\mathbf{R}^n_+)$, which permits the application of the Fourier transform. Definitions (4.53) and (4.58) will not be used.

Since pseudodifferential operators are applied to functions defined on the whole space $\mathbf{R}^n$, we can apply them to a function $f \in H_s(\mathbf{R}^n_+)$ only after extending f onto $\mathbf{R}^n$. We note the following lemma.

LEMMA 4.6. *Suppose* $A_-(\xi', \xi_n + i\tau)$, $\tau \leqslant 0$, *satisfies the conditions of Theorem* 4.4, A_- *is the p.o. with symbol* $A_-(\xi', x_n)$ *and* p *is the operator of restriction of a function to* $\mathbf{R}^n_+$. *Then the map* $f \to pA_- lf$, *where* $f \in H_s(\mathbf{R}^n_+)$ *and* $lf \in H_s$ *is an arbitrary extension of* f *from* $\mathbf{R}^n_+$ *onto* $\mathbf{R}^n$, *does not depend on the choice of the extension* lf *and is a bounded operator from* $H_s(\mathbf{R}^n_+)$ *into* $H_{s-\alpha}(\mathbf{R}^n_+)$.

PROOF. Suppose $l_1 f \in H_s$ is another extension of f. Then $f_- = lf - l_1 f \in H_s^-$. By Theorem 4.4, $A_- f_- \in H^-_{s-\alpha}$, so that $pA_- f_- = 0$. Consequently, $pA_- lf = pA_- l_1 f$.

Let us estimate $\|pA_- lf\|^+_{s-\alpha}$. Suppose lf is an extension of f such that

$$\|lf\|_s \leqslant 2\|f\|^+_s. \quad (4.59)$$

Then, by Lemma 4.4,

$$\|pA_- lf\|^+_{s-\alpha} \leqslant \|A_- lf\|_{s-\alpha} \leqslant C\|lf\|_s \leqslant 2C\|f\|^+_s. \quad (4.60)$$

Lemma 4.6 is proved.

5. *Dual spaces*. Let H^*_s, where s is an arbitrary number, denote the *dual space* of H_s, i.e. the space of continuous linear functionals on $H_s(\mathbf{R}^n)$. Since $H_s(\mathbf{R}^n)$ is a Hilbert space with scalar product (4.2), it follows by virtue of Riesz's theorem that any continuous linear functional $\Phi(u)$ on H_s is defined by an element $v \in H_s$ and its norm $\|\Phi\| = \sup_{\|u\|_s=1}|\Phi(u)|$ is equal to $\sqrt{\langle v, v\rangle_s} = \|v\|_s$.

Let

$$\tilde{w}(\xi) = (1 + |\xi|)^{2s}\tilde{v}(\xi), \qquad w = F^{-1}\tilde{w}. \tag{4.61}$$

Then $w \in H_{-s}$, $\|w\|_{-s} = \sqrt{\langle v, v\rangle_s}$ and $\langle u, v\rangle_s = \langle u, w\rangle_0$, where

$$\langle u, w\rangle_0 = \int_{-\infty}^{\infty} \tilde{u}(\xi)\,\overline{\tilde{w}(\xi)}\,d\xi, \quad u \in H_s, \quad w \in H_{-s}. \tag{4.62}$$

Thus (4.61) establishes an isomorphism between H_s^* and H_{-s}, with the value of a functional $w \in H_{-s}$ at an element $u \in H_s$ being given by (4.62). Form (4.62) multiplied by $(2\pi)^{-n}$ can be regarded as the extension by continuity of form (4.3), which was defined for $u \in H_s$ and $\varphi \in S$.

LEMMA 4.7. *Suppose* $u_+ \in H_s^+$ *and* $v_- \in H_{-s}^-$. *Then* $(u_+, v_-) = (2\pi)^{-n}\langle u_+, v_-\rangle_0 = 0$. *Conversely, if* $v_- \in H_{-s}$ *and* $(u_+, v_-) = 0$ *for all* $u_+ \in H_s^+$, *then* $v_- \in H_{-s}^-$.

PROOF. By definition of the support of a generalized function, $(u_+, \varphi) = 0$ $\varphi \in C_0^\infty(\mathbf{R}_-^n)$. By Lemma 4.3, the functions of class $C_0^\infty(\mathbf{R}_-^n)$ are dense in H_{-s}^-. It follows from (4.4) that $(u_+, v_-) = 0$ for $u_+ \in H_s^+$ and $v_- \in H_{-s}^-$. Suppose now $v_- \in H_{-s}(\mathbf{R}^n)$ and $(u_+, v_-) = 0$ for $u_+ \in H_s^+$. Then, in particular, $(v_-, \varphi) = \overline{(\varphi, v_-)} = 0$ for $\varphi \in C_0^\infty(\mathbf{R}_+^n)$. By definition of supp v_-, this means that supp $v_- \subset \overline{\mathbf{R}_-^n}$, i.e. $v_- \in H_{-s}^-$. Lemma 4.7 is proved.

Suppose $f \in H_{-s}(\mathbf{R}_+^n)$, $u_+ \in H_s^+$, and $lf \in H_{-s}(\mathbf{R}^n)$ is an arbitrary extension of f. Then the form

$$\langle u_+, lf\rangle_0 = \int_{-\infty}^{\infty} \tilde{u}_+(\xi)\,\overline{\widetilde{lf}}(\xi)d\xi \tag{4.63}$$

does not depend on the choice of the extension lf. In fact, if $l_1 f \in H_{-s}(\mathbf{R}^n)$ is another extension of f, then $lf - l_1 f \in H_{-s}^-(\mathbf{R}^n)$. Consequently, $(u_+, lf) = (u_+, l_1 f)$, since $(u_+, lf - l_1 f) = 0$ by virtue of Lemma 4.7.

The estimate $|\langle u_+, lf\rangle_0| \leqslant \|u_+\|_s\|lf\|_{-s}$ follows from (4.63) analogously to (4.4), and since $\langle u_+, lf\rangle_0$ does not depend on the choice of the extension lf,

$$|\langle u_+, lf\rangle_0| \leqslant \|u_+\|_s \inf_l \|lf\|_{-s} = \|u_+\|_s\|f\|_{-s}^+. \tag{4.64}$$

Consequently, to any element $f \in H_{-s}(\mathbf{R}_+^n)$ there corresponds a continuous functional on H_s^+ by (4.63), with distinct functionals corresponding to distinct elements of $H_{-s}(\mathbf{R}_+^n)$ by virtue of Lemma 4.7. Suppose now $\Phi(u_+)$ is an arbitrary continuous functional on H_s^+. The space H_s^+ is a subspace of the Hilbert space H_s with scalar product (4.2). Consequently, there exists according to Riesz's theorem an element $\varphi_+ \in H_s^+$ such that $\Phi(u_+) = \langle u_+, \varphi_+\rangle_s$. Let $\tilde{f}_0(\xi) = (1 + |\xi|)^{2s}\tilde{\varphi}_+(\xi)$ and $f_0 = F^{-1}\tilde{f}_0(\xi)$. Then $f_0 \in H_{-s}(\mathbf{R}^n)$, $pf_0 = f \in H_{-s}(\mathbf{R}_+^n)$, $\Phi(u_+) = \langle u_+, \varphi_+\rangle_s = \langle u_+, f_0\rangle_0$ and $\|\Phi\| = \|\varphi_+\|_s = \|f_0\|_{-s} \geqslant \|f\|_{-s}^+$. Consequently, an arbitrary continuous linear functional f

can be realized on H_s^+ by (4.63), with $\|\Phi\| \geqslant \|f\|_{-s}^+$. On the other hand,

$$\|\Phi\| = \sup_{\|u_+\|_s = 1} |\Phi(u_+)| \leqslant \|f\|_{-s}^+$$

by virtue of (4.64). Thus $\|\Phi\| = \|f\|_{-s}^+$, and we have proved the following lemma.

LEMMA 4.8. *Suppose $(H_s^+)^*$, where s is an arbitrary number, is the dual space of H_s^+. Then $(H_s^+)^*$ is isomorphic to $H_{-s}(\mathbf{R}_+^n)$, with the value of a functional $f \in H_{-s}(\mathbf{R}_+^n)$ at an element $u_+ \in H_s^+$ being given by* (4.63).

It is proved analogously that $(H_{-s}(\mathbf{R}_+^n))^*$ is isomorphic to H_s^+ for any s, with the value of a functional $u_+ \in H_s^+$ at an element $f \in H_{-s}(\mathbf{R}_+^n)$ being given by

$$\langle lf, u_+ \rangle_0 = \int_{-\infty}^{\infty} \widetilde{lf}(\xi)\,\overline{\tilde{u}_+}(\xi)\,d\xi = \overline{\langle u_+, lf \rangle_0}.$$

6. EXAMPLE 4.5. Let $\Lambda_+^s(\xi', \xi_n + i\tau) = (\xi_n + i|\xi'| + i\tau)^s$ for $\tau \geqslant 0$, let $\Lambda_+^s(\xi', \xi_n + i\tau)$ denote the regular functional corresponding to the function $\Lambda_+^s(\xi', \xi_n + i\tau)$ for $\tau > 0$, and let $\Lambda_+^s(\xi', \xi_n + i0)$ denote the limit in $S'(\mathbf{R}^n)$ of the regular functionals $\Lambda_+^s(\xi', \xi_n + i\tau)$ for $\tau \to +0$. The existence of a limit in S' is proved in the same way as in Example 1.7. Let $\Lambda_+^s(D', D_n + i\tau)$ denote the p.o. with symbol $\Lambda_+^s(\xi', \xi_n + i\tau)$ for $\tau \geqslant 0$. Analogously, let

$$\Lambda_-^s(D', D_n + i\tau)$$

denote the p.o. with symbol

$$\Lambda_-^s(\xi', \xi_n - i\tau) = (\xi_n - i|\xi'| - i\tau)^s$$

for $\tau \geqslant 0$, where $\Lambda_-^s(\xi', \xi_n - i0)$ is the limit in $S'(\mathbf{R}^n)$ of the regular functionals $\Lambda_-^s(\xi', \xi_n - i\tau)$ for $\tau \to +0$. We will find representations for the p.o.'s $\Lambda_\pm^s(D', D_n \pm i\tau)$ in the form of integrodifferential operators (see §3).

Suppose first that $s < 0$. Then, by (2.35),

$$F_\xi^{-1}\Lambda_+^s(\xi', \xi_n + i\tau) = \frac{e^{i(\pi/2)s}}{\Gamma(-s)}\theta(x_n)x_n^{-s-1}F_{\xi'}^{-1}e^{-x_n(|\xi'|+\tau)}. \qquad (4.64')$$

It can be shown that

$$F_{\xi'}^{-1}e^{-x_n|\xi'|} = \frac{\Gamma(n/2)}{\pi^{n/2}}\,\frac{x_n}{(x_n^2 + |x'|^2)^{n/2}}, \qquad x_n > 0. \qquad (4.65)$$

In fact, according to (4.64′),

$$F_\xi^{-1}\frac{1}{\xi_n + i|\xi'|} = -i\theta(x_n)F_{\xi'}^{-1}e^{-x_n|\xi'|}.$$

Analogously,

$$F_\xi^{-1}\frac{1}{\xi_n - i|\xi'|} = i\theta(-x_n)F_{\xi'}^{-1}e^{x_n|\xi'|}.$$

Further,

$$F_\xi^{-1}\frac{1}{\xi_n + i|\xi'|} + F_\xi^{-1}\frac{1}{\xi_n - i|\xi'|} = F_\xi^{-1}\frac{2\xi_n}{\xi_n^2 + |\xi'|^2}. \tag{4.66}$$

Since

$$\operatorname{supp} F_\xi^{-1}\frac{1}{\xi_n \pm i|\xi'|} \subset \overline{\mathbf{R}^n_\pm},$$

we have

$$pF_\xi^{-1}\frac{1}{\xi_n + i|\xi'|} = pF_\xi^{-1}\frac{2\xi_n}{\xi_n^2 + |\xi'|^2},$$

where p is the operator of restriction to $\mathbf{R}^n_+$. It follows by virtue of (3.16) and (3.42) that

$$F_{\xi'}^{-1}e^{-x_n|\xi'|} = ipF_\xi^{-1}\frac{1}{\xi_n + i|\xi'|} = \frac{\Gamma(n/2)}{\pi^{n/2}}\,\frac{x_n}{(x_n^2 + |x'|^2)^{n/2}}$$

for $x_n > 0$. Thus, substituting (4.65) in (4.64′),

$$F_\xi^{-1}\Lambda_+^s(\xi', \xi_n + i\tau) = \frac{e^{i(\pi/2)s}\Gamma(n/2)}{\pi^{n/2}\Gamma(-s)}\,\frac{\theta(x_n)x_n^{-s}e^{-x_n\tau}}{(x_n^2 + |x'|^2)^{n/2}}, \qquad s < 0. \tag{4.67}$$

Consequently, for $\tau \geqslant 0$ and $s < 0$

$$\Lambda_+^s(D', D_n + i\tau)u = \frac{e^{i(\pi/2)s}\Gamma(n/2)}{\pi^{n/2}\Gamma(-s)} \times \int_{-\infty}^{x_n}\int_{-\infty}^{\infty}\frac{(x_n - y_n)^{-s}e^{-(x_n - y_n)\tau}}{\left[(x_n - y_n)^2 + |x' - y'|^2\right]^{n/2}}u(y', y_n)\,dy'\,dy_n, \quad \forall u \in S. \tag{4.68}$$

Analogously, for $\tau \geqslant 0$ and $s < 0$

$$\Lambda_-^s(D', D_n - i\tau)u = -\frac{e^{-i(\pi/2)s}\Gamma(n/2)}{\pi^{n/2}\Gamma(-s)} \times \int_{x_n}^{\infty}\int_{-\infty}^{\infty}\frac{(y_n - x_n)^{-s}e^{(x_n - y_n)\tau}}{\left[(x_n - y_n)^2 + |x' - y'|^2\right]^{n/2}}u(y', y_n)\,dy'\,dy_n, \quad \forall u \in S. \tag{4.69}$$

Now suppose $s = m$, where m is a positive integer. Then

$$\Lambda_{\pm}^{m}(\xi', \xi_n \pm i\tau) = \Lambda_{m1}(|\xi'|^2, \xi_n \pm i\tau) \pm \Lambda_{m2}(|\xi'|^2, \xi_n \pm i\tau)|\xi'|, \quad (4.70)$$

where $\Lambda_{m1}(|\xi'|^2, \xi_n + i\tau)$, $\Lambda_{m2}(|\xi'|^2, \xi_n + i\tau)$ are polynomials in $|\xi'|^2$ and $\xi_n + i\tau$. For $n > 2$ the p.o. Λ' in $\mathbf{R}^{n-1}$ with symbol $|\xi'|$ can be represented in the form (see (3.17′))

$$\Lambda' u = -\frac{\Gamma((n-2)/2)}{2\pi^{n/2}} \Delta' \int_{-\infty}^{\infty} \frac{u(y', x_n)\,dy'}{|x' - y'|^{n-2}}, \qquad n > 2, \quad (4.71)$$

where $\Delta' = \sum_{k=1}^{n-1} \frac{\partial^2}{\partial x_k^2}$. If $n = 2$, then

$$|\xi'| = \xi_1 \operatorname{sgn} \xi_1 = \xi_1(\theta(\xi_1) - \theta(-\xi_1))$$

and hence

$$\Lambda' u = \frac{1}{\pi} \frac{\partial}{\partial x_1} \text{ p.v.} \int_{-\infty}^{\infty} \frac{u(y_1, x_2)}{x_1 - y_1}\, dy_1. \quad (4.71')$$

Thus

$$\Lambda_{\pm}^{m}(D', D_n \pm i\tau)u = \Lambda_{m1}(-\Delta', D_n \pm i\tau)u \pm \Lambda_{m2}(-\Delta', D_n \pm i\tau)\Lambda' u, \quad (4.72)$$

where $\Lambda' u$ has the form (4.71) for $n > 2$ and the form (4.71′) for $n = 2$.

Now suppose s is positive and nonintegral. Then there exists a positive integer m such that $s = m + \gamma$, where $-1 < \gamma < 0$. Consequently,

$$\Lambda_{+}^{s}(\xi', \xi_n + i\tau) = \Lambda_{m1}(|\xi|^2, \xi_n + i\tau)\Lambda_{+}^{\gamma}(\xi', \xi_n + i\tau) + \Lambda_{m2}(|\xi'|^2, \xi_n + i\tau)|\xi'|\Lambda_{+}^{\gamma}(\xi', \xi_n + i\tau). \quad (4.73)$$

Analogously to (4.64′),

$$F_{\xi}^{-1}|\xi'|\Lambda_{+}^{\gamma}(\xi', \xi_n + i\tau) = \frac{e^{i(\pi/2)\gamma}}{\Gamma(-\gamma)} \theta(x_n) x_n^{-\gamma-1} F_{\xi'}^{-1}|\xi'| e^{-x_n(|\xi'|+\tau)}. \quad (4.74)$$

In particular,

$$F_{\xi}^{-1} \frac{|\xi'|}{\xi_n + i|\xi'|} = -i\theta(x_n) F_{\xi'}^{-1}|\xi'| e^{-x_n|\xi'|}.$$

Further,

$$F_{\xi}^{-1} \frac{|\xi'|}{\xi_n - i|\xi'|} = i\theta(-x_n) F_{\xi'}^{-1}|\xi'| e^{x_n|\xi'|}.$$

Since

$$F_{\xi}^{-1} \frac{|\xi'|}{\xi_n + i|\xi'|} - F_{\xi}^{-1} \frac{|\xi'|}{\xi_n - i|\xi'|} = F_{\xi}^{-1} \frac{-2i|\xi'|^2}{\xi_n^2 + |\xi'|^2}$$

and

$$\operatorname{supp} F_{\xi}^{-1} \frac{|\xi'|}{\xi_n \pm i|\xi'|} \subset \overline{\mathbf{R}_{\pm}^{n}},$$

we have for $x_n > 0$

$$F_{\xi'}^{-1}|\xi'| e^{-x_n|\xi'|} = piF_{\xi}^{-1} \frac{|\xi'|}{\xi_n + i|\xi'|} = 2pF^{-1} \frac{|\xi'|^2}{\xi_n^2 + |\xi'|^2}. \quad (4.75)$$

Thus for $n > 2$ the p.o. $\Lambda'\Lambda_+^\gamma(D', D_n + i\tau)$ with symbol $|\xi'|\Lambda_+^\gamma(\xi', \xi_n + i\tau)$ has by virtue of (4.74), (4.75) and (3.15″) the form

$$\Lambda'\Lambda_+^\gamma u = -\frac{e^{i(\pi/2)\gamma}\Gamma((n-2)/2)}{2\Gamma(-\gamma)\pi^{n/2}}\Delta'$$
$$\times \int_{-\infty}^{x_n}\int_{-\infty}^{\infty}\frac{(x_n - y_n)^{-\gamma-1}e^{-(x_n-y_n)\tau}u(y', y_n)\,dy'\,dy_n}{\left[(x_n - y_n)^2 + |x' - y'|^2\right]^{(n-2)/2}}, \qquad \forall u \in S. \tag{4.76}$$

When $n = 2$, the p.o. $\Lambda'\Lambda_+^\gamma(D', D_n + i\tau)$ has by virtue of (4.74), (4.75) and (3.41) the form

$$\Lambda'\Lambda_+^\gamma u = \frac{e^{i(\pi/2)\gamma}}{2\pi\Gamma(-\gamma)}\frac{\partial^2}{\partial x_1^2}\int_{-\infty}^{x_2}\int_{-\infty}^{\infty}(x_2 - y_2)^{-\gamma-1}e^{-(x_2-y_2)\tau}\ln\left[(x_2 - y_2)^2\right.$$
$$\left. + (x_1 - y_1)^2\right]u(y_1, y_2)\,dy_1\,dy_2. \tag{4.76′}$$

Consequently, for $s = m + \gamma$

$$\Lambda_+^s(D', D_n + i\tau)u = \Lambda_{m1}(-\Delta', D_n + i\tau)\Lambda_+^\gamma(D', D_n + i\tau)u$$
$$+\Lambda_{m2}(-\Delta', D_n + i\tau)\Lambda'\Lambda_+^\gamma u, \tag{4.77}$$

where the p.o. Λ_+^γ has the form (4.68) and the p.o. $\Lambda'\Lambda_+^\gamma$ has the form (4.76) for $n > 2$ and the form (4.76′) for $n = 2$.

It is shown analogously that for $s = m + \gamma$ the p.o. $\Lambda_-^s(D', D_n - i\tau)$ is representable in the form

$$\Lambda_-^s(D', D_n - i\tau)u = \Lambda_{m1}(-\Delta', D_n - i\tau)\Lambda_-^\gamma(D', D_n - i\tau)u$$
$$-\Lambda_{m2}(-\Delta', D_n - i\tau)\Lambda'\Lambda_-^\gamma u, \tag{4.78}$$

where the p.o. $\Lambda_-^\gamma(D', D_n - i\tau)$ is given by (4.69) and the p.o. $\Lambda'\Lambda_-^\gamma$ has the forms

$$\Lambda'\Lambda_-^\gamma(D', D_n - i\tau)u = \frac{e^{-i(\pi/2)\gamma}\Gamma((n-2)/2)}{2\Gamma(-\gamma)\pi^{n/2}}\Delta'$$
$$\int_{x_n}^{\infty}\int_{-\infty}^{\infty}\frac{(y_n - x_n)^{-\gamma-1}e^{(x_n-y_n)\tau}u(y', y_n)}{\left[(x_n - y_n)^2 + |x' - y'|^2\right]^{n-2/2}}\,dy'\,dy_n,$$
$$\forall u \in S,\ n > 2, \tag{4.79}$$

$$\Lambda'\Lambda_-^\gamma u = \frac{-e^{-i(\pi/2)\gamma}\partial^2}{2\pi\Gamma(-\gamma)\partial x_1^2}$$
$$\cdot\int_{x_2}^{\infty}\int_{-\infty}^{\infty}(y_2 - x_2)^{-\gamma-1}e^{(x_2-y_2)\tau}\ln\left[(x_2 - y_2)^2 + (x_1 - y_1)^2\right]$$
$$\times u(y_1, y_2)\,dy_1\,dy_2, \qquad n = 2. \tag{4.79′}$$

We note that (4.79) and (4.79′), like the preceding formulas, remain in force for $\tau = 0$.

REMARK 4.3. Let us find the restriction of the functional

$$F^{-1}\Lambda_+^s(\xi', \xi_n + i\tau)$$

to the halfspace $\mathbf{R}^n_+$ when $\tau > 0$ and $s = m + \gamma$ is nonintegral. Since

$$F^{-1}_{\xi_n}\Lambda^s_+(\xi', \xi_n + i\tau) = (D_n + i|\xi'| + i\tau)^m F^{-1}_{\xi_n}\Lambda^\gamma_+(\xi', \xi_n + i\tau)$$
$$= (D_n + i|\xi'| + i\tau)^m \frac{e^{i(\pi/2)\gamma}}{\Gamma(-\gamma)}\theta(x_n)x_n^{-\gamma-1}e^{-x_n(|\xi'|+\tau)}, \tag{4.80}$$

$pF^{-1}_{\xi}\Lambda^s_+$ is a regular functional corresponding to a function that is infinitely differentiable for $x_n > 0$. Consequently, differentiating with respect to x_n in (4.80) and using (4.65), we get (cf. (2.41′))

$$pF^{-1}\Lambda^s_+(\xi', \xi_n + i\tau) = \frac{e^{i(\pi/2)s}\Gamma(n/2)}{\Gamma(-s)\pi^{n/2}} \frac{x_n^{-s}e^{-x_n\tau}}{(x_n^2 + |x'|^2)^{n/2}}, \qquad x_n > 0. \tag{4.81}$$

REMARK 4.4. When $n = 2$, the integrals (4.65) and (4.75) are easily calculated directly. In fact, for $x_2 > 0$

$$F^{-1}_{\xi_1}e^{-x_2|\xi_1|} = \frac{1}{2\pi}\left(\int_{-\infty}^0 e^{\xi_1(x_2 - ix_1)}\,d\xi_1 + \int_0^\infty e^{-\xi_1(x_2+ix_1)}\,d\xi_1\right)$$
$$= \frac{1}{\pi}\frac{x_2}{x_1^2 + x_2^2}, \quad F^{-1}_{\xi_1}|\xi_1|e^{-x_2|\xi_1|} = \frac{\partial}{\partial x_1}\frac{1}{\pi}\frac{x_1}{x_1^2 + x_2^2}.$$

CHAPTER II

BOUNDARY VALUE PROBLEMS FOR AN ELLIPTIC PSEUDODIFFERENTIAL OPERATOR IN A HALFSPACE

§5. Integral of Cauchy type

1. Suppose $\tilde{f}(\xi', \xi_n) \in S$ and

$$F_+(\xi', \xi_n + i\tau) = \frac{i}{2\pi}\int_{-\infty}^{\infty} \frac{\tilde{f}(\xi', \eta_n)}{\xi_n + i\tau - \eta_n}\, d\eta_n, \qquad \tau > 0. \tag{5.1}$$

Clearly, $F_+(\xi', \xi_n + i\tau)$ is an analytic function of $\xi_n + i\tau$ in the halfplane $\tau > 0$.

LEMMA 5.1. *For any $\tilde{f}(\xi', \xi_n) \in S$ the limit*

$$F_+(\xi', \xi_n + i0) = \lim_{\tau \to +0} F_+(\xi', \xi_n + i\tau)$$

exists, and

$$F_+(\xi', \xi_n + i0) = \frac{1}{2}\tilde{f}(\xi', \xi_n) + \text{p.v. } \frac{i}{2\pi}\int_{-\infty}^{\infty} \frac{\tilde{f}(\xi', \eta_n)}{\xi_n - \eta_n}\, d\eta_n, \tag{5.2}$$

where

$$\text{p.v. } \int_{-\infty}^{\infty} \frac{\tilde{f}(\xi', \eta_n)}{\xi_n - \eta_n}\, d\eta_n = \lim_{\varepsilon \to 0} \int_{|\xi_n - \eta_n| > \varepsilon} \frac{\tilde{f}(\xi', \eta_n)}{\xi_n - \eta_n}\, d\eta_n$$

is an integral in the sense of Cauchy's principal value.

PROOF. We have

$$F_+(\xi', \xi_n + i\tau) = F_1(\xi', \xi_n, \tau) + F_2(\xi', \xi_n, \tau), \tag{5.3}$$

where

$$F_1(\xi', \xi_n, \tau) = \frac{i}{2\pi}\int_{|\xi_n - \eta_n|<1} \frac{\tilde{f}(\xi', \eta_n) - \tilde{f}(\xi', \xi_n)}{\xi_n + i\tau - \eta_n}\, d\eta_n$$

$$+ \frac{i}{2\pi}\int_{|\xi_n - \eta_n|>1} \frac{\tilde{f}(\xi', \eta_n)}{\xi_n + i\tau - \eta_n}\, d\eta_n, \tag{5.4}$$

$$F_2(\xi', \xi_n, \tau) = \tilde{f}(\xi', \xi_n)\frac{i}{2\pi}\int_{|\xi_n - \eta_n|<1} \frac{d\eta_n}{\xi_n + i\tau - \eta_n}. \tag{5.5}$$

Clearly,

$$F_1(\xi', \xi_n, \tau) \underset{\tau\to+0}{\to} F_1(\xi', \xi_n, 0).$$

Let us find $\lim_{\tau\to+0} F_2(\xi', \xi_n, \tau)$. Setting $\xi_n - \eta_n = t$, we get

$$\begin{aligned} F_2(\xi', \xi_n, \tau) &= \tilde{f}(\xi', \xi_n)\frac{i}{2\pi}\int_{|t|<1} \frac{dt}{t + i\tau} \\ &= \tilde{f}(\xi', \xi_n)\frac{i}{2\pi}\left(\int_{-1}^{1} \frac{t\,dt}{t^2 + \tau^2} - i\tau\int_{-1}^{1} \frac{dt}{t^2 + \tau^2}\right) \\ &= \tilde{f}(\xi', \xi_n)\frac{1}{\pi}\operatorname{arctg}\frac{1}{\tau} \to \frac{1}{2}\tilde{f}(\xi', \xi_n) \qquad \text{as } \tau \to +0. \end{aligned}$$

Thus

$$\begin{aligned} F_+(\xi', \xi_n + i0) &= \frac{1}{2}\tilde{f}(\xi', \xi_n) + F_1(\xi', \xi_n, 0) \\ &= \frac{1}{2}\tilde{f}(\xi', \xi_n) + \frac{i}{2\pi}\int_{|\xi_n - \eta_n|<1} \frac{\tilde{f}(\xi', \eta_n) - \tilde{f}(\xi', \xi_n)}{\xi_n - \eta_n}\, d\eta_n \\ &\quad + \frac{i}{2\pi}\int_{|\xi_n - \eta_n|>1} \frac{\tilde{f}(\xi', \eta_n)}{\xi_n - \eta_n}\, d\eta_n. \end{aligned} \tag{5.6}$$

Since

$$\int_{\varepsilon<|\xi_n - \eta_n|<1} \frac{\tilde{f}(\xi', \xi_n)}{\xi_n - \eta_n}\, d\eta_n = 0$$

(by virtue of the oddness of the integrand), it follows that

$$F_1(\xi', \xi_n, 0) = \lim_{\varepsilon\to0} \frac{i}{2\pi}\int_{|\xi_n - \eta_n|>\varepsilon} \frac{\tilde{f}(\xi', \eta_n)}{\xi_n - \eta_n}\, d\eta_n = \text{p.v.}\ \frac{i}{2\pi}\int_{-\infty}^{\infty} \frac{\tilde{f}(\xi', \eta_n)\, d\eta_n}{\xi_n - \eta_n}.$$

Lemma 5.1 is proved.

Let

$$F_-(\xi', \xi_n + i\tau) = -\frac{i}{2\pi}\int_{-\infty}^{\infty} \frac{\tilde{f}(\xi', \eta_n)}{\xi_n + i\tau - \eta_n}\, d\eta_n, \qquad \tau < 0, \tag{5.7}$$

where $\tilde{f}(\xi', \xi_n) \in S$. The function $F_-(\xi', \xi_n + i\tau)$ is analytic with respect to $\xi_n + i\tau$ in the halfplane $\tau < 0$ and, analogously to Lemma 5.1, it is proved that the limit

$$F_-(\xi', \xi_n - i0) = \lim_{\tau \to -0} F_-(\xi', \xi_n + i\tau)$$

exists, and

$$F_-(\xi', \xi_n - i0) = \frac{1}{2}\tilde{f}(\xi', \xi_n) - \text{p.v.}\ \frac{i}{2\pi}\int_{-\infty}^{\infty} \frac{\tilde{f}(\xi', \eta_n)}{\xi_n - \eta_n}\, d\eta_n. \tag{5.8}$$

Let

$$\Pi^0\tilde{f} = \text{p.v.}\ \frac{i}{2\pi}\int_{-\infty}^{\infty} \frac{\tilde{f}(\xi', \eta_n)}{\xi_n - \eta_n}\, d\eta_n, \tag{5.9}$$

$$\Pi^+\tilde{f} = F_+(\xi', \xi_n + i0) = \frac{i}{2\pi}\int_{-\infty}^{\infty} \frac{\tilde{f}(\xi', \eta_n)}{\xi_n + i0 - \eta_n}\, d\eta_n, \tag{5.10}$$

$$\Pi^-\tilde{f} = F_-(\xi', \xi_n - i0) = -\frac{i}{2\pi}\int_{-\infty}^{\infty} \frac{\tilde{f}(\xi', \eta_n)}{\xi_n - i0 - \eta_n}\, d\eta_n. \tag{5.11}$$

By virtue of (5.2) and (5.8),

$$\Pi^+\tilde{f} = \tfrac{1}{2}\tilde{f} + \Pi^0\tilde{f}, \qquad \Pi^-\tilde{f} = \tfrac{1}{2}\tilde{f} - \Pi^0\tilde{f} \tag{5.12}$$

and hence

$$\tilde{f} = \Pi^+\tilde{f} + \Pi^-\tilde{f} \tag{5.13}$$

for any $\tilde{f} \in S$.

2. Let θ^+ denote the operator of multiplication by the function $\theta(x_n)$ that is equal to 1 for $x_n > 0$ and vanishes for $x_n < 0$. The following lemma holds.

LEMMA 5.2. *For any $f(x', x_n) \in S$*

$$F(\theta^+ f(x', x_n)) = \Pi^+\tilde{f}, \tag{5.14}$$

where $\tilde{f} = Ff$ is the Fourier transform of $f(x', x_n)$.

PROOF. We have

$$F(\theta(x_n)e^{-x_n\tau}) = \int_0^{\infty} e^{-x_n\tau + ix_n\xi_n} dx_n = \frac{i}{\xi_n + i\tau} \tag{5.15}$$

for $\tau > 0$. We note the following property of the Fourier transform (cf. (2.4)): if $\varphi(x)$ and $\psi(x)$ are absolutely integrable functions,

$$F(\varphi(x)\psi(x)) = \frac{1}{(2\pi)^n}\int_{-\infty}^{\infty} \tilde{\varphi}(\xi - \eta)\tilde{\psi}(\eta)\, d\eta. \tag{5.16}$$

Consequently, taking the Fourier transform of the function $\theta(x_n)e^{-x_n\tau}f(x', x_n)$ with respect to x and applying (5.16) with respect to x_n, we get

$$F(\theta(x_n)e^{-x_n\tau}f(x', x_n)) = \frac{i}{2\pi}\int_{-\infty}^{\infty}\frac{\tilde{f}(\xi', \eta_n)}{\xi_n + i\tau - \eta_n}\,d\eta_n. \tag{5.17}$$

But

$$\int_{-\infty}^{\infty}\theta(x_n)e^{-x_n\tau}f(x', x_n)e^{i(x,\xi)}\,dx \to \int_{-\infty}^{\infty}\theta(x_n)f(x', x_n)e^{i(x,\xi)}\,dx$$

as $\tau \to +0$ (cf. Example 1.5), so that, upon passing to the limit in (5.17) for $\tau \to +0$, relation (5.14) follows. Lemma 5.2 is proved.

Let θ^- denote the operator of multiplication by $(1 - \theta(x_n))$. It is proved analogously to (5.14) that

$$F(\theta^- f) = \Pi^-\tilde{f}, \qquad \forall f \in S, \tag{5.18}$$

where $\tilde{f}(\xi)$ is the Fourier transform of $f(x)$. Thus, by virtue of (5.14) and (5.18), the operator Π^+ (Π^-) is the Fourier image of the operator θ^+ (θ^-).

LEMMA 5.3. *For any* $|\delta| < \frac{1}{2}$ *and any integer* $p \geqslant 0$ *the operator*[1]

$$\Pi_\delta\tilde{f} = \frac{i}{2\pi}\int_{-\infty}^{\infty}\left(\frac{|\xi_n| + |\xi'| + 1}{|\eta_n| + |\xi'| + 1}\right)^{\delta}\ln^p\frac{|\xi_n| + |\xi'| + 1}{|\eta_n| + |\xi'| + 1}\,\frac{\tilde{f}(\xi', \eta_n)}{\xi_n + i0 - \eta_n}\,d\eta_n \tag{5.19}$$

satisfies the estimate

$$\|\Pi_\delta\tilde{f}\|_0 \leqslant C\|\tilde{f}\|_0, \qquad \forall\tilde{f} \in S. \tag{5.20}$$

PROOF. We divide (5.19) into four summands:

$$\begin{aligned}\Pi_\delta\tilde{f} = {}& \theta(\xi_n)\Pi_\delta\theta(\eta_n)\tilde{f} + \theta(\xi_n)\Pi_\delta(1 - \theta(\eta_n))\tilde{f}\\ &+ (1 - \theta(\xi_n))\Pi_\delta\theta(\eta_n)\tilde{f} + (1 - \theta(\xi_n))\Pi_\delta(1 - \theta(\eta_n))\tilde{f}.\end{aligned} \tag{5.21}$$

Let us show that $\theta(\xi_n)\Pi_\delta\theta(\eta_n)\tilde{f}$ satisfies the estimate

$$\int_{-\infty}^{\infty}\int_0^{\infty}|\theta(\xi_n)\Pi_\delta\theta(\eta_n)\tilde{f}|^2\,d\xi_n\,d\xi' \leqslant C\int_{-\infty}^{\infty}\int_0^{\infty}\theta(\eta_n)|\tilde{f}(\xi', \eta_n)|^2\,d\eta_n\,d\xi'. \tag{5.22}$$

Making the change of variables

$$\xi_n = (1 + |\xi'|)(t - 1), \qquad \eta_n = (1 + |\xi'|)(\tau - 1) \tag{5.23}$$

in $\theta(\xi_n)\Pi_\delta\theta(\eta_n)\tilde{f}$, we get

$$\theta(\xi_n)\Pi_\delta\theta(\eta_n)\tilde{f} = \theta(t - 1)\int_1^{\infty}\left(\frac{t}{\tau}\right)^{\delta}\ln^p\frac{t}{\tau}\,\frac{\tilde{f}(\xi', (|\xi'| + 1)(\tau - 1))}{t + i0 - \tau}\,d\tau. \tag{5.24}$$

---

[1] The case $p > 0$ will be needed only in Chapter IV.

Let $\Pi_\delta^{(1)}$ denote the operator defined by

$$\Pi_\delta^{(1)} g = \int_0^\infty \left(\frac{t}{\tau}\right)^\delta \ln^p \frac{t}{\tau} \frac{g(\tau)}{t + i0 - \tau}\, d\tau. \tag{5.25}$$

Estimate (5.22) will follow from

$$\int_0^\infty |\Pi_\delta^{(1)} g|^2\, dt \leqslant C \int_0^\infty |g(\tau)|^2\, d\tau, \tag{5.26}$$

if one puts $g(\tau) = \tilde{f}(\xi', (1 + |\xi'|)(\tau - 1))$ for $\tau \geqslant 1$ and $g(\tau) = 0$ for $0 \leqslant \tau < 1$, multiplies (5.26) by $(1 + |\xi'|)$ and then integrates with respect to ξ'.

A natural approach to the task of proving (5.26) is to apply the Mellin transform, since the kernel

$$M_0(t, \tau) = \left(\frac{t}{\tau}\right)^\delta \ln^p \frac{t}{\tau} \frac{1}{t + i0 - \tau}$$

is a homogeneous function of t and τ of degree -1. We take the Mellin transform of (5.25):

$$\widehat{\Pi_\delta^{(1)} g} = \lim_{\varepsilon \to +0} \int_0^\infty t^{s-1} \left(\int_0^\infty M_\varepsilon(t, \tau) g(\tau) d\tau\right) dt, \qquad -\delta < \operatorname{Re} s < 1 - \delta, \tag{5.27}$$

where

$$M_\varepsilon(t, \tau) = \left(\frac{t}{\tau}\right)^\delta \ln^p \frac{t}{\tau} \frac{1}{t + i\varepsilon - \tau}.$$

Varying the order of integration in (5.27) and making the change of variables $t = \tau t'$, we get $\hat{\Pi}_\delta^{(1)} g = \hat{M}_0(s) \hat{g}(s)$, where $\hat{g}(s) = \int_0^\infty g(\tau) \tau^{s-1} d\tau$ is the Mellin transform of $g(\tau)$ and

$$\hat{M}_0(s) = \lim_{\varepsilon \to 0} \int_0^\infty t^{s+\delta-1} \ln^p t \frac{dt}{t + i\varepsilon - 1}.$$

By Example 2.1,

$$\lim_{\varepsilon \to 0} \int_0^\infty t^{s+\delta-1} \frac{dt}{t + i\varepsilon - 1} = 2\pi i \frac{e^{2\pi i(s+\delta)}}{1 - e^{2\pi i(s+\delta)}}. \tag{5.28}$$

Differentiating (5.28) with respect to s, we get

$$\hat{M}_0(s) = 2\pi i \frac{d^p}{ds^p} \frac{e^{2\pi i(s+\delta)}}{1 - e^{2\pi i(s+\delta)}}.$$

Since $\hat{M}_0(s)$ is bounded on the straight line $s = \frac{1}{2} + i\eta$ for $|\delta| < \frac{1}{2}$, it follows by virtue of Parseval's equality (2.16) that

$$\int_0^\infty |\Pi_\delta^{(1)} g|^2\, dt = \frac{1}{2\pi} \int_{\frac{1}{2}-i\infty}^{\frac{1}{2}+i\infty} |\hat{M}_0(s)\hat{g}(s)|^2\, d\eta$$
$$\leqslant C \int_{\frac{1}{2}-i\infty}^{\frac{1}{2}+i\infty} |\hat{g}(s)|^2\, d\eta = C \int_0^\infty |g(t)|^2\, dt.$$

Estimate (5.26) is proved. The remaining summands in (5.21) are estimated analogously. The proof of their boundedness in $\tilde{H}_0$ reduces to a proof of estimate (5.26) for an integral operator of form (5.25) in which the denominator $t + i0 - \tau$ has been replaced by $t - i0 - \tau$ or $t + \tau$. Lemma 5.3 is proved.

THEOREM 5.1. *The operators $\Pi^\pm$ defined on $S(\mathbf{R}^n)$ by* (5.10) *and* (5.11) *are bounded in the norm of $\tilde{H}_\delta$ for $|\delta| < \frac{1}{2}$ and hence can be extended by continuity onto $\tilde{H}_\delta$ as bounded operators acting from $\tilde{H}_\delta$ into $\tilde{H}_\delta^\pm$.*

PROOF. The boundedness of Π^+ on $S(\mathbf{R}^n)$ in the norm of $\tilde{H}_\delta$ follows directly from Lemma 5.3 for $p = 0$. The boundedness of Π^- follows, for example, from the boundedness of Π^+ and (5.13). By (5.14) and (5.18), the operators θ^+ and θ^- are bounded in the norm of H_δ and hence can be extended by continuity to bounded operators in H_δ. For $f(x) \in S$ we have $\operatorname{supp} \theta^\pm f \subset \overline{\mathbf{R}}^n_\pm$, so that $\theta^\pm f \in H_\delta^\pm$. Since the subspaces $H_\delta^\pm$ are closed, the ranges of the operators $\theta^\pm$ lie in $H_\delta^\pm$. Consequently, $\Pi^\pm$ are bounded operators acting from $\tilde{H}_\delta$ into $\tilde{H}_\delta^\pm$.

REMARK 5.1. Suppose $f \in H_\delta(\mathbf{R}^n)$, $|\delta| < \frac{1}{2}$, and let

$$f_+ = \theta^+ f, \tilde{f}_+(\xi', \xi_n + i\tau) = F(f_+ e^{-x_n \tau}), \qquad \tau > 0.$$

We will show that for any $\tau > 0$

$$\tilde{f}_+(\xi', \xi_n + i\tau) = \frac{i}{2\pi} \int_{-\infty}^{\infty} \frac{\tilde{f}(\xi', \eta_n)\, d\eta_n}{\xi_n + i\tau - \eta_n}, \tag{5.29}$$

i.e. the function $\tilde{f}_+(\xi', \xi_n + i\tau)$ is representable in the halfplane $\tau > 0$ by an integral of Cauchy type. Since $1/(\xi_n + i\tau) \in \tilde{H}_{-\delta}(\mathbf{R}^1)$ for $\tau > 0$ and $|\delta| < \frac{1}{2}$, it follows by virtue of the inequality

$$\left| \int_{-\infty}^{\infty} \left| \frac{\tilde{f}(\xi', \eta_n)}{\xi_n + i\tau - \eta_n} \right| d\eta_n \right|^2 < \int_{-\infty}^{\infty} \frac{(1 + |\xi'| + |\eta_n|)^{-2\delta}}{|\xi_n + i\tau - \eta_n|^2}\, d\eta_n$$
$$\times \int_{-\infty}^{\infty} (1 + |\xi'| + |\eta_n|)^{2\delta} |\tilde{f}(\xi', \eta_n)|^2\, d\eta_n \tag{5.30}$$

that the integral (5.29) converges for almost all $\xi' \in \mathbf{R}^{n-1}$ and any $\xi_n + i\tau$, $\tau > 0$, and is an analytic function of $\xi_n + i\tau$. If $f \in S$, (5.29) coincides with (5.17). Suppose $f \in H_\delta(\mathbf{R}^n)$. By Theorem 4.1, there exists a sequence $f_m \in S$ that converges to f in the norm of $H_\delta(\mathbf{R}^n)$. From (5.30) it follows that

$$\int_{-\infty}^{\infty}\frac{\tilde f_m(\xi',\eta_n)}{\xi_n+i\tau-\eta_n}\,d\eta_n\to\int_{-\infty}^{\infty}\frac{\tilde f(\xi',\eta_n)}{\xi_n+i\tau-\eta_n}\,d\eta_n$$

in $S'(\mathbf{R}^n)$. Further, $f_m^+=\theta^+ f_m\to f^+=\theta^+ f$ in $H_\delta^\pm$ by virtue of Theorem 5.1, and

$$\tilde f_m^+(\xi',\xi_n+i\tau)=F(f_m^+e^{-x_n\tau})\to\tilde f_+(\xi',\xi_n+i\tau)$$

in S' (see Theorem 4.5). Thus, upon passing to the limit in (5.17) for $m\to\infty$, we get that (5.29) is valid for $f\in H_\delta(\mathbf{R}^n)$. From Theorem 4.5 it follows that $f_+(\xi',\xi_n+i\tau)\in\tilde H_\delta$ for $\tau>0$ and that

$$\tilde f_+(\xi',\xi_n+i\tau)\to\tilde f_+(\xi',\xi_n)=\Pi^+\tilde f$$

in the norm of $\tilde H_\delta$ as $\tau\to 0$. Consequently,

$$\frac{1}{2\pi}\int_{-\infty}^{\infty}\frac{\tilde f(\xi',\eta_n)\,d\eta_n}{\xi_n+i\tau-\eta_n}\to\Pi^+\tilde f$$

in the norm of $\tilde H_\delta(\mathbf{R}^n)$ as $\tau\to 0$ by virtue of (5.29).

LEMMA 5.4. *For $|\delta|<\frac{1}{2}$ any function $\tilde f\in\tilde H_\delta$ is uniquely representable in the form*

$$\tilde f=\tilde f_+ +\tilde f_-,\tag{5.31}$$

where $\tilde f_+=\Pi^+\tilde f\in\tilde H_\delta^+$ and $\tilde f_-=\Pi^-\tilde f\in\tilde H_\delta^-$.

PROOF. Suppose $\tilde f\in\tilde H_\delta$. Then, by Theorem 5.1, $\Pi^+\tilde f\in\tilde H_\delta^+$ and $\Pi^-\tilde f\in\tilde H_\delta^-$. Extending (5.13) by continuity from S onto $\tilde H_\delta$, we get

$$\tilde f=\Pi^+\tilde f+\Pi^-\tilde f,\qquad\forall\tilde f\in\tilde H_\delta.\tag{5.32}$$

Let us now prove the uniqueness of the decomposition (5.31). Suppose

$$\tilde f_+ +\tilde f_-=0,\tag{5.33}$$

where $\tilde f_+\in\tilde H^+_{-1/2+\varepsilon}$ and $\tilde f_-\in\tilde H^-_{-1/2+\varepsilon}$, $0<\varepsilon<1$. We will show that $\tilde f_+=\tilde f_-=0$. Suppose $g_\pm(x)\in C_0^\infty(\mathbf{R}^n_\pm)$. Then, clearly, $\theta(x_n)g_+(x)=g_+(x)$ and $\theta(x_n)g_-(x)=0$. By virtue of the boundedness of the operator θ^+ from $H_{-1/2+\varepsilon}$ into $H^+_{-1/2+\varepsilon}$ and Lemmas 4.2 and 4.3, we get $\theta^+f_+=f_+$ and $\theta^+f_-=0$ for any $f_\pm\in H^\pm_{-1/2+\varepsilon}$. Consequently,

$$\Pi^+\tilde f_+=\tilde f_+,\qquad\Pi^+\tilde f_-=0,\qquad\forall\tilde f_\pm\in\tilde H^\pm_{-1/2+\varepsilon}.\tag{5.34}$$

It is proved analogously that

$$\Pi^-\tilde f_+=0,\qquad\Pi^-\tilde f_-=\tilde f_-,\qquad\forall\tilde f_\pm\in\tilde H^\pm_{-1/2+\varepsilon}.\tag{5.35}$$

Thus, upon applying the operators Π^+ and Π^- to (5.33), we get $\tilde f_+=\tilde f_-=0$.

LEMMA 5.5 (expansion formula for an integral of Cauchy type). *Suppose* $\tilde{f} \in \tilde{H}_{m+\delta}(\mathbf{R}^n)$, *where m is a positive integer and* $|\delta| < \frac{1}{2}$. *Then*

$$\Pi^+\tilde{f} = \sum_{k=1}^{m} \frac{i\Pi'\hat{\Lambda}_+^{k-1}\tilde{f}}{\hat{\Lambda}_+^k} + \frac{1}{\hat{\Lambda}_+^m}\Pi^+\hat{\Lambda}_+^m \tilde{f}, \tag{5.36}$$

where

$$\hat{\Lambda}_+(\xi', \xi_n) = (\xi_n + i|\xi'| + i), \qquad \Pi'\tilde{g} = \frac{1}{2\pi}\int_{-\infty}^{\infty} \tilde{g}(\xi', \eta_n)d\eta_n.$$

PROOF. Expanding $(\xi_n - \eta_n)^{-1}$ by the formula for a geometric progression, we get

$$\frac{1}{\xi_n - \eta_n} = \frac{1}{(\xi_n + i|\xi'| + i)\left(1 - \dfrac{\eta_n + i|\xi'| + i}{\xi_n + i|\xi'| + i}\right)}$$

$$= \sum_{k=1}^{m} \frac{(\eta_n + i|\xi'| + i)^{k-1}}{(\xi_n + i|\xi'| + i)^k} + \frac{(\eta_n + i|\xi'| + i)^m}{(\xi_n + i|\xi'| + i)^{m+1}\left(1 - \dfrac{\eta_n + i|\xi'| + i}{\xi_n + i|\xi'| + i}\right)}$$

$$= \sum_{k=1}^{m} \frac{\hat{\Lambda}_+^{k-1}(\xi', \eta_n)}{\hat{\Lambda}_+^k(\xi', \xi_n)} + \frac{\hat{\Lambda}_+^m(\xi', \eta_n)}{\hat{\Lambda}_+^m(\xi', \xi_n)(\xi_n - \eta_n)}. \tag{5.37}$$

Suppose $f \in S$. Multiplying (5.37) by $(i/2\pi)\tilde{f}(\xi', \eta_n)$ and integrating with respect to η_n, we get

$$\Pi^0\tilde{f} = \sum_{k=1}^{m} \frac{i\Pi'\hat{\Lambda}^{k-1}\tilde{f}}{\hat{\Lambda}_+^k} + \frac{1}{\hat{\Lambda}_+^m}\Pi^0\hat{\Lambda}_+^m \tilde{f}. \tag{5.38}$$

Adding $\frac{1}{2}\tilde{f}(\xi', \xi_n)$ to both sides of (5.38), we obtain (5.36) for $\tilde{f} \in S$ by virtue of (5.12). It follows from Theorems 4.1, 4.2 and 5.1 that (5.36) remains valid for any $\tilde{f} \in \tilde{H}_{m+\delta}$ after completing the space S in the norm of $\tilde{H}_{m+\delta}$.

The validity for any $\tilde{f} \in \tilde{H}_{m+\delta}$ of the formula

$$\Pi^-\tilde{f} = \sum_{k=0}^{m} \frac{i\Pi'\hat{\Lambda}_-^{k-1}\tilde{f}}{\hat{\Lambda}_-^k} + \frac{1}{\hat{\Lambda}_-^m}\Pi^-\hat{\Lambda}_-^m\tilde{f}, \tag{5.39}$$

where $\hat{\Lambda}_- = \xi_n - i|\xi'| - i$ is proved analogously.

REMARK 5.2. *Suppose* $\delta = \pm\frac{1}{2}$. *Then* $\Pi^\pm$ *are unbounded operators in* $\tilde{H}_\delta$ *with a dense domain of definition.*

PROOF. Let $\tilde{L}_0(\mathbf{R}^n)$ denote the totality of all square integrable functions with compact support. For an arbitrary function $\tilde{f}(\xi) \in \tilde{H}_s$ we denote by $\tilde{f}_N$ the function that equals $\tilde{f}(\xi)$ for $|\xi| < N$ and vanishes for $|\xi| > N$. Then $\tilde{f}_N(\xi) \in \tilde{L}_0$ and $\|\tilde{f} - \tilde{f}_N\|_s \to 0$ as $N \to \infty$ for any s. Consequently, $\tilde{L}_0(\mathbf{R}^n)$ is dense in $\tilde{H}_s$.

Suppose $\delta = \frac{1}{2}$ and $\tilde{g} \in \tilde{L}_0$. Applying (5.36) for $m = 1$, we get

$$\Pi^+\tilde{g} = \frac{1}{\hat{\Lambda}_+} i\Pi'\tilde{g} + \hat{\Lambda}_+^{-1}\Pi^+\hat{\Lambda}_+\tilde{g}.$$

Clearly, $\hat{\Lambda}_+^{-1}\Pi^+\hat{\Lambda}_+\tilde{g} \in \tilde{H}_{1/2}$, since $\Pi^+\hat{\Lambda}_+\tilde{g} \in \tilde{H}_0$. On the other hand, $\hat{\Lambda}_+^{-1}i\Pi'\tilde{g} \in \tilde{H}_{1/2}$ if and only if

$$\Pi'\tilde{g} = \frac{1}{2\pi}\int_{-\infty}^{\infty}\tilde{g}(\xi', \xi_n)\,d\xi_n = 0 \tag{5.40}$$

for almost all ξ'. In fact,

$$\|\hat{\Lambda}_+^{-1}\Pi'\tilde{g}\|_{1/2}^2 = C\int_{-\infty}^{\infty}(1 + |\xi'| + |\xi_n|)^{-1}|\Pi'\tilde{g}|^2\,d\xi'd\xi_n = +\infty$$

if $\Pi'\tilde{g} \neq 0$ on a set of positive measure. The set of functions in $\tilde{L}_0(\mathbf{R}^n)$ satisfying (5.40) is denoted by $\tilde{L}_0^{(1)}$. We will show that $\tilde{L}_0^{(1)}$ is dense in $\tilde{L}_0(\mathbf{R}^n)$ and hence in $\tilde{H}_{1/2}$.

Let $\tilde{\varphi}_N(\xi_n) = 1/C_N(1 + |\xi_n|)$ for $|\xi_n| < N$ and $\tilde{\varphi}_N(\xi_n) = 0$ for $|\xi_n| > N$, where $C_N = \ln(1 + N)^2$. Clearly, $\int_{-\infty}^{\infty}\tilde{\varphi}_N(\xi_n)\,d\xi_n = 1$, and

$$\int_{-\infty}^{\infty}(1 + |\xi_n|)|\tilde{\varphi}_N(\xi_n)|^2\,d\xi_n = \frac{1}{C_N} \to 0$$

as $N \to \infty$. Let $\tilde{g}_N(\xi) = \tilde{g}(\xi) - 2\pi\tilde{\varphi}_N(\xi_n)\Pi'\tilde{g}(\xi)$, where $\tilde{g}(\xi) \in \tilde{L}_0(\mathbf{R}^n)$. Then $\tilde{g}_N(\xi) \in \tilde{L}_0^{(1)}$, since $\tilde{g}_N(\xi) \in \tilde{L}_0$ and $\Pi'\tilde{g}_N = 0$. But

$$\|\tilde{g} - \tilde{g}_N\|_{1/2}^2 = C\int_{-\infty}^{\infty}(1 + |\xi'| + |\xi_n|)|\tilde{\varphi}_N(\xi_n)|^2|\Pi'\tilde{g}|^2\,d\xi'd\xi_n$$

$$< C\int_{-\infty}^{\infty}(1+|\xi_n|)|\tilde{\varphi}_N(\xi_n)|^2|\Pi'\tilde{g}|^2\,d\xi'd\xi_n = \frac{C}{C_N}\int_{-\infty}^{\infty}|\Pi'\tilde{g}|^2\,d\xi' \to 0$$

as $N \to \infty$. Thus $\tilde{L}_0^{(1)}$ is dense in $\tilde{L}_0(\mathbf{R}^n)$.

Therefore Π^+ is unbounded on $\tilde{L}_0^{(1)}$, since otherwise $\Pi^+\tilde{f}$ would belong to $\tilde{H}_{1/2}$ for any $\tilde{f} \in \tilde{H}_{1/2}$, which is not the case. Further, the set $\tilde{\Omega}_{1/2}$ of functions $\tilde{f} \in \tilde{H}_{1/2}$ such that $\Pi^+\tilde{f} \in \tilde{H}_{1/2}$ is dense in $\tilde{H}_{1/2}$ since $\tilde{L}_0^{(1)} \subset \tilde{\Omega}_{1/2}$. Thus Π^+ is a densely defined unbounded operator in $\tilde{H}_{1/2}$.

Suppose now $\delta = -\frac{1}{2}$. Then $\Pi^+\tilde{f} \in \tilde{H}_{-1/2}$ for $\tilde{f} \in \tilde{L}_0(\mathbf{R}^n)$. Let

$$\tilde{h}_0(\xi', \xi_n) = \frac{\tilde{\psi}_0(\xi')\theta(\xi_n)}{\ln(\xi_n + 2)},$$

where $\tilde{\psi}_0(\xi') = 1$ for $|\xi'| < 1$ and $\tilde{\psi}_0(\xi') = 0$ for $|\xi'| > 1$. Then $\tilde{h}_0(\xi', \xi_n) \in \tilde{H}_{-1/2}$, since

$$\|\tilde{h}_0\|_{-1/2}^2 = \int_{-\infty}^{\infty}\int_0^{\infty}(1 + |\xi|)^{-1}\frac{|\tilde{\psi}_0(\xi')\theta(\xi_n)|^2}{\ln^2(\xi_n + 2)}\,d\xi'd\xi_n < \infty.$$

We put $\tilde{h}_N(\xi', \xi_n) = \tilde{h}_0(\xi', \xi_n)$ for $|\xi_n| < N$ and $\tilde{h}_N(\xi', \xi_n) = 0$ for $|\xi_n| > N$. Clearly, $\tilde{h}_N(\xi', \xi_n) \in \tilde{L}_0(\mathbf{R}^n)$, and $\|\tilde{h}_0 - \tilde{h}_N\|_{-1/2}^2 \to 0$ as $N \to \infty$. On the other hand,

$$\Pi^+\tilde{h}_N = \frac{i}{2\pi}\int_0^N\frac{\tilde{\psi}_0(\xi')\,d\eta_n}{(\xi_n + i0 - \eta_n)\ln(\eta_n + 2)} \to \infty$$

as $N \to \infty$ for $|\xi'| < 1$ and $\xi_n \neq 0$. Consequently, the sequence $\Pi^+\tilde{h}_N$ obviously does not converge in $\tilde{H}_{-1/2}$, since it would otherwise be possible to select a subsequence from $\Pi^+\tilde{h}_N$ that converges for almost all (ξ', ξ_n). Thus Π^+ is an unbounded operator in $\tilde{H}_{-1/2}$ with a dense domain of definition.

The unboundedness of Π^- in $\tilde{H}_{\pm 1/2}$ follows from the relation $\Pi^-\tilde{g} = \tilde{g} - \Pi^+\tilde{g}$, $\forall \tilde{g} \in \tilde{L}_0$.

We note that the function $\tilde{M}_0(s)$ is unbounded on the straight line $s = \frac{1}{2} + i\eta$ for $\delta = \frac{1}{2}$ (see the proof of Lemma 5.3), so that the unboundedness of Π^+ in $\tilde{H}_{1/2}$ can also be proved from the unboundedness of the operator $\Pi^{(1)}_{1/2}$ in $L_2(\mathbf{R}^1_+)$.

THEOREM 5.2 (Liouville's theorem). *Suppose $\tilde{f} \in \tilde{H}_s^+ \cap \tilde{H}_s^-$ for $s = -m + \delta$, where m is a positive integer and $-\frac{1}{2} \leqslant \delta < \frac{1}{2}$. Then for almost all ξ'*

$$\tilde{f}(\xi) = \sum_{k=1}^{m} \tilde{f}_k(\xi')\xi_n^{k-1}, \tag{5.41}$$

where $\tilde{f}_k(\xi') \in \tilde{H}_{s+k-1/2}(\mathbf{R}^{n-1})$.

PROOF. We take a function $\alpha(x_n) \in C_0^\infty(\mathbf{R}^1)$ such that $\alpha(x_n) = 1$ for $|x_n| \leqslant 1$ and $\alpha(x_n) = 0$ for $|x_n| \geqslant 2$. Since $f = F^{-1}\tilde{f} \in H_s^+ \cap H_s^-$, we have

$$(f, \varphi) = (f, \alpha(x_n)\varphi), \qquad \forall \varphi \in S. \tag{5.42}$$

Expanding $\varphi(x', x_n)$ in a Taylor series about $x_n = 0$ and multiplying this expansion by $\alpha(x_n)$, we get

$$\alpha(x_n)\varphi(x', x_n) = \sum_{k=1}^{m+1} \frac{(-ix_n)^{k-1}\alpha(x_n)}{(k-1)!}\varphi_k(x') + x_n^{m+1}\varphi_{m+1}(x', x_n), \tag{5.43}$$

where $\varphi_{m+1}(x', x_n) \in S(\mathbf{R}^n)$, $\varphi_k(x') = D_n^{k-1}\varphi(x', 0)$.

Let us show that

$$(f, x_n^{m+1}\varphi_{m+1}(x')) = 0. \tag{5.44}$$

Since $f \in H_s^+ \cap H_s^-$, it follows analogously to (5.42) that

$$(f, x_n^{m+1}\varphi_{m+1}(x)) = \left(f, \alpha\left(\frac{x_n}{\varepsilon}\right)x_n^{m+1}\varphi_{m+1}(x)\right), \qquad \forall \varepsilon > 0. \tag{5.45}$$

By virtue of (4.4) and the equivalence of the norms (4.1) and (4.5), we get

$$\begin{aligned} \left|\left(f, \alpha\left(\frac{x_n}{\varepsilon}\right)x_n^{m+1}\varphi_{m+1}\right)\right| &\leqslant \frac{1}{(2\pi)^n}\|f\|_s\left\|\alpha\left(\frac{x_n}{\varepsilon}\right)x_n^{m+1}\varphi_{m+1}\right\|_{-s} \\ &\leqslant C\|f\|_s\left\|\alpha\left(\frac{x_n}{\varepsilon}\right)x_n^{m+1}\varphi_{m+1}\right\|'_{m+1}. \end{aligned} \tag{5.46}$$

Since

$$\left\|\alpha\left(\frac{x_n}{\varepsilon}\right)x_n^{m+1}\varphi_{m+1}\right\|'^2_{m+1} = \sum_{|k| \leqslant m+1} \int_{|x_n| \leqslant 2\varepsilon} \left|D^k\left(\alpha\left(\frac{x_n}{\varepsilon}\right)x_n^{m+1}\varphi_{m+1}(x)\right)\right|^2 dx \to 0$$

as $\varepsilon \to 0$, (5.44) follows from (5.45) and (5.46).

Thus

$$(f, \varphi) = \sum_{k=1}^{m+1} (f, \alpha_k(x_n)\varphi_k(x')), \tag{5.47}$$

where

$$\alpha_k(x_n) = \frac{(-ix_n)^{k-1}\alpha(x_n)}{(k-1)!}.$$

From (5.47) we get by virtue of (4.3) that

$$(\tilde{f}, \tilde{\varphi}) = \sum_{k=1}^{m+1} (\tilde{f}, \tilde{\alpha}_k(\xi_n)\tilde{\varphi}_k(\xi')) = \sum_{k=1}^{m+1} 2\pi \int_{-\infty}^{\infty} \tilde{f}_k(\xi')\, \overline{\tilde{\varphi}_k(\xi')}\, d\xi', \tag{5.48}$$

where

$$\tilde{f}_k(\xi') = \frac{1}{2\pi} \int_{-\infty}^{\infty} \tilde{f}(\xi', \xi_n)\, \overline{\tilde{\alpha}_k(\xi_n)}\, d\xi_n.$$

Since

$$\tilde{\varphi}_k(\xi') = \frac{1}{2\pi} \int_{-\infty}^{\infty} \xi_n^{k-1} \tilde{\varphi}(\xi', \xi_n)\, d\xi_n,$$

it follows from (5.48) that

$$(\tilde{f}, \tilde{\varphi}) = \left(\sum_{k=1}^{m+1} \tilde{f}_k(\xi')\xi_n^{k-1}, \tilde{\varphi}(\xi', \xi_n) \right), \qquad \forall \tilde{\varphi} \in S(\mathbf{R}^n), \tag{5.49}$$

which implies that $\tilde{f} = \sum_{k=1}^{m+1} \tilde{f}_k(\xi')\xi_n^{k-1}$ almost everywhere. Since $\tilde{f} \in \tilde{H}_s$, it follows by virtue of Fubini's theorem that

$$\int_{-\infty}^{\infty} \left| \sum_{k=1}^{m+1} \tilde{f}_k(\xi')\xi_n^{k-1} \right|^2 (1 + |\xi'| + |\xi_n|)^{-2m+2\delta}\, d\xi_n < \infty \tag{5.50}$$

for almost all ξ'. Consequently, $\tilde{f}_{m+1}(\xi') = 0$ almost everywhere, since the integral

$$\int_{-\infty}^{\infty} |\xi_n|^p (1 + |\xi'| + |\xi_n|)^{-2m+2\delta}\, d\xi_n$$

diverges for $p = 2m$ and converges for $p \leqslant 2m - 2$.

Making the change of variables $\xi_n = (1 + |\xi'|)\eta_n$ in (5.50), we get

$$\int_{-\infty}^{\infty} \left| \sum_{k=1}^{m} \tilde{f}_k(\xi')\xi_n^{k-1} \right|^2 (1 + |\xi'| + |\xi_n|)^{2s}\, d\xi_n$$

$$= \sum_{i,k=1}^{m} a_{ik} \tilde{f}_i(\xi')(1 + |\xi'|)^{s+i-1/2} \tilde{f}_k(\xi')(1 + |\xi'|)^{s+k-1/2}, \tag{5.51}$$

where $a_{ik} = \int_{-\infty}^{\infty} \xi_n^{i+k-2}(1 + |\xi_n|)^{2s}\, d\xi_n$. The quadratic form (5.51) is positive definite. Consequently,

$$\int_{-\infty}^{\infty} \left| \sum_{k=1}^{m} \tilde{f}_k(\xi')\xi_n^{k-1} \right|^2 (1 + |\xi'| + |\xi_n|)^{2s}\, d\xi_n \geqslant C \sum_{k=1}^{m} |\tilde{f}_k(\xi')|^2 (1 + |\xi'|)^{2s+2k-1}. \tag{5.52}$$

Integrating both sides of (5.52), we get

$$\sum_{k=1}^{m} \left[\tilde{f}_k \right]^2_{s+k-1/2} \leqslant C \|\tilde{f}\|_s^2,$$

i.e. $\tilde{f}_k \in \tilde{H}_{s+k-1/2}(\mathbf{R}^{n-1})$. Theorem 5.2 is proved.

§6. Factorization of an elliptic symbol

1. Suppose $A(\xi) \in O^\infty_{\alpha+i\beta}$, i.e. $A(\xi) \in C^\infty$ for $\xi \neq 0$ and $A(\xi)$ is a homogeneous function of degree $\alpha + i\beta$:

$$A(t\xi) = t^{\alpha+i\beta}A(\xi), \qquad \forall t > 0. \tag{6.1}$$

A symbol $A(\xi)$ is said to be *elliptic* if

$$A(\xi) \neq 0, \qquad \forall \xi \neq 0. \tag{6.2}$$

Suppose $A(\xi) \in O^\infty_{\alpha+i\beta}$ is an elliptic symbol. Then $A^{-1}(\xi) \in O^\infty_{-\alpha-i\beta}$, and since $A(\xi) = |\xi|^{\alpha+i\beta}A\left(\frac{\xi}{|\xi|}\right)$, we have

$$C_1|\xi|^\alpha \leqslant |A(\xi)| \leqslant C_2|\xi|^\alpha. \tag{6.3}$$

By a *homogeneous factorization* of an elliptic symbol $A(\xi) \in O_{\alpha+i\beta}$ with respect to the variable ξ_n is meant a representation of $A(\xi', \xi_n)$ in the form

$$A(\xi', \xi_n) = A_-(\xi', \xi_n)A_+(\xi', \xi_n), \tag{6.4}$$

where $A_+(\xi', \xi_n)$ and $A_-(\xi', \xi_n)$ have the following properties:

a) $A_+(\xi', \xi_n)$ for $\xi' \neq 0$ admits an analytic continuation with respect to ξ_n into the upper halfplane $\tau > 0$.

b) $A_+(\xi', \xi_n + i\tau)$ is continuous with respect to the totality of variables (ξ', ξ_n, τ) for $\tau \geqslant 0$, $|\xi'| + |\xi_n| + \tau > 0$.

c) $A_+(\xi', \xi_n + i\tau)$ is a homogeneous function of degree $\kappa = \kappa_1 + i\kappa_2$:

$$A_+(t\xi', t(\xi_n + i\tau)) = t^\kappa A_+(\xi', \xi_n + i\tau), \qquad \forall t > 0. \tag{6.5}$$

d) $A_+(\xi', \xi_n + i\tau) \neq 0$ for $\tau \geqslant 0$, and $|\xi'| + |\xi_n| + \tau > 0$.

e) Analogously, the function $A_-(\xi', \xi_n)$ admits an analytic continuation with respect to ξ_n into the halfplane $\tau < 0$ for $\xi' \neq 0$, is continuous and different from zero for $\tau \leqslant 0$, $|\xi'| + |\xi_n| + |\tau| > 0$ and is homogeneous of degree $\alpha + i\beta - \kappa$.

If we replace A_+ by cA_+ and A_- by $c^{-1}A_-$, where c is an arbitrary constant, in (6.4), we will obtain another factorization of $A(\xi', \xi_n)$ with all of the above properties. We therefore assume in the sequel that

$$A_+(0, +1) = 1. \tag{6.6}$$

The class $O'_{\alpha+i\beta}$. We will write $\beta(\xi', \xi_n) \in O'_{\alpha+i\beta}$ if the function $B(\xi, \xi_n)$ is continuous for $\xi' \neq 0$ and homogeneous of degree $\alpha + i\beta$. Clearly, the $O^{\infty}_{\alpha+i\beta} \subset O'_{\alpha+i\beta}$.

The following lemma holds.

LEMMA 6.1. *Suppose* $b(\xi', \xi_n) \in O'_0$ *is a continuously differentiable function for* $\xi' \neq 0$, *and suppose*

$$|b(\xi', \xi_n)| \leqslant \frac{C|\xi'|}{|\xi'| + |\xi_n|}, \qquad \left|\frac{\partial b}{\partial \xi_k}\right| \leqslant \frac{C}{|\xi_n| + |\xi'|}, \qquad 1 \leqslant k \leqslant n-1. \tag{6.7}$$

Then the function

$$B_+(\xi', \xi_n + i\tau) = \frac{i}{2\pi}\int_{-\infty}^{\infty} \frac{b(\xi', \eta_n)}{\xi_n + i\tau - \eta_n}\, d\eta_n$$

is analytic with respect to $\xi_n + i\tau$ *in the halfplane* $\tau > 0$, *is continuous with respect to* $(\xi', \xi_n + i\tau)$ *for* $|\xi'| + |\xi_n| + \tau > 0, \tau \geqslant 0$, *and satisfies the estimate*

$$|B_+(\xi', \xi_n + i\tau)| \leqslant \frac{C_\varepsilon |\xi'|^{1-\varepsilon}}{(|\xi'| + |\xi_n| + \tau)^{1-\varepsilon}}, \qquad \forall \varepsilon > 0. \tag{6.8}$$

PROOF. The analyticity of $B_+(\xi', \xi_n + i\tau)$ for $\tau > 0$ follows from (6.7). Since

$$B_+(t\xi', t(\xi_n + i\tau)) = \frac{i}{2\pi}\int_{-\infty}^{\infty} \frac{b(t\xi', \eta_n)}{t\xi_n + it\tau - \eta_n}\, d\eta_n \tag{6.9}$$

and since $b(\xi', \eta_n)$ is homogeneous of degree zero, we get upon making the change of variables $\eta_n = t\eta'_n$ in (6.9) that $B_+(\xi', \xi_n + i\tau)$ is homogeneous of degree zero:

$$B_+(t\xi', t(\xi_n + i\tau)) = B_+(\xi', \xi_n + i\tau), \qquad \forall t > 0. \tag{6.10}$$

Let us now show that (6.7) implies the estimate

$$\left|\frac{\partial b(\xi', \xi_n)}{\partial \xi_n}\right| \leqslant \frac{C|\xi'|}{(|\xi'| + |\xi_n|)^2}. \tag{6.11}$$

From Euler's formula for homogeneous functions we have

$$\xi_n \frac{\partial b}{\partial \xi_n} + \sum_{k=1}^{n-1} \xi_k \frac{\partial b}{\partial \xi_k} \equiv 0.$$

Consequently,

$$\left|\frac{\partial b}{\partial \xi_n}\right| \leqslant \frac{C|\xi'|}{(|\xi'| + |\xi_n|)|\xi_n|} \leqslant \frac{2C|\xi'|}{(|\xi'| + |\xi_n|)^2}, \qquad |\xi'| \leqslant |\xi_n|, \tag{6.12}$$

and, since $\partial b/\partial\xi_n \in O'_{-1}$,

$$\left|\frac{\partial b}{\partial \xi_n}\right| \leqslant \frac{C}{|\xi'| + |\xi_n|} \leqslant \frac{2C|\xi'|}{(|\xi'| + |\xi_n|)^2}, \qquad |\xi'| \geqslant |\xi_n|. \tag{6.13}$$

Thus (6.11) is valid.

Suppose $\xi' = \omega$, $|\omega| = 1$. Then $b(\omega, \eta_n)$ and $\partial b(\omega, \eta_n)/\partial\eta_n$ are continuous with respect to (ω, η_n) and, by virtue of (6.7) and (6.11), satisfy the estimates

$$|b(\omega, \eta_n)| \leqslant \frac{C}{1 + |\eta_n|}, \qquad \left|\frac{\partial b(\omega, \eta_n)}{\partial \eta_n}\right| \leqslant \frac{C}{(1 + |\eta_n|)^2}. \tag{6.14}$$

Let

$$B_+(\omega, \xi_n + i\tau) = B_1(\omega, \xi_n, \tau) + B_2(\omega, \xi_n, \tau), \tag{6.15}$$

where

$$B_1(\omega, \xi_n, \tau) = \frac{i}{2\pi}\int_{|\xi_n - \eta_n| < 1/2} \frac{b(\omega, \eta_n) - b(\omega, \xi_n)}{\xi_n + i\tau - \eta_n}\, d\eta_n$$
$$+ \frac{i}{2\pi} b(\omega, \xi_n)\int_{|\xi_n - \eta_n| < 1/2} \frac{d\eta_n}{\xi_n + i\tau - \eta_n}, \tag{6.16}$$

$$B_2(\omega, \xi_n, \tau) = \frac{i}{2\pi}\int_{|\xi_n - \eta_n| > 1/2} \frac{b(\omega, \eta_n)}{\xi_n + i\tau - \eta_n}\, d\eta_n, \tag{6.17}$$

and let

$$b_1(\omega, \xi_n, \xi_n - \eta_n) = \frac{b(\omega, \eta_n) - b(\omega, \xi_n)}{\xi_n - \eta_n}. \tag{6.18}$$

Clearly, $b_1(\omega, \xi_n, \xi_n - \eta_n)$ is a continuous function with respect to the totality of variables. Estimating $b_1(\omega, \xi_n, \xi_n - \eta_n)$ for $|\xi_n - \eta_n| < \frac{1}{2}$ by Lagrange's formula for the remainder in Taylor's theorem, we get

$$|b_1(\omega, \xi_n, \xi_n - \eta_n)| \leqslant \frac{C}{(1 + |\xi_n + \theta(\eta_n - \xi_n)|)^2}$$
$$\leqslant \frac{C}{(1 + |\xi_n| - |\xi_n - \eta_n|)^2} \leqslant \frac{C}{(1 + |\xi_n|)^2}, \qquad 0 < \theta < 1. \tag{6.19}$$

On the other hand, making the change of variables $\xi_n - \eta_n = t$ in (6.16), we get

$$B_1(\omega, \xi_n, \tau) = \frac{i}{2\pi}\int_{|t| < \frac{1}{2}} \frac{t b_1(\omega, \xi_n, t)}{t + i\tau}\, dt + \frac{1}{\pi} b(\omega, \xi_n) \operatorname{arc\,tg} \frac{1}{2\tau}. \tag{6.19'}$$

To complete the proof of Lemma 6.1 we need

PROPOSITION 6.1. *Suppose the function $f(x, y)$ is continuous with respect to $x \in U$ for almost all $y \in V$ and integrable with respect to y for $x \in U$, where $U \subset \mathbf{R}^{n_1}$, $V \subset \mathbf{R}^{n_2}$. Suppose $|f(x, y)| \leqslant g(y)$, $y \in V$, where $\int_V g(y)\,dy < \infty$. Then $F(x) = \int_V f(x, y)\,dy$ is a continuous function in U.*

A proof of Proposition 6.1 can be obtained, for example, from Lebesgue's convergence theorem (see Example 1.5). For suppose $x_n \to x_0 \in U$. Then $\varphi_n(y) = f(x_n, y) \to \varphi_0(y) = f(x_0, y)$ for almost all $y \in V$ and $|\varphi_n(y)| \leqslant g(y)$. Consequently, by virtue of Lebesgue's convergence theorem,

$$F(x_n) = \int_V \varphi_n(y)\,dy \to F(x_0) = \int_V \varphi_0(y)\,dy,$$

i.e. $F(x)$ is a continuous function.

The function $tb_1(\omega, \xi_n, t)/(t + i\tau)$ is continuous with respect to (ω, ξ_n, τ) for any $t \neq 0$, and

$$\left| \frac{tb_1(\omega, \xi_n, t)}{t + i\tau} \right| \leqslant \frac{C|t|}{(|t| + \tau)(1 + |\xi_n|^2)}$$

for $|t| < \frac{1}{2}$. Consequently, the integral of (6.19′) satisfies all of the conditions of Proposition 6.1, so that $B_1(\omega, \xi_n, \tau)$ is continuous with respect to (ω, ξ_n, τ) for $\tau \geqslant 0$. In addition, by virtue of (6.7), (6.19) and (6.19′), we have

$$\begin{aligned} |B_1(\omega, \xi_n, \tau)| &\leqslant C \int_{|t| < \frac{1}{2}} \frac{|t|\,dt}{(|t| + \tau)(1 + |\xi_n|)^2} + \frac{C}{1 + |\xi_n|} \operatorname{arc\,tg} \frac{1}{2\tau} \\ &\leqslant \frac{C}{(1 + |\xi_n|)(1 + \tau)} \leqslant \frac{C}{1 + |\xi_n| + \tau}. \end{aligned} \tag{6.20}$$

Analogously, $B_2(\omega, \xi_n, \tau)$ is continuous with respect to (ω, ξ_n, τ) for $\tau \geqslant 0$. To estimate $B_2(\omega, \xi_n, \tau)$ we first note that

$$\begin{aligned} |B_2(\omega, \xi_n, \tau)| &\leqslant C \int_{|\xi_n - \eta_n| > \frac{1}{2}} \frac{d\eta_n}{(1 + |\eta_n|)|\xi_n + i\tau - \eta_n|} \\ &\leqslant C \int_{-\infty}^{\infty} \frac{d\eta_n}{|\xi_n + i\tau - \eta_n|^{1-\varepsilon/2} |\eta_n|^{1-\varepsilon/2}} \end{aligned} \tag{6.21}$$

for any $\varepsilon > 0$. Making the change of variables $\eta_n = (|\xi_n| + \tau)t$ in (6.21), we get

$$|B_2(\omega, \xi_n, \tau)| \leqslant \frac{C}{(|\xi_n| + \tau)^{1-\varepsilon}} \int_{-\infty}^{\infty} \frac{dt}{|t|^{1-\varepsilon/2} \left| \dfrac{\xi_n + i\tau}{|\xi_n| + \tau} - t \right|^{1-\varepsilon/2}} \leqslant \frac{C_\varepsilon}{(|\xi_n| + \tau)^{1-\varepsilon}}. \tag{6.22}$$

Since

$$|B_2(\omega, \xi_n, \tau)| \leqslant C \tag{6.23}$$

for $|\xi_n| + \tau \leqslant 1$, it follows by virtue of (6.22) and (6.23) that

$$|B_2(\omega, \xi_n, \tau)| \leqslant \frac{C_\varepsilon}{(1 + |\xi_n| + \tau)^{1-\varepsilon}}, \qquad \forall \varepsilon > 0. \tag{6.24}$$

From (6.15), (6.20) and (6.24) we obtain

$$|B_+(\omega, \xi_n + i\tau)| \leqslant \frac{C_\varepsilon}{(1 + |\xi_n| + \tau)^{1-\varepsilon}}, \qquad \forall \varepsilon > 0. \tag{6.25}$$

But the homogeneity of $B_+(\xi', \xi_n + i\tau)$ implies that

$$B_+(\xi', \xi_n + i\tau) = B_+\left(\omega, \frac{\xi_n + i\tau}{|\xi'|}\right),$$

where $\omega = \xi'/|\xi'|$, and hence (6.25) implies (6.8).

Finally, the continuity of $B_+(\xi', \xi_n + i\tau)$ for $\xi' \neq 0$ follows from the continuity of $B_1(\omega, \xi_n, \tau)$ and $B_2(\omega, \xi_n, \tau)$, while the continuity of $B_+(\xi', \xi_n + i\tau)$ for $\xi' = 0$ and $|\xi_n| + \tau > 0$ follows from (6.8) since $B_+(0, \xi_n + i\tau) = 0$. Lemma 6.1 is proved.

THEOREM 6.1. *Suppose $A(\xi) \in O^\infty_{\alpha + i\beta}$ is an elliptic symbol. Then $A(\xi)$ admits a unique (under the normalization condition (6.6)) homogeneous factorization.*

PROOF. A function $A(\xi) \in O^\infty_{\alpha + i\beta}$ can be represented in the form

$$A(\xi) = |\xi|^{\alpha + i\beta} A_0(\xi), \tag{6.26}$$

where $A_0(\xi) \in O_0^\infty$. The function $|\xi|^{\alpha + i\beta}$ is factored as follows:

$$|\xi|^{\alpha + i\beta} = (\xi_n - i|\xi'|)^{(\alpha + i\beta)/2}(\xi_n + i|\xi'|)^{(\alpha + i\beta)/2}, \tag{6.27}$$

where

$$(\xi_n \pm i|\xi'|)^{(\alpha + i\beta)/2} = e^{(\alpha + i\beta)\ln(\xi_n \pm i|\xi'|)/2},$$

with the branch of the logarithm being chosen, as in the preceding sections, so that it is real on the positive real axis.

Suppose $\xi' \neq 0$, and let

$$\alpha_1 = A_0(0, +1), \qquad \alpha_2 = A_0(0, -1). \tag{6.28}$$

By virtue of the homogeneity of the function $A_0(\xi', \xi_n)$,

$$\lim_{\xi_n \to +\infty} A_0(\xi', \xi_n) = \lim_{\xi_n \to +\infty} A_0\left(\frac{\xi'}{\xi_n}, +1\right) = A_0(0, +1) = \alpha_1, \tag{6.29}$$

$$\lim_{\xi_n \to -\infty} A_0(\xi', \xi_n) = \lim_{\xi_n \to -\infty} A_0\left(\frac{\xi'}{|\xi_n|}, \frac{\xi_n}{|\xi_n|}\right) = A_0(0, -1) = \alpha_2. \tag{6.30}$$

Let $\gamma = (1/2\pi i)\ln(\alpha_2/\alpha_1)$, with the branch of the logarithm being chosen arbitrarily, and consider the function

$$\alpha_1 \frac{(\xi_n + i|\xi'|)^\gamma}{(\xi_n - i|\xi'|)^\gamma}$$

for $\xi' \neq 0$. We have

$$\alpha_1 \frac{(\xi_n + i|\xi'|)^\gamma}{(\xi_n - i|\xi'|)^\gamma} = \alpha_1 \frac{e^{\gamma(\ln|\xi_n + i|\xi'|| + i\arg(\xi_n + i|\xi'|))}}{e^{\gamma(\ln|\xi_n - i|\xi'|| + i\arg(\xi_n - i|\xi'|))}} \to \alpha_1$$

as $\xi_n \to +\infty$, since $\ln|\xi_n + i|\xi'|| = \ln|\xi_n - i|\xi'||$ and $\arg(\xi_n \pm i|\xi'|) \to 0$ as $\xi_n \to +\infty$. When ξ_n varies through real values from $+\infty$ to $-\infty$, $\arg(\xi_n + i|\xi'|)$ varies from 0 to π while $\arg(\xi_n - i|\xi'|)$ varies from 0 to $-\pi$. Consequently,

$$\alpha_1 \frac{(\xi_n + i|\xi'|)^\gamma}{(\xi_n - i|\xi'|)^\gamma} \to \alpha_1 \frac{e^{i\gamma\pi}}{e^{-i\gamma\pi}} = \alpha_1 e^{2\pi i\gamma} = \alpha_2$$

as $\xi_n \to -\infty$.

Thus the function

$$A_1(\xi', \xi_n) = \frac{(\xi_n - i|\xi'|)^\gamma}{\alpha_1(\xi_n + i|\xi'|)^\gamma} A_0(\xi', \xi_n) \tag{6.31}$$

tends to 1 as $\xi_n \to \pm\infty$.

Let $2\pi m$ denote the increment in the argument of $A_1(\xi', \xi_n)$ for fixed $\xi' \neq 0$ and ξ_n varying from $+\infty$ to $-\infty$:

$$2\pi m = \Delta \arg A_1(\xi', \xi_n) \Big|_{\xi_n = +\infty}^{\xi_n = -\infty}.$$

We note that m is an integer, since $A_1(0, +1) = A_1(0, -1) = 1$. For $n \geqslant 3$ the sphere $|\xi'| = 1$ is connected, and since the quantity

$$\Delta \arg A_1(\xi', \xi_n) \Big|_{\xi_n = +\infty}^{\xi_n = -\infty}$$

clearly continuously depends on ξ', the integer m is one and the same for all $\xi' \neq 0$. For $n = 2$ the integer m can be different for $\xi' > 0$ and $\xi' < 0$. In this case we will assume in addition that m does not depend on ξ'.

Let

$$A_2(\xi', \xi_n) = \frac{(\xi_n - i|\xi'|)^m A_1(\xi', \xi_n)}{(\xi_n + i|\xi'|)^m}. \tag{6.32}$$

Since, when $\xi' \neq 0$,

$$\Delta \arg \frac{(\xi_n - i|\xi'|)^m}{(\xi_n + i|\xi'|)^m} \Big|_{\xi_n = +\infty}^{\xi_n = -\infty} = -2\pi m$$

and

$$\Delta \arg A_2(\xi', \xi_n)\Big|_{\xi_n = -\infty}^{\xi_n = +\infty} = \Delta \arg\left(\frac{\xi_n - i|\xi'|}{\xi_n + i|\xi'|}\right)^m\Big|_{\xi_n = -\infty}^{\xi_n = +\infty} + \Delta \arg A_1(\xi', \xi_n)\Big|_{\xi_n = -\infty}^{\xi_n = +\infty},$$

we have

$$\Delta \arg A_2(\xi', \xi_n)\Big|_{\xi_n = -\infty}^{\xi_n = +\infty} = 0. \tag{6.33}$$

In addition, $A_2(0, \pm 1) = 1$.

Let

$$b_2(\xi', \xi_n) = \ln A_2(\xi', \xi_n) = \ln|A_2(\xi', \xi_n)| + i \arg A_2(\xi', \xi_n), \tag{6.34}$$

with the branch of the logarithm being chosen so that $b_2(0, +1) = 0$. Then $b_2(0, -1) = 0$ by virtue of (6.33). We will show that $b_2(\xi', \xi_n)$ satisfies the conditions of Lemma 6.1. Clearly, $b_2(\xi', \xi_n)$ is continuously differentiable for $\xi' \neq 0$, $b_2(\xi', \xi_n) \in O_0'$ and its derivatives satisfy

$$\left|\frac{\partial b_2(\xi', \xi_n)}{\partial \xi_k}\right| \leqslant \frac{C}{|\xi'| + |\xi_n|}, \qquad 1 \leqslant k \leqslant n-1. \tag{6.35}$$

By means of Lagrange's formula for the remainder in Taylor's theorem, we get for $\xi_n > |\xi'|$

$$\begin{aligned} b_2(\xi', \xi_n)| &= \left|b_2\left(\frac{\xi'}{\xi_n}, +1\right) - b_2(0, +1)\right| \\ &\leqslant \frac{C}{\left|\theta\frac{\xi'}{\xi_n}\right| + 1}\,\frac{|\xi'|}{|\xi_n|} \leqslant \frac{C|\xi'|}{|\xi_n|} \leqslant \frac{2C|\xi'|}{|\xi'| + |\xi_n|}. \end{aligned} \tag{6.36}$$

Analogously, for $\xi_n < -|\xi'|$

$$\begin{aligned} |b_2(\xi', \xi_n)| &= \left|b_2\left(\frac{\xi'}{|\xi_n|}, -1\right) - b_2(0, -1)\right| \\ &\leqslant \frac{C}{\left|\theta\frac{\xi'}{\xi_n}\right| + 1}\,\frac{|\xi'|}{|\xi_n|} \leqslant \frac{2C|\xi'|}{|\xi'| + |\xi_n|}. \end{aligned} \tag{6.37}$$

Since $b_2(\xi', \xi_n)$ belongs to O_0' and is continuous for $|\xi'| + |\xi_n| > 0$, we have

$$|b_2(\xi', \xi_n)| \leqslant C \leqslant \frac{2C|\xi'|}{|\xi'| + |\xi_n|} \tag{6.38}$$

for $|\xi_n| \leqslant |\xi'|$. It follows from (6.36)–(6.38) that

$$|b_2(\xi', \xi_n)| \leqslant \frac{C|\xi'|}{|\xi'| + |\xi_n|}. \tag{6.39}$$

Thus $b_2(\xi', \xi_n)$ satisfies the conditions of Lemma 6.1, and hence the function

$$B_2^+(\xi', \xi_n + i\tau) = \frac{i}{2\pi}\int_{-\infty}^{\infty} \frac{b_2(\xi', \eta_n)}{\xi_n + i\tau - \eta_n}\, d\eta_n$$

is analytic with respect to $\xi_n + i\tau$ for $\tau > 0$, is continuous for $|\xi'| + |\xi_n| + \tau > 0$, $\tau \geqslant 0$, and satisfies the inequality

$$|B_2^+(\xi', \xi_n + i\tau)| \leqslant \frac{C_\varepsilon |\xi'|^{1-\varepsilon}}{(|\xi'| + |\xi_n| + \tau)^{1-\varepsilon}}, \qquad \forall \varepsilon > 0. \tag{6.40}$$

Analogously, the function

$$B_2^-(\xi', \xi_n + i\tau) = -\frac{i}{2\pi}\int_{-\infty}^{\infty} \frac{b_2(\xi', \eta_n)}{\xi_n + i\tau - \eta_n}\, d\eta_n$$

is analytic with respect to $\xi_n + i\tau$ for $\tau < 0$, is continuous for $|\xi'| + |\xi_n| + |\tau| > 0$, $\tau \leqslant 0$, and satisfies the estimate of form (6.40) with τ replaced by $|\tau|$. We note that both $B_2^+(\xi', \xi_n + i\tau)$ and $B_2^-(\xi', \xi_n + i\tau)$ are homogeneous of degree zero. By (5.31),

$$b_2(\xi', \xi_n) = \ln A_2(\xi', \xi_n) = B_2^+(\xi', \xi_n + i0) + B_2^-(\xi', \xi_n - i0). \tag{6.41}$$

Consequently,

$$A_2(\xi', \xi_n) = e^{B_2^-(\xi', \xi_n - i0)} e^{B_2^+(\xi', \xi_n + i0)}. \tag{6.42}$$

It therefore follows by (6.26), (6.27), (6.31), (6.32) and (6.42) that

$$A(\xi', \xi_n) = A_-(\xi', \xi_n) A_+(\xi', \xi_n), \tag{6.43}$$

where

$$A_+(\xi', \xi_n) = (\xi_n + i|\xi'|)^{(\alpha + i\beta)/2 + m + \gamma} e^{B_2^+(\xi', \xi_n + i0)}, \tag{6.44}$$

$$A_-(\xi', \xi_n) = \alpha_1(\xi_n - i|\xi'|)^{(\alpha + i\beta)/2 - m - \gamma} e^{B_2^-(\xi', \xi_n - i0)}. \tag{6.45}$$

The functions $A_\pm(\xi', \xi_n)$ satisfy all of the homogeneous factorization conditions, with $A_+(0, +1) = 1$ and the degree of homogeneity κ of $A_+(\xi', \xi_n)$ being equal to $2^{-1}(\alpha + i\beta) + m + \gamma$.

Let us now prove the uniqueness of the factorization under the normalization condition (6.6). Suppose

$$A(\xi', \xi_n) = A_-^{(1)}(\xi', \xi_n) A_+^{(1)}(\xi', \xi_n) \tag{6.46}$$

is another homogeneous factorization of $A(\xi', \xi_n)$, with $A_+^{(1)}(0, +1) = 1$ and the degree of homogeneity of $A_+^{(1)}(\xi', \xi_n)$ being equal to $\kappa^{(1)}$. Then for $\xi' \neq 0$

the function

$$\frac{A_+(\xi', \xi_n)}{A_+^{(1)}(\xi', \xi_n)} = \frac{A_-^{(1)}(\xi', \xi_n)}{A_-(\xi', \xi_n)} \tag{6.47}$$

is analytic with respect to $\xi_n + i\tau$ in the whole $\xi_n + i\tau$ plane, behaves like a power of $\xi_n + i\tau$ as $|\xi_n + i\tau| \to \infty$, and does not have any zeros. Consequently, according to Liouville's theorem,

$$\frac{A_+(\xi', \xi_n)}{A_+^{(1)}(\xi', \xi_n)} = C(\xi') \neq 0.$$

Letting $\xi_n \to +\infty$, we get

$$C(\xi') = \lim_{\xi_n \to +\infty} \frac{A_+(\xi', \xi_n)}{A_+^{(1)}(\xi', \xi_n)} = \lim_{\xi_n \to +\infty} \frac{\xi_n^{\kappa} A_+\left(\dfrac{\xi'}{\xi_n}, +1\right)}{\xi_n^{\varkappa^{(1)}} A_+^{(1)}\left(\dfrac{\xi'}{\xi_n}, +1\right)}. \tag{6.48}$$

Since the limit in (6.48) exists and is different from zero and infinity, it follows that $\kappa = \kappa^{(1)}$, and hence

$$C(\xi') = \frac{A_+(0, +1)}{A_+^{(1)}(0, +1)} = 1.$$

Thus

$$A_+(\xi', \xi_n) \equiv A_+^{(1)}(\xi', \xi_n) \qquad \text{and} \qquad A_-(\xi', \xi_n) \equiv A_-^{(1)}(\xi', \xi_n).$$

Theorem 6.1 is proved.

2. *Formula for the factorization index*. The degree of homogeneity $\kappa = \kappa_1 + i\kappa_2$ of the function $A_+(\xi', \xi_n)$ is called the *factorization index* of an elliptic symbol $A(\xi', \xi_n) \in O^{\infty}_{\alpha+i\beta}$. By virtue of (6.44), $\kappa = (\alpha + i\beta)/2 + m + \gamma$, where $\gamma = (1/2\pi i)\ln(\alpha_2/\alpha_1)$ is generally a complex number: $\gamma = \gamma_1 + i\gamma_2$. We note that $\gamma_2 = -(1/2\pi)\ln|\alpha_2/\alpha_1|$. Since

$$A(\xi', \xi_n) = |\xi|^{\alpha+i\beta}\left(\frac{\xi_n + i|\xi'|}{\xi_n - i|\xi'|}\right)^{m+\gamma} \alpha_1 A_2(\xi', \xi_n) \tag{6.49}$$

and

$$\frac{1}{2\pi}\Delta \arg\left(\frac{\xi_n + i|\xi'|}{\xi_n - i|\xi'|}\right)^{m+\gamma_1+i\gamma_2}\Bigg|_{\xi_n = +\infty}^{\xi_n = -\infty} = m + \gamma_1,$$

it follows by virtue of (6.33) that

$$\frac{1}{2\pi}\Delta \arg A(\xi', \xi_n)|\xi|^{-\alpha-i\beta}\Bigg|_{\xi_n = +\infty}^{\xi_n = -\infty} = m + \gamma_1. \tag{6.50}$$

Thus

$$\kappa = \frac{\alpha + i\beta}{2} + \frac{1}{2\pi}\Delta \arg A(\xi', \xi_n)|\xi|^{-\alpha - i\beta}\Bigg|_{\xi_n = +\infty}^{\xi_n = -\infty} - \frac{i}{2\pi}\ln\left|\frac{A(0, -1)}{A(0, +1)}\right|. \tag{6.51}$$

We cite some examples in which the factorization index of an elliptic symbol is calculated.

EXAMPLE 6.1. A symbol $A(\xi) \in O_\alpha^\infty$, where α is a real number, is said to be *strongly elliptic* if

$$A(\xi) = A_1(\xi) + iA_2(\xi), \tag{6.52}$$

where the functions $A_i(\xi) \in O_\alpha^\infty$ $(i = 1, 2)$ are real symbols and

$$A_1(\xi) \neq 0, \qquad \forall \xi \neq 0. \tag{6.53}$$

For any fixed $\xi' \neq 0$ and ξ_n varying from $+\infty$ to $-\infty$ the function $A(\xi)|\xi|^{-\alpha}$ describes a contour in the complex plane that lies entirely in the right halfplane and joins the points $A(0, +1)$ and $A(0, -1)$. Consequently,

$$\frac{1}{2\pi}\Delta \arg A(\xi)|\xi|^{-\alpha}\Bigg|_{\xi_n = +\infty}^{\xi_n = -\infty} = \frac{1}{2\pi}(\arg A(0, -1) - \arg A(0, +1)).$$

Thus, by virtue of (6.51),

$$\begin{aligned}\kappa &= \frac{\alpha}{2} + \frac{1}{2\pi}(\arg A(0, -1) - \arg A(0, +1)) - \frac{i}{2\pi}\ln\left|\frac{A(0, -1)}{A(0, +1)}\right| \\ &= \frac{\alpha}{2} + \frac{1}{2\pi i}\ln\frac{A(0, -1)}{A(0, +1)},\end{aligned} \tag{6.54}$$

where $-\pi/2 < \arg A(0, \pm 1) < \pi/2$. If, in particular, $A(0, -1) = A(0, +1)$, then $\kappa = \alpha/2$.

EXAMPLE 6.2. Suppose $A(\xi) \in O_{\alpha+i\beta}^\infty$ is an even elliptic symbol, i.e. $A(\xi) = A(-\xi)$. We will show that the factorization index of $A(\xi', \xi_n)$ is equal to $(\alpha + i\beta)/2$ for $n \geqslant 3$. Since the sphere $|\xi'| = 1$ is connected for $n \geqslant 3$, we have

$$\Delta \arg A(\xi', \xi_n)|\xi|^{-\alpha - i\beta}\Bigg|_{\xi_n = +\infty}^{\xi_n = -\infty} = \Delta \arg A(-\xi', \xi_n)|\xi|^{-\alpha - i\beta}\Bigg|_{\xi_n = +\infty}^{\xi_n = -\infty}. \tag{6.54'}$$

while the evenness of $A(\xi', \xi_n)$ implies

$$\Delta \arg A(-\xi', \xi_n)|\xi|^{-\alpha - i\beta}\Bigg|_{\xi_n = +\infty}^{\xi_b = -\infty} = \Delta \arg A(\xi', -\xi_n)|\xi|^{-\alpha - i\beta}\Bigg|_{\xi_n = +\infty}^{\xi_n = -\infty}. \tag{6.55}$$

And since the increment in the argument clearly changes sign if one reverses the direction in which ξ_n varies, it follows from (6.54′) and (6.55) that

$$\begin{aligned}\Delta \arg A(\xi', \xi_n)|\xi|^{-\alpha - i\beta}\Bigg|_{\xi_n = +\infty}^{\xi_n = -\infty} &= \Delta \arg A(\xi', -\xi_n)|\xi|^{-\alpha - i\beta}\Bigg|_{\xi_n = +\infty}^{\xi_n = -\infty} \\ &= -\Delta \arg A(\xi', \xi_n)|\xi|^{-\alpha - i\beta}\Bigg|_{\xi_n = +\infty}^{\xi_n = -\infty}.\end{aligned} \tag{6.56}$$

Thus

$$\Delta \arg A(\xi', \xi_n)|\xi|^{-\alpha - i\beta} \Bigg| \begin{matrix} \xi_n = -\infty \\ \xi_n = +\infty \end{matrix} = 0,$$

and hence $\kappa = (\alpha + i\beta)/2$ by virtue of (6.51).

EXAMPLE 6.3. *Factorization of a homogeneous elliptic polynomial*. A homogeneous polynomial $P(\xi) = \Sigma_{|k|=r} a_k \xi^k$ is said to be *elliptic* if $P(\xi) \neq 0$ for $\xi \neq 0$. Consequently, for $\xi' \neq 0$ the polynomial $P(\xi', \xi_n)$ has no real zeros with respect to the variable ξ_n. Let $P_+(\xi', \xi_n + i\tau)$ denote the polynomial in $\xi_n + i\tau$ whose zeros are precisely those zeros of $P(\xi', \xi_n + i\tau)$ that lie in the halfplane $\tau < 0$ for $\xi' \neq 0$ and which is such that $P_+(0, +1) = 1$. Analogously, let $P_-(\xi', \xi_n + i\tau)$ denote the polynomial whose zeros are precisely those zeros of $P(\xi', \xi_n + i\tau)$ that lie in the halfplane $\tau > 0$ for $\xi' \neq 0$ and which is such that $P(0, +1) = P_-(0, +1)$. Then

$$P(\xi', \xi_n) = P_-(\xi', \xi_n) P_+(\xi', \xi_n) \tag{6.57}$$

is the homogeneous factorization of $P(\xi', \xi_n)$. Suppose $n \geqslant 3$. Since $P(-\xi) = (-1)^r P(\xi)$, we can prove, as in Example 6.2, that the degree of homogeneity κ of $P_+(\xi)$ is equal to $r/2$. It follows that r is an even integer: $r = 2m$, and that the degree of homogeneity of both $P_+(\xi)$ and $P_-(\xi)$ is equal to m.

EXAMPLE 6.4. Suppose $n = 2$ and $P(\xi) = \xi_2 - i\xi_1$, i.e. $P(\xi)$, is the symbol of the Cauchy-Riemann operator $\partial/\partial x_1 + i\partial/\partial x_2$. Then the factorization of $\xi_2 - i\xi_1$ has the form $P_+(\xi_1, \xi_2) = \xi_2 - i\xi_1$, $P_-(\xi_1, \xi_2) = 1$ for $\xi_1 < 0$ and the form $P_+(\xi_1, \xi_2) = 1$, $P_-(\xi_1, \xi_2) = \xi_2 - i\xi_1$ for $\xi_1 > 0$. Thus the factorization index is equal to 1 for $\xi_1 < 0$ and vanishes for $\xi_1 > 0$. In the sequel we will consider for the case $n = 2$ only those elliptic symbols whose factorization index does not depend on ξ_1.

§7. Pseudodifferential equations in a halfspace

1. Let A be a pseudodifferential operator (p.o.) with symbol $A(\xi) \in S_\alpha^0$ (see §3):

$$Au = \frac{1}{(2\pi)^n} \int_{-\infty}^{\infty} A(\xi)\tilde{u}(\xi) e^{-i(x,\xi)}\, d\xi, \tag{7.1}$$

where $u(x) \in S(\mathbf{R}^n)$. In this section, as throughout Chapters I–IV, only symbols $A(\xi)$ not depending on x are considered. We assume that the symbol $A(\xi)$ can be represented in the form

$$A(\xi) = A_0(\xi) + A_1(\xi), \tag{7.2}$$

where $A_0(\xi) \in O_{\alpha + i\beta}^\infty$ and

$$|A_1(\xi)| \leqslant C|\xi|^{\alpha - \varepsilon} \quad \text{for } |\xi| \geqslant N \tag{7.3}$$

in which $\varepsilon > 0$ and N are given numbers. As will be shown in Chapter VI, the p.o. A_1 with symbol $A_1(\xi)$ plays a subordinate role in the theory of boundary value problems in a bounded domain in comparison with the p.o. A_0 in the sense that the statement of a boundary value problem for the p.o. A and the conditions for its normal solvability do not depend on A_1 but only on $A_0(\xi)$.

Therefore, in this and the following chapter, we specially choose $A_1(\xi)$ so as to simplify the properties of the p.o. A with symbol $A(\xi) = A_0(\xi) + A_1(\xi)$ in comparison with those of $A_0(\xi)$. The choice $A_1(\xi) = 0$, i.e. $A(\xi) = A_0(\xi)$ for all ξ, is not really suitable, since the function $A_0(\xi)$ is unbounded in a neighborhood of $\xi = 0$ and hence A_0 is an unbounded p.o. from $H_s(\mathbf{R}^n)$ into $H_{s-\alpha}(\mathbf{R}^n)$ for any s when $\alpha < 0$, while the p.o. A_0^{-1} with symbol $A_0^{-1}(\xi)$ becomes unbounded in the spaces $H_s(\mathbf{R}^n)$ when $\alpha > 0$ and $A_0(\xi) \neq 0$ for $\xi \neq 0$.

Let

$$\hat{A}_0(\xi', \xi_n) = A_0((1 + |\xi'|)\omega, \xi_n). \tag{7.4}$$

where $\omega = \xi'/|\xi'|$ and $A_0(\xi) \in O^{\infty}_{\alpha+i\beta}$, and let $A_1(\xi', \xi_n) = \hat{A}_0(\xi', \xi_n) - A_0(\xi', \xi_n)$. We will show that $A_1(\xi', \xi_n)$ satisfies (7.3). According to Lagrange's formula for the remainder in Taylor's theorem, we have

$$\begin{aligned} |A_1(\xi', \xi_n)| &= |A_0(\xi' + \omega, \xi_n) - A_0(\xi', \xi_n)| \\ &\leqslant \sum_{k=1}^{n-1} \left| \frac{\partial A_0(\xi' + \theta\omega, \xi_n)}{\partial \xi_k} \right| \leqslant C(|\xi' + \theta\omega| + |\xi_n|)^{\alpha-1}, \end{aligned} \tag{7.5}$$

where $0 < \theta < 1$. Suppose $|\xi'| + |\xi_n| \geqslant 2$. Then

$$\begin{aligned} \tfrac{1}{2}(|\xi'| + |\xi_n|) &\leqslant |\xi'| + |\xi_n| - 1 \leqslant |\xi' + \theta\omega| + |\xi_n| \\ &\leqslant |\xi'| + 1 + |\xi_n| \leqslant 2(|\xi'| + |\xi_n|). \end{aligned}$$

Consequently,

$$(|\xi' + \theta\omega| + |\xi_n|)^{\alpha-1} \leqslant C(|\xi'| + |\xi_n|)^{\alpha-1}.$$

for either sign of $\alpha - 1$. Thus

$$|A_1(\xi', \xi_n)| \leqslant C(|\xi'| + |\xi_n|)^{\alpha-1} \quad \text{for} \quad |\xi'| + |\xi_n| \geqslant 2. \tag{7.6}$$

We note the following simple lemma on an elliptic p.o. with symbol $\hat{A}(\xi', \xi_n)$.

LEMMA 7.1. *Suppose $A_0(\xi) \in O^{\infty}_{\alpha+i\beta}$ is an elliptic symbol, and*

$$\hat{A}_0(\xi', \xi_n) = A_0((1 + |\xi'|)\omega, \xi_n).$$

Then the pseudodifferential equation (p.e.)

$$\hat{A}_0 u = f, \qquad \forall f \in H_{s-\alpha}(\mathbf{R}^n), \tag{7.7}$$

is uniquely solvable in $H_s(\mathbf{R}^n)$ for any s.

PROOF. Since $A_0(\xi', \xi_n)$ satisfies (6.3), $\hat{A}_0(\xi', \xi_n)$ satisfies

$$C_2(1 + |\xi'| + |\xi_n|)^{\alpha} \leqslant |\hat{A}_0(\xi', \xi_n)| \leqslant C_1(1 + |\xi'| + |\xi_n|)^{\alpha}. \tag{7.8}$$

By Lemma 4.4, the operator $\hat{A}_0$ defined on the functions of class S by the formula of form (7.1) can be extended to a bounded operator from $H_s(\mathbf{R}^n)$ into $H_{s-\alpha}(\mathbf{R}^n)$ for any s. Taking the Fourier transform with respect to x of equation (7.7), we get

$$\hat{A}_0(\xi)\tilde{u}(\xi) = \tilde{f}(\xi), \qquad \tilde{f}(\xi) \in \tilde{H}_{s-\alpha}(\mathbf{R}^n). \tag{7.9}$$

It clearly follows from (7.8) that $\tilde{u}(\xi) = \tilde{f}(\xi)/\hat{A}_0(\xi) \in \tilde{H}_s(\mathbf{R}^n)$ is the only solution of (7.9) such that

$$\|u\|_s^2 = \int_{-\infty}^{\infty} (1 + |\xi|)^{2s} \frac{|\tilde{f}(\xi)|^2}{|\hat{A}_0(\xi)|^2}\, d\xi \leqslant C\|f\|_{s-\alpha}^2. \tag{7.10}$$

Thus the p.o. $\hat{A}_0^{-1}$ with symbol $\hat{A}_0^{-1}(\xi)$ is a bounded operator from $H_{s-\alpha}(\mathbf{R}^n)$ into $H_s(\mathbf{R}^n)$ and is the inverse of $\hat{A}_0$.

2. We now go over to the investigation of elliptic pseudodifferential equations in a halfspace. Let $\hat{A}_0$ be a p.o. with symbol $\hat{A}_0(\xi)$, where $A_0(\xi) \in O_{\alpha+i\beta}^{\infty}$. By a *pseudodifferential equation* in the halfspace $\mathbf{R}_+^n$ is meant an equation of the form

$$p\hat{A}_0 u_+ = f, \tag{7.11}$$

where $u_+ \in H_s^+(\mathbf{R}^n)$, i.e. $u_+ = 0$ for $x_n < 0$, $f \in H_{s-\alpha}(\mathbf{R}_+^n)$ and p is the operator of restriction to the halfspace $\mathbf{R}_+^n$. Thus the equation $\hat{A}_0 u_+ = f$ is satisfied for $x_n > 0$. The operator defined by (7.11) is a bounded operator from H_s^+ into $H_{s-\alpha}(\mathbf{R}_+^n)$ for any s. In fact, if $u_+ \in H_s^+$, then $\hat{A}_0 u_+ \in H_{s-\alpha}(\mathbf{R}^n)$ by virtue of Lemma 4.4. Consequently, by definition of $H_{s-\alpha}(\mathbf{R}_+^n)$,

$$p\hat{A}_0 u_+ \in H_{s-\alpha}(\mathbf{R}_+^n), \quad \text{with } \|p\hat{A}_0 u_+\|_{s-\alpha}^+ \leqslant \|\hat{A}_0 u_+\|_{s-\alpha} \leqslant C\|u_+\|_s.$$

Suppose $A_0(\xi', \xi_n) \in O_{\alpha+i\beta}^{\infty}$ is an elliptic symbol with homogeneous factorization

$$A_0(\xi', \xi_n) = A_-(\xi', \xi_n)A_+(\xi', \xi_n) \tag{7.12}$$

and let κ denote the factorization index of $A_0(\xi', \xi_n)$. By applying the factorization (or Wiener-Hopf) method, we will find a solution of (7.11) belonging to H_s^+ under the assumption that $f \in H_{s-\alpha}(\mathbf{R}_+^n)$.

Suppose there exists a solution $u_+ \in H_s^+$ of equation (7.11) for $f \in H_{s-\alpha}(\mathbf{R}_+^n)$ and let

$$u_- = lf - \hat{A}_0 u_+, \tag{7.13}$$

where $lf \in H_{s-\alpha}(\mathbf{R}^n)$ is an arbitrary extension of $f \in H_{s-\alpha}(\mathbf{R}^n)$ from $\mathbf{R}_+^n$ onto $\mathbf{R}^n$. Since $lf \in H_{s-\alpha}(\mathbf{R}^n)$ and $\hat{A}_0 u_+ \in H_{s-\alpha}(\mathbf{R}^n)$, we have $u_- \in H_{s-\alpha}(\mathbf{R}^n)$. In addition, $u_- = 0$ for $x_n > 0$ by virtue of (7.11). Consequently, $u_- \in H_{s-\alpha}^-(\mathbf{R}^n)$.

The Fourier transform with respect to x of (7.13) is

$$\hat{A}_0(\xi)\tilde{u}_+(\xi) + \tilde{u}_-(\xi) = \widetilde{lf}(\xi). \tag{7.14}$$

Substituting

$\hat{A}_0(\xi', \xi_n) = \hat{A}_-(\xi', \xi')\hat{A}_+(\xi', \xi_n)$, where $\hat{A}_\pm(\xi', \xi_n) \coloneqq A_\pm((1 + |\xi'|)\omega, \xi_n)$

in (7.14) and multiplying both sides of (7.14) by $\hat{A}_-^{-1}(\xi', \xi_n)$, we get

$$\hat{A}_+(\xi)\tilde{u}_+(\xi) + \hat{A}_-^{-1}(\xi)\tilde{u}_-(\xi) = \hat{A}_-^{-1}(\xi)\widetilde{lf}(\xi). \tag{7.15}$$

Let

$$\tilde{v}_+(\xi) = \hat{A}_+(\xi)\tilde{u}_+(\xi), \qquad \tilde{v}_-(\xi) = \hat{A}_-^{-1}(\xi)\tilde{u}_-(\xi), \tag{7.16}$$

$$\tilde{g}(\xi) = \hat{A}_-^{-1}(\xi)\widetilde{lf}(\xi). \tag{7.17}$$

By Theorem 4.4 and Lemma 4.4,

$$\tilde{v}_+(\xi) \in \tilde{H}^+_{s-\operatorname{Re}\kappa}, \qquad \tilde{v}_-(\xi) \in \tilde{H}^-_{s-\operatorname{Re}\kappa}, \qquad \tilde{g}(\xi) \in \tilde{H}_{s-\operatorname{Re}\kappa}, \tag{7.18}$$

since the degrees of homogeneity of $A_+(\xi)$ and $A_-^{-1}(\xi)$ are equal to κ and $\kappa - \alpha - i\beta$ respectively. In the new notation (7.15) acquires the form

$$\tilde{v}_+(\xi) + \tilde{v}_-(\xi) = \tilde{g}(\xi). \tag{7.19}$$

Suppose

$$\operatorname{Re}\kappa - s - \tfrac{1}{2} \in \mathbf{R} \setminus \mathbf{Z}, \tag{7.20}$$

where $\mathbf{Z}$ denotes the set of all integers. Then there exists an integer m such that

$$\operatorname{Re}\kappa - s = m + \delta, \tag{7.21}$$

where $|\delta| < \frac{1}{2}$. We will separately consider in subsections 3, 4 and 6 below the three cases $m = 0$, $m > 0$ and $m < 0$.

3. Suppose $m = 0$, i.e. $\operatorname{Re}\kappa - s = \delta$, $|\delta| < \frac{1}{2}$. Then $\tilde{v}_+(\xi) \in \tilde{H}^\pm_{s-\operatorname{Re}\kappa} = \tilde{H}^\pm_{-\delta}$, $\tilde{g}(\xi) \in \tilde{H}_{-\delta}$ in (7.19). Since $|\delta| < \frac{1}{2}$, we have by virtue of Lemma 5.4 that the decomposition (7.19) is unique, with

$$\tilde{v}_+ = \Pi^+\tilde{g}, \tilde{v}_- = \Pi^-\tilde{g}. \tag{7.22}$$

It follows from (7.16), (7.17) and (7.22) that the Fourier transform of any solution $u_+(x) \in H_s^+$ of (7.11) has the form

$$\tilde{u}_+(\xi) = \hat{A}_+^{-1}\Pi^+\hat{A}_-^{-1}\widetilde{lf}. \tag{7.23}$$

We note that (7.23) does not depend on the choice of the extension lf. For suppose $l_1 f \in H_{s-\alpha}(\mathbf{R}^n)$ is another extension of $f \in H_{s-\alpha}(\mathbf{R}^n)$, i.e. $pl_1 f = f$. Since $f_- = lf - l_1 f \in H^-_{s-\alpha}(\mathbf{R}^n)$, it follows by virtue of Theorem 4.4 that $\hat{A}_-^{-1}(\xi)\tilde{f}_-(\xi) \in \tilde{H}^-_{s-\operatorname{Re}\kappa} = \tilde{H}^-_{-\delta}$; while Lemma 5.4 implies that $\Pi^+\hat{A}_-^{-1}(\xi)\tilde{f}_- = 0$. Thus

$$\Pi^+\hat{A}_-^{-1}\widetilde{lf} = \Pi^+\hat{A}_-^{-1}\widetilde{l_1 f}. \tag{7.24}$$

If, in particular, $f = 0$, we can choose $lf = 0$, and hence $u_+ = 0$ by virtue of (7.23). Thus the solution of (7.11) is unique in the case $\operatorname{Re}\kappa - s = \delta$, $|\delta| < \frac{1}{2}$.

In the sequel we will always choose an extension lf such that

$$\|lf\|_{s-\alpha} \leqslant 2\|f\|^+_{s-\alpha}. \tag{7.25}$$

Relations (7.23), (7.25), Lemma 4.4 and the boundedness of the operator Π^+ in $\tilde{H}_{-\delta} = \tilde{H}_{s-\operatorname{Re}\kappa}$ imply

$$\|u_+\|_s = \|\tilde{u}_+\|_s \leqslant C\|\Pi^+\hat{A}_-^{-1}\widetilde{lf}\|_{s-\operatorname{Re}\kappa} \leqslant C\|\hat{A}_-^{-1}\widetilde{lf}\|_{s-\operatorname{Re}\kappa} \leqslant C\|lf\|_{s-\alpha}$$
$$\leqslant C\|f\|^+_{s-\alpha}. \tag{7.26}$$

We note that $u_+ = F^{-1}\hat{A}_+^{-1}\Pi^+\hat{A}_-^{-1}\widetilde{lf}$ is a solution of (7.11) for any $f \in H_{s-\alpha}(\mathbf{R}^n_+)$. In fact, since $\Pi^+\hat{A}_-^{-1}\widetilde{lf} = \hat{A}_-^{-1}\widetilde{lf} - \Pi^-\hat{A}_-^{-1}\widetilde{lf}$, we have

$$\hat{A}_0(\xi)\tilde{u}_+(\xi) = \hat{A}_-(\xi)\Pi^+\hat{A}_-^{-1}\widetilde{lf} = \widetilde{lf} - \hat{A}_-\Pi^-\hat{A}_-^{-1}\widetilde{lf}.$$

By Theorems 4.4 and 5.1, $\hat{A}_-(\xi)\Pi^-\hat{A}_-^{-1}\widetilde{lf} \in \tilde{H}^-_{s-\alpha}$. Consequently,

$$p\hat{A}_0u_+ = p\big(lf - F^{-1}\hat{A}_-(\xi)\Pi^-\hat{A}_-^{-1}\widetilde{lf}\big) = f,$$

i.e. $u_+ = F^{-1}\hat{A}_+^{-1}(\xi)\Pi^+\hat{A}_-^{-1}\widetilde{lf}$ is a solution of (7.11) belonging to H_s^+. We have thus proved

THEOREM 7.1. *Suppose* $\operatorname{Re}\kappa - s = \delta$, $|\delta| < \frac{1}{2}$. *Then* (7.11) *has a unique solution* $u_+ \in H_s^+$ *for any* $f \in H_{s-\alpha}(\mathbf{R}^n_+)$.

EXAMPLE 7.1. Let $\Lambda^{2\kappa}(D', D_n, \tau)$ denote the p.o. with symbol $\Lambda^{2\kappa}(\xi', \xi_n, \tau) = |\xi_n^2 + (|\xi'| + \tau)^2|^\kappa$, where $\tau > 0$ and κ is an arbitrary real number. The factorization of $\Lambda^{2\kappa}(\xi', \xi_n, \tau)$ has the form

$$\Lambda^{2\kappa}(\xi', \xi_n, \tau) = \Lambda_-^\kappa(\xi', \xi_n, -i\tau)\Lambda_+^\kappa(\xi', \xi_n + i\tau),$$

$$\text{where } \Lambda_\pm^\kappa(\xi', \xi_n \pm i\tau) = (\xi_n \pm i|\xi'| \pm i\tau)^\kappa$$

(see Example 4.5). Thus the factorization index of $\Lambda^{2\kappa}(\xi', \xi_n, \tau)$ is equal to κ. Consider the p.e. in $\mathbf{R}^n_+$

$$p\Lambda^{2\kappa}(D', D_n, \tau)u_+ = f(x), \qquad f(x) \in H_{\kappa-2\kappa}(\mathbf{R}^n_+), \tag{7.27}$$

under the assumption that $|s - \kappa| < \frac{1}{2}$. By Theorem 7.1, equation (7.27) has a unique solution $u_+ \in H_s^+$, which has the form

$$u_+(x) = F^{-1}\Lambda_+^{-\kappa}(\xi', \xi_n + i\tau)\Pi^+\Lambda_-^{-\kappa}(\xi', \xi_n - i\tau)\widetilde{lf}$$
$$= \Lambda_+^{-\kappa}(D', D_n + i\tau)\theta^+\Lambda_-^{-\kappa}(D', D_n - i\tau)lf, \tag{7.28}$$

where $\Lambda_\pm^{-\kappa}(D', D_n \pm i\tau)$ is the p.o. with symbol $\Lambda_\pm^{-\kappa}(\xi', \xi_n \pm i\tau)$. By making use of Example 4.5, one can obtain an expression for (7.28) in the form of integrodifferential operators. Suppose e.g. $\kappa > 0$ and $lf \in S(\mathbf{R}^n)$. Then (4.68) and (4.69) imply

$$u_+(x) = -\frac{\Gamma^2\left(\frac{n}{2}\right)}{\pi^n\Gamma^2(\kappa)}\int_0^{x_n}\int_{-\infty}^{\infty}\frac{(x_n - y_n)^\kappa e^{-(x_n-y_n)\tau}}{\left[(x - y_n)^2 + |x' - y'|^2\right]^{n/2}}$$
$$\times\left(\int_{y_n}^{\infty}\int_{-\infty}^{\infty}\frac{(y_n - z_n)^\kappa e^{(z_n-y_n)\tau}f(z', z_n)}{\left[(y_n - z_n)^2 + |y' - z'|^2\right]^{n/2}}\,dz'\,dz_n\right)dy'\,dy_n. \tag{7.28'}$$

We note that (7.28) and (7.28′) continue to hold when $\tau = 0$ and $lf \in S(\mathbf{R}^n)$.

4. Suppose $\operatorname{Re}\kappa - s = m + \delta$, where m is a positive integer and $|\delta| < \frac{1}{2}$. Then

$$\tilde{v}_{\pm}(\xi) \in \tilde{H}^{\pm}_{s-\operatorname{Re}\kappa} = \tilde{H}^{\pm}_{-m-\delta}, \qquad \tilde{g}(\xi) \in \tilde{H}_{-m-\delta}$$

in (7.19). Since $m + \delta > \frac{1}{2}$, we cannot apply Lemma 5.4 to $\tilde{g}(\xi)$.

Let $Q(\xi', \xi_n)$ denote an arbitrary polynomial in ξ_n of degree m:

$$Q(\xi', \xi_n) = \xi_n^m + \sum_{k=1}^{m} q_k(\xi')\xi_n^{k-1},$$

where $q_k(\xi')$ is a homogeneous function of degree $m - k + 1$ and $Q(\xi', \xi_n) \neq 0$ for $|\xi'| + |\xi_n| > 0$, and let $\hat{Q}(\xi', \xi_n) = Q((1 + |\xi'|)\omega, \xi_n)$. By Lemma 4.4, $\hat{Q}^{-1}(\xi)\tilde{g}(\xi) \in \tilde{H}_{-\delta}$. Applying Lemma 5.4 to $\hat{Q}^{-1}(\xi)\tilde{g}(\xi)$, we get

$$\hat{Q}^{-1}(\xi)\tilde{g}(\xi) = \Pi^{+}\hat{Q}^{-1}(\xi)\tilde{g}(\xi) + \Pi^{-}\hat{Q}^{-1}(\xi)\tilde{g}(\xi), \tag{7.29}$$

where $\Pi^{\pm}\hat{Q}^{-1}\tilde{g} \in \tilde{H}^{\pm}_{-\delta}$, from which, upon multiplying both sides by $\hat{Q}(\xi)$, we get

$$\tilde{g}(\xi) = \hat{Q}(\xi)\Pi^{+}\hat{Q}^{-1}\tilde{g} + \hat{Q}(\xi)\Pi^{-}\hat{Q}^{-1}\tilde{g}, \tag{7.30}$$

with $\hat{Q}(\xi)\Pi^{\pm}\hat{Q}^{-1}\tilde{g} \in \tilde{H}^{\pm}_{-m-\delta}$ by virtue of Theorem 4.4.

Subsitituting (7.30) in (7.19) and transferring the "plus" functions to one side of the equality and the "minus" functions to the other, we get

$$\tilde{v}_{+}(\xi) - \hat{Q}(\xi)\Pi^{+}\hat{Q}^{-1}\tilde{g} = \hat{Q}(\xi)\Pi^{-}\hat{Q}^{-1}\tilde{g} - \tilde{v}_{-}(\xi). \tag{7.31}$$

The left side of (7.31) belongs to $\tilde{H}^{+}_{-m-\delta}$, while the right side belongs to $\tilde{H}^{-}_{-m-\delta}$. Consequently, Theorem 5.2 (Liouville's theorem) implies (7.31) is equal to a polynomial in ξ_n of degree $m - 1$:

$$\tilde{v}_{+}(\xi) - \hat{Q}(\xi)\Pi^{+}\hat{Q}^{-1}\tilde{g} = \sum_{k=1}^{m} \tilde{c}_k(\xi')\xi_n^{k-1}, \tag{7.32}$$

where $\tilde{c}_k(\xi') \in \tilde{H}_{s_k}(\mathbf{R}^{n-1})$, $s_k = -m - \delta + k - \frac{1}{2} = s - \operatorname{Re}\kappa + k - \frac{1}{2}$, $1 \leqslant k \leqslant m$. Thus, by (7.32), (7.16) and (7.17), the solution $u_{+}(x)$ of (7.11) has the form

$$\tilde{u}_{+}(\xi) = \hat{A}_{+}^{-1}\hat{Q}\Pi^{+}\hat{Q}^{-1}\hat{A}_{-}^{-1}lf + \frac{\sum_{k=1}^{m}\tilde{c}_k(\xi')\xi_n^{k-1}}{\hat{A}_{+}(\xi)}. \tag{7.33}$$

It can be verified analogously to (7.24) that (7.33) does not depend on the choice of the extension $lf(x)$.

Let us estimate (7.33). By (7.25), Lemma 4.4 and Theorem 5.1,

$$\begin{aligned} \|\hat{A}_{+}^{-1}\hat{Q}\Pi^{+}\hat{Q}^{-1}\hat{A}_{-}^{-1}\widetilde{lf}\|_s &\leqslant C\|\Pi^{+}\hat{Q}^{-1}\hat{A}_{-}^{-1}\widetilde{lf}\|_{-\delta} \\ &\leqslant C\|\hat{Q}^{-1}\hat{A}_{-}^{-1}\widetilde{lf}\|_{-\delta} \leqslant C\|\widetilde{lf}\|_{s-\alpha} \leqslant C\|f\|^{+}_{s-\alpha}. \end{aligned} \tag{7.34}$$

Since $|\hat{A}_+^{-1}(\xi', \xi_n)| \leqslant C(1 + |\xi'| + |\xi_n|)^{-\mathrm{Re}\,\kappa}$, we get upon making the change of variables $\xi_n = (1 + |\xi'|)\eta_n$ that

$$\left\| \frac{\sum_{k=1}^m \tilde{c}_k(\xi')\xi_n^{k-1}}{\hat{A}_+(\xi', \xi_n)} \right\|_s^2 \leqslant C \sum_{k=1}^m \int_{-\infty}^{\infty} \frac{|\tilde{c}_k(\xi')\xi_n^{k-1}|^2(1 + |\xi'| + |\xi_n|)^{2s}}{|\hat{A}(\xi', \xi_n)|^2} \, d\xi' \, d\xi_n$$

$$\leqslant C \sum_{k=1}^m \int_{-\infty}^{\infty} |\tilde{c}_k(\xi')|^2 (1 + |\xi'|)^{2s_k} \, d\xi' = C \sum_{k=1}^m [c_k]_{s_k}^2, \tag{7.35}$$

where $s_k = s - \mathrm{Re}\,\kappa + k - \frac{1}{2}$, $1 \leqslant k \leqslant m$. It follows from (7.34) and (7.35) that

$$\|u_+\|_s \leqslant C\left(\|f\|_{s-\alpha}^+ + \sum_{k=1}^m [c_k]_{s_k} \right), \qquad s_k = s - \mathrm{Re}\,\kappa + k - \tfrac{1}{2}. \tag{7.36}$$

Let us verify that the function $u_+(x) = F^{-1}\tilde{u}_+(\xi)$, where $\tilde{u}_+(\xi)$ is given by (7.33), satisfies (7.11) for $f \in H_{s-\alpha}(\mathbf{R}_+^n)$ and arbitrary $c_k(x') \in H_{s_k}(\mathbf{R}^{n-1})$. We have

$$\begin{aligned} \hat{A}_0 \tilde{u}_+(\xi) &= \hat{A}_- \hat{Q} \Pi^+ \hat{Q}^{-1} \hat{A}_-^{-1} \widetilde{lf} + \hat{A}_- \sum_{k=1}^m \tilde{c}_k(\xi')\xi_n^{k-1} \\ &= \hat{A}_- \hat{Q}(\hat{Q}^{-1}\hat{A}_-^{-1}\widetilde{lf} - \Pi^- \hat{Q}^{-1}\hat{A}_-^{-1}\widetilde{lf}) + \hat{A}_- \sum_{k=1}^m \tilde{c}_k(\xi')\xi_n^{k-1} \\ &= \widetilde{lf}(\xi) - \tilde{w}_-(\xi), \end{aligned} \tag{7.37}$$

where

$$\tilde{w}_-(\xi) = -\hat{A}_- \hat{Q} \Pi^- \hat{Q}^{-1} \hat{A}_-^{-1} \widetilde{lf} + \hat{A}_- \sum_{k=1}^m \tilde{c}_k(\xi')\xi_n^{k-1}. \tag{7.38}$$

By Theorems 4.4 and 5.1, $\tilde{w}_-(\xi) \in \tilde{H}_{s-\alpha}^-$. Consequently, $p\hat{A}_0 u_+ = f$ since $pw_- = 0$. We have thus proved

THEOREM 7.2. *Suppose* $\mathrm{Re}\,\kappa - s = m + \delta$, *where* m *is a positive integer and* $|\delta| < \frac{1}{2}$. *Then equation* (7.11) *for any* $f \in H_{s-\alpha}(\mathbf{R}_+^n)$ *has a solution* $u_+(x) = F^{-1}\tilde{u}_+(\xi) \in H_s^+$, *where* $\tilde{u}_+(\xi)$ *is given by* (7.33). *This solution is not unique and depends on* m *arbitrary functions* $c_k(x') \in H_{s_k}(\mathbf{R}^{n-1})$, $s_k = s - \mathrm{Re}\,\kappa + k - \frac{1}{2}$, $1 \leqslant k \leqslant m$. *The a priori estimate* (7.36) *holds.*

5. *Pseudodifferential operators of potential type* (*coboundary operators*). Before proceeding to the case $m < 0$, let us study a class of operators acting from a space of functions defined on $\mathbf{R}^{n-1}$ into a space of functions defined on $\mathbf{R}_+^n$ or $\mathbf{R}^n$.

By a *pseudodifferential operator of potential type* or a *coboundary operator* is meant an operator C defined by a formula of the form

$$C(v(x') \times \delta(x_n)) = F^{-1}(C(\xi', \xi_n)\tilde{v}(\xi')), \qquad \forall v(x') \in C_0^\infty(\mathbf{R}^{n-1}), \tag{7.39}$$

where $C(\xi) \in S_\alpha^0$ and the right side is the inverse Fourier transform of the regular functional $C(\xi', \xi_n)\tilde{v}(\xi')$. Since

$$F(v(x') \times \delta(x_n)) = \tilde{v}(\xi') \text{ and } \int_{-\infty}^{\infty} |\tilde{v}(\xi')|^2(1 + |\xi'| + |\xi_n|)^{-1-2\varepsilon}\, d\xi'\, d\xi_n < \infty,$$

we have

$$v(x') \times \delta(x_n) \in H_{-1/2-\varepsilon}(\mathbf{R}^n), \qquad \forall \varepsilon > 0.$$

Thus (7.39) is the result of applying a p.o. C to a generalized function $v(x') \times \delta(x_n) \in H_{-1/2-\varepsilon}(\mathbf{R}^n)$. Analogously to (3.7), if $\mathcal{Q}$ is an arbitrary functional on $S(\mathbf{R}^n)$ and $a = F^{-1}\mathcal{Q}$, then the p.o. of potential type A with symbol $\mathcal{Q} \in S'$ is defined by

$$A(v(x') \times \delta(x_n)) = F^{-1}(\mathcal{Q}\tilde{v}(\xi')) = a * (v(x') \times \delta(x_n)),$$
$$\forall v(x') \in C_0^\infty(\mathbf{R}^{n-1}). \tag{7.40}$$

We note that the convolution of a functional $a \in S'$ and a functional $v(x') \times \delta(x_n)$ with compact support is defined in §2.7.

EXAMPLE 7.2. Let P_n be the p.o. with symbol $\xi_n/|\xi|^2$, $n > 2$. Then (3.16) and (3.42) imply

$$P_n(v(x') \times \delta(x_n)) = -\frac{i\Gamma\left(\frac{n}{2}\right)}{2\pi^{n/2}} \int_{\mathbf{R}^{n-1}} \frac{x_n v(y')\, dy'}{\left[x_n^2 + (x' - y')^2\right]^{n/2}}, \tag{7.41}$$

i.e. $iP_n(v(x') \times \delta(x_n))$ is a double layer potential for Laplace's equation.

If $\Lambda_+^\kappa(D', D_n + i\tau)$ is the p.o. with symbol $\Lambda_+^\kappa(\xi', \xi_n + i\tau)$, $\tau > 0$, $\kappa < 0$, then (4.68) implies

$$\Lambda_+^\kappa(D', D_n + i\tau)(v(x') \times \delta(x_n))$$
$$= \frac{e^{i(\pi/2)\kappa}\Gamma\left(\frac{n}{2}\right)}{\pi^{n/2}\Gamma(-\kappa)} \int_{\mathbf{R}^{n-1}} \frac{\theta(x_n) x_n^{|\kappa|} e^{-x_n\tau} v(y')}{\left[x_n^2 + (x' - y')^2\right]^{n/2}}\, dy', \qquad \kappa < 0. \tag{7.42}$$

6. Suppose $\operatorname{Re}\kappa - s = m + \delta$, where m is a negative integer and $|\delta| < \frac{1}{2}$. Then $s - \alpha = \operatorname{Re}\kappa - \delta - m - \alpha > \operatorname{Re}\kappa - \delta - \alpha$. Consequently, $f \in H_{\operatorname{Re}\kappa-\delta-\alpha}(\mathbf{R}_+^n)$ if $f \in H_{s-\alpha}(\mathbf{R}_+^n)$. By Theorem 7.1, there exists a unique solution $w_+ \in \overset{+}{H}_{\operatorname{Re}\kappa-\delta}$ of the equation $p\hat{A}_0 w_+ = f$, with $\tilde{w}_+(\xi)$ having the form

$$\tilde{w}_+(\xi) = \hat{A}_+^{-1}(\xi)\Pi^+\hat{A}_-^{-1}(\xi)\tilde{lf}(\xi). \tag{7.43}$$

Since $\hat{A}_-^{-1}(\xi)\widetilde{lf}(\xi) \in \tilde{H}_{s-\operatorname{Re}\kappa} = \tilde{H}_{|m|-\delta}$, it follows by virtue of the expansion formula (5.36) for an integral of Cauchy type that

$$\tilde{w}_+(\xi) = \sum_{k=1}^{|m|} \frac{i\Pi'\hat{\Lambda}_+^{k-1}\hat{A}_-^{-1}\widetilde{lf}}{\hat{A}_+(\xi)\hat{\Lambda}_+^k(\xi)} + \frac{1}{\hat{A}_+(\xi)\hat{\Lambda}_+^{|m|}}\Pi^+\hat{\Lambda}_+^{|m|}\hat{A}_-^{-1}\widetilde{lf}.$$

Let

$$\tilde{v}_k(\xi') = i\Pi'\hat{\Lambda}_+^{k-1}\hat{A}_-^{-1}\widetilde{lf},\ \tilde{u}_+(\xi) = \hat{A}_+^{-1}\hat{\Lambda}_+^{-|m|}\Pi^+\hat{\Lambda}_+^{|m|}\hat{A}_-^{-1}\widetilde{lf}, \tag{7.44}$$

where

$$\hat{\Lambda}_+(\xi) = \xi_n + i|\xi'| + i \quad \text{and} \quad \Pi'\tilde{g} = \frac{1}{2\pi}\int_{-\infty}^{\infty} \tilde{g}(\xi', \xi_n)\, d\xi_n.$$

Then

$$\tilde{w}_+(\xi) = \sum_{k=1}^{|m|} \frac{1}{\hat{A}_+(\xi)\hat{\Lambda}_+^k}\tilde{v}_k(\xi') + \tilde{u}_+(\xi), \tag{7.45}$$

where Lemma 4.4 and Theorems 4.2, 4.4 and 5.1 imply that $\tilde{u}_+(\xi) \in \tilde{H}_s^+$, $\tilde{v}_k(\xi') \in \tilde{H}_{m_k}(\mathbf{R}^{n-1})$, $m_k = |m| - \delta - k + \frac{1}{2}$, with

$$\|u_+\|_s \leqslant C\|f\|_{s-\alpha}^+, \qquad \sum_{k=1}^{|m|}[v_k]_{m_k} \leqslant C\|f\|_{s-\alpha}^+. \tag{7.46}$$

Let us show that a function $\tilde{w}_+(\xi) \in \tilde{H}^+_{\operatorname{Re}\kappa-\delta}$ is uniquely representable in the form (7.45). Suppose

$$\tilde{w}_+(\xi) = \sum_{k=1}^{|m|} \frac{1}{\hat{A}_+(\xi)\hat{\Lambda}_+^k}\tilde{v}_k^{(1)}(\xi') + \tilde{u}_+^{(1)}(\xi)$$

is another representation. Then

$$\sum_{k=1}^{|m|} \frac{1}{\hat{A}_+(\xi)\hat{\Lambda}_+^k}\tilde{v}_k^{(2)}(\xi') = \tilde{u}_+^{(1)}(\xi) - \tilde{u}_+(\xi) \in \tilde{H}_s^+,$$

where $\tilde{v}_k^{(2)}(\xi') = \tilde{v}_k(\xi') - \tilde{v}_k^{(1)}(\xi)$. We will show that

$$\sum_{k=1}^{|m|} \frac{1}{\hat{A}_+(\xi)\hat{\Lambda}_+^k}\tilde{v}_k^{(2)}(\xi') \in \tilde{H}_s^+$$

only when $\tilde{v}_k^{(2)}(\xi') = 0$ for almost all ξ' and $1 \leqslant k \leqslant m$. By Fubini's theorem,

$$\int_{-\infty}^{\infty}\left|\sum_{k=1}^{|m|} \hat{A}_+^{-1}\hat{\Lambda}_+^{-k}\tilde{v}_k^{(2)}(\xi')\right|^2 (1 + |\xi'| + |\xi_n|)^{2s}\, d\xi_n < \infty \tag{7.47}$$

for almost all ξ'. On the other hand, if not all of the $\tilde{v}_k^{(2)}$ in (7.47) are equal to zero, then the integral (7.47) diverges. Suppose e.g. $\tilde{v}_1^{(2)} = \ldots = \tilde{v}_{p-1}^{(2)} = 0$, $\tilde{v}_p^{(2)} \neq 0$, $1 \leqslant p \leqslant |m|$. Then for $|\xi_n| \geqslant N$, where N is sufficiently large, we get

$$\left| \sum_{k=1}^{|m|} \frac{\tilde{v}_k^{(2)}}{\hat{\Lambda}_+^k} \right| \geqslant \frac{C_p}{(1 + |\xi'| + |\xi_n|)^p}$$

and hence

$$\int_{-\infty}^{\infty} \left| \sum_{k=1}^{|m|} \hat{A}_+^{-1} \hat{\Lambda}_+^{-k} \tilde{v}_k^{(2)} \right|^2 (1 + |\xi'| + |\xi_n|)^{2s} \, d\xi_n$$

$$\geqslant C_p^{(1)} \int_{|\xi_n| \geqslant N} (1 + |\xi'| + |\xi_n|)^{2(|m| - p - \delta)} \, d\xi_n = +\infty.$$

Thus the representation (7.45) is unique.

In particular, $\tilde{w}_+(\xi) \in \tilde{H}_s^+$ if and only if

$$\tilde{v}_k(\xi') = \frac{i}{2\pi} \int_{-\infty}^{\infty} (\xi_n + i|\xi'| + i)^{k-1} \frac{\widetilde{lf}(\xi)}{\hat{A}_-(\xi)} \, d\xi_n = 0, \qquad 1 \leqslant k \leqslant |m|, \tag{7.48}$$

for almost all ξ' or, equivalently,

$$\int_{-\infty}^{\infty} \xi_n^{k-1} \frac{\widetilde{lf}(\xi', \xi_n)}{\hat{A}_-(\xi', \xi_n)} \, d\xi_n = 0, \qquad 1 \leqslant k \leqslant |m|, \tag{7.49}$$

for almost all ξ'.

We note that (7.49) does not depend on the choice of the extension lf. For suppose $l_1 f \in H_{s-\alpha}(\mathbf{R}^n)$ is another extension of $f(x)$. Then $f_- = lf - l_1 f \in H_{s-\alpha}^-$ and

$$\tilde{f}_k(\xi', \xi_n) = \frac{\xi_n^{k-1} \tilde{f}_-(\xi', \xi_n)}{\hat{A}_-(\xi', \xi_n)} \in \tilde{H}_{|m|-k+1-\delta}^- \subset \tilde{H}_{1-\delta}^-, \qquad 1 \leqslant k \leqslant |m|.$$

By Theorem 4.2,

$$F_k(\xi', x_n) = \frac{1}{2\pi} \int_{-\infty}^{\infty} \tilde{f}_k(\xi', \xi_n) e^{-ix_n\xi_n} \, d\xi_n$$

is a continuous function of x_n with values in $\tilde{H}_{1/2-\delta}(\mathbf{R}^{n-1})$. Since $F_k(\xi', x_n) = 0$ for almost all ξ' when $x_n > 0$, it follows by continuity that

$$F_k(\xi', 0) = \frac{1}{2\pi} \int_{-\infty}^{\infty} \tilde{f}_k(\xi', \xi_n) \, d\xi_n = 0, \qquad 1 \leqslant k \leqslant |m|.$$

Consequently, (7.49) is satisfied for either all of the extensions lf or none of them. We have thus proved

THEOREM 7.3. *Suppose* $\operatorname{Re}\kappa - s = m + \delta$, *where* m *is a negative integer and* $|\delta| < \frac{1}{2}$. *Then for any* $f \in H_{s-\alpha}(\mathbf{R}^n_+)$ *there exists a unique collection of functions* $u_+ \in H_s^+$, $v_k(x') \in H_{m_k}(\mathbf{R}^{n-1})$, $m_k = |m| - \delta - k + \frac{1}{2}$, $1 \leqslant k \leqslant |m|$, *such that*

$$p\hat{A}_0\left[u_+ + \sum_{k=1}^{|m|} \hat{A}_+^{-1}\hat{\Lambda}_+^{-k}(v_k(x') \times \delta(x_n))\right] = f, \tag{7.50}$$

where the operators

$$\hat{A}_+^{-1}\hat{\Lambda}_+^{-k}(v_k(x') \times \delta(x_n)) = F_\xi^{-1}\left[\hat{A}_+^{-1}(\xi', \xi_n)\hat{\Lambda}_+^{-k}(\xi', \xi_n)\tilde{v}_k(\xi')\right]$$

are pseudodifferential operators of potential type. In particular, the equation $p\hat{A}_0 u_+ = f$ *has a solution belonging to* H_s^+ *if and only if* $f(x)$ *satisfies* (7.49) *for almost all* ξ'. *Estimates* (7.46) *hold.*

EXAMPLE 7.3. *Integral equation of the first kind.* Consider the equation

$$p\Lambda^{2\kappa}(D', D_n, \tau)w_+ = f, \qquad lf \in S(\mathbf{R}^n) \tag{7.51}$$

which is an integral equation of the first kind when $\kappa < 0$.

According to Theorem 7.1, if $|s - \kappa| < \frac{1}{2}$, then (7.51) has a unique solution $w_+ \in H_s^+$, with

$$\tilde{w}_+(\xi) = \Lambda_+^{-\kappa}(\xi', \xi_n + i\tau)\Pi^+\Lambda_-^{-\kappa}(\xi', \xi_n - i\tau)\widetilde{lf}.$$

When $\kappa \leqslant -\frac{1}{2}$, this solution generally does not belong to L_2.

Suppose $\kappa = m + \gamma$, where m is a negative integer and $0 < \gamma < \frac{1}{2}$. Then, by (5.36),

$$\tilde{w}_+(\xi) = \sum_{k=1}^{|m|} \Lambda_+^{|m|-k-\gamma}(\xi', \xi_n + i\tau)\tilde{v}_k(\xi') + \tilde{u}_+(\xi), \tag{7.52}$$

where

$$\tilde{v}_k(\xi') = i\Pi'\Lambda_+^{k-1}(\xi', \xi_n + i\tau)\Lambda_-^{|m|-\gamma}(\xi', \xi_n - i\tau)\widetilde{lf}, \tag{7.53}$$

$$\tilde{u}_+(\xi) = \Lambda_+^{-\gamma}(\xi', \xi_n + i\tau)\Pi^+\Lambda_+^{|m|}(\xi', \xi_n + i\tau)\Lambda_-^{|m|-\gamma}(\xi', \xi_n - i\tau)\widetilde{lf}.$$

We note that $u_+ \in H_0^+$, $v_k(x') \in H_N(\mathbf{R}^{n-1})$, $\forall N$. According to Theorem 7.3, (7.51) has a unique solution of the form

$$w_+(x) = \sum_{k=1}^{|m|} \Lambda_+^{|m|-k-\gamma}(D', D_n + i\tau)(v_k(x') \times \delta(x_n)) + u_+(x), \tag{7.52'}$$

where $u_+ \in H_0^+$, $v_k(x') \in H_{|m|-\gamma-k+1/2}(\mathbf{R}^{n-1})$. We note that (4.81) implies that the restriction of the operator

$$\Lambda_+^{|m|-k-\gamma}(D', D_n + i\tau)(v_k(x') \times \delta(x_n))$$

to $\mathbf{R}^n_+$ has the form

$$p\Lambda_+^{|m|-k-\gamma}(D', D_n + i\tau)(v_k(x') \times \delta(x_n))$$
$$= \frac{e^{-i(\pi/2)(m+k+\gamma)}\Gamma\left(\frac{n}{2}\right)}{\Gamma(m+k+\gamma)\pi^{n/2}} \int_{-\infty}^{\infty} \frac{x_n^{-|m|+k+\gamma}e^{-x_n\tau}v_k(y')\,dy'}{[x_n^2 + |x' - y'|^2]^{n/2}}. \tag{7.54}$$

Thus, if the right side $f(x)$ is sufficiently smooth, Theorem 7.3 enables us to give a more precise form of the generalized function w_+ satisfying (7.51): w_+ decomposes into the sum of a square integrable function $u_+(x)$ and the potentials

$$\Lambda_+^{|m|-k-\gamma}(D', D_n + i\tau)(v_k(x') \times \delta(x_n)),$$

which contain the principal singularities of the solution.

EXAMPLE 7.4. Consider (7.51) when $\kappa = m$, where m is a negative integer. Then, setting $\gamma = 0$ in (7.52) and (7.53), we get

$$\begin{aligned}\tilde{w}_+(\xi) &= \sum_{k=1}^{|m|} \Lambda_+^{|m|-k}(\xi', \xi_n + i\tau)\tilde{v}_k(\xi') + \tilde{u}_+(\xi)\\ &= \sum_{k=1}^{|m|} \tilde{w}_k(\xi')(\xi_n + i\tau)^{k-1} + \tilde{u}_+(\xi),\end{aligned}$$

where

$$\tilde{u}_+(\xi) = \Pi^+\Lambda^{2|m|}(\xi', \xi_n, \tau)\tilde{lf}, \qquad \tilde{w}_k(\xi') = \sum_{r=k}^{|m|} c_{kr}|\xi'|^{r-k}\tilde{v}_{|m|-r}(\xi')$$

and the c_{kr} are certain numerical constants. Thus for any $f \in H_{2|m|}(\mathbf{R}^n_+)$ there exist unique functions $u_+ \in H_0^+(\mathbf{R}^n)$ and $w_k(x') \in H_{k-1/2}(\mathbf{R}^{n-1})$ such that

$$p\Lambda^{2m}\left(u_+ + \sum_{k=1}^{|m|} w_k(x') \times (D_n + i\tau)^{k-1}\delta(x_n)\right) = f(x). \tag{7.55}$$

Since m is a negative integer, $\Lambda^{2|m|}(D', D_n, \tau)$ is a differential operator with respect to x_n of order $2|m|$. Applying $\Lambda^{2|m|}(D', D_n, \tau)$ to (7.55), we get

$$\Lambda^{2|m|}(D', D_n, \tau)f = u_+ \quad \text{for } x_n > 0, \tag{7.56}$$

$$\Lambda^{2|m|}(D', D_n, \tau) \sum_{k=1}^{|m|} \Lambda^{2m}(D', D_n, \tau)\big(w_k(x')(D_n + i\tau)^{k-1}\delta(x_n)\big) = 0 \quad \text{for } x_n > 0. \tag{7.57}$$

Thus f is a solution of (7.56) in $\mathbf{R}^n_+$ with right side u_+, while

$$\sum_{k=1}^{|m|} \Lambda^{2m}(D', D_n, \tau)\big(w_k(x') \times (D_n + i\tau)^{k-1}\delta(x_n)\big)$$

is a solution of the homogeneous equation $\Lambda^{2|m|}(D', D_n, \tau)g = 0$ in $\mathbf{R}^n_+$. Since

$$\begin{aligned}&\Pi'\Lambda_+^{k-1}(\xi', \xi_n + i\tau)\Lambda_-^{|m|}(\xi', \xi_n - i\tau)\tilde{lf}\\ &\qquad = (D_n + i|\xi'| + i\tau)^{k-1}(D_n - i|\xi'| - i\tau)^{|m|}F_{\xi'}lf|_{x_n=0},\end{aligned}$$

the functions $w_k(x')$ have the form

$$w_k(x') = B_k lf|_{x_n=0}, \qquad 1 < k < |m|, \tag{7.58}$$

where the B_k are certain p.o.'s that are differential operators with respect to x_n. Consequently, the solution of (7.55) can be interpreted as the solution of problem inverse to the boundary value problem (7.56), (7.58). In fact, whereas in the boundary value problem (7.56), (7.58) we seek the solution f in terms of the given right side $u_+(x)$ and boundary data $w_k(x')$, in (7.55) we seek to recover the right side $u_+(x)$ and boundary data $w_k(x')$ of problem

(7.56), (7.58) from the solution f or, equivalently, we seek the volume potential density $u_+(x)$ and surface potential densities $w_k(x')$, $1 < k < |m|$, that determine f.

EXAMPLE 7.5. Let us consider in more detail a special case of the preceding example. Suppose $\kappa = -1$ and $n > 3$. Then (7.51) takes the form

$$p\Lambda^{-2}(D', D_n, \tau)u_+ = f(x). \tag{7.59}$$

We note that $\Lambda^{-2}(D', D_n, 0)u_+$ is a Newtonian potential (see (3.15″)). Suppose f is a solution of the equation $[D_n^2 + (|D'| + \tau)^2]f = 0$ that decreases sufficiently rapidly in $\mathbf{R}_+^n$ but does not identically vanish. Then (7.59) does not have a solution of class H_0^+. In fact, applying the operator $D_n^2 + (|D'| + \tau)^2$ to both sides of (7.59), we get $u_+ = [D_n^2 + (|D'| + \tau)^2]f = 0$, which is impossible since f does not identically vanish. On the other hand, the equation

$$p\Lambda^{-2}(D', D_n, \tau)(u_+(x) + v(x') \times \delta(x_n)) = f(x)$$

has a solution $u_+(x) \in H_0^+$, $v(x') \in H_{1/2}(\mathbf{R}^{n-1})$ by virtue of Theorem 7.3.

REMARK 7.1. Let κ be the factorization index of an elliptic symbol $A(\xi', \xi_n)$, and let $Z(\kappa)$ denote the set of points on the real line of the form $\operatorname{Re}\kappa - m + \frac{1}{2}$, where m ranges over the set of all integers. The set $Z(\kappa)$ divides the real line into a countable set of intervals

$$I_m = \left(\operatorname{Re}\kappa - m - \tfrac{1}{2}, \operatorname{Re}\kappa - m + \tfrac{1}{2}\right), \qquad m = 0, \pm 1, \pm 2, \ldots.$$

By Theorem 7.1, when $s \in I_0$, equation (7.11) for $f \in H_{s-\alpha}(\mathbf{R}_+^n)$ has a unique solution in H_s^+. By Theorem 7.2, when $s \in I_m$ for $m > 0$, equation (7.11) for $f \in H_{s-\alpha}(\mathbf{R}_+^n)$ has a solution in H_s^+ that is not unique and depends on m arbitrary functions $c_k(x') \in H_{s-\operatorname{Re}\kappa+k-1/2}(\mathbf{R}^{n-1})$, $1 < k < m$. Finally, when $s \in I_m$ for $m < 0$, equation (7.11) for $f \in H_{s-\alpha}(\mathbf{R}_+^n)$ has at most one solution in H_s^+ that exists only under the fulfilment of $|m|$ relations of form (7.49).

Thus (since $H_{s_1}^+ \supset H_{s_2}^+$ for $s_1 < s_2$), as s decreases subject to the restrictions $s \notin Z(\kappa)$ and $s < \operatorname{Re}\kappa - \frac{1}{2}$, the space H_s^+ in which the solutions of (7.11) are sought becomes larger and the number of solutions of (7.11) increases; as s increases subject to the restrictions $s \notin Z(\kappa)$ and $s > \operatorname{Re}\kappa + \frac{1}{2}$, the space H_s^+ becomes smaller and (7.11) has a solution in H_s^+ only under the fulfilment of condition (7.49). As s traverses a point of the set $Z(\kappa)$, there occurs a variation in either the number of arbitrary functions $c_k(x')$ on which the solution of (7.11) depends or the number of relations in (7.49).

§8. Formulation of boundary value problems for pseudodifferential equations

1. Consider the pseudodifferential equation

$$p\hat{A}_0 u_+ = f, \tag{8.1}$$

where $\hat{A}_0$ is the elliptic p.o. with symbol

$$\hat{A}_0(\xi', \xi_n),\ A_0(\xi', \xi_n) \in O_{\alpha+i\beta}^\infty, \qquad f \in H_{s-\alpha}(\mathbf{R}_+^n)$$

and the solution $u_+(x)$ is sought in H_s^+. It is known by Theorem 7.1 that (8.1) has a unique solution for $\operatorname{Re}\kappa - s = \delta$, where κ is the factorization index of $A_0(\xi', \xi_n)$ and $|\delta| < \frac{1}{2}$. We will therefore assume that $\operatorname{Re}\kappa - s = m + \delta$, where m is a nonzero integer and $|\delta| < \frac{1}{2}$.

Let us first consider the case $m > 0$. Here it is known by Theorem 7.2 that (8.1) has an infinite number of solutions depending on m arbitrary functions $c_k(x') \in H_{s_k}(\mathbf{R}^{n-1})$, $s_k = s - \operatorname{Re}\kappa + k - \frac{1}{2}$. In order to determine the functions $c_k(x')$ $(1 \leqslant k \leqslant m)$ and thereby remove the nonuniqueness of the solution of (8.1), we add the following m pseudodifferential boundary conditions to (8.1):

$$p'\hat{B}_j u_+ = g_j(x'), \qquad 1 \leqslant j \leqslant m, \tag{8.2}$$

where $\hat{B}_j$ $(1 \leqslant j \leqslant m)$ is the p.o. with symbol $\hat{B}_j(\xi', \xi_n)$, $B_j(\xi', \xi_n) \in O'_{\beta_j}$, and p' is the operator of restriction of a function to the hyperplane $x_n = 0$.

In this section we will adopt the assumption

$$|B_j(\xi', \xi_n)| \leqslant C|\xi'|^{\operatorname{Re}\beta_j - \beta_{j1}}(|\xi'| + |\xi_n|)^{\beta_{j1}}, \qquad \beta_{j1} < s - \tfrac{1}{2}, \tag{8.3}$$

which implies by virtue of Lemma 4.4 and Theorem 4.2 that the restriction of $B_j u$ to the hyperplane $x_n = 0$ exists and belongs to $H_{s-\operatorname{Re}\beta_j - 1/2}(\mathbf{R}^{n-1})$.

We proceed to solve the boundary value problem (8.1), (8.2) under the assumption that the functions $f \in H_{s-\alpha}(\mathbf{R}^n_+)$ and $g_j(x') \in H_{s-\operatorname{Re}\beta_j - 1/2}(\mathbf{R}^{n-1})$ are given. In doing so, we will determine conditions on the symbols $B_j(\xi', \xi_n)$ that ensure the unique solvability of problem (8.1), (8.2). On account of (4.32′), the Fourier transform with respect to x' of the boundary conditions (8.2) is

$$\Pi'\hat{B}_j(\xi', \xi_n)\tilde{u}_+(\xi', \xi_n) = \tilde{g}_j(\xi'), \qquad 1 \leqslant j \leqslant m, \tag{8.4}$$

where

$$\Pi'\hat{B}_j(\xi', \xi_n)\tilde{u}_+(\xi', \xi_n) = \frac{1}{2\pi}\int_{-\infty}^{\infty} \hat{B}_j(\xi', \xi_n)\tilde{u}_+(\xi', \xi_n)\, d\xi_n;$$

while, according to (7.33), the Fourier transform of the general solution of (8.1) has the form

$$\tilde{u}_+(\xi) = \frac{\hat{Q}(\xi)}{\hat{A}_+(\xi)}\Pi^+\frac{\widetilde{lf}(\xi)}{\hat{Q}(\xi)\hat{A}_-(\xi)} + \frac{\sum_{k=1}^m \tilde{c}_k(\xi')\xi_n^{k-1}}{\hat{A}_+(\xi)}, \tag{8.5}$$

where $Q(\xi', \xi_n)$ is an arbitrary homogeneous elliptic polynomial in ξ_n of degree m. Substituting (8.5) in (8.4), we get

$$\sum_{k=1}^{m} \hat{b}_{jk}(\xi')\tilde{c}_k(\xi') = \tilde{g}_j(\xi') - \tilde{f}_j(\xi'), \qquad 1 \leqslant j \leqslant m, \tag{8.6}$$

where

$$\hat{b}_{jk}(\xi') = \frac{1}{2\pi}\int_{-\infty}^{\infty} \frac{\hat{B}_j(\xi', \xi_n)\xi_n^{k-1}}{\hat{A}_+(\xi', \xi_n)}\, d\xi_n, \qquad 1 \leqslant j, k \leqslant m, \tag{8.7}$$

$$\tilde{f}_j(\xi') = \Pi' \frac{\hat{B}_j(\xi)\hat{Q}(\xi)}{\hat{A}_+(\xi)} \Pi^+ \frac{\widetilde{lf}}{A_- Q}, \qquad 1 \leqslant j \leqslant m. \tag{8.7'}$$

We note that the integral in (8.7) converges, since

$$\left|\frac{\hat{B}_j(\xi)\xi_n^{k-1}}{\hat{A}_+(\xi)}\right| \leqslant C(1 + |\xi'|)^{\operatorname{Re}\beta_j - \beta_{j1}}(1 + |\xi'| + |\xi_n|)^{\beta_{j1} + k - 1 - \operatorname{Re}\kappa}$$

and $\beta_{j1} + k - 1 - \operatorname{Re}\kappa < s - \frac{1}{2} + m - 1 - \operatorname{Re}\kappa = -\frac{3}{2} - \delta < -1$.

To ensure the unique solvability of (8.6) for the $\tilde{c}_k(\xi')$, we require that

$$\det\|b_{jk}(\xi')\|_{j,k=1}^m \neq 0 \quad \text{for } \xi' \neq 0, \tag{8.8}$$

and hence that $\det\|\hat{b}_{jk}(\xi')\|_{j,k=1}^m \neq 0$ for any $\xi' \in \mathbf{R}^{n-1}$. Condition (8.8) is called the *Šapiro-Lopatinskiĭ condition* in analogy with the case of boundary value problems for elliptic differential equations.

The solution of (8.6) then has the form

$$\tilde{c}_k(\xi') = \sum_{j=1}^m \hat{d}_{kj}\big(\tilde{g}_j(\xi') - \tilde{f}_j(\xi')\big), \qquad 1 \leqslant k \leqslant m, \tag{8.9}$$

where $\|d_{ij}(\xi')\|_{i,j=1}^m$ is the inverse of the matrix $\|b_{jk}(\xi')\|_{j,k=1}^m$. Substituting (8.9) in (8.5), we obtain the following formula for the Fourier transform of the solution u_+ of problem (8.1), (8.2):

$$\tilde{u}_+(\xi) = \frac{\hat{Q}(\xi)}{\hat{A}_+(\xi)} \Pi^+ \frac{\widetilde{lf}}{\hat{Q}\hat{A}_-} + \sum_{j,k=1}^m \frac{\hat{d}_{kj}(\xi')\xi_n^{k-1}}{\hat{A}_+(\xi)}\big(\tilde{g}_j(\xi') - \tilde{f}_j(\xi')\big), \tag{8.10}$$

where the $\tilde{f}_j(\xi')$ are defined by (8.7').

Let us estimate u_+. By (8.7), the matrix element $b_{jk}(\xi')$ is a homogeneous function of degree $\beta_j + k - \kappa$, while the matrix element $d_{kj}(\xi')$ is a homogeneous function of degree $-(\beta_j + k - \kappa)$. Consequently, (8.9) implies

$$\begin{aligned}[c_k]_{s_k} &\leqslant C\sum_{j=1}^m [g_j - f_j]_{s_k - \operatorname{Re}\beta_j - k + \operatorname{Re}\kappa} \\ &\leqslant C\sum_{j=1}^m [g_j]_{s - \operatorname{Re}\beta_j - 1/2} + C\sum_{j=1}^m [f_j]_{s - \operatorname{Re}\beta_j - 1/2},\end{aligned} \tag{8.11}$$

where $s_k = s - \operatorname{Re}\kappa + k - \frac{1}{2}$. Estimating (8.7'), we get by virtue of Theorems 4.2, 5.1 and Lemma 4.4 that

$$[f_j]_{s-\operatorname{Re}\beta_j-1/2} \leqslant C\left\| \frac{\hat{B}_j\hat{Q}}{\hat{A}_+}\Pi^+\frac{\widetilde{lf}}{\hat{Q}\hat{A}_-}\right\|_{s-\operatorname{Re}\beta_j} \leqslant C\left\|\Pi^+\frac{\widetilde{lf}}{\hat{Q}\hat{A}_-}\right\|_{s+m-\operatorname{Re}\kappa}$$
$$\leqslant C\left\|\frac{\widetilde{lf}}{\hat{Q}\hat{A}_-}\right\|_{s+m-\operatorname{Re}\kappa} \leqslant C\|\widetilde{lf}\|_{s-\alpha} \leqslant C\|f\|^+_{s-\alpha}. \tag{8.12}$$

It therefore follows by (7.36), (8.11) and (8.12) that

$$\|u_+\|_s \leqslant C\left(\|f\|^+_{s-\alpha} + \sum_{j=1}^m [g_j]_{s-\operatorname{Re}\beta_j-1/2}\right). \tag{8.13}$$

We have thus proved

THEOREM 8.1. *Suppose* $A_0(\xi', \xi_n) \in O^\infty_{\alpha+i\beta}$ *is an elliptic symbol with factorization index* κ *such that* $\operatorname{Re}\kappa - s = m + \delta$, *where* m *is a positive integer and* $|\delta| < \frac{1}{2}$. *Suppose* $B_j(\xi) \in O'_{\beta_j}$, $1 \leqslant j \leqslant m$, *and conditions* (8.3) *and* (8.8) *are satisfied. Then the boundary value problem* (8.1), (8.2) *has a unique solution* $u_+ \in H_s^+$ *for any* $f(x) \in H_{s-\alpha}(\mathbf{R}^n_+)$ *and* $g_j(x') \in H_{s-\operatorname{Re}\beta_j-1/2}(\mathbf{R}^{n-1})$, $1 \leqslant j \leqslant m$. *The a priori estimate* (8.13) *holds.*

REMARK 8.1. We note that the pseudodifferential operators $\hat{B}_j$ satisfying the conditions of Theorem 8.1 cannot be differential operators, and what is more, their symbols $\hat{B}_j(\xi', \xi_n)$ cannot be "plus" symbols, i.e. admit an analytic continuation with respect to $\xi_n + i\tau$ into the halfplane $\tau > 0$ with preservation of the estimate

$$\hat{B}_j(\xi', \xi_n + i\tau)| \leqslant C(1 + |\xi'|)^{\beta_j - \beta_{j1}}(1 + |\xi'| + |\xi_n| + \tau)^{\beta_{j1}}.$$

For then the integrand in (8.7) would admit for any k an analytic continuation into the halfplane $\tau > 0$ that satisfies

$$\left|\frac{\hat{B}_j(\xi', \xi_n + i\tau)(\xi_n + i\tau)^{k-1}}{\hat{A}_+(\xi', \xi_n + i\tau)}\right|$$
$$\leqslant C(1 + |\xi'|)^{\operatorname{Re}\beta_j - \beta_{j1}}(1 + |\xi'| + |\xi_n| + \tau)^{-3/2-\delta}, \qquad |\delta| < \tfrac{1}{2}, \tag{8.14}$$

and it would consequently be possible to close the contour in (8.7) at infinity and get, by Cauchy's theorem, $\hat{b}_{jk}(\xi') \equiv 0$ for $1 < k < m$, so that (8.8) would not be satisfied.

It can be verified by calculating the integrals of (8.7) with the use of the residue theorem that the conditions of Theorem 8.1 are satisfied, for example, by a p.o. with one of the following symbols:

$$B_j(\xi', \xi_n) = \frac{A_+(\xi)\xi_n^{j-1}}{|\xi|^{2m}} \quad \text{or} \quad B_j(\xi', \xi_n) = \frac{A_0(\xi)\xi_n^{j-1}}{|\xi|^{2m+m_1}}, \qquad m_1 > \alpha - \operatorname{Re}\kappa.$$

In §11, by restricting the class of p.o.'s $\hat{A}_0$, we will be able to widen the class of admissible boundary operators so that it will include differential operators.

2. *Operators of potential type in* $\mathbf{R}^n_+$. Suppose $C(\xi', \xi_n) \in O'_{\alpha+i\beta}$ and

$$|C(\xi', \xi_n)| \leqslant C|\xi'|^{\alpha-\alpha_1}(|\xi'| + |\xi_n|)^{\alpha_1}. \tag{8.15}$$

If $s < \alpha - \alpha_1 - \frac{1}{2}$, then the operator of potential type $\hat{C}(v(x') \times \delta(x_n))$ can be extended by continuity to a bounded operator from $H_{s+1/2}(\mathbf{R}^{n-1})$ into $H_{s-\alpha}(\mathbf{R}^n)$. For suppose $v(x') \in C_0^\infty(\mathbf{R}^{n-1})$. Then (8.15) implies

$$\|\hat{C}(\xi', \xi_n)\tilde{v}(\xi')\|^2_{s-\alpha} \leqslant C\int_{-\infty}^{\infty}(1 + |\xi'|)^{2(\alpha-\alpha_1)} \times (1 + |\xi'| + |\xi_n|)^{2(s+\alpha_1-\alpha)}|\tilde{v}(\xi')|^2\, d\xi'\, d\xi_n. \tag{8.16}$$

The integral in (8.16) converges since $s + \alpha_1 - \alpha < -\frac{1}{2}$. Making the change of variables $\xi_n = (1 + |\xi'|)\eta_n$ in (8.16), we get

$$\|\hat{C}(\xi', \xi_n)\tilde{v}(\xi')-^2_{s-\alpha} \leqslant C\int_{-\infty}^{\infty}|\tilde{v}(\xi')|^2(1 + |\xi'|)^{2(s+1/2)}\, d\xi' = C[v]^2_{s+1/2}. \tag{8.17}$$

Since the functions of class $C_0^\infty(\mathbf{R}^{n-1})$ are dense in $H_{s+1/2}(\mathbf{R}^{n-1})$, (8.17) remains valid for any $v \in H_{s+1/2}(\mathbf{R}^{n-1})$.

We note that if $s < \alpha - \alpha_1 - \frac{1}{2}$, then the operator $p\hat{C}(v(x') \times \delta(x_n))$ is bounded from $H_{s+1/2}(\mathbf{R}^{n-1})$ into $H_{s-\alpha}(\mathbf{R}^n_+)$ since the restriction operator p is bounded from $H_{s-\alpha}(\mathbf{R}^n)$ into $H_{s-\alpha}(\mathbf{R}^n_+)$. If, on the other hand, $s \geqslant \alpha - \alpha_1 - \frac{1}{2}$, then, in general,

$$p\hat{C}(v(x') \times \delta(x_n)) \notin H_{s-\alpha}(\mathbf{R}^n_+).$$

We introduce here a class of symbols $C(\xi', \xi_n)$ such that for any s

$$\|p\hat{C}(v(x') \times \delta(x_n))\|^+_{s-\alpha} \leqslant C[v]_{s+1/2}, \qquad \forall v(x') \in H_{s+1/2}(\mathbf{R}^{n-1}). \tag{8.18}$$

Suppose a symbol $C(\xi', \xi_n) \in O'_{\alpha+i\beta}$ has for any $N > 0$ the decomposition

$$C(\xi', \xi_n) = C_N^-(\xi', \xi_n) + R_N(\xi', \xi_n), \tag{8.19}$$

where $C_N^-(\xi', \xi_n)$ admits an analytic continuation with respect to $\xi_n + i\tau$ into the halfplane $\tau < 0$ that is continuous for $\xi' \neq 0$ and satisfies

$$|C_N^-(\xi', \xi_n + i\tau)| \leqslant C_N(|\xi'| + |\xi_n| + |\tau|)^{\alpha_1}|\xi'|^{\alpha-\alpha_1}, \tag{8.20}$$

while the function $R_N(\xi', \xi_n)$ satisfies

$$|R_N(\xi', \xi_n)| \leqslant C_N\frac{|\xi'|^N}{(|\xi'| + |\xi_n|)^{N-\alpha}}. \tag{8.21}$$

The following lemma holds.

LEMMA 8.1. *Suppose a function* $C(\xi', \xi_n) \in O'_{\alpha+i\beta}$ *satisfies* (8.19). *Then* $p\hat{C}(v(x') \times \delta(x_n))$ *is a bounded operator from* $H_{s+1/2}(\mathbf{R}^{n-1})$ *into* $H_{s-\alpha}(\mathbf{R}^n_+)$ *for any* s.

PROOF. We choose $N > s + \frac{1}{2}$ in (8.19). Then

$$p\hat{C}(v(x') \times \delta(x_n)) = p\hat{C}_N^-(v(x') \times \delta(x_n)) + p\hat{R}_N(v(x') \times \delta(x_n)).$$

If $v(x') \in C_0^\infty(\mathbf{R}^{n-1})$, then

$$v(x') \times \delta(x_n) \in H^-_{-1/2-\varepsilon}(\mathbf{R}^n)$$

and, by Theorem 4.4, $\hat{C}_N^-(\xi', \xi_n)\tilde{v}(\xi') \in \tilde{H}^-_{-\alpha_1-1/2-\varepsilon}$. Consequently,

$$p\hat{C}_N^-(v(x') \times \delta(x_n)) = 0,$$

and hence

$$p\hat{C}(v(x') \times \delta(x_n)) = p\hat{R}_N(v(x') \times \delta(x_n)). \tag{8.22}$$

But the boundedness of $p\hat{R}_N(v(x') \times \delta(x_n))$ from $H_{s+1/2}(\mathbf{R}^{n-1})$ into $H_{s-\alpha}(\mathbf{R}^n_+)$ follows from (8.17), since $s < N - \frac{1}{2}$. Lemma 8.1 is proved.

3. We now consider (8.1) under the assumption that $f \in H_{s-\alpha}(\mathbf{R}^n_+)$ and $s = \operatorname{Re} \kappa - m - \delta$, where m is a negative integer and $|\delta| < \frac{1}{2}$.

By Theorem 7.3, there exists a unique collection of functions $u_+ \in H_s^+$, $v_k(x') \in H_{m_k}(\mathbf{R}^{n-1})$, $m_k = |m| - \delta - k + \frac{1}{2}$, $1 \leqslant k \leqslant |m|$, satisfying the equation

$$p\hat{A}_0\left(u_+ + \sum_{k=1}^{|m|} \hat{A}_+^{-1}\hat{\Lambda}_+^{-k}(v_k(x') \times \delta(x_n))\right) = f(x). \tag{8.23}$$

which can be written in the form

$$p\hat{A}_0 u_+ + \sum_{k=1}^{|m|} p\hat{G}_k(v_k(x') \times \delta(x_n)) = f(x), \tag{8.24}$$

where $\hat{G}_k$ is the p.o. with symbol

$$\hat{G}_k(\xi', \xi_n) = \hat{A}_0(\xi)\hat{A}_+^{-1}\hat{\Lambda}_+^{-k} = \hat{A}_-(\xi)\hat{\Lambda}_+^{-k}(\xi).$$

Let us show that the symbols $\hat{G}_k(\xi', \xi_n)$ satisfy (8.19). Expanding $\hat{\Lambda}_+^{-k}$ in powers of Λ_-^{-1}, we get for any N

$$\hat{\Lambda}_+^{-k} = (\xi_n + i|\xi'| + i)^{-k} = \hat{\Lambda}_-^{-k}\left(1 + \frac{2i(1 + |\xi'|)}{\hat{\Lambda}_-}\right)^{-k}$$

$$= \sum_{r=0}^{N} \frac{a_r(1 + |\xi'|)^r}{\hat{\Lambda}_-^{k+r}} + \hat{\Lambda}_N(\xi', \xi_n), \tag{8.25}$$

where the a_r are certain numbers and

$$|\hat{\Lambda}_N(\xi', \xi_n)| \leqslant \frac{C(1 + |\xi'|)^{N+1}}{(1 + |\xi'| + |\xi_n|)^{N+k+1}}.$$

Since N is arbitrary, it follows from (8.25) that $\hat{G}_k(\xi', \xi_n) = \hat{A}_-\hat{\Lambda}_+^{-k}$ satisfies (8.19).

Consequently, Lemma 8.1 implies $\|p\hat{G}_k(v_k \times \delta(x_n))\|_{s-\alpha}^{+} \leqslant C[v_k]_{m_k}$, since the degree of homogeneity of $G_k(\xi', \xi_n)$ is equal to $\alpha + i\beta - \kappa - k$. Thus, since it is equivalent to (8.23), equation (8.24) has a unique solution $u_+ \in H_s^+$, $v_k(x') \in H_{m_k}(\mathbf{R}^{n-1})$ for any $f \in H_{s-\alpha}(\mathbf{R}_+^n)$.

Let us now consider a more general class of equations than (8.24). Suppose

$$p\left(\hat{A}_0 u_+ + \sum_{k=1}^{|m|} \hat{C}_k(v_k \times \delta(x_n))\right) = f(x), \tag{8.26}$$

where $f \in H_{s-\alpha}(\mathbf{R}_+^n)$, $\hat{C}_k$ is the p.o. with symbol $\hat{C}_k(\xi', \xi_n)$ and $C_k(\xi', \xi_n) \in O'_{\alpha_k + i\beta_k}$. If $s \geqslant \alpha - \alpha_k - \frac{1}{2}$, it is assumed that each $C_k(\xi', \xi_n)$ admits a decomposition of form (8.19):

$$C_k(\xi', \xi_n) = C_k^-(\xi', \xi_n) + R_k(\xi', \xi_n), \tag{8.27}$$

where $C_k^-(\xi', \xi_n)$ is a "minus" function and

$$|R_k(\xi', \xi_n)| \leqslant C|\xi'|^{\alpha_k - \alpha_{k1}}(|\xi'| + |\xi_n|)^{\alpha_{k1}}, \qquad \alpha_{k1} < \alpha - s - \tfrac{1}{2}. \tag{8.28}$$

We wish to find functions $u_+ \in H_s^+$ and $v_k(x') \in H_{s_k}(\mathbf{R}^{n-1})$, $s_k = s - \alpha + \alpha_k + \frac{1}{2}$, $1 \leqslant k \leqslant |m|$.

In contrast to (8.24), it is not assumed that the symbols $C_k(\xi)$ have the form $A_-(\xi)A_k^+(\xi)$, where $A_k^+(\xi)$ is a "plus" function. Consequently, (8.26) is not equivalent, in general, to an equation of the form $p\hat{A}_0 w_+ = f$ for some w_+ expressed in terms of $u_+(x)$ and $v_k(x')$, $1 \leqslant k \leqslant |m|$. Equation (8.26) is converted into the equation $p\hat{A}_0 u_+ = f$ only when $v_k \equiv 0$, which, as will be shown below, is equivalent to the fulfilment of condition (7.49).

Let us solve (8.26). In doing so, we will determine conditions on the symbols $C_k(\xi)$ that ensure the unique solvability of (8.26). We write (8.26) in the form

$$p\hat{A}_0 u_+ = f - p\sum_{k=1}^{|m|} \hat{C}_k(v_k \times \delta(x_n)). \tag{8.29}$$

Since $p\hat{C}_k(v_k \times \delta(x_n)) = p\hat{R}_k(v_k \times \delta(x_n))$, we can replace $\hat{C}_k$ by $\hat{R}_k$ in (8.29). By Theorem 7.3, a necessary and sufficient condition for the existence of a solution $u_+ \in H_s^+$ of (8.29) is the fulfilment of the following condition of form (7.49) in which $\widetilde{lf}$ has been replaced by $\widetilde{lf} - \sum_{j=1}^{|m|}\hat{R}_j(\xi)\tilde{v}_j(\xi')$:

$$\int_{-\infty}^{\infty} \frac{\xi_n^{k-1}}{\hat{A}_-(\xi)}\left(\widetilde{lf}(\xi) - \sum_{j=1}^{|m|} \hat{R}_j(\xi)\tilde{v}_j(\xi')\right) d\xi_n = 0. \tag{8.30}$$

If we let

$$\hat{c}_{kj}(\xi') = \int_{-\infty}^{\infty} \frac{\hat{R}_j(\xi', \xi_n)\xi_n^{k-1}}{\hat{A}_-(\xi', \xi_n)}\, d\xi_n, \qquad 1 \leqslant k, j \leqslant |m|, \tag{8.31}$$

$$\tilde{\varphi}_k(\xi') = \int_{-\infty}^{\infty} \frac{\xi_n^{k-1}\widetilde{lf}(\xi', \xi_n)}{\hat{A}_-(\xi', \xi_n)}\, d\xi_n, \qquad 1 \leqslant k \leqslant |m|, \tag{8.32}$$

condition (8.30) assumes the form

$$\sum_{j=1}^{|m|} \hat{c}_{kj}(\xi')\tilde{v}_j(\xi') = \tilde{\varphi}_k(\xi'), \qquad 1 \leqslant k \leqslant |m|. \tag{8.33}$$

To ensure the unique solvability of (8.33) for any right sides $\tilde{\varphi}_k(\xi')$, we require that

$$\det\|c_{kj}(\xi')\|_{k,j=1}^{|m|} \neq 0 \qquad \text{for } \xi' \neq 0. \tag{8.34}$$

This condition is the analog for equation (8.26) of the Šapiro-Lopatinskiĭ condition.

The solution of (8.33) is

$$\tilde{v}_j(\xi') = \sum_{k=1}^{|m|} \hat{c}_{jk}^{(1)}(\xi')\tilde{\varphi}_k(\xi'), \tag{8.35}$$

where $\|\hat{c}_{rk}^{(1)}(\xi')\|_{r,k=1}^{|m|}$ is the inverse of the matrix $\|\hat{c}_{kj}(\xi')\|_{k,j=1}^{|m|}$. Under the fulfilment of condition (8.30), $\tilde{u}_+(\xi)$ has, by Theorem 7.3, the form

$$\tilde{u}_+(\xi) = \frac{1}{\hat{A}_+(\xi)\hat{\Lambda}_+^{|m|}}\Pi^+\frac{\hat{\Lambda}_+^{|m|}}{\hat{A}_-}\left(\widetilde{lf} - \sum_{k=1}^{|m|} \hat{R}_k(\xi)\tilde{v}_k(\xi')\right). \tag{8.36}$$

Therefore, substituting (8.35) in (8.36), we conclude that

$$\tilde{u}_+(\xi) = \frac{1}{A_+\hat{\Lambda}_+^{|m|}}\Pi^+\frac{\hat{\Lambda}_+^{|m|}}{\hat{A}_-}\left(\widetilde{lf} - \sum_{j,k=1}^{|m|} \hat{R}_k(\xi)\hat{c}_{jk}^{(1)}(\xi')\tilde{\varphi}_k(\xi')\right), \tag{8.37}$$

where $\tilde{\varphi}_k(\xi')$ is defined by (8.32). Formulas (8.37) and (8.35) determine the unique solution $(u_+(x), v_1(x'), \ldots, v_{|m|}(x'))$ of (8.26).

Let us estimate this solution. Since $\deg_\xi R_j(\xi', \xi_n) = \alpha_j + i\beta_j$ and $\deg_\xi A_-(\xi) = \alpha + i\beta - \kappa$, where $\deg_\xi$ denotes the degree of homogeneity with respect to ξ of a function, we have $\deg_{\xi'} c_{kj}(\xi') = \alpha_j + i\beta_j + k - \alpha - i\beta - \kappa$. Consequently,

$$\deg_{\xi'} c_{jk}^{(1)}(\xi') = -(\alpha_j + i\beta_j + k - \alpha - i\beta - \kappa). \tag{8.37$'$}$$

We note that the integral (8.31) converges by virtue of (8.28). Moreover, $\hat{c}_{kj}(\xi')$ is completely determined by $\hat{C}_k(\xi)$ and $\hat{A}_-(\xi)$ and does not depend on the representation of $\hat{C}_k$ in the form (8.27).

In fact, if $C_k(\xi) = C_{k1}^- + R_{k1}$ is another decomposition of form (8.27), then $R_{k1}(\xi) - R_k(\xi) = C_{k2}^-(\xi', \xi_n)$, where $C_{k2}^-(\xi', \xi_n)$ is a "minus" function satisfying

(8.28). Consequently, the function

$$\frac{\hat{C}_{j2}^{-}(\xi', \xi_n)\xi_n^{k-1}}{\hat{A}_{-}(\xi', \xi_n)}$$

is a "minus" function satisfying

$$\left|\frac{\hat{C}_{j2}^{-}(\xi', \xi_n + i\tau)(\xi_n + i\tau)^{k-1}}{\hat{A}_{-}(\xi', \xi_n + i\tau)}\right| < \frac{C}{(1 + |\xi'| + |\xi_n| + |\tau|)^{1+\varepsilon}}$$

for $\tau \leqslant 0$ and $\varepsilon > 0$. Therefore Cauchy's theorem implies that

$$\int_{-\infty}^{\infty} \frac{\hat{C}_{j2}^{-}\xi_n^{k-1}}{\hat{A}_{-}(\xi)}\, d\xi_n = 0,$$

i.e.

$$\int_{-\infty}^{\infty} \frac{\hat{R}_k(\xi)\xi_n^{k-1}}{\hat{A}_{-}}\, d\xi_n = \int_{-\infty}^{\infty} \frac{\hat{R}_{k1}(\xi)\xi_n^{k-1}}{\hat{A}_{-}}\, d\xi_n.$$

From (8.35) we get by virtue of (8.37′) and Lemma 4.4 that

$$[v_j]_{s-\alpha+\alpha+\alpha_j+1/2} \leqslant C \sum_{k=1}^{|m|} [\varphi_k]_{s-\operatorname{Re}\kappa-k+1/2}. \tag{8.38}$$

On the other hand, since $s - \operatorname{Re}\kappa - k + \frac{1}{2} = |m| - \delta - k + \frac{1}{2} > 0$, Theorem 4.2 implies

$$[\varphi_k]_{s-\operatorname{Re}\kappa-k+1/2} \leqslant C\|lf\|_{s-\alpha} \leqslant C\|f\|_{s-\alpha}^{+}. \tag{8.39}$$

Therefore

$$\sum_{j=1}^{|m|} [v_j]_{s-\alpha+\alpha_j+1/2} \leqslant C\|f\|_{s-\alpha}^{+} \tag{8.40}$$

and hence, by (8.36),

$$\|u\|_s^{+} \leqslant C\left(\|f\|_s^{+} + \sum_{k=1}^{|m|} [v_k]_{s-\alpha+\alpha_j+1/2}\right) \leqslant C\|f\|_{s-\alpha}^{+}. \tag{8.41}$$

We have thus proved

THEOREM 8.2. *Suppose $A_0(\xi)$ is an elliptic symbol with factorization index κ such that* $\operatorname{Re}\kappa - s = m + \delta$, *where m is a negative integer and $|\delta| < \frac{1}{2}$, and suppose $C_k(\xi) \in O'_{\alpha_k+i\beta_k}$ satisfies* (8.27). *Then, under the fulfilment of the analog* (8.34) *of the Šapiro-Lopatinskiĭ condition, equation* (8.26) *has for any $f \in H_{s-\alpha}(\mathbf{R}_+^n)$ a unique solution $u_+ \in H_s^+$, $v_k(x') \in H_{s_k}(\mathbf{R}^{n-1})$, $s_k = s - \alpha + \alpha_k + \frac{1}{2}$, $1 \leqslant k \leqslant |m|$. Estimates* (8.40) *and* (8.41) *hold.*

REMARK 8.2. Condition (7.49) can be written in the form

$$D_n^{k-1}\hat{A}_-^{-1}lf|_{x_n=0} = 0, \qquad f \in H_{s-\alpha}(\mathbf{R}_+^n), \tag{8.42}$$

where $\hat{A}_-^{-1}$ is the p.o. with symbol $A_-^{-1}(\xi', \xi_n)$. Under the fulfilment of condition (7.49) or (8.42), equation (8.1) has a unique solution $u_+ \in H_s^+$ for $s = \operatorname{Re}\kappa - m - \delta$, where m is a negative integer and $|\delta| < \frac{1}{2}$. In the case of a p.o. in a bounded domain with smooth boundary, the condition on f ensuring the membership of the solution in H_s^+ is not as visible as condition (8.42). Nevertheless, as will be shown in Chapter VI, the formulation of a problem for (8.23) or (8.26) can be extended without difficulty to the case of a p.o. in a bounded domain.

REMARK 8.3. Let $A_0(\xi)$ be an elliptic symbol of zero degree of homogeneity. Then the p.o. A_0 is a singular integral operator. Theorems 7.1, 7.3, 8.1 and 8.2 yield the formulation and solution of boundary value problems for the p.o. $\hat{A}_0$ in a half-space. We note that, according to (6.51), the factorization index of an elliptic symbol $A_0(\xi) \in O_{\alpha+i\beta}^{\infty}$ is generally not equal to $(\alpha + i\beta)/2$. Consequently, for $\alpha + i\beta = 0$ or $\alpha < 0$, i.e. for singular integral equations or integral equations of the first kind, it can turn out that $\operatorname{Re}\kappa > \frac{1}{2}$. For the unique solvability of such equations in H_0^+ it is necessary to either assign boundary conditions on the hyperplane $x_n = 0$ or seek the solution in a space H_s^+, $|s - \operatorname{Re}\kappa| < \frac{1}{2}$, which essentially corresponds to imposing zero boundary conditions. On the other hand, for $\alpha > 0$, i.e. in the case of integrodifferential equations, it can happen that $\operatorname{Re}\kappa < -\frac{1}{2}$, and then a pseudodifferential equation generally does not have a solution in H_0^+. In this case one can consider an equation with potentials of form (8.23) and thereby reduce the problem to the determination of a function in H_0^+ and potential densities containing the singularities of the solution.

REMARK 8.4. Let us consider the existence of a solution of (8.1) in H_s^+ under the assumption that $f \in H_{s-\alpha}(\mathbf{R}_+^n)$ and $s \in Z(\kappa)$.

Suppose $s = \operatorname{Re}\kappa - m + \frac{1}{2}$, where m is a nonpositive integer. If $0 < \varepsilon < 1$ and (7.49) is satisfied, then, by Theorem 7.3, equation (8.1) has a solution $u_+ \in H_{s-\varepsilon}^+$, with

$$\tilde{u}_+ = \frac{1}{\hat{A}_+(\xi)\hat{\Lambda}_+^{|m|}} \Pi^+ \hat{\Lambda}_+^{|m|} \frac{\widetilde{lf}}{\hat{A}_-}.$$

Since $lf \in H_{s-\alpha}(\mathbf{R}^n)$, we have $\tilde{g} = \hat{\Lambda}_+^{|m|}\widetilde{lf}/\hat{A}_- \in \tilde{H}_{1/2}$ by Lemma 4.4. According to Remark 5.2, $\Pi^+\tilde{g}$ does not belong to $\tilde{H}_{1/2}^+$ for all $\tilde{g} \in \tilde{H}_{1/2}$ but only for $\tilde{g}$ in a dense subset of $\tilde{H}_{1/2}$. Consequently, (8.1) has a solution $u_+ \in H_s^+$, $s = \operatorname{Re}\kappa - m + \frac{1}{2}$, only for f in a dense subset of $H_{s-\alpha}(\mathbf{R}_+^n)$. Estimate (7.46) does not hold due to the unboundedness of Π^+ in $\tilde{H}_{1/2}$.

Suppose now $s = \operatorname{Re}\kappa - m - \frac{1}{2}$, where m is a nonnegative integer. By Theorem 5.2 (cf. the proof of Theorem 7.2), the homogeneous equation $p\hat{A}_0u_+ = 0$ has a solution $u_+(x) \in H_s^+$, with

$$\tilde{u}_+(\xi) = Fu_+ = \sum_{k=1}^{m} \frac{\tilde{c}_k(\xi')\xi_n^{k-1}}{\hat{A}_+(\xi)},$$

that depends on m arbitrary functions $c_k(x') \in H_{-m+k-1}(\mathbf{R}^{n-1})$.

Let us now show that the inhomogeneous equation $p\hat{A}_0 u_+ = f$ has a solution $u_+ \in H_s^+$ only for f in a dense subset of $H_{s-\alpha}(\mathbf{R}^n_+)$. Since $H_{s-\alpha+\varepsilon}(\mathbf{R}^n_+)$ is dense in $H_{s-\alpha}(\mathbf{R}^n_+)$ and since Theorem 7.2 implies the existence of a solution $u_+ \in H^+_{s+\varepsilon} \subset H_s^+$ for $f \in H_{s-\alpha+\varepsilon}(\mathbf{R}^n_+)$, it remains to show that a solution does not exist in H_s^+ for all $f \in H_{s-\alpha}(\mathbf{R}^n_+)$. Suppose $\widetilde{lf} = \hat{A}_- \hat{\Lambda}_+^{m+1} \tilde{f}_0(\xi)$, where

$$\tilde{f}_0(\xi', \xi_n) = \frac{\theta(\xi_n)\tilde{\varphi}_0(\xi')}{(2 + \xi_n)\ln(2 + \xi_n)},$$

with $\tilde{\varphi}_0(\xi') = 1$ for $|\xi'| < 1$ and $\tilde{\varphi}_0(\xi') = 0$ for $|\xi'| \geqslant 1$. Then $\widetilde{lf} \in H_{s-\alpha}(\mathbf{R}^n)$. Suppose $0 < \varepsilon < 1$. Since $\widetilde{lf} \in \tilde{H}_{s-\alpha-\varepsilon}(\mathbf{R}^n)$, it follows by virtue of Theorem 7.2 that (8.1) has a solution $u_+ \in H^+_{s-\varepsilon}$, with

$$\tilde{u}_+(\xi) = \frac{\hat{\Lambda}_+^{m+1}}{\hat{A}_+} \Pi^+ \tilde{f}_0 + \sum_{k=1}^{m+1} \frac{\tilde{c}_k(\xi')\xi_n^{k-1}}{\hat{A}_+}. \tag{8.43}$$

We will show that $\tilde{u}_+(\xi) \notin \tilde{H}_s^+$. If it were true that $\tilde{u}_+(\xi) \in \tilde{H}_s^+$, we would have

$$\int_{-\infty}^{\infty} |\tilde{u}_+(\xi)|^2 (1 + |\xi'| + |\xi_n|)^{2s}\, d\xi_n < \infty, \qquad s = \operatorname{Re}\kappa - m - \tfrac{1}{2}, \tag{8.44}$$

for almost all ξ'. But

$$\begin{aligned} \left| \frac{\hat{\Lambda}_+^{m+1}}{\hat{A}_+} \Pi^+ \tilde{f}_0 \right| &\geqslant C \left| |\xi_n|^{m+1-\operatorname{Re}\kappa} \int_0^\infty \frac{d\eta_n}{(|\xi_n| + \eta_n)(2 + \eta_n)\ln(2 + \eta_n)} \right| \\ &\geqslant \frac{1}{2} C |\xi_n|^{m-\operatorname{Re}\kappa} \int_0^{|\xi_n|} \frac{d\eta_n}{(2 + \eta_n)\ln(2 + \eta_n)} \\ &= \tfrac{1}{2} C |\xi_n|^{m-\operatorname{Re}\kappa} (\ln\ln(2 + |\xi_n|) - C_1), \end{aligned} \tag{8.45}$$

for $|\xi'| \leqslant 1$ and $\xi_n < -N$. And since

$$\left| \sum_{k=1}^{m+1} \frac{\tilde{c}_k(\xi')\xi_n^{k-1}}{\hat{A}_+} \right| = O\big(|\xi_n|^{m-\operatorname{Re}\kappa}\big),$$

we actually have

$$\int_{-\infty}^{\infty} |\tilde{u}_+(\xi)|^2 (1 + |\xi'| + |\xi_n|)^{2s}\, d\xi_n = \infty$$

for $|\xi'| \leqslant 1$. Consequently, $\tilde{u}_+ \notin \tilde{H}_s^+$.

CHAPTER III

SMOOTHNESS OF SOLUTIONS OF PSEUDODIFFERENTIAL EQUATIONS

§9. Asymptotic behavior of a solution of a pseudodifferential equation in a halfspace

Suppose $A(\xi', \xi_n) \in O^{\infty}_{\alpha+i\beta}$ is an elliptic symbol with factorization $A(\xi', \xi_n) = A_-(\xi', \xi_n)A_+(\xi', \xi_n)$ with respect to ξ_n and factorization index $\kappa = \deg_\xi A_+(\xi', \xi_n)$. Consider the p.e.

$$p\hat{A}u_+ = f, \qquad lf \in S(\mathbf{R}^n). \tag{9.1}$$

By Theorem 7.1, (9.1) has a unique solution $u_+ \in H_s^+$ for $|s - \operatorname{Re}\kappa| < \frac{1}{2}$. We will find the asymptotic behavior of this solution in the vicinity of the hyperplane $x_n = 0$. The following assertion is valid.

LEMMA 9.1. *The function $A_+^{-1}(\xi', \xi_n)$ admits for any positive integer N the expansion*

$$A_+^{-1}(\xi', \xi_n) = \Lambda_+^{-\kappa}(\xi', \xi_n) + \Lambda_+^{-\kappa-1}\left(c_{11}(\xi')\ln\frac{\Lambda_+}{|\xi'|} + c_{10}(\xi')\right)$$
$$+ \ldots + \Lambda_+^{-\kappa-N}\sum_{p=0}^{N} c_{Np}(\xi')\ln^p\frac{\Lambda_+}{|\xi'|} + R_N(\xi', \xi_n), \tag{9.2}$$

where $c_{kp}(\xi')$ is a homogeneous function of degree k and

$$|R_N(\xi', \xi_n)| \leqslant C_{N\varepsilon}|\xi'|^{N+1-\varepsilon}(|\xi'| + |\xi_n|)^{-\operatorname{Re}\kappa - N - 1 + \varepsilon}, \qquad \forall\varepsilon > 0. \tag{9.3}$$

PROOF. From (6.44) we have

$$A_+(\xi', \xi_n) = (\xi_n + i|\xi'|)^{\kappa}e^{B_2^+(\xi', \xi_n + i0)},$$

where

$$B_2^+(\xi', \xi_n + i\tau) = \frac{i}{2\pi}\int_{-\infty}^{\infty} \frac{b_2(\xi', \eta_n)}{\xi_n + i\tau - \eta_n}\, d\eta_n, \tag{9.4}$$

$$b_2(\xi', \xi_n) = \ln\left[A(\xi', \xi_n)A^{-1}(0, +1)(\xi_n + i|\xi'|)^{-\kappa}(\xi_n - i|\xi'|)^{\kappa-\alpha-i\beta}\right]. \tag{9.5}$$

Let $r = |\xi'|$, $\omega = \xi'/|\xi'|$ and let $B_2(\omega, r, \xi_n)$ denote $b_2(\xi', \xi_n)$ as a function of the coordinates (ω, r, ξ_n). By (9.5), $B_2(\omega, r, \xi_n)$ is a homogeneous function of degree zero with respect to (r, ξ_n) and belongs to C^∞ with respect to the totality of variables (ω, r, ξ_n) for $r + |\xi_n| > 0$.

Suppose $\xi_n > 0$. Since $B_2(\omega, 0, +1) = b_2(0, +1) = 0$, and since

$$B_2(\omega, r, \xi_n) = B_2\left(\omega, \frac{r}{r+\xi_n}, \frac{\xi_n}{r+\xi_n}\right),$$

by virtue of the homogeneity, it follows from Taylor's theorem that

$$B_2(\omega, r, \xi_n) = B_2\left(\omega, \frac{r}{r+\xi_n}, \frac{\xi_n}{r+\xi_n}\right)$$
$$= \sum_{k_1+k_2=1}^{N} \frac{(-1)^{k_2}}{k_1!k_2!} \frac{\partial^{k_1+k_2}B_2(\omega, 0, +1)}{\partial r^{k_1}\partial\xi_n^{k_2}} \frac{r^{k_1+k_2}}{(r+\xi_n)^{k_1+k_2}} + B_{3N}(\omega, r, \xi_n), \tag{9.6}$$

where

$$|B_{3N}(\omega, r, \xi_n)| \leqslant \frac{C_N r^{N+1}}{(r+\xi_n)^{N+1}}. \tag{9.7}$$

Suppose now $\xi_n < 0$. Then

$$B_2(\omega, r, \xi_n) = B_2\left(\omega, \frac{r}{r+|\xi_n|}, \frac{\xi_n}{r+|\xi_n|}\right),$$

and Taylor's theorem implies

$$B_2(\omega, r, \xi_n) = \sum_{k_1+k_2=1}^{N} \frac{1}{k_1!k_2!}$$
$$\times \frac{\partial^{k_1+k_2}B_2(\omega, 0, -1)}{\partial r^{k_1}\partial\xi_n^{k_2}} \frac{r^{k_1+k_2}}{(r+|\xi_n|)^{k_1+k_2}} + B_{4N}(\omega, r, \xi_n), \tag{9.8}$$

where

$$|B_{4N}(\omega, r, \xi_n)| \leqslant \frac{C_N r^{N+1}}{(r+|\xi_n|)^{N+1}}. \tag{9.9}$$

Let

$$b_{k1}(\omega) = \sum_{k_1+k_2=k} \frac{(-1)^{k_2}}{k_1!k_2!} \frac{\partial^{k_1+k_2}B_2(\omega, 0, +1)}{\partial r^{k_1}\partial\xi_n^{k_2}},$$

$$b_{k2}(\omega) = \sum_{k_1+k_2=k} \frac{1}{k_1!k_2!} \frac{\partial^{k_1+k_2}B_2(\omega, 0, -1)}{\partial r^{k_1}\partial\xi_n^{k_2}}. \tag{9.10}$$

Then

$$\begin{aligned} B_2^+(\xi', \xi_n + i\tau) &= \frac{i}{2\pi}\int_{-\infty}^{\infty} \frac{B_2(\omega, r, \eta_n)}{\xi_n + i\tau - \eta_n}\, d\eta_n \\ &= \sum_{k=1}^{N} \frac{i}{2\pi}\int_0^{\infty} \frac{b_{k1}(\omega)r^k d\eta_n}{(r + \eta_n)^k(\xi_n + i\tau - \eta_n)} \\ &\quad + \sum_{k=1}^{N} \frac{i}{2\pi}\int_{-\infty}^{0} \frac{b_{k2}(\omega)r^k d\eta_n}{(r + |\eta_n|)^k(\xi_n + i\tau - \eta_n)} + B_{5N}^+(\xi', \xi_n + i\tau), \end{aligned} \tag{9.11}$$

where

$$B_{5N}^+(\xi', \xi_n + i\tau) = \frac{i}{2\pi}\int_{-\infty}^{0} \frac{B_{4N}(\omega, r, \eta_n)}{\xi_n + i\tau - \eta_n}\, d\eta_n + \frac{i}{2\pi}\int_0^{\infty} \frac{B_{3N}(\omega, r, \eta_n)}{\xi_n + i\tau - \eta_n}\, d\eta_n. \tag{9.12}$$

We have

$$\int_0^{\infty} \frac{d\eta_n}{(r + \eta_n)(\xi_n + i\tau - \eta_n)} = \frac{\ln(-(\xi_n + i\tau)/r)}{\xi_n + i\tau + r},$$

$$\int_{-\infty}^{0} \frac{d\eta_n}{(r - \eta_n)(\xi_n + i\tau - \eta_n)} = \frac{\ln(\xi_n + i\tau)/r}{\xi_n + i\tau - r}. \tag{9.13}$$

Here the branch of the logarithm is taken so that it is real on the positive real axis. Differentiating (9.13) $k - 1$ times with respect to r, we get

$$\int_0^{\infty} \frac{d\eta_n}{(r + \eta_n)^k(\xi_n + i\tau - \eta_n)} = \frac{1}{(\xi_n + i\tau + r)^k} \ln\left(-\frac{\xi_n + i\tau}{r}\right) + \sum_{p=1}^{k-1} \frac{1}{pr^p(\xi_n + i\tau + r)^{k-p}}, \tag{9.14}$$

$$\int_{-\infty}^{0} \frac{d\eta_n}{(r - \eta_n)^k(\xi_n + i\tau - \eta_n)} = \frac{(-1)^{k-1}}{(\xi_n + i\tau - r)^k} \ln\frac{\xi_n + i\tau}{r} + \sum_{p=1}^{k-1} \frac{(-1)^{k-p-1}}{pr^p(\xi_n + i\tau - r)^{k-p}}. \tag{9.14'}$$

Suppose $|\xi_n| \geqslant 2r$. Then

$$\frac{1}{(\xi_n + i\tau \pm r)^k} = \frac{1}{(\xi_n + i\tau + ir)^k}\left(1 - \frac{(i \mp 1)r}{\xi_n + i\tau + ir}\right)^{-k}$$

$$= \sum_{p=0}^{N} \frac{C_p^{\pm} r^p}{(\xi_n + i\tau + ir)^{p+k}} + B_{6N}^{\pm}(\xi_n + i\tau, r), \qquad (9.15)$$

where the $C_p^{\pm}$ are numerical constants, $C_0^{\pm} = 1$ and for $|\xi_n| \geqslant 2r$

$$|B_{6N}^{\pm}(\xi_n + i\tau, r)| \leqslant \frac{C_N r^{N+1}}{(|\xi_n| + r + \tau)^{N+1+k}}. \qquad (9.16)$$

Analogously,

$$\ln(\xi_n + i\tau) = \ln(\xi_n + i\tau + ir) + \ln\left(1 - \frac{ir}{\xi_n + i\tau + ir}\right)$$

$$= \ln(\xi_n + i\tau + ir) - \sum_{p=1}^{N} \frac{i^p r^p}{p(\xi_n + i\tau + ir)^p} + B_{7N}(\xi_n + i\tau, r), \qquad (9.17)$$

where $B_{7N}(\xi_n + i\tau, r)$ for $|\xi_n| \geqslant 2r$ satisfies the estimate of form (9.16) with $k = 0$. Thus, expanding the right sides of (9.13), (9.14) and (9.14′) with the use of (9.15) and (9.17) and substituting the results in (9.11), we get

$$B_2^+(\xi', \xi_n + i\tau) = \sum_{k=1}^{N} \left(b_{k1}(\omega) + (-1)^{k-1} b_{k2}(\omega)\right) \frac{r^k}{(\xi_n + i\tau + ir)^k} \ln \frac{\xi_n + i\tau + ir}{r}$$

$$+ \sum_{k=1}^{N} \frac{C_k(\omega) r^k}{(\xi_n + i\tau + ir)^k} + B_{8N}(\omega, r, \xi_n + i\tau) + B_{5N}^+(\xi', \xi_n + i\tau), \qquad (9.18)$$

where $B_{8N}(\omega, r, \xi_n + i\tau)$ for $|\xi_n| \geqslant 2r$ satisfies the estimate of form (9.16) with $k = 0$. Applying the expansion formula for an integral of Cauchy type (see (5.36)) to $B_{5N}^+(\xi', \xi_n + i\tau)$, we get

$$B_{5N}^+(\xi', \xi_n + i\tau) = \sum_{k=1}^{N} \frac{d_k(\omega) r^k}{(\xi_n + i\tau + ir)^k} + B_{9N}(\omega, r, \xi_n + i\tau), \qquad (9.19)$$

with $B_{9N}(\omega, r, \xi_n + i\tau)$ satisfying by virtue of Lemma 6.1 the estimate

$$|B_{9N}(\omega, r, \xi_n + i\tau)| \leqslant \frac{C_{N\varepsilon} r^{N+1-\varepsilon}}{(|\xi_n| + \tau + r)^{N+1-\varepsilon}}, \qquad \forall \varepsilon > 0. \qquad (9.20)$$

It therefore follows from (9.18) and (9.19) that

$$B_2^+(\xi', \xi_n + i\tau) = B_3^+(\xi', \xi_n + i\tau) + B_{10N}(\xi', \xi_n + i\tau), \tag{9.21}$$

where

$$B_3^+(\xi', \xi_n + i\tau) = \sum_{k=1}^{N} \sum_{p=0}^{1} \frac{d_{pk}(\omega) r^k}{(\xi_n + i\tau + ir)^k} \ln^p \frac{\xi_n + i\tau + ir}{r} \tag{9.22}$$

and $B_{10N}(\xi', \xi_n + i\tau)$ satisfies (9.20) for $|\xi_n| \geqslant 2r$.

Suppose now $|\xi_n| \leqslant 2r$. Then Lemma 6.1 implies

$$|B_2^+(\xi', \xi_n + i\tau)| \leqslant \frac{C_\varepsilon r^{1-\varepsilon}}{(|\xi_n| + \tau + r)^{1-\varepsilon}} \leqslant \frac{C_{N\varepsilon} r^{N+1-\varepsilon}}{(|\xi_n| + \tau + r)^{N+1-\varepsilon}}$$

while (9.22) implies that for $|\xi_n| \leqslant 2r$

$$|B_3^+(\xi', \xi_n + i\tau)| \leqslant \frac{C_\varepsilon r^{1-\varepsilon}}{(|\xi_n| + \tau + r)^{1-\varepsilon}} \leqslant \frac{C_{N\varepsilon} r^{N+1-\varepsilon}}{(|\xi_n| + \tau + r)^{N+1-\varepsilon}}, \qquad \forall \varepsilon > 0.$$

Consequently, $B_{10N}(\xi', \xi_n + i\tau)$ satisfies (9.20) for any $|\xi_n| + r > 0$. Substituting (9.21) in (6.44), we get (9.2). Lemma 9.1 is proved.

By Theorem 7.1, $\tilde{u}_+(\xi) = \hat{A}_+^{-1}\Pi^+\hat{A}_-^{-1}\tilde{lf}$. Applying (5.36) to $\Pi^+\hat{A}_-^{-1}\tilde{lf}$, we get

$$\tilde{u}_+(\xi) = \sum_{k=1}^{N} \hat{A}_+^{-1}\hat{\Lambda}_+^{-k}\tilde{f}_k(\xi') + \tilde{u}_N^+(\xi), \tag{9.23}$$

where

$$\tilde{f}_k(\xi') = i\Pi'\hat{\Lambda}_+^{k-1}\hat{A}_-^{-1}\tilde{lf}, \qquad \tilde{u}_N^+(\xi) = \hat{A}_+^{-1}\hat{\Lambda}_+^{-N}\Pi^+\hat{\Lambda}_+^{N}\hat{A}_-^{-1}\tilde{lf}.$$

Since $\tilde{lf} \in S(\mathbf{R}^n)$, we have

$$|\tilde{f}_k(\xi')| \leqslant C_{mk}(1 + |\xi'|)^{-m}, \qquad \forall m, \tag{9.24}$$

$$|\hat{\Lambda}_+^N\hat{A}_-^{-1}\tilde{lf}| + \left|\frac{\partial}{\partial\xi_n}\hat{\Lambda}_+^N\hat{A}_-^{-1}\tilde{lf}\right| \leqslant C_{Nm}\frac{(1 + |\xi'|)^{-m}}{(1 + |\xi'| + |\xi_n|)^2}, \qquad \forall m. \tag{9.25}$$

(9.25) and Lemma 6.1 imply

$$|\tilde{u}_N^+(\xi', \xi_n + i\tau)| \leqslant C_{mN}(1 + |\xi'| + |\xi_n| + \tau)^{-\operatorname{Re}\kappa - N - 1 + \varepsilon}(1 + |\xi'|)^{-m}, \qquad \forall \varepsilon > 0, \quad \forall m. \tag{9.26}$$

Thus, substituting (9.2) in (9.23), we get

$$\tilde{u}_+(\xi', \xi_n) = \hat{\Lambda}_+^{-\kappa-1}\tilde{v}_{01}(\xi') + \sum_{k=2}^{N} \Lambda_+^{-\kappa-k} \sum_{p=0}^{k-1} \tilde{v}_{pk}(\xi')\ln^p \frac{\hat{\Lambda}_+}{1 + |\xi'|} + \tilde{u}_{N1}^+(\xi', \xi_n), \tag{9.27}$$

where the $\tilde{v}_{pk}(\xi')$ satisfy the estimates of form (9.24), $\tilde{u}^+_{N1}(\xi', \xi_n)$ satisfies the estimate of form (9.26) and

$$\tilde{v}_{01}(\xi') = i\Pi'\hat{A}_-^{-1}\widetilde{lf}. \tag{9.28}$$

Suppose $N > -\operatorname{Re}\kappa + 1$ and let

$$u^+_{N1}(x', x_n) = F^{-1}\tilde{u}^+_{N1}(\xi', \xi_n) = \frac{1}{(2\pi)^n}\int_{-\infty}^{\infty} \tilde{u}^+_{N1}(\xi', \xi_n)e^{-i(x,\xi)}\,d\xi. \tag{9.29}$$

It follows from (9.26) that $D^r_{x'}u^+_{N1}(x', x_n)$ for any r is continuously differentiable $\operatorname{Re}\kappa + N - 1$ times. Displacing the contour of integration with respect to ξ_n in (9.29) in the direction perpendicular to the real axis, we get

$$u^+_{N1}(x', x_n) = \frac{1}{(2\pi)^n}\int_{-\infty}^{\infty} \tilde{u}^+_{N1}(\xi', \xi_n + i\tau)e^{-i(x',\xi')-ix_n(\xi_n+i\tau)}\,d\xi'\,d\xi_n, \qquad \forall\tau \geqslant 0. \tag{9.30}$$

This and (9.26) imply

$$\begin{aligned} |D^r_{x'}u^+_{N1}(x', x_n)| &\leqslant C\int_{-\infty}^{\infty}(1 + |\xi'| + |\xi_n| + \tau)^{-\operatorname{Re}\kappa - N - 1 + \varepsilon}e^{x_n\tau}(1 + |\xi'|)^{-n}\,d\xi' d\xi_n \\ &\leqslant C^{(1)}\tau^{-\operatorname{Re}\kappa - N + 2\varepsilon}e^{x_n\tau}\int_{-\infty}^{\infty}(1 + |\xi_n|)^{-1-\varepsilon}\,d\xi_n \\ &= C^{(2)}\tau^{-\operatorname{Re}\kappa - N + 2\varepsilon}e^{x_n\tau}, \qquad \forall\tau > 0. \end{aligned} \tag{9.31}$$

Since τ is arbitrary, we get upon passing to the limit in (9.31) for $\tau \to +\infty$ that $u^+_{N1}(x', x_n) = 0$ for $x_n < 0$. For $x_n > 0$, we get upon setting $\tau = 1/x_n$ that

$$|D^r_{x'}u^+_{N1}(x', x_n)| \leqslant C_{\varepsilon Nr}x_n^{\operatorname{Re}\kappa + N - 2\varepsilon}, \quad x_n > 0, \quad \forall\varepsilon > 0. \tag{9.32}$$

Let k_0 be the least positive integer such that

$$\operatorname{Re}\kappa + k_0 > 0. \tag{9.33}$$

According to Example 2.3, for $k \geqslant k_0$ the inverse Fourier transform of the function $\hat{\Lambda}_+^{-\kappa-k}\ln^p\hat{\Lambda}_+$ has the form

$$F^{-1}_{\xi_n}\hat{\Lambda}_+^{-\kappa-k}\ln^p\hat{\Lambda}_+ = \theta(x_n)x_n^{\kappa+k-1}e^{-x_n(1+|\xi'|)}\sum_{r=0}^{p} C_{pr}(-\kappa - k)\ln^r x_n. \tag{9.34}$$

If $k < k_0$, it has the form (see (2.38))

$$F^{-1}_{\xi_n}\hat{\Lambda}_+^{-\kappa-k}\ln^p\hat{\Lambda}_+ = e^{-x_n(1+|\xi'|)}D_n^{k_0-k}\sum_{r=0}^{p} C_{pr}(-\kappa - k_0)\theta(x_n)x_n^{\kappa+k_0-1}\ln^r x_n. \tag{9.35}$$

Thus, letting $u_+(x', x_n) = F^{-1}\tilde{u}_+(\xi', \xi_n)$ and expanding $e^{-x_n(1+|\xi'|)}$ in a series, we get from (9.24), (9.27), (9.34) and (9.35) that

$$u_+(x', x_n) = \sum_{k=1}^{k_0-1} D_n^{k_0-k} \sum_{p=0}^{k-1} d_{pk}(x')\theta(x_n)x_n^{\kappa+k_0-1}\ln^p x_n$$
$$+ \sum_{k=k_0}^{N} \sum_{p=0}^{k-1} d_{pk}(x')\theta(x_n)x_n^{\kappa+k-1}\ln^p x_n + u_{N2}^+(x', x_n), \qquad (9.36)$$

where $u_{N2}^+(x', x_n)$ satisfies the estimate of form (9.32), $d_{pk}(x') \in H_m(\mathbf{R}^{n-1})$ for any m, which implies $d_{pk}(x') \in C^\infty(\mathbf{R}^{n-1})$, and

$$d_{01}(x') = \frac{e^{-i(\pi/2)(\kappa+k_0-1)}}{\Gamma(\kappa + k_0)} p'\hat{A}_-^{-1}(D', D_n)lf. \qquad (9.37)$$

It follows from (9.36) that $u_+(x', x_n)$ is continuously differentiable $\operatorname{Re}\kappa + N - 1$ times for $x_n \neq 0$ and, since N is arbitrary, $u_+(x', x_n) \in C^\infty$ for $x_n \neq 0$. We have thus proved

THEOREM 9.1. *Suppose $u_+(x', x_n)$ is the solution of (9.1) in H_s^+ for $|s - \operatorname{Re}\kappa| < 1/2$, where κ is the factorization index of the elliptic symbol $A(\xi', \xi_n)$. Then $u_+(x', x_n) \in C^\infty$ for $x_n \neq 0$ and the asymptotic expansion (9.36), where $u_{N2}^+(x', x_n)$ satisfies the estimate of form (9.32), holds for any positive integer N.*

We note that if $\operatorname{Re}\kappa > -1$, then $k_0 = 1$ and hence the first sum in (9.36) vanishes. In this case

$$u_+(x', x_n) = d_{01}(x')\theta(x_n)x_n^\kappa + O\big(\theta(x_n)x_n^{\operatorname{Re}\kappa+1-\varepsilon}\big), \qquad \forall\varepsilon > 0.$$

If κ is a negative integer, then $\delta(x_n)$ or its derivatives appear in (9.36) (cf. Example 7.3).

EXAMPLE 9.1. Consider the equation $p\Lambda^{2\kappa}(D', D_n, \tau)u_+ = f$ for $u_+ \in H_s^+$ when $lf \in S(\mathbf{R}^n)$, $\kappa > -1$ and $|s - \kappa| < 1/2$. Then

$$u_+(x', x_n) = d_{01}(x')\theta(x_n)x_n^\kappa + O(\theta(x_n)x_n^{\kappa+1}), \qquad (9.38)$$

with

$$d_{01}(x') = \frac{e^{-i(\pi/2)\kappa}}{\Gamma(\kappa+1)} p'\Lambda_-^{-\kappa}(D', D_n - i\tau)lf$$
$$= \frac{-\Gamma(n/2)}{\pi^{n/2}\Gamma(\kappa)\Gamma(\kappa+1)} \int_0^\infty \int_{-\infty}^\infty \frac{y_n^\kappa e^{-y_n\tau} f(y', y_n)\,dy'\,dy_n}{[y_n^2 + |x' - y'|^2]^{n/2}}. \qquad (9.39)$$

REMARK 9.1. It follows from (9.36) that, when κ is nonintegral, $u_+(x', x_n) \notin C^\infty(\overline{\mathbf{R}_+^n})$, i.e. is not infinitely differentiable in the closed halfspace $\overline{\mathbf{R}_+^n}$.

When κ is a nonnegative integer, u_+ can also fail to belong to $C^\infty(\overline{\mathbf{R}_+^n})$ because of the presence of logarithms in (9.36). By (9.18), if

$$b_{k1}(\omega) + (-1)^{k-1}b_{k2}(\omega) = 0, \qquad k = 1, 2, \ldots, \qquad (9.40)$$

then (9.2) has the form

$$A_+^{-1}(\xi', \xi_n) = \sum_{k=0}^{N} \tilde{C}_k(\xi')\Lambda_+^{-\kappa-k} + R_N(\xi', \xi_n), \tag{9.41}$$

where $\tilde{C}_0(\xi') = 1$ and $R_N(\xi', \xi_n)$ satisfies (9.3). In this case the asymptotic expansion of $u_+(x', x_n)$ does not contain logarithms and, when $\operatorname{Re}\kappa > -1$, has the form

$$u_+(x', x_n) = \sum_{k=1}^{N} d_k(x')\theta(x_n)x_n^{\kappa+k-1} + u_N^+(x', x_n), \tag{9.42}$$

where $d_k(x') \in C^\infty(\mathbf{R}^{n-1})$ and $u_N^+(x', x_n)$ satisfies the estimate of form (9.32). Thus $u_+(x', x_n) \in C^\infty(\overline{\mathbf{R}_+^n})$ if (9.40) is satisfied and κ is a nonnegative integer.

Suppose κ is arbitrary and (9.40) is not necessarily satisfied. We pose the problem of increasing the smoothness of the solution $u_+ \in H_s^+$ of (9.1) when $|s - \operatorname{Re}\kappa| < 1/2$. According to Theorem 7.3, for this purpose it is necessary that $f(x)$ satisfy a condition of the form

$$\int_{-\infty}^{\infty} \frac{\xi_n^{k-1}\widetilde{lf}(\xi', \xi_n)}{\hat{A}_-(\xi', \xi_n)}\, d\xi_n = 0, \qquad 1 \leq k \leq L. \tag{9.43}$$

A condition imposed on the right side $f(x', x_n)$ of (9.1) is said to be *local* if its fulfilment depends only on the values of $f(x', x_n)$ in an arbitrarily small neighborhood of the hyperplane $x_n = 0$, and to be *nonlocal* otherwise. Clearly, condition (9.43) is local only when $\hat{A}_-^{-1}(\xi', \xi_n)$ is a polynomial in ξ_n:

$$\hat{A}_-^{-1}(\xi', \xi_n) = \sum_{p=0}^{\kappa-\alpha-i\beta} a_p(\xi')\xi_n^p.$$

Then

$$\frac{1}{2\pi}\int_{-\infty}^{\infty} \frac{\xi_n^{k-1}\widetilde{lf}(\xi', \xi_n)}{\hat{A}_-(\xi', \xi_n)}\, d\xi_n = \sum_{p=0}^{\kappa-\alpha-i\beta} D_n^{p+k-1}a_p(\xi')F_{\xi'}lf(x)\big|_{x_n=0} = 0. \tag{9.44}$$

If $\hat{A}_-^{-1}(\xi', \xi_n)$ is not a polynomial in ξ_n, condition (9.43) is nonlocal, i.e. its fulfilment depends on the values of $f(x', x_n)$ for all $x \in \mathbf{R}_+^n$. Even e.g. if $f(x', x_n) \in C_0^\infty(\mathbf{R}_+^n)$, i.e. $f(x', x_n) = 0$ in a neighborhood of $x_n = 0$, condition (9.43) is generally not satisfied. We note that in Example 7.1 condition (9.43) is local when κ is a nonnegative integer and nonlocal otherwise.

If the L relations of condition (9.43) are satisfied, then $\tilde{f}_k(\xi') = 0$ for $1 \leq k \leq L$ in (9.23). In this case (9.36) takes the form

$$u_+(x', x_n) = \sum_{k=L+1}^{N} \sum_{p=0}^{k-L-1} d_{pk}(x')\theta(x_n)x_n^{\kappa+k-1}\ln^p x_n + u_N^+(x', x_n), \tag{9.45}$$

i.e.

$$u_+(x', x_n) = O(\theta(x_n)x_n^{\operatorname{Re}\kappa+L}),$$

for $\operatorname{Re}\kappa + L > 0$.

§10. Smooth pseudodifferential operators in a halfspace

1. As follows from Theorem 9.1, the solution $u_+ \in H_s^+$ of the p.e. (9.1) when $lf \in S(\mathbf{R}^n)$ and $|s - \operatorname{Re} \kappa| < 1/2$ is generally not infinitely differentiable in the closed halfspace $\overline{\mathbf{R}^n_+}$. For example, in the case of the p.e. $p\Lambda^{2\kappa}(D', D_n, \tau)u_+ = f$ when $\kappa > 0$ the solution $u_+(x) \in C^\infty(\overline{\mathbf{R}^n_+})$ if and only if κ is an integer. We distinguish below a class of p.e.'s whose solution belongs to $C^\infty(\overline{\mathbf{R}^n_+})$. A member of this class has the following property: if $u(x) \in S(\mathbf{R}^n)$ and $u_+ = \theta(x_n)u(x)$, then $p\hat{A}u_+ \in C^\infty(\overline{\mathbf{R}^n_+})$. Such a p.o. will be said to be *smooth* in the halfspace $\mathbf{R}^n_+$.

The class $D^{(0)}_{\alpha+i\beta}$. Suppose $A(\xi', \xi_n) \in O^\infty_{\alpha+i\beta}$, i.e. $A(\xi', \xi_n) \in C^\infty$ for $|\xi'| + |\xi_n| > 0$ and is homogeneous of degree $\alpha + i\beta$. We will write $A(\xi', \xi_n) \in D^{(0)}_{\alpha+i\beta}$, if

$$\frac{\partial^k A(0, -1)}{\partial \xi'^k} = (-1)^{|k|} e^{-(\alpha+i\beta)\pi i} \frac{\partial^k A(0, +1)}{\partial \xi'^k}, \qquad 0 \leqslant |k| < \infty. \quad (10.1)$$

It will be shown below that a p.o. with symbol of class $D^{(0)}_{\alpha+i\beta}$ is smooth in $\mathbf{R}^n_+$.

EXAMPLE 10.1. Let $P(\xi', \xi_n)$ be a homogeneous polynomial of degree p. Clearly,

$$P(0, -1) = (-1)^p P(0, +1), \quad \frac{\partial^k P(0, -1)}{\partial \xi'^k} = (-1)^{p-|k|} \frac{\partial^k P(0, +1)}{\partial \xi'^k}, \quad \text{i.e. } P(\xi', \xi_n) \in D^{(0)}_p.$$

It is verified analogously that a rational symbol $P(\xi', \xi_n)/Q(\xi', \xi_n)$, where $Q(\xi', \xi_n)$ is a homogeneous elliptic polynomial of degree q, belongs to the class $D^{(0)}_{p-q}$.

The class $D^{(0)}_{\alpha+i\beta}$ is somewhat restrictive for the applications because of the requirement of infinite differentiability of a symbol for $|\xi'| + |\xi_n| > 0$. Even the factorization factors of an elliptic symbol $A(\xi', \xi_n) \in O^\infty_{\alpha+i\beta}$ need not belong to C^∞ for $|\xi'| + |\xi_n| > 0$. For example, the factors $\xi_n \pm i|\xi'|$ of the symbol $A(\xi', \xi_n) = |\xi|^2$ are not infinitely differentiable at the point $(\xi', \xi_n) = (0, +1)$.

Suppose $A(\xi', \xi_n) \in C^\infty$ for $\xi' \neq 0$ and is homogeneous of degree $\alpha + i\beta$. Let $r = |\xi'|$, $\omega = \xi'/|\xi'|$ and let $\mathcal{A}(\omega, r, \xi_n)$ denote $A(\xi', \xi_n)$ as a function of the coordinates (ω, r, ξ_n). Suppose $\mathcal{A}(\omega, r, \xi_n) \in C^\infty$ with respect to the totality of variables (ω, r, ξ_n) for $|\omega| = 1$, $0 \leqslant r < \infty$, $-\infty < \xi_n < +\infty$, $|\xi_n| + r > 0$. We will say that $A(\xi', \xi_n)$ is of class $D_{\alpha+i\beta}$ if

$$A(0, -1) = e^{-(\alpha+i\beta)\pi i} A(0, +1), \quad (10.2)$$

$$\frac{\partial^k \mathcal{A}(\omega, 0, -1)}{\partial r^k} = (-1)^k e^{-(\alpha+i\beta)\pi i} \frac{\partial^k \mathcal{A}(\omega, 0, +1)}{\partial r^k}, \qquad 1 \leqslant k < \infty. \quad (10.2')$$

Clearly, $\Lambda_-^{\alpha+i\beta}(\xi', \xi_n) = (\xi_n - i|\xi'|)^{\alpha+i\beta} \in D_{\alpha+i\beta}$ for arbitrary $\alpha + i\beta$, and $\Lambda_+^m(\xi', \xi_n) = (\xi_n + i|\xi'|)^m \in D_m$ for an arbitrary integer m. We note that $D_{\alpha+i\beta}^{(0)} \subset D_{\alpha+i\beta}$. In fact, if $A(\xi', \xi_n) \in D_{\alpha+i\beta}^{(0)}$, then

$$\frac{\partial^k A(\xi', \xi_n)}{\partial r^k} = \left(\sum_{i=1}^{n-1} \omega_i \frac{\partial}{\partial \xi_i} \right)^k A(\xi', \xi_n),$$

where the $\omega_i = \xi_i / r$ are fixed. Consequently, (10.2) and (10.2′) follow from (10.1).

LEMMA 10.1. *For any positive integer N a symbol $A(\xi', \xi_n) \in D_{\alpha+i\beta}$ admits the decomposition*

$$A(\xi', \xi_n) = A_N^-(\xi', \xi_n) + R_N(\xi', \xi_n), \tag{10.3}$$

where $A_N^-(\xi', \xi_n)$ has an analytic continuation $A_N^-(\xi', \xi_n + i\tau)$ with respect to $\xi_n + i\tau$ into the halfplane $\tau < 0$ that is continuous for $|\xi'| + |\xi_n| + |\tau| > 0$ and satisfies

$$|A_N^-(\xi', \xi_n + i\tau)| \leqslant C_N(|\xi'| + |\xi_n| + |\tau|)^\alpha,$$

while $R_N(\xi', \xi_n)$ is a homogeneous function of degree $\alpha + i\beta$ that satisfies

$$|R_N(\xi', \xi_n)| \leqslant C_N |\xi'|^{N+1} (|\xi'| + |\xi_n|)^{\alpha - N - 1}. \tag{10.4}$$

PROOF. Let $\mathcal{A}(\omega, r, \xi_n) = A(\xi', \xi_n)$, where $\omega = \xi'/|\xi'|$ and $r = |\xi'|$. Expanding $\mathcal{A}(\omega, r, \xi_n)$ with the use of Taylor's theorem in a neighborhood of the point $(\omega, r, \xi_n) = (\omega, 0, +1)$, we get for $\xi_n > 0$

$$\mathcal{A}(\omega, r, \xi_n) = \xi_n^{\alpha+i\beta} \mathcal{A}\left(\omega, \frac{r}{\xi_n}, +1\right)$$
$$= \sum_{k=0}^{N} \xi_n^{\alpha+i\beta} \frac{1}{k!} \frac{\partial^k \mathcal{A}(\omega, 0, +1)}{\partial r^k} \left(\frac{r}{\xi_n}\right)^k + R_{N1}(\omega, r, \xi_n), \tag{10.5}$$

where

$$|R_{N1}(\omega, r, \xi_n)| \leqslant \frac{C_N r^{N+1}}{|\xi_n|^{N+1-\alpha}} \tag{10.6}$$

for $0 \leqslant r \leqslant \xi_n$ by virtue of Lagrange's formula for the remainder in Taylor's theorem. Analogously, expanding $\mathcal{A}(\omega, r, \xi_n)$ with the use of Taylor's theorem in a neighborhood of the point $(\omega, r, \xi_n) = (\omega, 0, -1)$, we get for $\xi_n < 0$

$$\mathscr{A}(\omega, r, \xi_n) = |\xi_n|^{\alpha+i\beta}\mathscr{A}\left(\omega, \frac{r}{|\xi_n|}, -1\right)$$

$$= \sum_{k=0}^{N} |\xi_n|^{\alpha+i\beta}\frac{1}{k!}\,\frac{\partial^k \mathscr{A}(\omega, 0, -1)}{\partial r^k}\,\frac{r^k}{|\xi_n|^k} + R_{N2}(\omega, r, \xi_n), \qquad (10.7)$$

where $R_{N2}(\omega, r, \xi_n)$ satisfies (10.6) for $0 \leqslant r \leqslant |\xi_n|$, $\xi_n < 0$.

As in §1, let

$$(\xi_n - i0)^{\alpha+i\beta-k} = \lim_{\tau\to -0}(\xi_n + i\tau)^{\alpha+i\beta-k}.$$

Then

$$(\xi_n - i0)^{\alpha+i\beta-k} = \xi_n^{\alpha+i\beta-k}$$

for $\xi_n > 0$ and

$$(\xi_n - i0)^{\alpha+i\beta-k} = e^{(\alpha+i\beta-k)\ln(\xi_n - i0)} = e^{-(\alpha+i\beta-k)\pi i}|\xi_n|^{\alpha+i\beta-k}$$

for $\xi_n < 0$. If (10.2) and (10.2′) are satisfied, then (10.5) and (10.7) can be consolidated:

$$A(\xi', \xi_n) = \sum_{k=0}^{N} a_k(\omega) r^k (\xi_n - i0)^{\alpha+i\beta-k} + R_{N3}(\omega, r, \xi_n), \qquad (10.8)$$

where $a_0 = A(0, +1)$,

$$a_k(\omega) = \frac{1}{k!}\,\frac{\partial^k A(\omega, 0, +1)}{\partial r^k}$$

for $1 \leqslant k \leqslant N$ and $R_{N3}(\omega, r, \xi_n)$ satisfies (10.6) for $0 \leqslant r < |\xi_n|$.

From the binomial theorem we have for $r/|\xi_n| < 1$

$$(\xi_n - i0)^{\alpha+i\beta-k} = (\xi_n - ir)^{\alpha+i\beta-k}\left(1 + \frac{ir}{\xi_n - ir}\right)^{\alpha+i\beta-k}$$

$$= (\xi_n - ir)^{\alpha+i\beta-k}\sum_{p=0}^{N-k} c_p \frac{r^p}{(\xi_n - ir)^p} + R_{N4}(r, \xi_n), \qquad (10.9)$$

where the c_p are certain numbers and $R_{N4}(r, \xi_n)$ satisfies (10.6) for $0 \leqslant r < |\xi_n|$. Therefore, substituting (10.9) in (10.8) and collecting the coefficients of like powers of $\xi_n - ir$, we get

$$A(\xi', \xi_n) = \sum_{k=0}^{N} b_k(\omega) r^k (\xi_n - ir)^{\alpha+i\beta-k} + R_{N5}(\omega, r, \xi_n), \qquad (10.10)$$

where $b_0 = A(0, +1)$, $b_k(\omega) \in C^\infty$ for $1 \leqslant k \leqslant N$ and

$$|R_{N5}(\omega, r, \xi_n)| \leqslant C_N r^{N+1}|\xi_n|^{\alpha-N-1}, \qquad 0 \leqslant r < |\xi_n|. \qquad (10.11)$$

We now note that the function

$$A_N^-(\xi', \xi_n) = \sum_{k=0}^{N} b_k(\omega) r^k (\xi_n - ir)^{\alpha + i\beta - k} \tag{10.12}$$

in (10.10) satisfies all of the requirements of Lemma 10.1.

Let us show that $R_{N5}(\omega, r, \xi_n)$ satisfies (10.4) for $r + |\xi_n| > 0$. For $0 \leqslant r < |\xi_n|$, (10.4) follows from (10.11), since

$$|R_{N5}(\omega, r, \xi_n)| \leqslant C_N r^{N+1} |\xi_n|^{\alpha - N - 1} \leqslant 2^{|\alpha - N - 1|} C_N r^{N+1} (|\xi_n| + r)^{\alpha - N - 1}.$$

For $r \geqslant |\xi_n|$, inequality (10.4) follows from the estimate

$$|R_{N5}| \leqslant |A(\xi', \xi_n)| + |A_N^-(\xi', \xi_n)| \leqslant C(|\xi_n| + r)^{\alpha} \leqslant 2^{N+1} C r^{N+1} (|\xi_n| + r)^{\alpha - N - 1}.$$

Lemma 10.1 is proved.

REMARK 10.1. Suppose $\mathcal{Q}(\omega, r, \xi_n)$ is a homogeneous function of degree $\alpha + i\beta$ that belongs to C^∞ with respect to the totality of variables (ω, r, ξ_n) for $r + |\xi_n| > 0$, and suppose (10.10) holds for any positive integer N. Then $A(\xi', \xi_n) \in D_{\alpha + i\beta}$. In fact, since N is an arbitrary positive integer, it follows from (10.10) that

$$A(0, -1) = e^{-(\alpha + i\beta)\pi i} A(0, +1)$$

and the derivatives of $\mathcal{Q}(\omega, r, \xi_n)$ with respect to r for $r = 0$, $\xi_n = \pm 1$ satisfy (10.2′).

REMARK 10.2. Suppose $A(\xi', \xi_n) \in D_\alpha$, where α is an integer. Then

$$A(\xi', \xi_n) = \sum_{k=0}^{N} c_k(\omega) r^k (\xi_n + ir)^{\alpha - k} + R_N(\omega, r, \xi_n), \tag{10.13}$$

where $c_0 = A(0, +1)$, $c_k(\omega) \in C^\infty$ for $1 \leqslant k \leqslant N$ and

$$|R_N(\omega, r, \xi_n)| \leqslant \frac{c_N r^{N+1}}{(|\xi_n| + r)^{N+1-\alpha}}.$$

For suppose α is an integer. Then

$$e^{-(\alpha - k)\pi i} = e^{(\alpha - k)\pi i} \quad \text{and} \quad (\xi_n - i0)^{\alpha - k} = (\xi_n + i0)^{\alpha - k} = \xi_n^{\alpha - k},$$

so that (10.8) can be written in the form

$$A(\xi', \xi_n) = \sum_{k=0}^{N} a_k(\omega) r^k \xi_n^{\alpha - k} + R_{N3}(\omega, r, \xi_n).$$

Replacing now $\xi_n - ir$ by $\xi_n + ir$ and repeating the proof of Lemma 10.1, we obtain (10.13).

We note that if, in particular, α is a nonnegative integer, then (10.13) implies

$$A(\xi', \xi_n) = P(\xi', \xi_n) + R(\xi', \xi_n), \tag{10.14}$$

where $P(\xi', \xi_n)$ is a polynomial in ξ_n of degree α, while $R(\xi', \xi_n) \in D_\alpha$ and

$$|R(\xi', \xi_n)| \leqslant \frac{C r^{\alpha + 1}}{|\xi_n| + r}.$$

LEMMA 10.2. *Suppose* $A(\xi', \xi_n) \in D_{\alpha_1+i\beta_1}$ *and* $B(\xi', \xi_n) \in D_{\alpha_2+i\beta_2}$. *Then* $A(\xi', \xi_n)B(\xi', \xi_n) \in D_{\alpha_3+i\beta_3}$, *where* $\alpha_3 + i\beta_3 = \alpha_1 + \alpha_2 + i(\beta_1 + \beta_2)$.

In fact, $A(\xi', \xi_n)B(\xi', \xi_n)$ satisfies (10.2) and (10.2′) by virtue of Leibniz's formula for differentiating a product.

LEMMA 10.3. *Suppose* $A(\xi', \xi_n) \in O^{\infty}_{\alpha+i\beta}$ *is an elliptic symbol with factorization*

$$A(\xi', \xi_n) = A_-(\xi', \xi_n)A_+(\xi', \xi_n)$$

and factorization index κ. *If* $A(\xi', \xi_n) \in D^{(0)}_{\alpha+i\beta}$, *then*

a) κ *is an integer, and*

b) $A_+(\xi', \xi_n) \in D_\kappa$ *and* $A_-(\xi', \xi_n) \in D_{\alpha+i\beta-\kappa}$.

PROOF. Suppose $A(\xi', \xi_n) \in D^{(0)}_{\alpha+i\beta}$ is an elliptic symbol. By Lemma 10.2,

$$A_0(\xi', \xi_n) = (\xi_n - i|\xi'|)^{-\alpha-i\beta}A(\xi', \xi_n) \in D_0.$$

Since $(\xi_n - i|\xi'|)^{\alpha+i\beta}$ is a "minus" function, the factorization index of $A(\xi', \xi_n)$ is equal to the factorization index of $A_0(\xi', \xi_n)$. It follows from (10.2) that $A_0(0, -1) = A_0(0, +1)$. Consequently, the quantity

$$\frac{1}{2\pi}\Delta \arg A_0(\xi', \xi_n)\Big|_{\xi_n=+\infty}^{\xi_n=-\infty} = m$$

is an integer for $\xi' \neq 0$. But $\kappa = m$ by virtue of (6.51), so that κ is an integer.

Let

$$A_1(\xi', \xi_n) = \frac{(\xi_n - i|\xi'|)^m A_0(\xi', \xi_n)}{(\xi_n + i|\xi'|)^m A_0(0, +1)} \qquad \text{and} \qquad A_2(\xi', \xi_n) = \ln A_1(\xi', \xi_n).$$

Then, clearly, $A_2(\xi', \xi_n) \in D_0$ and $A_2(0, \pm 1) = 0$. The factorization of $A(\xi', \xi_n)$ has the form

$$A(\xi', \xi_n) = A_-(\xi', \xi_n)A_+(\xi', \xi_n),$$

where

$$A_-(\xi', \xi_n) = A(0, +1)(\xi_n - i|\xi'|)^{\alpha+i\beta-m}e^{\Pi^- A_2},$$

$$A_+(\xi', \xi_n) = (\xi_n + i|\xi'|)^m e^{\Pi^+ A_2}.$$

Consequently, Lemma 10.3 will follow if we show that $\Pi^{\pm}A_2 \in D_0$.

By Lemma 10.1,

$$A_2(\xi', \xi_n) = \sum_{k=1}^{N} \frac{a_k(\xi')}{(\xi_n - i|\xi'|)^k} + R_N(\omega, r, \xi_n), \tag{10.15}$$

where $a_k(\xi')$ is a homogeneous function of degree k that belongs to C^∞ for $\xi' \neq 0$ and

$$|R_N(\omega, r, \xi_n)| \leqslant \frac{c_N r^{N+1}}{(r + |\xi_n|)^{N+1}}. \tag{10.16}$$

Since

$$\Pi^+ \sum_{k=1}^{N} \frac{a_k(\xi')}{(\xi_n - i|\xi'|)^k} = 0,$$

we have $\Pi^+ A_2 = \Pi^+ R_N$. Applying the expansion formula for an integral of Cauchy type, we get

$$\Pi^+ A_2(\xi', \xi_n) = \sum_{k=1}^{N} \frac{a_{k1}(\xi')}{(\xi_n - i|\xi'|)^k} + R_{N1}(\xi', \xi_n), \tag{10.17}$$

where $a_{k1}(\xi') = i\Pi' \Lambda_-^{k-1} R_N$ and $R_{N1} = \Lambda_-^{-N} \Pi^+ \Lambda_-^N R_N$.

Let us show that

$$|\Lambda_-^N R_N| \leqslant \frac{c_N r^{N+1}}{r + |\xi_n|}, \qquad \left|\frac{\partial}{\partial \xi_n}(\Lambda_-^N R_N)\right| \leqslant \frac{c_N r^{N+1}}{(r + |\xi_n|)^2}. \tag{10.18}$$

The function $\partial R_N / \partial \xi_n \in C^\infty$ with respect to the totality of variables (ω, r, ξ_n) and is homogeneous of degree -1. By (10.16),

$$\frac{\partial^{k+1} R_N(\omega, 0, \pm 1)}{\partial r^k \partial \xi_n} = 0$$

for $0 \leqslant k \leqslant N$. Consequently, expanding $\partial R_N(\omega, r, \xi_n)/\partial \xi_n$ by Taylor's theorem in a neighborhood of the points $(\omega, 0, \pm 1)$, we get

$$\left|\frac{\partial R_N}{\partial \xi_n}\right| \leqslant \frac{c_N r^{N+1}}{|\xi_n|^{N+2}} \leqslant \frac{2^{N+2} c_N r^{N+1}}{(r + |\xi_n|)^{N+2}}$$

for $r \leqslant |\xi_n|$. If $r \geqslant |\xi_n|$, then

$$\left|\frac{\partial R_N}{\partial \xi_n}\right| \leqslant \frac{c_N}{r + |\xi_n|} \leqslant \frac{2^{N+1} c_N r^{N+1}}{(|\xi_n| + r)^{N+2}}.$$

Thus (10.18) is proved.

It follows from Lemma 6.1 that $R_{N1} = \Lambda_-^{-N} \Pi^+ \Lambda_-^N R_N$ satisfies the estimate

$$|R_{N1}(\xi', \xi_n)| \leqslant \frac{c_N r^{N+1-\varepsilon}}{(r + |\xi_n|)^{N+1-\varepsilon}}, \qquad \forall \varepsilon > 0. \tag{10.19}$$

It is proved analogously that

$$\left|D_\omega^p D_r^q D_{\xi_n}^l R_N\right| \leq C_{pql} r^{N+1-q}(r+|\xi_n|)^{-l-N-1}$$

for $0 \leq q \leq N$, $0 \leq |p| + l < \infty$. As a result, Lemma 6.1 implies that R_{N1} can be continuously differentiated N times with respect to (ω, r, ξ_n) for $r + |\xi_n| > 0$. Since N is arbitrary, it follows from (10.17) and (10.19) that $\Pi^+ A_2 \in C^\infty$ with respect to (ω, r, ξ_n) for $r + |\xi_n| > 0$, and hence that $\Pi^+ A_2 \in D_0$ (see Remark 10.1). Clearly, $\Pi^- A_2 = A_2 - \Pi^+ A_2$ also belongs to D_0. Lemma 10.3 is proved.

2. THEOREM 10.1. *Suppose $A(\xi', \xi_n) \in D_{\alpha+i\beta}$ and $u(x) \in H_r(\mathbf{R}^n_+)$, where r is any nonnegative number. Let $u_+ = u$ for $x_n > 0$ and $u_+ = 0$ for $x_n < 0$. Then $p\hat{A}u_+$ is a bounded operator from $H_r(\mathbf{R}^n_+)$ into $H_{r-\alpha}(\mathbf{R}^n_+)$, i.e.*

$$\|p\hat{A}u_+\|^+_{r-\alpha} \leq c\|u\|^+_r. \tag{10.20}$$

PROOF. Let $lu \in H_r(\mathbf{R}^n)$ be an arbitrary extension of $u(x)$ from $\mathbf{R}^n_+$ onto $\mathbf{R}^n$ such that

$$\|lu\|_r \leq 2\|u\|^+_r. \tag{10.21}$$

Then $u_+(x) = \theta(x_n)lu$. Let $u_- = (1-\theta(x_n))lu = lu - u_+$. By Lemma 10.1, for any positive integer N

$$\hat{A}(\xi', \xi_n) = \hat{A}^-_N(\xi', \xi_n) + \hat{R}_N(\xi', \xi_n), \tag{10.22}$$

where

$$|\hat{A}^-_N(\xi', \xi_n + i\tau)| \leq c_N(1+|\xi'|+|\xi_n|+|\tau|)^\alpha, \qquad \tau \leq 0,$$

$$|\hat{R}_N(\xi', \xi_n)| \leq \frac{c_N(1+|\xi'|)^N}{(1+|\xi'|+|\xi_n|)^{N-\alpha}}.$$

Consequently,

$$p\hat{A}u_+ = p\hat{A}^-_N u_+ + p\hat{R}_N u_+. \tag{10.23}$$

Since $\hat{A}^-_N(\xi', \xi_n)$ is a "minus" function,

$$p\hat{A}^-_N u_+ = p\hat{A}^-_N lu. \tag{10.24}$$

In fact, $p\hat{A}^-_N u_+ = p\hat{A}^-_N lu - p\hat{A}^-_N u_-$. Since $u_- \in H^-_0$, it follows by virtue of Theorem 4.4 that $\hat{A}^-_N u_- \in H^-_{-\alpha}$ and hence that $p\hat{A}^-_N u_- = 0$. Therefore, by (10.21) and Lemma 4.4,

$$\|pA^-_N lu\|^+_{r-\alpha} \leq \|A^-_N lu\|_{r-\alpha} \leq c\|lu\|_r \leq 2c\|u\|^+_r. \tag{10.25}$$

Choosing $N \geqslant r$ in (10.22), we get

$$\left(\|p\hat{R}_N u_+\|^+_{r-\alpha}\right)^2 \leqslant \|\hat{R}_N u_+\|^2_{r-\alpha}$$
$$\leqslant c\int_{-\infty}^{\infty} (1+|\xi'|+|\xi_n|)^{2(r-\alpha)} \frac{(1+|\xi'|)^{2N}|\tilde{u}_+(\xi',\xi_n)|^2}{(1+|\xi'|+|\xi_n|)^{2(N-\alpha)}}\, d\xi'\, d\xi_n$$
$$\leqslant c\int_{-\infty}^{\infty} (1+|\xi'|)^{2r}|\tilde{u}_+(\xi',\xi_n)|^2\, d\xi'\, d\xi_n. \tag{10.26}$$

Since $lu(x) = u_+(x) + u_-(x)$, $u_+(x) = \theta(x_n)lu$, $u_- = (1-\theta(x_n))lu$, it follows by virtue of Parseval's equality that

$$\int_{-\infty}^{\infty} |\tilde{u}_+(\xi',\xi_n)|^2\, d\xi_n + \int_{-\infty}^{\infty} |\tilde{u}_-(\xi',\xi_n)|^2\, d\xi_n = \int_{-\infty}^{\infty} |\widetilde{lu}(\xi',\xi_n)|^2\, d\xi_n \tag{10.27}$$

for almost all ξ'. Multiplying (10.27) by $(1+|\xi'|)^{2r}$ and integrating with respect to ξ', we get

$$\int_{-\infty}^{\infty} (1+|\xi'|)^{2r}|\tilde{u}_+(\xi',\xi_n)|^2\, d\xi'\, d\xi_n \leqslant \int_{-\infty}^{\infty} (1+|\xi'|)^{2r}|\widetilde{lu}|^2\, d\xi'\, d\xi_n$$
$$\leqslant \int_{-\infty}^{\infty} (1+|\xi'|+|\xi_n|)^{2r}|\widetilde{lu}|^2\, d\xi'\, d\xi_n. \tag{10.28}$$

Therefore, (10.26), (10.28) and (10.21) imply

$$\|p\hat{R}_N u_+\|^+_{r-\alpha} \leqslant c\|u\|^+_r. \tag{10.29}$$

Theorem 10.1 is proved.

REMARK 10.3. If $lu \in S(\mathbf{R}^n)$ and $A(\xi',\xi_n) \in \hat{D}_{\alpha+i\beta}$, then Theorem 10.1 implies $p\hat{A}u_+ \in H_{r-\alpha}(\mathbf{R}^n_+)$, $\forall r > 0$, and hence $p\hat{A}u_+ \in C^\infty(\overline{\mathbf{R}^n_+})$. If, on the other hand, (10.2) and (10.2′) are not satisfied for at least one k, then, in general, $p\hat{A}u_+ \notin C^\infty(\overline{\mathbf{R}^n_+})$.

For suppose (10.2) and (10.2′) are satisfied for $k < m$ but

$$\frac{\partial^m \mathcal{A}(\omega, 0, -1)}{\partial r^m} \not\equiv e^{-(\alpha+i\beta-m)\pi i}\frac{\partial^m \mathcal{A}(\omega, 0, +1)}{\partial r^m} \tag{10.30}$$

and suppose first $\alpha + i\beta$ is not an integer. We choose $c_{k1}(\omega)$ and $c_{k2}(\omega)$ so that

$$c_{k1}(\omega)(\xi_n + i0)^{\alpha+i\beta-k} + c_{k2}(\omega)(\xi_n - i0)^{\alpha+i\beta-k}$$
$$= \begin{cases} a_{k1}(\omega)\xi_n^{\alpha+i\beta-k} & \text{for } \xi_n > 0, \\ a_{k2}(\omega)|\xi_n|^{\alpha+i\beta-k} & \text{for } \xi_n < 0. \end{cases} \tag{10.31}$$

where

$$a_{k1}(\omega) = \frac{1}{k!}\,\frac{\partial^k \mathcal{A}(\omega, 0, +1)}{\partial r^k}, \quad a_{k2}(\omega) = \frac{1}{k!}\,\frac{\partial^k \mathcal{A}(\omega, 0, -1)}{\partial r^k}.$$

Then c_{k1} and c_{k2} satisfy the system of equations

$$c_{k1} + c_{k2} = a_{k1}, \quad e^{(\alpha + i\beta - k)\pi i}c_{k1} + e^{-(\alpha + i\beta - k)\pi i}c_{k2} = a_{k2}.$$

Consequently,

$$c_{k1}(\omega) = (-1)^k\left(e^{-(\alpha+i\beta)\pi i} - e^{(\alpha+i\beta)\pi i}\right)^{-1}\left(e^{-(\alpha+i\beta-k)\pi i}a_{k1}(\omega) - a_{k2}(\omega)\right),$$

$$c_{k2}(\omega) = (-1)^k\left(e^{(\alpha+i\beta)\pi i} - e^{-(\alpha+i\beta)\pi i}\right)^{-1}\left(e^{(\alpha+i\beta-k)\pi i}a_{k1}(\omega) - a_{k2}(\omega)\right).$$

We note that $c_{k1}(\omega) \equiv 0$ for $k < m$ and $c_{m1}(\omega) \not\equiv 0$ by virtue of (10.30). Combining (10.5) and (10.7), we get

$$\mathcal{A}(\omega, r, \xi_n) = \sum_{k=m}^{N} c_{k1}(\omega)r^k(\xi_n + i0)^{\alpha+i\beta-k}$$

$$+ \sum_{k=0}^{N} c_{k2}(\omega)r^k(\xi_n - i0)^{\alpha+i\beta-k} + R_N(\omega, r, \xi_n), \tag{10.32}$$

where $R_N(\omega, r, \xi_n)$ satisfies (10.6) for $0 < r < |\xi_n|$ and N is sufficiently large ($N > \alpha + 1$). Analogously to (10.10), we get from (10.32) that

$$\mathcal{A}(\omega, r, \xi_n) = A_N^-(\omega, r, \xi_n) + \sum_{k=m}^{N} b_{k1}(\omega)r^k(\xi_n + ir)^{\alpha+i\beta-k} + R_{N1}(\omega, r, \xi_n), \tag{10.33}$$

where $A_N^-(\omega, r, \xi_n)$ has the form (10.12), $R_{N1}(\omega, r, \xi_n)$ satisfies the estimate of form (10.4) and $b_{m1}(\omega) = c_{m1}(\omega) \not\equiv 0$.

It follows from Theorem 10.1 that $p\hat{A}_N^- u_+ \in C^\infty(\overline{\mathbf{R}^n_+})$.

We apply the expansion formula for an integral of Cauchy type to $\tilde{u}_+ = \Pi^+\widetilde{lu}$:

$$\tilde{u}_+ = \Pi^+\widetilde{lu} = \sum_{k=1}^{N} \hat{\Lambda}_+^{-k}i\Pi'\hat{\Lambda}_+^{k-1}\widetilde{lu} + \hat{\Lambda}_+^{-N}\Pi^+\hat{\Lambda}_+^N\widetilde{lu}.$$

Suppose $u_+(x)$ is such that $b_{m1}(\omega)\Pi'\widetilde{lu} \not\equiv 0$. As in the proof of Theorem 9.1, we get

$$F_\xi^{-1}\Bigg[\left(\sum_{k=m}^{N} b_{k1}(\omega)(r+1)^k(\xi_n + ir + i)^{\alpha+i\beta-k} + R_{N1}(\omega, r+1, \xi_n)\right)$$

$$\times\left(\sum_{k=1}^{N} \hat{\Lambda}_+^{-k}i\Pi'\hat{\Lambda}_+^{k-1}\widetilde{lu} + \hat{\Lambda}_+^{-N}\Pi^+\hat{\Lambda}_+^N\widetilde{lu}\right)\Bigg]$$

$$= \sum_{k=m}^{N} d_k(x')x_{n,+}^{-\alpha-i\beta+k-1} + u_N(x', x_n), \tag{10.34}$$

where $d_k(x') \in C^\infty(\mathbf{R}^{n-1})$, $d_m(x') \not\equiv 0$, $x_{n,+}^{-\alpha-i\beta+k-1} = \theta(x_n)x_n^{-\alpha-i\beta+k-1}$ for $k > \alpha$, $x_{n,+}^{-\alpha-i\beta+k-1}$ is defined by (1.23) for $k < \alpha$, and $u_N(x', x_n)$ can be continuously differentiated $N - \alpha - 1$ times. Since $\alpha + i\beta$ is not an integer and $d_m(x') \not\equiv 0$, it follows from (10.34) that $p\hat{A}u_+ \notin C^\infty(\overline{\mathbf{R}^n_+})$.

In the case when α is an integer and $\beta = 0$, we choose $c_{k1}(\omega)$ and $c_{k2}(\omega)$ so that

$$a_{k1}\xi_n^{\alpha-k} + c_{k1}\xi_n^{\alpha-k}\ln(\xi_n + i0) + c_{k2}\xi_n^{\alpha-k}\ln(\xi_n - i0)$$

$$= \begin{cases} a_{k1}\xi_n^{\alpha-k} & \text{for } \xi_n > 0, \\ a_{k2}|\xi_n|^{\alpha-k} & \text{for } \xi_n < 0. \end{cases} \tag{10.35}$$

Then c_{k1} and c_{k2} satisfy the system of equations

$$c_{k1} + c_{k2} = 0, \qquad i\pi c_{k1} - i\pi c_{k2} = (-1)^{\alpha-k} a_{k2} - a_{k1}, \qquad (10.36)$$

and hence are uniquely determined. We note that neither $c_{k1}(\omega)$ nor $c_{k2}(\omega)$ identically vanishes if (10.2) or (10.2′) is not satisfied. The rest of the proof is not different from the case of nonintegral $\alpha + i\beta$.

3. Suppose $A(\xi', \xi_n) \in D_{\alpha+i\beta}$ and consider the operator of potential type $p\hat{A}(v(x') \times \delta(x_n))$. The estimate

$$\|p\hat{A}(v(x') \times \delta(x_n))\|^{+}_{r-\alpha} \leq c[v]_{r+1/2}$$

holds for any r by virtue of Lemmas 10.1 and 8.1.

Suppose $f(x', x_n) \in H_r(\mathbf{R}^n_+)$ and $lf \in H_r(\mathbf{R}^n)$ is an arbitrary extension of f from $\mathbf{R}^n_+$ onto $\mathbf{R}^n$. If $r > 1/2$, then lf has, by Theorem 4.2, a restriction $p'lf \in H_{r-1/2}(\mathbf{R}^{n-1})$ to the hyperplane $x_n = 0$. We note that if $l_1 f \in H_r(\mathbf{R}^n)$ is another extension of f, then $lf - l_1 f \in H_r^-(\mathbf{R}^n)$ and hence $p'(lf - l_1 f) = 0$. Thus $p'lf$ does not depend on the choice of the extension of f. We will denote $p'lf$ below by $p'_+ f$. The restriction $p'_+ f$ can also be defined as follows: restrict f to the hyperplane $x_n = \varepsilon > 0$ and then let ε tend to zero.

THEOREM 10.2. *Suppose $A(\xi', \xi_n) \in D_{\alpha+i\beta}$ and $v(x') \in S(\mathbf{R}^{n-1})$. Then the restriction $p'_+\hat{A}(D', D_n)(v(x') \times \delta(x_n))$ exists, and*

$$p'_+\hat{A}(D', D_n)(v(x') \times \delta(x_n)) = \hat{a}(D')v(x'), \qquad (10.37)$$

where $\hat{a}(D')$ is the p.o. in $\mathbf{R}^{n-1}$ with symbol

$$\hat{a}(\xi') = \frac{1}{2\pi}\int_{-\infty}^{\infty} \hat{R}_N(\xi', \xi_n)\,d\xi_n, \qquad N > \alpha, \qquad (10.38)$$

in which $\hat{R}_N(\xi', \xi_n)$ is the same as in (10.3).

PROOF. We take the decomposition (10.3) with $N > \alpha$. Then

$$p\hat{A}(v(x') \times \delta(x_n)) = p\hat{A}_N^-(v(x') \times \delta(x_n)) + p\hat{R}_N(v(x') \times \delta(x_n)).$$

As in the proof of Lemma 8.1, we get $p\hat{A}_N^-(v(x') \times \delta(x_n)) = 0$. Since

$$|\hat{R}_N(\xi', \xi_n)| \leq \frac{c(1 + |\xi'|)^{N+1}}{(1 + |\xi'| + |\xi_n|)^{N+1-\alpha}},$$

with $N - \alpha = 2\delta > 0$, we have $\hat{R}_N(\xi', \xi_n) \times \tilde{v}(\xi') \in \tilde{H}_{1/2+\delta}$. Consequently, Theorem 4.2 implies the existence of the restriction

$$p'_+\hat{R}_N(v(x') \times \delta(x_n)) = p'\hat{R}_N(v(x') \times \delta(x_n)).$$

Taking the Fourier transform with respect to x', we get, by (4.32′), that

$$F_{x'}p'\hat{R}_N(v(x') \times \delta(x_n)) = \frac{1}{2\pi}\int_{-\infty}^{\infty} \hat{R}_N(\xi', \xi_n)\tilde{v}(\xi')\,d\xi_n. \qquad (10.39)$$

We note that the integral here converges since $N > \alpha$. Let

$$\hat{a}(\xi') = \frac{1}{2\pi}\int_{-\infty}^{\infty} \hat{R}_N(\xi', \xi_n)d\xi_n.$$

As in the proof of Theorem 8.2, it can be shown that $\hat{a}(\xi')$ does not depend on the choice of the decomposition (10.3) with $N > \alpha$. Thus

$$p'_+\hat{A}(v(x') \times \delta(x_n)) = p'\hat{R}_N(v(x') \times \delta(x_n)) = F_{\xi'}^{-1}\hat{a}(\xi')\tilde{v}(\xi'),$$

which proves Theorem 10.2.

Since $S(\mathbf{R}^{n-1})$ is dense in $H_r(\mathbf{R}^{n-1})$ for any r, we get, by completing $S(\mathbf{R}^{n-1})$ in the norm of $H_r(\mathbf{R}^{n-1})$, that (10.37) holds for any $v(x') \in H_r(\mathbf{R}^{n-1})$.

REMARK 10.4. In order to note a property of the symbols of class D_0 we make the change of variables

$$z = \frac{\xi_n + ir}{\xi_n - ir}, \qquad r \neq 0. \tag{10.40}$$

When ξ_n ranges over the real axis, $z = e^{i\varphi}$ ranges over the unit circle c_1 with deleted point $z = 1$: $0 < \varphi = \arg z < 2\pi$. By (10.40),

$$\xi_n = -ir\frac{1+z}{1-z} = -ir\frac{1+e^{i\varphi}}{1-e^{i\varphi}} = r \operatorname{ctg}\frac{\varphi}{2}.$$

Consequently,

$$\mathcal{A}(\omega, r, \xi_n) = \mathcal{A}\left(\omega, r, -ir\frac{1+z}{1-z}\right) = \mathcal{A}\left(\omega, 1, \operatorname{ctg}\frac{\varphi}{2}\right). \tag{10.41}$$

Since $A(\xi', \xi_n) \in C^\infty$ for $\xi' \neq 0$, we have $\mathcal{A}(\omega, 1, \operatorname{ctg}(\varphi/2)) \in C^\infty$ for $0 < \varphi < 2\pi$. We will show that (10.2) and (10.2′) imply that $\mathcal{A}(\omega, 1, \operatorname{ctg}(\varphi/2)) \in C^\infty$ on the circle c_1.

Let us study the behavior of $\mathcal{A}(\omega, 1, \operatorname{ctg}(\varphi/2))$ for $\varphi \to +0$ and $\varphi \to 2\pi - 0$. We note that $\varphi \to +0$ as $\xi_n \to +\infty$, while $\varphi \to 2\pi - 0$ as $\xi_n \to -\infty$. Since $\mathcal{A}(\omega, r, \xi_n)$ is homogeneous of degree zero and $\operatorname{ctg}(\varphi/2) > 0$ for $0 < \varphi < \pi$, we have

$$\mathcal{A}\left(\omega, 1, \operatorname{ctg}\frac{\varphi}{2}\right) = \mathcal{A}\left(\omega, \operatorname{tg}\frac{\varphi}{2}, 1\right) \quad \text{for } 0 < \varphi < \pi. \tag{10.42}$$

Analogously, since $\operatorname{ctg}(\varphi/2) < 0$ for $\pi < \varphi < 2\pi$,

$$\mathcal{A}\left(\omega, 1, \operatorname{ctg}\frac{\varphi}{2}\right) = \mathcal{A}\left(\omega, -\operatorname{tg}\frac{\varphi}{2}, -1\right) \quad \text{for } \pi < \varphi < 2\pi. \tag{10.43}$$

By (10.2),

$$\mathcal{A}(\omega, 0, +1) = \mathcal{A}(\omega, 0, -1) = A_0(0, +1), \tag{10.44}$$

i.e. $\mathcal{A}(\omega, 1, \operatorname{ctg}(\varphi/2))$ is continuous on c_1. Further, by (10.2′),

$$\frac{\partial^k}{\partial\varphi^k}\mathcal{A}\left(\omega, \operatorname{tg}\frac{\varphi}{2}, +1\right)\Bigg|_{\varphi=+0} = \frac{\partial^k}{\partial\varphi^k}\mathcal{A}\left(\omega, -\operatorname{tg}\frac{\varphi}{2}, -1\right)\Bigg|_{\varphi=2\pi-0}, \tag{10.45}$$

i.e. $\mathcal{A}(\omega, 1, \operatorname{ctg}(\varphi/2)) \in C^\infty$ on the circle c_1.

Clearly, we also have the converse assertion: if $B(\omega, \operatorname{ctg}(\varphi/2)) \in C^\infty$ for $|\omega| = 1$, $\varphi \in c_1$ and $B(\omega, \infty)$ does not depend on ω, then $B(\omega, \xi_n/r) \in D_0$.

§11. Boundary value problems for smooth pseudodifferential operators in a halfspace

Boundary value problems for a p.o. with symbol in $D^{(0)}_{\alpha+i\beta}$ can be considered in classes of functions that are smooth in $\mathbf{R}^n_+$ and vanish for $x_n < 0$.

Assuming that $A(\xi', \xi_n) \in D^{(0)}_{\alpha+i\beta}$ and $A(\xi', \xi_n) \neq 0$ for $|\xi'| + |\xi_n| > 0$, consider the equation

$$p\hat{A}u_+ = f(x), \tag{11.1}$$

where $f(x) \in H_{r-\alpha}(\mathbf{R}^n_+)$, $r \geqslant 0$, $u_+ = u(x) \in H_r(\mathbf{R}^n_+)$ for $x_n > 0$ and $u_+(x) = 0$ for $x_n < 0$. By Theorem 10.1, $p\hat{A}u_+$ is a bounded operator from $H_r(\mathbf{R}^n_+)$ into $H_{r-\alpha}(\mathbf{R}^n_+)$. The factorization index κ of $A(\xi', \xi_n)$ is an integer (see Lemma 10.3). Consequently, in the notation of §7 one can put $s = 0$, $m = \kappa$ and $\delta = 0$, since a function $u(x) \in H_r(\mathbf{R}^n_+)$ that vanishes for $x_n < 0$ belongs to $H_0^+(\mathbf{R}^n)$.

We will separately consider the three cases a) $\kappa = 0$, b) $\kappa > 0$, c) $\kappa < 0$.

a) When $\kappa = 0$, the following theorem is valid.

THEOREM 11.1. *Suppose $A(\xi', \xi_n) \in D^{(0)}_{\alpha+i\beta}$ is an elliptic symbol with factorization index $\kappa = 0$. Then for any $f(x) \in H_{r-\alpha}(\mathbf{R}^n_+)$ there exists a unique solution of* (11.1) *belonging to the space $H_r(\mathbf{R}^n_+)$.*

PROOF. Since $r \geqslant 0$, we see that $f(x) \in H_{r-\alpha}(\mathbf{R}^n_+) \subset H_{-\alpha}(\mathbf{R}^n_+)$. By Theorem 7.1, there exists a unique solution $u_+ \in H_0^+(\mathbf{R}^n)$ of (11.1), with

$$\tilde{u}_+(\xi) = \frac{1}{\hat{A}_+(\xi)} \Pi^+ \frac{\widetilde{lf}}{\hat{A}_-(\xi)}, \tag{11.2}$$

where $lf \in H_{r-\alpha}(\mathbf{R}^n)$ and

$$\|lf\|_{r-\alpha} \leqslant 2\|f\|^+_{r-\alpha}. \tag{11.3}$$

Let us show that $u(x) = u_+(x)$ for $x_n > 0$ satisfies

$$\|u\|^+_r \leqslant C\|f\|^+_{r-\alpha} \tag{11.4}$$

and hence belongs to $H_r(\mathbf{R}^n_+)$. By Lemma 10.3, $A_+(\xi) \in D_0$. Consequently, on the basis of Theorem 10.1,

$$\|u\|^+_r \leqslant C\|\hat{A}_-^{-1} lf\|^+_r. \tag{11.5}$$

But by Lemma 4.4,

$$\|\hat{A}_-^{-1} lf\|^+_r \leqslant \|\hat{A}_-^{-1} lf\|_r \leqslant C\|lf\|_{r-\alpha} \leqslant 2C\|f\|^+_{r-\alpha} \tag{11.6}$$

and (11.4) follows. Theorem 11.1 is proved.

b) Suppose $\kappa > 0$. We will consider the boundary value problem

$$p\hat{A}u_+ = f(x), \tag{11.7}$$

$$p'_+ \hat{B}_j u_+ = g_j(x'), \qquad 1 \leqslant j \leqslant \kappa, \tag{11.8}$$

where p'_+ is the operator of restriction to the hyperplane $x_n = \varepsilon > 0$ with a subsequent passage to the limit for $\varepsilon \to 0$ (see §10.3), and $B_j(\xi', \xi_n) \in D_{\beta_j}$.

Suppose $r \geqslant 0$, $r > \max_{1 \leqslant j \leqslant \kappa}(\operatorname{Re}\beta_j + 1/2)$. If $u(x) \in H_r(\mathbf{R}^n_+)$, Theorem 10.1 implies

$$p\hat{A}u_+ \in H_{r-\alpha}(\mathbf{R}^n_+), \qquad p\hat{B}_j u_+ \in H_{r-\operatorname{Re}\beta_j}(\mathbf{R}^n_+).$$

Since $r - \operatorname{Re}\beta_j > 1/2$, the restriction $p'_+ B_j u_+$ exists and belongs to $H_{r-\operatorname{Re}\beta_j - 1/2}(\mathbf{R}^{n-1})$.

Now let $f \in H_{r-\alpha}(\mathbf{R}^n_+)$ and $g_j(x') \in H_{r-\operatorname{Re}\beta_j-1/2}(\mathbf{R}^{n-1})$, $1 \leqslant j \leqslant \kappa$ be given. We will find a solution of the boundary value problem (11.7), (11.8) that belongs to $H_r(\mathbf{R}^n_+)$ for $x_n > 0$.

By Theorem 7.2, the Fourier transform of the general solution $u_+ \in H_0^+$ of (11.7) has the form

$$\tilde{u}_+(\xi) = \frac{\hat{Q}(\xi)}{\hat{A}_+(\xi)} \Pi^+ \frac{\widetilde{lf}}{\hat{Q}\hat{A}_-} + \frac{\sum_{k=1}^{\kappa} \tilde{c}_k(\xi')\xi_n^{k-1}}{\hat{A}_+(\xi', \xi_n)}, \tag{11.9}$$

where $Q(\xi', \xi_n)$ is an arbitrary elliptic polynomial in ξ_n that is homogeneous of degree κ with respect to (ξ', ξ_n). The extension lf of f is chosen from the space $H_{r-\alpha}(\mathbf{R}^n)$, while the arbitrary functions $\tilde{c}_k(\xi')$ are respectively chosen from the spaces $\tilde{H}_{r-\kappa+k-1/2}(\mathbf{R}^{n-1})$.

Let us show that $u = pF_\xi^{-1}\tilde{u}_+ \in H_r(\mathbf{R}^n_+)$. By Lemmas 10.2 and 10.3,

$$\frac{Q(\xi)}{A_+(\xi)} \in D_0, \qquad \frac{B_j(\xi)Q(\xi)}{A_+(\xi)} \in D_{\beta_j}.$$

Consequently, analogously to (11.4), we get

$$\left\| pF_\xi^{-1} \frac{\hat{Q}(\xi)}{\hat{A}_+(\xi)} \Pi^+ \frac{\widetilde{lf}}{\hat{Q}\hat{A}_-} \right\|_r^+ \leqslant C\|f\|_{r-\alpha}^+, \tag{11.10}$$

$$\left\| pF_\xi^{-1} \frac{\hat{B}_j(\xi)\hat{Q}(\xi)}{\hat{A}_+(\xi)} \Pi^+ \frac{\widetilde{lf}}{\hat{Q}\hat{A}_-} \right\|_{r-\operatorname{Re}\beta_j}^+ \leqslant C\|f\|_{r-\alpha}^+, \qquad 1 \leqslant j \leqslant \kappa. \tag{11.11}$$

By Lemma 8.1,

$$\left\| pF_\xi^{-1} \frac{\sum_{k=1}^{\kappa} \tilde{c}_k(\xi')\xi_n^{k-1}}{\hat{A}_+(\xi', \xi_n)} \right\|_r^+ \leqslant C \sum_{k=1}^{\kappa} [c_k]_{r-\kappa+k-1/2} \tag{11.12}$$

which together with (11.10) implies

$$\|u\|_r^+ \leqslant C\left(\|f\|_{r-\alpha}^+ + \sum_{k=1}^{\kappa} [c_k]_{r-\kappa+k-1/2} \right). \tag{11.13}$$

Consequently, $u \in H_r(\mathbf{R}^n_+)$.

By way of preparation for substituting (11.9) in the boundary conditions (11.8), let

$$f_j(x') = p'_+ F_\xi^{-1} \frac{\hat{B}_j(\xi)\hat{Q}(\xi)}{\hat{A}_+} \Pi^+ \frac{\tilde{l}f}{\hat{Q}\hat{A}}, \qquad 1 \leqslant j \leqslant \kappa. \tag{11.14}$$

Since $r - \operatorname{Re} \beta_j > 1/2$, the restriction (11.14) exists and, by (11.11),

$$[f_j(x')]_{r-\operatorname{Re}\beta_j - 1/2} \leqslant C\|f\|^+_{r-\alpha}, \qquad 1 \leqslant j \leqslant \kappa. \tag{11.15}$$

Since $\dfrac{\hat{B}_j \xi_n^{k-1}}{\hat{A}_+} \in D_{\beta_j + k - 1 - \kappa}$, Lemma 10.1 implies

$$\frac{\hat{B}_j(\xi', \xi_n)\xi_n^{k-1}}{\hat{A}_+(\xi)} = \hat{B}_{jk}^-(\xi', \xi_n) + \hat{R}_{jk}(\xi', \xi_n), \tag{11.16}$$

where $\hat{B}_{jk}^-(\xi', \xi_n)$ is a "minus" function, i.e. it has an analytic continuation with respect to $\xi_n + i\tau$ into the halfplane $\tau < 0$ that satisfies the estimate

$$|\hat{B}_{jk}^-(\xi', \xi_n + i\tau)| \leqslant C(1 + |\xi'| + |\xi_n| + |\tau|)^{\operatorname{Re}\beta_j - \kappa + k - 1},$$

while $\hat{R}_{jk}(\xi', \xi_n)$ is a continuous function for $\xi' \neq 0$ that satisfies

$$|\hat{R}_{jk}(\xi', \xi_n)| \leqslant \frac{C(1 + |\xi'|)^{\operatorname{Re}\beta_j - \kappa + k + 1}}{(1 + |\xi'| + |\xi_n|)^2}. \tag{11.17}$$

By (11.17), $F_{\xi_n}^{-1}\hat{R}_{jk}(\xi', \xi_n)$ is a continuous function with respect to x_n and ξ' for $\xi' \neq 0$. Since $pF_{\xi_n}^{-1}\hat{B}_{jk}^-(\xi', \xi_n) = 0$, the restriction

$$p'_+ F_{\xi_n}^{-1} \frac{\hat{B}_j(\xi)\xi_n^{k-1}}{\hat{A}_+}$$

exists, and

$$p'_+ F_{\xi_n}^{-1} \frac{\hat{B}_j(\xi)\xi_n^{k-1}}{\hat{A}_+(\xi)} = p' F_{\xi_n}^{-1} \hat{R}_{jk}(\xi', \xi_n). \tag{11.18}$$

Let

$$\hat{b}_{jk}(\xi') = p' F_{\xi_n}^{-1} \hat{R}_{jk}(\xi', \xi_n) = \frac{1}{2\pi} \int_{-\infty}^{\infty} \hat{R}_{jk}(\xi', \xi_n) \, d\xi_n. \tag{11.18$'$}$$

Then $b_{jk}(\xi')$ is a continuous function for $\xi' \neq 0$ that is homogeneous of degree $\beta_j - \kappa + k$. As in the proof of Theorem 8.2, it can be verified that $b_{jk}(\xi')$ is independent of the lack of uniqueness in the decomposition (11.16). Thus, after substituting (11.9) in the boundary conditions (11.8), we get

$$\sum_{k=1}^{\kappa} \hat{b}_{jk}(\xi')\tilde{c}_k(\xi') = \tilde{g}_j(\xi') - \tilde{f}_j(\xi'). \tag{11.19}$$

As before, we require that

$$\det\|b_{jk}(\xi')\|_{j,k=1}^{\kappa} \neq 0 \quad \text{for } \xi' \neq 0. \tag{11.20}$$

Then (11.19) has the unique solution

$$\tilde{c}_k(\xi') = \sum_{j=1}^{\kappa} \hat{b}_{kj}^{(1)}(\xi')\big(\tilde{g}_j(\xi') - \tilde{f}_j(\xi')\big), \tag{11.21}$$

where $\|\hat{b}_{kj}^{(1)}(\xi')\|_{k,j=1}^{\kappa}$ is the inverse of the matrix $\|\hat{b}_{jk}(\xi')\|_{j,k=1}^{\kappa}$ and $b_{kj}^{(1)}(\xi')$ is a homogeneous function of degree $-(\beta_j - \kappa + k)$. From (11.21) and (11.15) we get

$$\begin{aligned}[c_k]_{r-\kappa+k-1/2} &\leqslant C\left(\sum_{j=1}^{\kappa}[g_j]_{r-\operatorname{Re}\beta_j-1/2} + \sum_{j=1}^{\kappa}[f_j]_{r-\operatorname{Re}\beta_j-1/2}\right)\\ &\leqslant C\left(\sum_{j=1}^{\kappa}[g_j]_{r-\operatorname{Re}\beta_j-1/2} + \|f\|_{r-\alpha}^{+}\right).\end{aligned} \tag{11.22}$$

Thus (11.9) and (11.21) determine the (clearly unique) solution of the boundary value problem (11.7), (11.8), with the estimate

$$\|u\|_r^{+} \leqslant C\left(\|f\|_{r-\alpha}^{+} + \sum_{k=1}^{\kappa}[g_j]_{r-\operatorname{Re}\beta_j-1/2}\right) \tag{11.23}$$

holding by virtue of (11.13) and (11.22). We consequently have

THEOREM 11.2. *Suppose* $A(\xi) \in D_{\alpha+i\beta}^{(0)}$ *is an elliptic symbol,* $\kappa > 0$, $B_j(\xi) \in D_{\beta_j}$, $1 \leqslant j \leqslant \kappa$, *and the Šapiro-Lopatinskiĭ condition* (11.20) *is satisfied. Then the boundary value problem* (11.7), (11.8) *has a unique solution* $u(x) \in H_r(\mathbf{R}_+^n)$ *for any* $f \in H_{r-\alpha}(\mathbf{R}_+^n)$, $g_j(x') \in H_{r-\operatorname{Re}\beta_j-1/2}(\mathbf{R}^{n-1})$ *when* $r \geqslant 0$, $r > \max_{1\leqslant j\leqslant\kappa}(\operatorname{Re}\beta_j + 1/2)$.

REMARK 11.1. If $r > 0$, then a particular solution of (11.7) does not have to be taken in the form

$$F_\xi^{-1}\frac{\hat{Q}}{\hat{A}_+}\Pi^+\frac{\tilde{lf}}{\hat{Q}\hat{A}_-},$$

where $Q(\xi', \xi_n)$ is a homogeneous elliptic polynomial of degree κ. Let k be a nonnegative integer such that $r + k > \kappa$ and let

$$Q_k(\xi', \xi_n) = \sum_{i=0}^{k} q_i(\xi')\xi_n^i$$

be a homogeneous elliptic polynomial of degree k. Then

$$\frac{\tilde{lf}}{\hat{Q}_k\hat{A}_-} \in \tilde{H}_{r-\kappa+k}(R^n) \subset H_0(\mathbf{R}^n).$$

Consequently, the operator Π^+ can be applied to $\dfrac{\widetilde{lf}}{\hat{Q}_k\hat{A}_-}$, and

$$F_\xi^{-1}\frac{\hat{Q}_k}{\hat{A}_+}\Pi^+\frac{\widetilde{lf}}{\hat{Q}_k\hat{A}_-}$$

is a particular solution of (11.7) that belongs to $H_r(\mathbf{R}^n_+)$ for $x_n > 0$. If $r > \kappa$, one can put $k = 0$ and $\hat{Q}_0 \equiv 1$.

EXAMPLE 11.1. In contrast to the boundary value problem (8.1), (8.2), the boundary conditions (11.8) can be differential conditions with respect to x_n. We will show that the Dirichlet data

$$u|_{x_n=+0} = g_1(x'), \qquad D_n u|_{x_n=+0} = g_2(x'), \ldots, D_n^{\kappa-1}u|_{x_n=+0} = g_\kappa(x'), \tag{11.24}$$

where

$$g_j(x') \in H_{r-j+1/2}(\mathbf{R}^{n-1}), \quad D_n^{j-1} = \left(i\frac{\partial}{\partial x_n}\right)^{j-1}, \quad 1 \leqslant j \leqslant \kappa, r > \kappa - 1/2,$$

can always be taken as the boundary conditions (11.8) for equation (11.7).

Let

$$\hat{B}_{jk}^+(\xi', \xi_n) = \frac{\xi_n^{j+k-2}}{\hat{A}_+(\xi', \xi_n)}, \qquad 1 \leqslant j, k \leqslant \kappa.$$

If $j + k \leqslant \kappa$, then

$$\left|\hat{B}_{jk}^+(\xi', \xi_n + i\tau)\right| \leqslant \frac{C}{(1 + |\xi'| + |\xi_n| + \tau)^2}.$$

Thus one can put $\hat{B}_{jk}^- = 0$ and $\hat{R}_{jk} = \hat{B}_{jk}^+$ in the decomposition (11.16), and hence

$$\hat{b}_{jk}(\xi') = \frac{1}{2\pi}\int_{-\infty}^{\infty}\hat{B}_{jk}^+(\xi', \xi_n)\,d\xi_n.$$

But Cauchy's theorem implies

$$\int_{-\infty}^{\infty}\hat{B}_{jk}^+(\xi', \xi_n)\,d\xi_n = 0. \tag{11.25}$$

Consequently, $\hat{b}_{jk}(\xi') = 0$ for $j + k \leqslant \kappa$.

Suppose $j + k = \kappa + 1$. Since $A_+(0, +1) = 1$, it follows by virtue of (10.13) that

$$\hat{B}_{jk}^+(\xi', \xi_n) = \frac{\xi_n^{\kappa-1}}{\hat{A}_+(\xi', \xi_n)} = \frac{1}{\xi_n + i|\xi'| + i} + \hat{A}_1^+(\xi', \xi_n), \qquad j + k = \kappa + 1,$$

where

$$|\hat{A}_1^+(\xi', \xi_n + i\tau)| \leqslant \frac{C(1 + |\xi'|)}{(|\xi_n| + |\xi'| + \tau + 1)^2}.$$

On the other hand, (11.18) and (11.18′) imply

$$\hat{b}_{jk}(\xi') = \lim_{x_n \to +0} F_{\xi_n}^{-1}\hat{B}_{jk}^+(\xi', \xi_n).$$

Since

$$F_{\xi_n}^{-1}\frac{1}{\xi_n + i|\xi'| + i} = i\theta(x_n)e^{-x_n(|\xi'|+1)}$$

(see Example 2.3) and, analogously to (11.25), $\int_{-\infty}^{\infty}\hat{A}_1^+(\xi', \xi_n)d\xi_n = 0$, we conclude that $\hat{b}_{jk}(\xi') = i$ for $j + k = \kappa + 1$.

Thus $\hat{b}_{jk}(\xi') = 0$ for $2 \leq j + k \leq \kappa$ and $\hat{b}_{jk}(\xi') = i$ for $j + k = \kappa + 1$. Consequently,

$$\det\left\|\hat{b}_{jk}(\xi')\right\|_{j,k=1}^{\kappa} = (-1)^{(\kappa+3)\kappa/2} i^{\kappa} \neq 0,$$

so that the Šapiro-Lopatinskiĭ condition (11.20) is satisfied. Thus, by Theorem 11.2, the boundary value problem (11.7), (11.24) is uniquely solvable.

EXAMPLE 11.2. *Boundary value problems for elliptic differential equations.* Consider the equation

$$\hat{L}(D', D_n)u(x', x_n) = f(x', x_n), \qquad x_n > 0, \tag{11.26}$$

where

$$\hat{L}(\xi', \xi_n) = L\left((1 + |\xi'|)\frac{\xi'}{|\xi'|}, \ \xi_n\right)$$

and $L(\xi', \xi_n)$ is a homogeneous elliptic polynomial in ξ' and ξ_n of degree $2m$. Although, as a result of the replacement of $L(\xi', \xi_n)$ by $\hat{L}(\xi', \xi_n)$, equation (11.26) is not a differential equation with respect to x', the most important fact in the case of the halfspace $\mathbf{R}^n_+$ is that (11.26) is nevertheless a differential equation with respect to x_n. The boundary conditions, which are differential conditions with respect to x_n, are given in the form

$$\hat{B}_j(D', D_n)u(x', x_n)\big|_{x_n = +0} = g_j(x'), \qquad 1 \leq j \leq m, \tag{11.27}$$

where $B_j(\xi', \xi_n)$ $(1 \leq j \leq m)$ is a homogeneous polynomial in (ξ', ξ_n) of degree m_j. It is assumed that

$$f \in H_{r-2m}(\mathbf{R}^n_+), \quad g_j(x') \in H_{r-m_j-1/2}(\mathbf{R}^{n-1}), \quad r > \max_{1 \leq j \leq m} (m_j + 1/2).$$

The boundary value problem (11.26), (11.27) is usually solved in the following way (see, for example, [2] and [3]). We take the Fourier transform with respect to x' and obtain the following boundary value problem for an ordinary differential equation:

$$\hat{L}(\xi', D_n)\tilde{w}(\xi', x_n) = \tilde{h}(\xi', x_n), \qquad x_n > 0, \tag{11.28}$$

$$\hat{B}_j(\xi', D_n)\tilde{w}(\xi', x_n)\big|_{x_n = +0} = \tilde{g}_j(\xi'), \qquad 1 \leq j \leq m, \tag{11.29}$$

where $\tilde{w}(\xi', x_n) = F_{x'}u(x', x_n)$ and $\tilde{h}(\xi', x_n) = F_{x'}f(x', x_n)$.

We next find a particular solution of (11.28). To this end we take an arbitrary extension of $f(x', x_n)$ from $\mathbf{R}^n_+$ onto $\mathbf{R}^n$ that belongs to $H_{r-2m}(\mathbf{R}^n)$. Then

$$\tilde{v}(\xi', x_n) = \frac{1}{2\pi}\int_{-\infty}^{\infty} \frac{\tilde{l}f(\xi', \xi_n)}{\hat{L}(\xi', \xi_n)} e^{-ix_n\xi_n}\, d\xi_n \tag{11.30}$$

is a particular solution of (11.28) for $x_n > 0$.

Let $L(\xi', \xi_n) = L_-(\xi', \xi_n)L_+(\xi', \xi_n)$ be the factorization of the elliptic polynomial $L(\xi', \xi_n)$ (see Example 6.3). The general solution of the homogeneous equation (11.28) contains $2m$ arbitrary functions of ξ'. For $x_n \to +\infty$, m of these functions are exponentially decreasing, while the remaining m are exponentially increasing. Therefore, if we confine ourselves to functions belonging to $H_r(\mathbf{R}^n_+)$, we get that the general solution of (11.28) for $x_n > 0$ has the form

$$\tilde{w}(\xi', x_n) = \tilde{v}(\xi', x_n) + \sum_{k=1}^{m} \frac{1}{2\pi}\int_{\Gamma_+} \frac{\tilde{c}_k(\xi')\zeta_n^{k-1}}{\hat{L}_+(\xi', \zeta_n)} e^{-ix_n\zeta_n}\, d\zeta_n, \tag{11.31}$$

where Γ_+ is a contour in the complex ζ_n plane that encloses all of the zeros of $\hat{L}_+(\xi', \zeta_n)$.

Substituting (11.31) in (11.29), we get

$$\sum_{k=1}^{m} \hat{b}_{jk}(\xi')\tilde{c}_k(\xi') = \tilde{g}_j(\xi') - \tilde{f}_j(\xi'), \tag{11.32}$$

where

$$\hat{b}_{jk}(\xi') = \frac{1}{2\pi}\int_{\Gamma_+} \frac{\hat{B}_j(\xi', \zeta_n)\zeta_n^{k-1}}{\hat{L}_+(\xi', \zeta_n)}\, d\zeta_n,$$

$$\tilde{f}_j(\xi') = \frac{1}{2\pi}\int_{-\infty}^{\infty} \frac{\hat{B}_j(\xi', \xi_n)\tilde{lf}(\xi', \xi_n)}{\hat{L}(\xi', \xi_n)}\, d\xi_n. \tag{11.33}$$

The fulfilment of the Šapiro-Lopatinskiĭ condition for the boundary value problem (11.26), (11.27) means that $\det\|\hat{b}_{jk}(\xi')\|_{j,k=1}^{m} \neq 0$. If this condition is satisfied, then (11.32) has the unique solution

$$\tilde{c}_k(\xi') = \sum_{j=1}^{m} \hat{b}_{kj}^{(1)}(\xi')\big(\tilde{g}_j(\xi') - \tilde{f}_j(\xi')\big). \tag{11.34}$$

Substituting (11.34) in (11.31), we get that the boundary value problem (11.26), (11.27) is uniquely solvable. In this connection, it is easily verified that

$$\|u\|_r^+ \leqslant C\Big(\|f\|_{r-2m}^+ + \sum_{j=1}^{m} [g_j]_{r-m_j-1/2}\Big).$$

Let us compare the above solution of (11.26), (11.27) with the solution obtained by applying the method of proof of Theorem 11.2 to (11.26), (11.27). If we put $\hat{A}(\xi', \xi_n) = \hat{L}(\xi', \xi_n)$ and $\kappa = m$ and choose $\hat{Q}(\xi', \xi_n) = \hat{L}_+(\xi', \xi_n)$ in (11.9), we get

$$\tilde{u}_+(\xi', \xi_n) = \Pi^+ \frac{\tilde{lf}(\xi)}{\hat{L}(\xi)} + \sum_{k=1}^{m} \frac{\tilde{c}_k(\xi')\xi_n^{k-1}}{\hat{L}_+(\xi', \xi_n)}. \tag{11.35}$$

On the other hand, by Jordan's lemma,

$$\frac{1}{2\pi}\int_{-\infty}^{\infty} \frac{\sum_{k=1}^{m} \tilde{c}_k(\xi')\xi_n^{k-1}}{\hat{L}_+(\xi', \xi_n)} e^{-ix_n\xi_n}\, d\xi_n = \frac{1}{2\pi}\int_{\Gamma_+} \frac{\sum_{k=1}^{m} \tilde{c}_k(\xi')\zeta_n^{k-1}e^{-ix_n\zeta_n}}{\hat{L}_+(\xi', \zeta_n)}\, d\zeta_n \tag{11.36}$$

for $x_n > 0$. Thus (11.35) can be obtained from (11.31) by extending all of the functions in (11.31) by (setting them equal to) zero for $x_n < 0$ and then taking the Fourier transform with respect to x_n.

c) Let us return to the study of (11.1) and consider the case $\kappa < 0$. Suppose $f \in H_{r-\alpha}(\mathbf{R}_+^n)$, $r \geqslant 0$. By Theorem 7.1, the unique solution $u_+ \in H_\kappa^+$ of (11.1) has the form

$$\tilde{u}_+(\xi) = \hat{A}_+^{-1}\Pi^+\hat{A}_-^{-1}\tilde{lf}, \tag{11.37}$$

where $lf \in H_{r-\alpha}(\mathbf{R}^n)$ is an arbitrary extension of f. Expanding $\Pi^+\hat{A}_-^{-1}\widetilde{lf}$ by (5.36), we get

$$\tilde{u}_+ = \sum_{k=1}^{|\kappa|} \hat{A}_+^{-1}\hat{\Lambda}_+^{-k}\tilde{u}_k(\xi') + \tilde{w}_+(\xi), \tag{11.38}$$

where

$$\tilde{u}_k(\xi') = i\Pi'\hat{\Lambda}_+^{k-1}\hat{A}_-^{-1}\widetilde{lf} \in H_{r+|\kappa|-k+1/2}(\mathbf{R}^{n-1}),$$

and

$$\tilde{w}_+(\xi) = \hat{A}_+^{-1}\hat{\Lambda}_+^{-|\kappa|}\Pi^+\hat{\Lambda}_+^{|\kappa|}\hat{A}_-^{-1}\widetilde{lf} \in \tilde{H}_0^+(\mathbf{R}^n).$$

Since $A_+^{-1}\Lambda_+^{-|\kappa|} \in D_0$, we have

$$\|w_+\|_r^+ \leqslant c\|lf\|_{r-\alpha}^+, \tag{11.39}$$

which is analogous to (11.4). By (10.4),

$$\hat{A}_+^{-1}\hat{\Lambda}_+^{-k} = \sum_{p=0}^{|\kappa|-k} \hat{c}_{pk}(\xi')\xi_n^p + \hat{R}_k^+(\xi', \xi_n), \tag{11.40}$$

where $c_{|\kappa|-k,k} = 1$, $c_{pk}(\xi')$ is a homogeneous function of degree $|\kappa| - k - p$, $R_k^+(\xi', \xi_n) \in D_{|\kappa|-k}$ and

$$|\hat{R}_k^+(\xi', \xi_n)| \leqslant c(1 + |\xi'|)^{|\kappa|-k+1}(1 + |\xi'| + |\xi_n|)^{-1}. \tag{11.41}$$

Therefore, substituting (11.40) in (11.38), we get

$$\tilde{u}_+(\xi) = \sum_{p=1}^{|\kappa|} \tilde{v}_p(\xi')\xi_n^{p-1} + \tilde{v}_+(\xi), \tag{11.42}$$

where

$$\tilde{v}_p(\xi') = \sum_{k=1}^{|\kappa|-k} \hat{c}_{pk}(\xi')\tilde{u}_k(\xi') \in \tilde{H}_{r+p+1/2}(\mathbf{R}^{n-1}), \tag{11.43}$$

$$\tilde{v}_+(\xi) = \sum_{k=1}^{|\kappa|} \hat{R}_k^+(\xi)\tilde{u}_k(\xi') + \tilde{w}_+(\xi) \in \tilde{H}_0^+(\mathbf{R}^n). \tag{11.44}$$

Expansion (11.42) is unique, with $\tilde{v}_p(\xi') = 0$ for almost all ξ' and $1 \leqslant p \leqslant |\kappa|$ if and only if for almost all ξ'

$$\Pi'\hat{\Lambda}_+^{k-1}\hat{A}_-^{-1}\widetilde{lf} = 0, \quad 1 \leqslant k \leqslant |\kappa|. \tag{11.45}$$

It follows from (11.44) by virtue of (11.39) and Lemma 8.1 that

$$\|v\|_r^+ \leqslant c\left(\sum_{k=1}^{|\kappa|} [u_k]_{r+|\kappa|-k+1/2} + \|lf\|_{r-\alpha}^+\right) \leqslant c\|lf\|_{r-\alpha}^+ \tag{11.46}$$

while (11.43) implies

$$\sum_{p=1}^{|\kappa|} [v_p]_{r+p+1/2} \leqslant c\|f\|_{r-\alpha}^{+}. \tag{11.47}$$

We have thus proved

THEOREM 11.3. *Suppose $A(\xi) \in D_{\alpha+i\beta}^{(0)}$ is an elliptic symbol with factorization index $\kappa < 0$. Then for any $f \in H_{r-\alpha}(\mathbf{R}_+^n)$, $r \geqslant 0$, there exists a unique solution $v(x) \in H_r(\mathbf{R}_+^n)$, $v_k(x') \in H_{r+k+1/2}(\mathbf{R}^{n-1})$, $1 \leqslant k \leqslant |\kappa|$, of the equation*

$$p\hat{A}\left(v_+ + \sum_{k=1}^{|\kappa|} v_k(x') \times D_n^{k-1}\delta(x_n)\right) = f(x). \tag{11.48}$$

The boundary value problem (11.48) is a special case (when $C_k(\xi', \xi_n) = A(\xi', \xi_n)\xi_n^{k-1}$) of the boundary value problem with potentials (coboundary value problem) of the form (8.26)

$$p\left(\hat{A}u_+ + \sum_{k=1}^{|\kappa|} \hat{C}_k(v_k(x') \times \delta(x_n))\right) = f(x). \tag{11.49}$$

Let us now consider (11.49) under the assumption that $f \in H_{r-\alpha}(\mathbf{R}_+^n)$ and each $C_k(\xi', \xi_n)$ is an arbitrary symbol of class $D_{\alpha_k+i\beta_k}$. By Lemma 10.1, each $\hat{C}_k(\xi', \xi_n)$ admits the decomposition

$$\hat{C}_k(\xi', \xi_n) = \hat{C}_k^-(\xi', \xi_n) + \hat{R}_k(\xi', \xi_n), \tag{11.50}$$

where $\hat{C}_k^-(\xi', \xi_n)$ is a "minus" function and $\hat{R}_k(\xi)$ satisfies

$$|\hat{R}_k(\xi', \xi_n)| \leqslant \frac{C(1 + |\xi'|)^N}{(1 + |\xi'| + |\xi_n|)^{N-\alpha_k}}, \qquad N - \alpha_k > r - \alpha + \frac{1}{2}. \tag{11.51}$$

If condition (8.34) is satisfied, then, as in the proof of Theorem 8.2, the unique solution $u_+ \in H_0^+(\mathbf{R}^n)$, $v_k(x') \in H_{r_k}(\mathbf{R}^{n-1})$, $r_k = r - \alpha + \alpha_k + 1/2$, $1 \leqslant k \leqslant |\kappa|$, of (11.49) is determined by (8.35), (8.37) and (8.32), in which m should be replaced by κ. Since $\hat{A}_+^{-1}\hat{\Lambda}_+^{-|\kappa|} \in D_0$, it follows from (8.35), (8.37) and (8.32) by virtue of Theorem 10.1 that

$$\|u\|_r^+ + \sum_{k=1}^{|\kappa|} [v_k]_{r_k} \leqslant C\|f\|_{r-\alpha}^{+}. \tag{11.52}$$

We thus have

THEOREM 11.4. *Suppose $A(\xi) \in D_{\alpha+i\beta}^{(0)}$ is an elliptic symbol with factorization index $\kappa < 0$, $C_k(\xi) \in D_{\alpha_k+i\beta_k}$, and condition (8.34) is satisfied. Then for any $f \in H_{r-\alpha}(\mathbf{R}_+^n)$ equation (11.49) has a unique solution*

$$u \in H_r(\mathbf{R}_+^n), \quad v_k(x') \in H_{r_k}(\mathbf{R}^{n-1}), \quad r_k = r - \alpha + \alpha_k + 1/2, \quad 1 \leqslant k \leqslant |\kappa|. \tag{11.53}$$

We note that Theorem 11.3 can be obtained from Theorem 11.4 by verifying the fulfilment of condition (8.34) for $C_k(\xi) = A(\xi)\xi_n^{k-1}$ analogously to the procedure in Example 11.1.

EXAMPLE 11.3. *Inhomogeneous Dirichlet problem*. Suppose $A(\xi) \in O_{\alpha+i\beta}^{\infty}$ is an elliptic symbol with factorization index κ with respect to ξ_n. It was shown in §7 that when $|s - \operatorname{Re}\kappa| < 1/2$, the equation

$$p\hat{A}u_+ = f \tag{11.54}$$

has a unique solution $u_+ \in H_s^+$ for any $f \in H_{s-\alpha}(\mathbf{R}_+^n)$. When $s > m - 1/2$, where m is a positive integer, it follows from the membership of $u_+(x)$ in H_s^+ (see Theorem 4.2) that $u_+(x', 0) = D_n^k u_+(x', 0) = 0$ for $1 \leqslant k \leqslant m-1$. Thus $u_+(x', x_n)$ satisfies the null Dirichlet data for $x_n = 0$. Suppose $A(\xi) \in D_{\alpha+i\beta}^{(0)}$ and $\kappa = m > 0$. Then (11.54) becomes uniquely solvable if $u_+(x', x_n) = u(x', x_n) \in H_s(\mathbf{R}_+^n)$ for $x_n > 0$, $s > m - 1/2$ and $u(x', x_n)$ satisfies the boundary conditions (11.24) (see Example 11.1). Thus the inhomogeneous Dirichlet problem is correct for a p.o. with symbol of class $D_{\alpha+i\beta}^{(0)}$. If, on the other hand, $A(\xi) \notin D_{\alpha+i\beta}$ then, as was noted in Remark 8.1, boundary conditions of the form (11.24) cannot be posed. We establish below an analog of the inhomogeneous Dirichlet problem for a p.o. with an arbitrary elliptic symbol of class $O_{\alpha+i\beta}^{\infty}$.

Let us consider the problem of finding a function $u(x) \in H_s(\mathbf{R}^n)$ that satisfies the equation

$$p\hat{A}u = f(x), \qquad f(x) \in H_{s-\alpha}(\mathbf{R}_+^n) \tag{11.55}$$

and coincides in $\mathbf{R}_-^n$ with a given function $g(x) \in H_s(\mathbf{R}_-^n)$:

$$u|_{\mathbf{R}_-^n} = g(x). \tag{11.56}$$

Equations (11.55) and (11.56) are a special case of the so-called *paired equations* considered in §13.4.

To solve problem (11.55), (11.56), we let

$$v_- = lf - \hat{A}u, \qquad v_+ = u - lg, \tag{11.57}$$

where $lf \in H_{s-\alpha}(\mathbf{R}^n)$ and $lg \in H_s(\mathbf{R}^n)$ are arbitrary extensions of f and g onto $\mathbf{R}^n$. Clearly, $v_- \in H_{s-\alpha}^-$, $v_+ \in H_s^+$. The Fourier transform of (11.57) with respect to all variables is

$$\hat{A}(\xi)\tilde{u}(\xi) + \tilde{v}_-(\xi) = \widetilde{lf}(\xi), \qquad \tilde{v}_+(\xi) = \tilde{u}(\xi) - \widetilde{lg}(\xi). \tag{11.58}$$

Eliminating $\tilde{u}(\xi)$ from (11.58), we obtain

$$\hat{A}(\xi)\tilde{v}_+(\xi) + \tilde{v}_-(\xi) = \widetilde{lf}(\xi) - \hat{A}(\xi)\widetilde{lg}(\xi), \tag{11.59}$$

which is analogous to (7.14). Suppose $|s - \operatorname{Re} \kappa| < 1/2$. Then, by Theorem 7.1, the function

$$\tilde{v}_+(\xi) = \hat{A}_+^{-1}\Pi^+\hat{A}_-^{-1}(\widetilde{lf}(\xi) - \hat{A}(\xi)\widetilde{lg}(\xi)) \tag{11.60}$$

is the only element of $\tilde{H}_s^+$ satisfying (11.59). Consequently (assuming still $|s - \operatorname{Re} \kappa| < 1/2$), for any $f \in H_{s-\alpha}(\mathbf{R}_+^n)$ and $g \in H_s(\mathbf{R}_-^n)$ problem (11.55), (11.56) has a unique solution $u \in H_s(\mathbf{R}^n)$, with

$$\tilde{u}(\xi) = \tilde{v}_+(\xi) + \widetilde{lg}(\xi) = \hat{A}_+^{-1}\Pi^+\hat{A}_-^{-1}\widetilde{lf} + \widetilde{lg}(\xi) - \hat{A}_+^{-1}\Pi^+\hat{A}_+\widetilde{lg}(\xi). \tag{11.61}$$

Expression (11.61) can be simplified. In fact, since

$$\hat{A}_+^{-1}\Pi^+\hat{A}_+\widetilde{lg} = \hat{A}_+^{-1}(\hat{A}_+\widetilde{lg} - \Pi^-\hat{A}_+\widetilde{lg}) = \widetilde{lg}(\xi) - \hat{A}_+^{-1}\Pi^-\hat{A}_+\widetilde{lg},$$

we have

$$\tilde{u}(\xi) = \hat{A}_+^{-1}\Pi^+\hat{A}_-^{-1}\widetilde{lf} + \hat{A}_+^{-1}\Pi^-\hat{A}_+\widetilde{lg}. \tag{11.62}$$

If $s > m - 1/2$, where m is a positive integer, then, by (11.56) and Theorem 4.2,

$$u(x', 0) = g(x', x_n)|_{x_n = -0}, \qquad D_n^k u(x', 0) = D_n^k g(x', x_n)|_{x_n = -0}. \tag{11.63}$$

Consequently, problem (11.55), (11.56) can be regarded as an analog of the inhomogeneous Dirichlet problem.

§12. General boundary value problems for elliptic pseudodifferential equations in a halfspace

Let $P(\xi)$ and $Q(\xi)$ be homogeneous elliptic polynomials of degrees $2m_1$ and $2m_2$ respectively. In the halfspace $\mathbf{R}_+^n$ we can correctly formulate for the operator $\hat{P}$ a boundary value problem with m_1 boundary conditions (see Example 11.2) and for the p.o. $\hat{Q}^{-1}$ a coboundary value problem with m_2 potentials (see Theorem 11.3).

Let $\hat{A}$ be the p.o. with symbol $\hat{A}(\xi) = \hat{P}(\xi)/\hat{Q}(\xi)$ and suppose, for definiteness, that $m_1 > m_2$. By Theorem 11.1, we can correctly formulate for the p.o. $\hat{A}$ in $\mathbf{R}_+^n$ a boundary value problem with $\kappa = m_1 - m_2$ boundary conditions. There naturally arises the question of whether for the p.o. $\hat{A}$ it is possible to increase the number of boundary conditions to m_1 and at the same time to add m_2 potentials in the equations, i.e. to consider a boundary value problem of the form

$$p\left(\hat{A}_+u_+ + \sum_{k=1}^{m_2} \hat{C}_k(v_k(x') \times \delta(x_n))\right) = f(x), \tag{12.1}$$

$$p'\hat{B}_j u_+ = g_j(x'), \qquad 1 \leqslant j \leqslant m_1. \tag{12.2}$$

It will be shown below that the boundary value problem (12.1), (12.2) is uniquely solvable under certain conditions on the symbols $\hat{B}_j(\xi)$ and $\hat{C}_k(\xi)$.

We now formulate a general boundary value problem for the elliptic p.o. $\hat{A}$ that generalizes the boundary value problem (12.1), (12.2). Let $A(\xi) \in O^{\infty}_{\alpha+i\beta}$ be an elliptic symbol with factorization index κ that is generally not of class $D^{(0)}_{\alpha+i\beta}$, let s be a real number satisfying the condition $\operatorname{Re}\kappa - s = m + \delta$, where m is an integer and $|\delta| < 1/2$, and let m_+ and m_- be arbitrary nonnegative integers such that $m_+ - m_- = m$. We will consider the boundary value problem

$$p\left(\hat{A}(D)u_+ + \sum_{k=1}^{m_-} \hat{C}_k(D)(v_k(x') \times \delta(x_n))\right) = f(x), \tag{12.3}$$

$$p'\hat{B}_j(D)u_+ + \sum_{k=1}^{m_-} \hat{E}_{jk}(D')v_k(x') = g_j(x'), \qquad 1 \leqslant j \leqslant m_+. \tag{12.4}$$

It is assumed that

$$B_j(\xi) \in O'_{\beta_j}, \qquad |B_j(\xi', \xi_n)| \leqslant C|\xi'|^{\operatorname{Re}\beta_j - \beta_{j1}}(|\xi'| + |\xi_n|)^{\beta_{j1}},$$
$$\beta_{j1} < s - 1/2, \qquad 1 \leqslant j \leqslant m_+, \tag{12.5}$$

$$C_k(\xi) \in O'_{\gamma_k}, \qquad |C_k(\xi', \xi_n)| \leqslant C|\xi'|^{\operatorname{Re}\gamma_k - \gamma_{k1}}(|\xi'| + |\xi_n|)^{\gamma_{k1}},$$
$$\gamma_{k1} < \alpha - s - 1/2, \qquad 1 \leqslant k \leqslant m_-, \tag{12.6}$$

and each symbol $E_{jk}(\xi')$ is continuous for $\xi' \neq 0$ and homogeneous of degree $\delta_{jk} = \beta_j + \gamma_k - \alpha - i\beta + 1$. In (12.3), (12.4) we are given functions $f(x) \in H_{s-\alpha}(\mathbf{R}^n_+)$, $g_j(x') \in H_{s-\operatorname{Re}\beta_j - 1/2}(\mathbf{R}^{n-1})$, $1 \leqslant j \leqslant m_+$, and we wish to find functions $u_+ \in H_s^+$, $v_k(x') \in H_{s_k}(\mathbf{R}^{n-1})$, $1 \leqslant k \leqslant m_-$, $s_k = s - \alpha + \operatorname{Re}\gamma_k + 1/2$.

a) Suppose $m < 0$. With the use of (12.5) and (12.6), we get that $f \in H_{s-\alpha}(\mathbf{R}^n_+)$, $g_j(x') \in H_{s-\operatorname{Re}\beta_j-1/2}(\mathbf{R}^{n-1})$ if $u_+ \in H_s^+$, $v_k(x') \in H_{s_k}(\mathbf{R}^{n-1})$, $1 \leqslant k \leqslant m_-$. Equation (12.3) can be written in the form

$$p\hat{A}u_+ = f - p\sum_{k=1}^{m_-} \hat{C}_k(v_k(x') \times \delta(x_n)), \tag{12.7}$$

which, by (7.49), has a solution $u_+ \in H_s^+$ if and only if

$$\sum_{j=1}^{m_-} \hat{d}_{kj}(\xi')\tilde{v}_j(\xi') = \tilde{\varphi}_k(\xi'), \qquad 1 \leqslant k \leqslant |m|, \tag{12.8}$$

where

$$\hat{d}_{kj}(\xi') = \int_{-\infty}^{\infty} \frac{\hat{C}_j(\xi', \xi_n)\xi_n^{k-1}}{\hat{A}_-(\xi', \xi_n)}\,d\xi_n, \qquad 1 \leqslant j \leqslant m_-, \quad 1 \leqslant k \leqslant |m|, \tag{12.9}$$

$$\tilde{\varphi}_k(\xi') = \int_{-\infty}^{\infty} \frac{\xi_n^{k-1} \widetilde{lf}(\xi', \xi_n)}{\hat{A}_-(\xi', \xi_n)} \, d\xi_n, \qquad 1 \leqslant k \leqslant |m|. \tag{12.10}$$

It follows from (12.6) and (12.9) that each $d_{kj}(\xi')$ is continuous for $\xi' \neq 0$ and homogeneous of degree $\gamma_j + k - \alpha - i\beta + \kappa$. By Theorem 4.2 and Lemma 4.4, $\tilde{\varphi}_k(\xi') \in H_{s-\operatorname{Re}\kappa-k+1/2}(\mathbf{R}^{n-1})$. Under the fulfilment of condition (12.8) the Fourier transform of the solution (12.7) is given by

$$\tilde{u}_+(\xi) = \frac{1}{\hat{A}_+(\xi)\hat{\Lambda}_+^{|m|}} \Pi^+ \frac{\hat{\Lambda}_+^{|m|}}{\hat{A}_-} \left(\widetilde{lf} - \sum_{k=1}^{m_-} \hat{C}_k(\xi', \xi_n) \tilde{v}_k(\xi') \right), \tag{12.11}$$

which is analogous to (8.36).

Analogously to (8.4), the Fourier transform with respect to x' of the boundary conditions (12.4) is

$$\Pi' \hat{B}_j(\xi) \tilde{u}_+(\xi) + \sum_{k=1}^{m_-} \hat{E}_{jk}(\xi') \tilde{v}_k(\xi') = \tilde{g}_j(\xi'), \qquad 1 \leqslant j \leqslant m_+. \tag{12.12}$$

Substituting (12.11) in (12.12), we get

$$\sum_{j=1}^{m_-} \hat{d}_{|m|+k,j}(\xi') \tilde{v}_j(\xi') = \tilde{\varphi}_{|m|+k}(\xi'), \qquad 1 \leqslant k \leqslant m_+, \tag{12.13}$$

where

$$\hat{d}_{|m|+k,j}(\xi') = -\Pi' \frac{\hat{B}_j(\xi)}{\hat{A}_+(\xi)\hat{\Lambda}_+^{|m|}} \Pi^+ \frac{\hat{\Lambda}_+^{|m|} \hat{C}_k(\xi)}{\hat{A}_-(\xi)} + \hat{E}_{kj}(\xi'),$$

$$1 \leqslant k \leqslant m_+, \qquad 1 \leqslant j \leqslant m_-, \tag{12.14}$$

$$\tilde{\varphi}_{|m|+k}(\xi') = \tilde{g}_k(\xi') - \Pi' \frac{\hat{B}_j(\xi)}{\hat{A}_+ \hat{\Lambda}_+^{|m|}} \Pi_+ \frac{\hat{\Lambda}_+^{|m|} \widetilde{lf}}{\hat{A}_-(\xi)}, \qquad 1 \leqslant k \leqslant m_+. \tag{12.15}$$

From (12.5), (12.6) and (12.14) it follows that the degree of homogeneity of $d_{|m|+k,j}(\xi')$ is $\delta_{jk} = \beta_j + \gamma_k - \alpha - i\beta + 1$.

Since each function $\Lambda_+^{|m|}(\xi) C_k(\xi)/A_-(\xi)$ is continuous for $\xi' \neq 0$ and, by (12.6), satisfies the estimate

$$|\Lambda_+^{|m|} C_k(\xi)/A_-(\xi)| \leqslant C|\xi'|^{\operatorname{Re}\gamma_k - \gamma_{k1}} (|\xi'| + |\xi_n|)^{\gamma_{k1} + |m| - \alpha + \operatorname{Re}\kappa},$$

in which $\gamma_{k1} - \alpha + |m| + \operatorname{Re}\kappa = \gamma_{k1} - \alpha + s + \delta < -1/2 + \delta$, it is a continuous function of ξ' for $\xi' \neq 0$ with values in $\tilde{H}_{-\delta}(\mathbf{R}^1)$. It follows from Theorem 5.1 that $\Pi^+(\Lambda_+^{|m|} C_k(\xi)/A_-(\xi))$ is also a continuous function of ξ' for $\xi' \neq 0$ with values in $\tilde{H}_{-\delta}(\mathbf{R}^1)$. Analogously, (12.5) implies that each $B_j(\xi)/A_+\Lambda_+^{|m|}$ is a continuous function of ξ' for $\xi' \neq 0$ with values in $\tilde{H}_\delta(\mathbf{R}^1)$. Consequently, by (12.14) and (4.4), each function $d_{|m|+k,j}(\xi')$ is continuous for $\xi' \neq 0$.

System (12.8), (12.13) consists of $|m| + m_+ = m_-$ equations for the m_- unknown functions $\tilde{v}_k$, $1 \leqslant k \leqslant m_-$. Under the fulfilment of the condition

$$\det\|d_{kj}(\xi')\|_{k,j=1}^{m_-} \neq 0 \quad \text{for } \xi' \neq 0 \tag{12.16}$$

it has the unique solution

$$\tilde{v}_j(\xi') = \sum_{k=1}^{m_-} \hat{d}_{jk}^{(1)}(\xi')\tilde{\varphi}_k(\xi'), \tag{12.17}$$

where $\|\hat{d}_{jk}^{(1)}(\xi')\|_{j,k=1}^{m_-}$ is the inverse of the matrix $\|\hat{d}_{kj}(\xi')\|_{j,k=1}^{m_-}$. Since each function $d_{jk}^{(1)}(\xi')$ is continuous for $\xi' \neq 0$ and homogeneous, we obtain from (12.17)

$$\sum_{j=1}^{m_-} [v_j]_{s_j} \leqslant C \sum_{k=1}^{|m|} [\varphi_k]_{s-\operatorname{Re}\kappa-k+1/2} + C \sum_{k=1}^{m_+} [\varphi_{|m|+k}]_{s-\operatorname{Re}\beta_k-1/2},$$

$$s_j = s - \alpha + \operatorname{Re}\gamma_j + 1/2, \tag{12.18}$$

which together with (12.10), (12.14), Lemma 4.4 and Theorems 4.2 and 5.1 implies

$$\sum_{j=1}^{m_-} [v_j]_{s_j} \leqslant C\left(\|lf\|_{s-\alpha}^{+} + \sum_{k=1}^{m_-} [g_k]_{s-\operatorname{Re}\beta_k-1/2}\right). \tag{12.19}$$

Finally, substituting (12.17) in (12.11), we get

$$\tilde{u}_+(\xi) = \frac{1}{\hat{A}_+\hat{\Lambda}_+^{|m|}}\Pi_+\frac{\hat{A}_+^{|m|}}{\hat{A}_-}\left(\widetilde{lf} - \sum_{j,k=1}^{m_-} \hat{C}_k(\xi)\hat{d}_{kj}^{(1)}(\xi')\tilde{\varphi}_j(\xi')\right), \tag{12.20}$$

which implies

$$\|u_+\|_s \leqslant C\left(\|lf\|_{s-\alpha}^{+} + \sum_{k=1}^{m_-} [g_k]_{s-\operatorname{Re}\beta_j-1/2}\right). \tag{12.21}$$

Thus the Fourier transform of the unique solution of the boundary value problem (12.3), (12.4) is given by (12.17), (12.20), (12.10) and (12.15). In addition, (12.19) and (12.21) hold. It is easily verified that (12.17), (12.20), (12.10) and (12.15) determine a solution $u_+ \in H_s^+$, $v_k(x') \in H_{s_k}(\mathbf{R}^{n-1})$, $s_k = s - \alpha + \operatorname{Re}\gamma_k + 1/2$, $1 \leqslant k \leqslant m_-$, of the boundary value problem (12.3), (12.4) for any $f \in H_{s-\alpha}(\mathbf{R}_+^n)$, $g_j(x') \in H_{s-\operatorname{Re}\beta_j-1/2}(\mathbf{R}^{n-1})$.

b) Let us now consider the case $m \geqslant 0$. Analogously to (7.33), the Fourier transform of the general solution of (12.7) has the form

$$\tilde{u}_+(\xi) = \frac{\hat{Q}(\xi)}{\hat{A}_+(\xi)}\Pi^+\frac{1}{\hat{Q}\hat{A}^-}\left(\widetilde{lf}(\xi) - \sum_{k=1}^{m_-} \hat{C}_k(\xi)\tilde{v}_k(\xi')\right) + \sum_{k=1}^{m} \frac{\tilde{v}_{m_-+k}(\xi')\xi_n^{k-1}}{\hat{A}_+(\xi)}, \tag{12.22}$$

where $Q(\xi)$ is an arbitrary elliptic polynomial in ξ_n that is homogeneous of degree m with respect to (ξ', ξ_n) and $\tilde{v}_{m_-+k}(\xi') \in \tilde{H}_{s-\operatorname{Re}\kappa+k-1/2}(\mathbf{R}^{n-1})$. When $m = 0$, the last sum in (12.22) should be omitted.

Substituting (12.22) in the boundary conditions (12.12), we get

$$\sum_{k=1}^{m_+} \hat{e}_{jk}(\xi')\tilde{v}_k(\xi') = \tilde{g}_j(\xi') - \tilde{f}_j(\xi'), \qquad 1 \leqslant j \leqslant m_+, \tag{12.23}$$

where

$$\tilde{f}_j(\xi') = \Pi' \frac{\hat{B}_j(\xi)\hat{Q}(\xi)}{\hat{A}_+(\xi)} \Pi^+ \frac{\tilde{lf}}{\hat{Q}\hat{A}_-}, \tag{12.24}$$

$$\hat{e}_{jk}(\xi') = -\Pi' \frac{\hat{B}_j\hat{Q}}{\hat{A}_+} \Pi^+ \frac{\hat{C}_k}{\hat{Q}\hat{A}_-} + \hat{E}_{jk}(\xi'), \qquad 1 \leqslant j \leqslant m_+, \qquad 1 \leqslant k \leqslant m_-, \tag{12.25}$$

$$\hat{e}_{j,k+m_-}(\xi') = \Pi' \frac{\hat{B}_j\xi_n^{k-1}}{\hat{A}_+}, \qquad 1 \leqslant j \leqslant m_+, \qquad 1 \leqslant k \leqslant m. \tag{12.26}$$

It can be shown analogously to (12.9) and (12.14) that each $e_{jk}(\xi')$ is a continuous function for $\xi' \neq 0$ that is homogeneous of degree $\beta_j + \gamma_k - \alpha - i\beta + 1$ for $1 \leqslant j \leqslant m_+$, $1 \leqslant k \leqslant m_-$, and that each $e_{j,k+m_-}(\xi')$ is a continuous function for $\xi' \neq 0$ that is homogeneous of degree $\beta_j + k - \kappa$.

Under the fulfilment of the condition

$$\det\|e_{jk}(\xi')\|_{j,k=1}^{m_+} \neq 0 \quad \text{for } \xi' \neq 0 \tag{12.27}$$

system (12.23) has the unique solution

$$\tilde{v}_k(\xi') = \sum_{j=1}^{m_+} \hat{e}_{kj}^{(1)}(\xi')\big(\tilde{g}_j(\xi') - \tilde{f}_j(\xi')\big), \qquad 1 \leqslant k \leqslant m_+, \tag{12.28}$$

where $\|\hat{e}_{kj}^{(1)}(\xi')\|_{k,j=1}^{m_+}$ is the inverse of the matrix $\|\hat{e}_{jk}(\xi')\|_{j,k=1}^{m_+}$. Since the matrix $\|\hat{e}_{kj}^{(1)}(\xi')\|_{k,j=1}^{m_+}$ is clearly continuous for $\xi' \neq 0$, it follows from (12.28) that

$$\sum_{k=1}^{m_-} [v_k]_{s-\alpha+\operatorname{Re}\gamma_k+1/2} + \sum_{k=1}^{m} [v_{m_-+k}]_{s-\operatorname{Re}\kappa+k-1/2} \leqslant C\left(\|f\|^{+}_{s-\alpha} + \sum_{j=1}^{m} [g_j]_{s-\operatorname{Re}\beta_j-1/2}\right). \tag{12.29}$$

Finally, substituting (12.28) in (12.22), we get

$$\tilde{u}_+(\xi) = \frac{\hat{Q}}{\hat{A}_+}\Pi^+\frac{1}{\hat{Q}\hat{A}_-}\left(\widetilde{lf} - \sum_{j=1}^{m_+}\sum_{k=1}^{m_-}\hat{C}_k(\xi)\hat{e}_{kj}^{(1)}\big(\tilde{g}_j(\xi') - \tilde{f}_j(\xi')\big)\right)$$

$$+\sum_{j=1}^{m_+}\sum_{k=1}^{m}\frac{\hat{e}_{m_-+k,j}^{(1)}(\xi')\xi_n^{k-1}}{\hat{A}_+(\xi)}\big(\tilde{g}_j(\xi') - \tilde{f}_j(\xi')\big). \qquad (12.30)$$

Consequently, when $m \geqslant 0$, the Fourier transform of the unique solution of the boundary value problem (12.3), (12.4) has the form (12.28), (12.30). When $m = 0$, the second double sum in (12.30) should be omitted. Also, as in the case $m < 0$, estimates (12.19) and (12.21) hold. It is easily verified that (12.28), (12.30) and (12.24) determine the solution of the boundary value problem (12.3), (12.4) for any $f \in H_{s-\alpha}(\mathbf{R}_+^n)$ and $g_j(x') \in H_{s-\operatorname{Re}\beta_j-1/2}(\mathbf{R}^{n-1})$, $1 \leqslant j \leqslant m_+$. We thus have

THEOREM 12.1. *Suppose $A(\xi) \in O_{\alpha+i\beta}^{\infty}$ is an elliptic symbol with factorization index κ, suppose $\operatorname{Re}\kappa - s = m + \delta$, where m is an integer and $|\delta| < 1/2$, suppose m_+ and m_- are nonnegative integers such that $m_+ - m_- = m$, suppose $B_j(\xi) \in O'_{\beta_j}$, $C_k(\xi) \in O'_{\gamma_k}$ and $E_{jk}(\xi')$ are continuous functions for $\xi' \neq 0$ that are homogeneous, with the degree of homogeneity of $E_{jk}(\xi')$ being equal to $\beta_j + \gamma_k - \alpha - i\beta + 1$, $1 \leqslant j \leqslant m_+$, $1 \leqslant k \leqslant m_-$, suppose* (12.5) *and* (12.6) *hold, and suppose* (12.16) *is satisfied for $m < 0$ or* (12.27) *is satisfied for $m \geqslant 0$. Then the boundary value problem* (12.3), (12.4) *has a unique solution*

$$u_+ \in H_s^+,\ v_k(x') \in H_{s-\alpha+\operatorname{Re}\gamma_k+1/2}(\mathbf{R}^{n-1}), \qquad 1 \leqslant k \leqslant m_-,$$

for any $f \in H_{s-\alpha}(\mathbf{R}_+^n)$, $g_j(x') \in H_{s-\operatorname{Re}\beta_j-1/2}(\mathbf{R}^{n-1})$, $1 \leqslant j \leqslant m_+$. In addition, (12.19) *and* (12.21) *are valid.*

Theorem 12.1 contains as special cases Theorems 8.1 and 8.2: if $m > 0$, $m_- = 0$, $m_+ = m$, we obtain the assertion of Theorem 8.1; while if $m < 0$, $m_- = |m|$, $m_+ = 0$, we obtain Theorem 8.2. We note that if the symbols $C_k(\xi)$ are of class D_{γ_k}, then, as in Theorem 8.2, the restrictions (12.6) can be dropped.

REMARK 12.1. Let

$$\mathcal{H}_s^{(1)} = H_s^+ \times \prod_{k=1}^{m_-} H_{s-\alpha+\operatorname{Re}\gamma_k+1/2}(\mathbf{R}^{n-1}),$$

$$\mathcal{H}_s^{(2)} = H_{s-\alpha}(\mathbf{R}_+^n) \times \prod_{j=1}^{m_+} H_{s-\operatorname{Re}\beta_j-1/2}(\mathbf{R}^{n-1}).$$

The operator $\mathfrak{A}$ defined by the left sides of (12.3) and (12.4) is a bounded operator from $\mathcal{H}_s^{(1)}$ into $\mathcal{H}_s^{(2)}$. Theorem 12.1 asserts that there exists (and can even be found in explicit form) a bounded operator R from $\mathcal{H}_s^{(2)}$ into $\mathcal{H}_s^{(1)}$

that is the inverse of $\mathfrak{A}$, i.e.

$$R\mathfrak{A} = I_1, \qquad \mathfrak{A}R = I_2, \tag{12.31}$$

where I_1 and I_2 are the identity operators in $\mathcal{H}_s^{(1)}$ and $\mathcal{H}_s^{(2)}$.

The following theorem, which is analogous to Theorem 12.1 and contains Theorems 11.1 and 11.3, holds for an elliptic symbol of class $D^{(0)}_{\alpha+i\beta}$.

THEOREM 12.2. *Suppose $A(\xi) \in D^{(0)}_{\alpha+i\beta}$ is an elliptic symbol with factorization index κ, suppose m_+ and m_- are arbitrary nonnegative integers such that $m_+ - m_- = \kappa$, suppose $B_j(\xi) \in D_{\beta_j}$, $C_k(\xi) \in D_{\gamma_k}$ and $E_{jk}(\xi')$ are continuous functions for $\xi' \neq 0$ and the degree of homogeneity of $E_{jk}(\xi')$ is equal to $\beta_j + \gamma_k - \alpha - i\beta + 1$, $1 \leqslant j \leqslant m_+$, $1 \leqslant k \leqslant m_-$, suppose the condition analogous to* (12.16) *is satisfied when $\kappa < 0$ and the condition analogous to* (12.27) *is satisfied with $\kappa \geqslant 0$, and suppose*

$$r \geqslant 0, \quad r > \max_{1 \leqslant j \leqslant m_+} (\operatorname{Re} \beta_j + 1/2).$$

Then the boundary value problem (12.3), (12.4) *has a unique solution $u(x) \in H_r(\mathbf{R}^n_+)$, $v_k(x') \in H_{r-\alpha+\operatorname{Re}\gamma_k+1/2}(\mathbf{R}^{n-1})$, $1 \leqslant k \leqslant m_-$, for any $f \in H_{r-\alpha}(\mathbf{R}^n_+)$, $g_j(x') \in H_{r-\operatorname{Re}\beta_j-1/2}(\mathbf{R}^{n-1})$, $1 \leqslant j \leqslant m_+$. In addition,*

$$\|u\|_r^+ + \sum_{k=1}^{m_-} [v_k]_{r-\alpha+\operatorname{Re}\gamma_k+1/2} \leqslant C\left(\|f\|^+_{r-\alpha} + \sum_{j=1}^{m_+} [g_j]_{r-\operatorname{Re}\beta_j-1/2}\right). \tag{12.32}$$

Theorem 12.2 is proved in basically the same way as Theorem 12.1. The variations in the proof are the same as the variations in the proofs of Theorems 11.1 and 11.3 in comparison with Theorems 8.1 and 8.2.

EXAMPLE 12.1. Let $\hat{\Delta}$ be the p.o. with symbol $-[\xi_n^2 + (|\xi'| + 1)^2]$, $G(\xi) \in D_\gamma$, $\operatorname{Re}\gamma < 0$, and consider the boundary value problem

$$p\big(\hat{\Delta}u_+ + G(v(x') \times \delta(x_n))\big) = f(x), \tag{12.33}$$

$$u(x', x_n)|_{x_n=+0} = g_1(x'), \qquad \left.\frac{\partial u(x)}{\partial x_n}\right|_{x_n=+0} = g_2(x'). \tag{12.34}$$

Here $\kappa = 1$, $m_+ = 2$, $m_- = 1$, so that $m_+ - m_- = \kappa$. It is assumed that

$$f(x) \in H_{r-2}(\mathbf{R}^n_+), \quad g_j(x') \in H_{r-j+1/2}(\mathbf{R}^{n-1}), \quad 1 \leqslant j \leqslant 2, \quad r > 3/2.$$

According to Example 11.2, the general solution $u(x) \in H_r(\mathbf{R}^n_+)$ of (12.33) has the form

$$u(x', x_n) = -F_\xi^{-1}\frac{\widetilde{lf}(\xi) - \hat{G}(\xi)\tilde{v}(\xi')}{\xi_n^2 + (|\xi'| + 1)^2} + F_\xi^{-1}\frac{\tilde{w}(\xi')}{\xi_n + i(|\xi'| + 1)}$$

$$= -F_\xi^{-1}\frac{\widetilde{lf}(\xi) - \hat{G}(\xi)\tilde{v}(\xi')}{\xi_n^2 + (|\xi'| + 1)^2} + iF_{\xi'}^{-1}e^{-x_n(|\xi'|+1)}\tilde{w}(\xi'), \qquad x_n > 0, \tag{12.35}$$

where $\tilde{w}(\xi') \in \tilde{H}_{r-1/2}(\mathbf{R}^{n-1})$. Substituting (12.35) in the boundary conditions (12.34) and taking the Fourier transform with respect to x', we obtain the following system of equations for $\tilde{w}(\xi')$ and $\tilde{v}(\xi')$:

$$\begin{gathered} i\tilde{w}(\xi') + \hat{c}_1(\xi')\tilde{v}(\xi') = \tilde{g}_1(\xi') - \tilde{f}_1(\xi'), \\ -i(1+|\xi'|)\tilde{w}(\xi') + \hat{c}_2(\xi')\tilde{v}(\xi') = \tilde{g}_2(\xi') - \tilde{f}_2(\xi'), \end{gathered} \tag{12.36}$$

where

$$\hat{c}_1(\xi') = \Pi' \frac{\hat{G}(\xi)}{\xi_n^2 + (|\xi'|+1)^2}, \qquad \hat{c}_2(\xi') = -i\Pi' \frac{\xi_n \hat{G}(\xi)}{\xi_n^2 + (|\xi'|+1)^2}, \tag{12.37}$$

$$\tilde{f}_1(\xi') = -\Pi' \frac{\tilde{lf}(\xi)}{\xi_n^2 + (|\xi'|+1)^2}, \qquad \tilde{f}_2(\xi') = i\Pi' \frac{\xi_n \tilde{lf}(\xi)}{\xi_n^2 + (|\xi'|+1)^2}. \tag{12.38}$$

If the determinant $\hat{c}_0(\xi')$ of (12.36) doesn't vanish for all ξ', then (12.36) determines unique functions $\tilde{w}(\xi') \in \tilde{H}_{r-1/2}(\mathbf{R}^{n-1})$, $\tilde{v}(\xi') \in \tilde{H}_{r-3/2+\operatorname{Re}\gamma}(\mathbf{R}^{n-1})$ and hence the unique solution $u(x) \in H_r(\mathbf{R}_+^n)$ of the boundary value problem (12.33), (12.34).

An example in which $\hat{c}_0(\xi') \neq 0$ for all ξ' is obtained by taking

$$\hat{G}(\xi) = \frac{1}{\xi_n^2 + (|\xi'|+1)^2},$$

since then $\hat{c}_1(\xi') \neq 0$ for all ξ' and $\hat{c}_2(\xi') \equiv 0$.

We note that, when $\tilde{v}(\xi') = 0$, $u(x', x_n)$ is a solution of a Cauchy problem for an elliptic equation of second order, i.e. a solution of an overdetermined problem. The addition of a potential in (12.33) removes this overdeterminateness. On the other hand, the conditions for the vanishing of $v(x')$, which are easily obtained from (12.36), yield a relation between $g_1(x')$, $g_2(x')$ and $f(x)$ that is necessary and sufficient for the existence of a solution of the Cauchy problem.

EXAMPLE 12.2. Let us consider the general boundary value problem (12.3), (12.4) in the case when $A(\xi) = P(\xi)/Q(\xi) \in D^{(0)}_{2(m_1-m_2)}$ is a rational function in which $P(\xi)$ and $Q(\xi)$ are homogeneous elliptic polynomials of degrees $2m_1$ and $2m_2$ respectively, and the $B_j(\xi)$ are also polynomials in ξ. We will show that in this case the proof of Theorem 12.2 can be obtained by reducing problem (12.3), (12.4) to a system of simultaneous differential equations.

Let $\hat{Q}^{-1}(D)u_+ = v$ and $v^\pm$ be the restriction of v to $\mathbf{R}^n_\pm$. Since $u_+ \in H_0^+$, we have $v \in H_{2m_2}(\mathbf{R}^n)$ and $v^\pm \in H_{2m_2}(\mathbf{R}^n_\pm)$. Consequently,

$$D_n^{k-1}v^+\big|_{x_n=+0} = D_n^{k-1}v^-\big|_{x_n=-0}, \qquad 1 \leqslant k \leqslant 2m_2. \tag{12.39}$$

While from the boundary value problem (12.3), (12.4) we obtain the relations

$$p\left(\hat{P}(D)v^+ + \sum_{k=1}^{m_-} \hat{C}_k(D)(v_k(x') \times \delta(x_n))\right) = f(x), \tag{12.40}$$

$$p'_+\hat{B}_j(D)\hat{Q}(D)v^+ + \sum_{k=1}^{m_-} \hat{E}_{jk}(D')v_k(x') = g_j(x'), \qquad 1 \leqslant j \leqslant m_+, \tag{12.41}$$

$$\hat{Q}(D)v^- = 0 \quad \text{for } x \in \mathbf{R}^n_-. \tag{12.42}$$

The general solution $v^+ \in H_{2m_2}(\mathbf{R}^n_+)$ of (12.40) depends on m_1 arbitrary functions $c_1^+(x'), \ldots, c_{m_1}^+(x')$ (see Example 11.2). Analogously, the general solution $v^- \in H_{2m_2}(\mathbf{R}^n_-)$ of (12.42) depends on m_2 arbitrary functions $c_1^-(x'), \ldots, c_{m_2}^-(x')$. Substituting the expressions for v^+ and v^- in (12.39) and (12.41), we obtain a system of $2m_2 + m_+$ equations in $\mathbf{R}^{n-1}$ for the determination of the $m_1 + m_2 + m_-$ $(= 2m_2 + m_+)$ unknown functions $c_1^+, \ldots, c_{m_1}^+, c_1^-, \ldots, c_{m_2}^-, v_1, \ldots, v_{m_-}$. This system is uniquely solvable under the fulfilment of the Šapiro-Lopatinskiĭ condition.

§13. Mixed boundary value problems for an elliptic equation of second order in a halfspace

1. We consider in $\mathbf{R}^n_+$ an elliptic differential equation of second order with constant coefficients

$$L(D)u(x) \equiv -\sum_{j,k=1}^{n} a_{jk}\frac{\partial^2 u(x)}{\partial x_j \partial x_k} = f(x), \tag{13.1}$$

where the a_{jk} are real numbers such that $\sum_{j,k=1}^n a_{jk}\xi_j\xi_k > 0$ for $\xi = (\xi_1, \ldots, \xi_n) \neq 0$.

Let $\mathbf{R}^{n-1}$ denote the $(n-1)$-dimensional hyperplane defined in $\mathbf{R}^n$ by the equation $x_n = 0$, where $x = (x', x_n) \in \mathbf{R}^n$ and $x' = (x_1, \ldots, x_{n-1})$, let $\mathbf{R}^{n-1}_+ \subset \mathbf{R}^{n-1}$ denote the $(n-1)$-dimensional halfspace $x_{n-1} > 0$, $x_n = 0$, and let $\mathbf{R}^{n-1}_- \subset \mathbf{R}^{n-1}$ denote the halfspace $x_{n-1} < 0$, $x_n = 0$. For equation (13.1) we pose mixed boundary conditions on the hyperplane $x_n = 0$, i.e. we assign one boundary condition on $\mathbf{R}^{n-1}_+$ and another on $\mathbf{R}^{n-1}_-$:

$$B_1(D)u(x)|_{\mathbf{R}^{n-1}_+} = g_1(x'), \qquad B_2(D)u(x)|_{\mathbf{R}^{n-1}_-} = g_2(x'), \tag{13.2}$$

where $B_1(D)$ and $B_2(D)$ are homogeneous differential operators of orders m_1 and m_2 with constant coefficients.

Let

$$\hat{\hat{L}}(\xi'', \xi_{n-1}, \xi_n) = L((1 + |\xi''|)\omega'', \xi_{n-1}, \xi_n),$$
$$\hat{\hat{B}}_i(\xi'', \xi_{n-1}, \xi_n) = B_i((1 + |\xi''|)\omega'', \xi_{n-1}, \xi_n),$$
$$i = 1, 2, \quad \xi'' = (\xi_1, \ldots, \xi_{n-2}), \quad \omega'' = \xi''/|\xi''|,$$

where $L(\xi'', \xi_{n-1}, \xi_n)$ and $B_i(\xi'', \xi_{n-1}, \xi_n)$ are the symbols of the differential operators $L(D)$, $B_i(D)$. The pseudodifferential operators $\hat{\hat{L}}$, $\hat{\hat{B}}_1$ and $\hat{\hat{B}}_2$ with symbols $\hat{\hat{L}}(\xi)$, $\hat{\hat{B}}_1(\xi)$ and $\hat{\hat{B}}_2(\xi)$ are differential operators with respect to x_{n-1} and x_n but not with respect to the remaining variables. Analogously to the way we considered the p.o. with symbol $\hat{A}_0(\xi)$ in place of a p.o. with homogeneous symbol $A_0(\xi)$ in §§7–12, it will be more convenient for us to consider in place of the mixed boundary value problem (13.1), (13.2) the mixed boundary value problem

$$\hat{\hat{L}}(D)u(x) = f(x), \qquad x \in \mathbf{R}^n_+, \tag{13.1'}$$

$$\hat{\hat{B}}_1(D)u(x)\Big|_{\mathbf{R}^{n-1}_+} = g_1(x'), \qquad \hat{\hat{B}}_2(D)u(x)\Big|_{\mathbf{R}^{n-1}_-} = g_2(x'). \tag{13.2'}$$

2. Even when $f(x)$, $g_1(x')$ and $g_2(x')$ are infinitely differentiable, the solution $u(x)$ of problem (13.1′), (13.2′) generally has a singularity at $x_n = x_{n-1} = 0$ (see §13.6 below) while being infinitely differentiable elsewhere. It is therefore natural to study problem (13.1′), (13.2′) in classes of functions $u(x)$ belonging to $H_s(\mathbf{R}^n_+)$ after multiplication by a weight factor that vanishes at $x_n = x_{n-1} = 0$. If s is sufficiently large, such functions $u(x)$ will be sufficiently smooth off the hyperline $x_n = x_{n-1} = 0$. Mixed boundary value problems will be studied in weighted function classes in Chapter VI. In the present section, however, for the sake of simplicity we will seek the solution of a mixed problem in somewhat less natural but simpler spaces of functions having one smoothness with respect to x_n and another with respect to x', the smoothness with respect to x_n being sufficient to ensure the existence of restrictions to the hyperplane $x_n = 0$ of the functions $\hat{\hat{B}}_1(D)u(x)$ and $\hat{\hat{B}}_2(D)u(x)$.

Let $H_{s,r}(\mathbf{R}^n)$ denote the space of generalized functions u in $\mathbf{R}^n$ with finite norm

$$\|u\|^2_{s,r} = \int_{-\infty}^{\infty} (1 + |\xi'| + |\xi_n|)^{2s}(1 + |\xi'|)^{2r}|\tilde{u}(\xi', \xi_n)|^2\, d\xi'\, d\xi_n,$$

$$\tilde{u}(\xi', \xi_n) = Fu. \tag{13.3}$$

Clearly,

$$H_{s+r_+}(\mathbf{R}^n) \subset H_{s,r}(\mathbf{R}^n) \subset H_{s+r_-}(\mathbf{R}^n),$$

$$\text{where } r_+ = \max(r, 0), \quad r_- = \min(r, 0).$$

The space $H_{s,r}(\mathbf{R}^n_+)$ consists, by definition, of the generalized functions f in $\mathbf{R}^n_+$ admitting extensions onto $\mathbf{R}^n$ that belong to $H_{s,r}(\mathbf{R}^n)$. The norm in $H_{s,r}(\mathbf{R}^n_+)$ is given by the following formula, which is analogous to (4.51):

$$\|f\|^+_{s,r} = \inf_l \|lf\|_{s,r}. \tag{13.4}$$

The infimum here is taken over all of the extensions of f that belong to $H_{s,r}(\mathbf{R}^n)$.

We note the following properties of $H_{s,r}(\mathbf{R}^n)$ and $H_{s,r}(\mathbf{R}^n_+)$.

LEMMA 13.1. *Suppose $A(\xi', \xi_n) \in S^0_\alpha$. Then the p.o. A with symbol $A(\xi', \xi_n)$ is a bounded operator from $H_{s,r}(\mathbf{R}^n)$ into $H_{s-\alpha,r}(\mathbf{R}^n)$ for any s and r.*

LEMMA 13.2. *Suppose $u(x', x_n) \in H_{s,r}(\mathbf{R}^n_+)$, $s > 1/2$. Then the restriction p'_+u of $u(x', x_n)$ to the hyperplane $x_n = 0$ is defined, belongs to $H_{s+r-1/2}(\mathbf{R}^{n-1})$ and satisfies*

$$[p'_+u]_{s+r-1/2} \leqslant c\|u\|^+_{s,r}. \tag{13.5}$$

LEMMA 13.3. *Suppose $A(\xi', \xi_n) \in D_{\alpha+i\beta}$. Then the operator of potential type $p\hat{A}(v(x') \times \delta(x_n))$ is a bounded operator from $H_{s+r+1/2}(\mathbf{R}^{n-1})$ into $H_{s-\alpha,r}(\mathbf{R}^n_+)$ for any s and r, i.e.*

$$\|p\hat{A}(v(x') \times \delta(x_n))\|^+_{s-\alpha,r} \leqslant c[v]_{s+r+1/2}, \qquad \forall v \in H_{s+r+1/2}(\mathbf{R}^{n-1}). \tag{13.6}$$

The proofs of Lemmas 13.1–13.3 do not differ from the proofs of Lemma 4.4, Theorem 4.2 and Lemma 8.1.

3. Suppose $u(x) \in H_{s,r}(\mathbf{R}^n_+)$, where $s > \max(m_1 + 1/2, m_2 + 1/2)$, is a solution of the boundary value problem (13.1′), (13.2′). Then, by Lemmas 13.1 and 13.2,

$$f(x) \in H_{s-2,r}(\mathbf{R}^n_+), \qquad g_1(x') \in H_{s+r-m_1-1/2}(\mathbf{R}^{n-1}_+),$$
$$g_2(x') \in H_{s+r-m_2-1/2}(\mathbf{R}^{n-1}_-), \tag{13.7}$$

with the following estimate holding:

$$\|f\|^+_{s-2,r} + [g_1]^+_{s+r-m_1-1/2} + [g_2]^-_{s+r-m_2-1/2} \leqslant C\|u\|^+_{s,r}, \tag{13.8}$$

in which $[\cdot]^\pm_s$ denotes the norm in $H_s(\mathbf{R}^{n-1}_\pm)$.

We will determine conditions on s, r, $B_1(D)$ and $B_2(D)$ that ensure the unique solvability of the boundary value problem (13.1′), (13.2′).

Let $L(\xi', \xi_n) = a_{nn}(\xi_n - \lambda_1(\xi'))(\xi_n - \lambda_2(\xi'))$ be the factorization of $L(\xi', \xi_n)$ with respect to ξ_n, with $\operatorname{Im}\lambda_1(\xi') > 0$ for $\xi' \neq 0$ and $\operatorname{Im}\lambda_2(\xi') < 0$ for $\xi' \neq 0$. The general solution of (13.1′) in $\mathbf{R}^n_+$ that belongs to $H_{s,r}(\mathbf{R}^n_+)$ has the form (cf. (12.35))

$$u(x', x_n) = F^{-1}\frac{\widetilde{lf}}{\hat{\hat{L}}(\xi)} + F^{-1}\frac{\tilde{v}(\xi')}{\xi_n - \hat{\lambda}_2(\xi')}$$
$$= F^{-1}\frac{\widetilde{lf}}{\hat{\hat{L}}(\xi)} + \frac{i}{(2\pi)^{n-1}}\int_{-\infty}^{\infty} e^{-ix_n\hat{\lambda}_2(\xi') - i(x',\xi')}\tilde{v}(\xi')\,d\xi', \quad x_n > 0, \tag{13.9}$$

where $\hat{\lambda}_2(\xi') = \lambda_2((1 + |\xi''|)\omega'', \xi_{n-1})$, $lf \in H_{s,r}(\mathbf{R}^n)$ is an extension of f and $v(x') \in H_{s+r-1/2}(\mathbf{R}^{n-1})$. Also, by Lemmas 13.1 and 13.3,

$$\|u\|^{+}_{s,r} \leqslant c\left(\|lf\|^{+}_{s-2,r} + [v]_{s+r-1/2}\right). \tag{13.10}$$

Let p_+ (p_-) denote the operator of restriction to $\mathbf{R}^{n-1}_+$ ($\mathbf{R}^{n-1}_-$) of a function defined in $\mathbf{R}^{n-1}$. Substituting (13.9) in the boundary conditions (13.2′), we get

$$p_+\hat{b}_1 v = g_1(x') - f_1(x'), \qquad x_{n-1} > 0, \tag{13.11}$$

$$p_-\hat{b}_2 v = g_2(x') - f_2(x'), \qquad x_{n-1} < 0, \tag{13.12}$$

where $\hat{b}_k$ is the p.o. in $\mathbf{R}^{n-1}$ with symbol

$$\hat{b}_k(\xi') = iB_k((1 + |\xi''|)\omega'', \xi_{n-1}, \hat{\lambda}_2(\xi')),$$

and

$$f_k(x') = \frac{1}{(2\pi)^n}\int_{-\infty}^{\infty} \frac{\hat{\hat{B}}_i(\xi)\widetilde{lf}(\xi)}{\hat{L}(\xi)} e^{-i(x',\xi')}\, d\xi'\, d\xi_n, \qquad k = 1, 2. \tag{13.13}$$

4. *Paired equations*. Equations of form (13.11), (13.12) are called *paired equations*. It is assumed that $B_1(D)$ and $B_2(D)$ satisfy the Šapiro-Lopatinskiĭ condition and hence that

$$\hat{b}_1(\xi') \neq 0, \qquad \hat{b}_2(\xi') \neq 0. \tag{13.14}$$

Let $lg_i(x') \in H_{s+r-m_i-1/2}(\mathbf{R}^{n-1})$, $i = 1, 2$, be extensions of the functions $g_1 \in H_{s+r-m_1-1/2}(\mathbf{R}^{n-1}_+)$ and $g_2 \in H_{s+r-m_2-1/2}(\mathbf{R}^{n-1}_-)$ onto $\mathbf{R}^{n-1}$ and let

$$v_-(x') = lg_1(x') - f_1(x') - \hat{b}_1 v, \tag{13.15}$$

$$v_+(x') = lg_2(x') - f_2(x') - \hat{b}_2 v, \tag{13.16}$$

which implies

$$v_+ \in H^{+}_{s+r-m_2-1/2}(\mathbf{R}^{n-1}), \quad v_- \in H^{-}_{s+r-m_1-1/2}(\mathbf{R}^{n-1}).$$

The Fourier transform with respect to $x' = (x'', x_{n-1})$ is

$$\hat{b}_1(\xi')\tilde{v}(\xi') + \tilde{v}_-(\xi') = \widetilde{lg_1}(\xi') - \tilde{f}_1(\xi'), \tag{13.17}$$

$$\hat{b}_2(\xi')\tilde{v}(\xi') + \tilde{v}_+(\xi') = \widetilde{lg_2}(\xi') - \tilde{f}_2(\xi'). \tag{13.18}$$

Eliminating $\tilde{v}(\xi')$ from (13.17) and (13.18), we obtain the following equation of form (7.14):

$$\hat{b}_1(\xi')\hat{b}_2^{-1}(\xi')\tilde{v}_+(\xi') - \tilde{v}_-(\xi') = \hat{b}_1(\xi')\hat{b}_2^{-1}(\xi')\left(\widetilde{lg_2} - \tilde{f}_2\right) - \widetilde{lg_1} + \tilde{f}_1. \tag{13.19}$$

Thus paired equations reduce to a p.e. of form (7.11).

Let

$$\hat{b}_1(\xi'', \xi_{n-1})\hat{b}_2^{-1}(\xi'', \xi_{n-1}) = \hat{b}_-(\xi'', \xi_{n-1})\hat{b}_+(\xi'', \xi_{n-1}) \tag{13.20}$$

be the factorization of $\hat{b}_1\hat{b}_2^{-1}$ with respect to ξ_n and let $\kappa = \deg_{\xi'} b_+(\xi'', \xi_{n-1})$ be its factorization index. Substituting (13.20) in (13.19) and dividing by $\hat{b}_-(\xi')$, we get

$$\hat{b}_+(\xi')\tilde{v}_+(\xi') - \frac{\tilde{v}_-(\xi')}{\hat{b}_-(\xi')} = \hat{b}_+(\xi')\big(lg_2 - \tilde{f}_2\big) - \frac{\widetilde{lg_1} - \tilde{f}_1}{\hat{b}_-(\xi)}. \tag{13.21}$$

Letting

$$\kappa_0 = \kappa + m_2, \tag{13.22}$$

we choose s and r so that

$$|s + r - 1/2 - \operatorname{Re}\kappa_0| < 1/2, \qquad s > \max(m_1 + 1/2, m_2 + 1/2). \tag{13.23}$$

If, in particular, $\operatorname{Re}\kappa_0 > \max(m_1 - 1/2, m_2 - 1/2)$, we can satisfy (13.23) by choosing $r = 0$, i.e. we can consider the mixed boundary value problem in an ordinary Sobolev-Slobodeckiĭ space $H_s(\mathbf{R}^n_+)$. But if

$$\operatorname{Re}\kappa_0 \leqslant \max(m_1 - 1/2, m_2 - 1/2),$$

we must choose $r < 0$.

In (13.21) we have

$$\hat{b}_+\tilde{v}_+ \in \tilde{H}_\delta^+(\mathbf{R}^{n-1}), \qquad \hat{b}_-^{-1}\tilde{v}_- \in \tilde{H}_\delta^-(\mathbf{R}^{n-1})$$

and hence the right side belongs to $\tilde{H}_\delta(\mathbf{R}^{n-1})$, where $\delta = s + r - 1/2 - \operatorname{Re}\kappa_0$. Since $|\delta| < 1/2$, the only function $\tilde{v}_+(\xi')$ satisfying (13.21) is (cf. the proof of Theorem 7.1)

$$\tilde{v}_+(\xi') = \hat{b}_+^{-1}(\xi')\Pi^+\big[\hat{b}_+\big(\widetilde{lg_2} - \tilde{f}_2\big) - \hat{b}_-^{-1}\big(\widetilde{lg_1} - \tilde{f}_1\big)\big]. \tag{13.24}$$

Consequently, substituting (13.24) in (13.18), we get

$$\begin{aligned}\tilde{v}(\xi') = -\hat{b}_2^{-1}\tilde{v}_+ + \hat{b}_2^{-1}\big(\widetilde{lg_2} - \tilde{f}_2\big) = -\hat{b}_2^{-1}\hat{b}_+^{-1}\Pi^+\big[\hat{b}_+\big(\widetilde{lg_2} - \tilde{f}_2\big)\\ -\hat{b}_-^{-1}\big(\widetilde{lg_1} - \tilde{f}_1\big)\big] + \hat{b}_2^{-1}\big(\widetilde{lg_2} - \tilde{f}_2\big).\end{aligned} \tag{13.25}$$

Further, since $\deg_{\xi'} b_2^{-1}(\xi')b_+^{-1}(\xi') = -m_2 - \kappa = -\kappa_0$, it follows from (13.25) by virtue of Lemmas 4.4, 13.1, 13.2 and Theorems 5.1, 4.2 that

$$\begin{aligned}[v]_{s+r-1/2} &\leqslant c\left([g_1]^+_{s+r-m_1-1/2} + [g_2]^-_{s+r-m_2-1/2} + \sum_{k=1}^{2}[f_k]_{s+r-m_k-1/2}\right)\\ &\leqslant c\big([g_1]^+_{s+r-m_1-1/2} + [g_2]^-_{s+r-m_2-1/2} + \|f\|^+_{s-2,r}\big).\end{aligned} \tag{13.26}$$

Formula (13.25) can be given a more symmetric form. Since $\Pi^+ + \Pi^- = I$, where I is the identity operator, we have

$$\hat{b}_2^{-1}(\widetilde{lg_2} - \tilde{f}_2) - \hat{b}_2^{-1}\hat{b}_+^{-1}\Pi^+\hat{b}_+(\widetilde{lg_2} - \tilde{f}_2) = \hat{b}_2^{-1}\hat{b}_+^{-1}\Pi^-\hat{b}_+(\widetilde{lg_2} - \tilde{f}_2), \tag{13.27}$$

while from (13.20) we have

$$\hat{b}_2^{-1}\hat{b}_+^{-1} = \hat{b}_1^{-1}\hat{b}_-. \tag{13.28}$$

Therefore (13.25), (13.27) and (13.28) imply

$$\tilde{v}(\xi') = \hat{b}_2^{-1}\hat{b}_+^{-1}\Pi^+\hat{b}_-^{-1}(\widetilde{lg_1} - \tilde{f}_1) + \hat{b}_1^{-1}\hat{b}_-\Pi^-\hat{b}_+(\widetilde{lg_2} - \tilde{f}_2). \tag{13.29}$$

Substituting (13.25) or (13.29) in (13.9), we obtain an expression for the unique solution of problem (13.1′), (13.2′) belonging to $H_{s,r}(\mathbf{R}^n_+)$. It is easily verified that (13.9) and (13.29) determine the solution of the mixed boundary value problem (13.1′), (13.2′) for any

$$f \in H_{s-2,r}(\mathbf{R}^n_+), \quad g_1(x') \in H_{s+r-m_1-1/2}(\mathbf{R}^{n-1}_+),$$
$$g_2(x') \in H_{s+r-m_2-1/2}(\mathbf{R}^{n-1}_-).$$

Also (13.10) and (13.26) imply the estimate

$$\overset{+}{\|u\|}_{s,r} \leqslant c\left(\overset{+}{\|f\|}_{s-2,r} + \overset{+}{[g_1]}_{s+r-m_1-1/2} + \overset{-}{[g_2]}_{s+r-m_2-1/2}\right). \tag{13.30}$$

We have thus proved

THEOREM 13.1. *Suppose the Šapiro-Lopatinskiĭ condition* (13.14) *is satisfied,* κ *is the factorization index of* $\hat{b}_1(\xi'', \xi_{n-1})\hat{b}_2^{-1}(\xi'', \xi_{n-1})$, $\kappa_0 = \kappa + m_2$, *and the numbers s and r satisfy* (13.23). *Then the mixed boundary value problem* (13.1′), (13.2′) *has a unique solution* $u(x) \in H_{s,r}(\mathbf{R}^n_+)$ *for any* $f \in H_{s-2,r}(\mathbf{R}^n_+)$, $g_1 \in H_{s+r-m_1-1/2}(\mathbf{R}^{n-1}_+)$, $g_2 \in H_{s+r-m_2-1/2}(\mathbf{R}^{n-1}_-)$. *In addition,* (13.30) *holds.*

5. *Examples of mixed boundary value problems.* Suppose we are given in $\mathbf{R}^n_+$ $(n > 3)$ the equation

$$-\frac{\partial^2 u}{\partial x_n^2} - \frac{\partial^2 u}{\partial x_{n-1}^2} + (\Lambda''(D'') + \tau)^2 u = 0, \qquad x_n > 0, \tag{13.31}$$

where $\Lambda''(D'')$ is the p.o. with symbol $|\xi''|$ and $\tau > 0$.

If one puts $\tau = 0$ in (13.31), one obtains Laplace's equation. The general solution of (13.31) belonging to $H_{s,r}(\mathbf{R}^n_+)$ has the form (see (13.9))

$$u(x', x_n) = iF_{\xi'}^{-1}e^{-x_n\sqrt{\xi_{n-1}^2 + (|\xi''| + \tau)^2}}\,\tilde{v}(\xi'), \qquad x_n > 0, \tag{13.32}$$

where $v(x') \in H_{s+r-1/2}(\mathbf{R}^{n-1})$. When $\tau = 0$ and $v(x') \in S(\mathbf{R}^{n-1})$, (13.32) is a double-layer potential for Laplace's equation (see (7.42)):

$$u(x', x_n) = \frac{\Gamma(n/2)}{\pi^{n/2}}\int_{\mathbf{R}^{n-1}}\frac{x_n v(y')dy'}{[x_n^2 + |x' - y'|^2]^{n/2}}, \qquad x_n > 0. \tag{13.32′}$$

We consider some examples of mixed boundary value problems for equation (13.31).

EXAMPLE 13.1. *Zaremba's problem.* We pose the following boundary conditions on the hyperplane $x_n = 0$:

$$i\frac{\partial u}{\partial x_n}\bigg|_{\mathbf{R}^{n-1}_+} = g_1(x'), \qquad u|_{\mathbf{R}^{n-1}_-} = 0. \tag{13.33}$$

In the notation of subsections 3 and 4 we have

$$\hat{b}_1(\xi') = \sqrt{\xi_{n-1}^2 + (|\xi''| + \tau)^2}\,, \quad m_1 = 1, \quad \hat{b}_2(\xi') = i, \quad m_2 = 0,$$

so that the symbol

$$\hat{b}_1(\xi')\hat{b}_2^{-1}(\xi') = -i\sqrt{\xi_{n-1}^2 + (|\xi''| + \tau)^2}$$

has the factorization

$$\hat{b}_1\hat{b}_2^{-1} = -i(\xi_{n-1} - i|\xi''| - i\tau)^{1/2}(\xi_{n-1} + i|\xi''| + i\tau)^{1/2}.$$

Consequently, $\kappa = 1/2$ and $\kappa_0 = \kappa + m_2 = 1/2$.

Suppose $lg_1 \in S(\mathbf{R}^{n-1})$. From (4.68), (4.69) and (13.29) we obtain (cf. (7.28′))

$$v(x'', x_{n-1}) = -\frac{\Gamma^2((n-1)/2)}{\pi^n}\theta(x_{n-1})$$
$$\times \int_0^{x_{n-1}}\int_{-\infty}^{\infty}\frac{(x_{n-1} - y_{n-1})^{1/2}e^{-(x_{n-1}-y_{n-1})\tau}}{\left[(x_{n-1} - y_{n-1})^2 + |x'' - y''|^2\right]^{(n-1)/2}}$$
$$\times\left(\int_{y_{n-1}}^{\infty}\int_{-\infty}^{\infty}\frac{(y_{n-1} - z_{n-1})^{1/2}e^{(z_{n-1}-y_{n-1})\tau}g_1(z')\,dz'}{\left[(y_{n-1} - z_{n-1})^2 + |y'' - z''|^2\right]^{(n-1)/2}}\right)dy', \tag{13.34}$$

which remains valid for $\tau = 0$. Inequalities (13.23) take the form

$$1/2 < s + r < 3/2, \qquad s > 3/2. \tag{13.35}$$

Consequently, the mixed boundary value problem (13.31), (13.33) has a solution $u(x) \in H_{3/2+\varepsilon,-2\varepsilon}(\mathbf{R}^n_+)$ for any $g_1 \in H_{-\varepsilon}(\mathbf{R}^{n-1}_+)$ $(0 < \varepsilon < 1)$. We note that when $\tau = 0$, problem (13.31), (13.33) is an example of the so-called contact problems of the theory of elasticity.

EXAMPLE 13.2. Let

$$l_0(D) = \sum_{k=1}^{n} a_k i\frac{\partial}{\partial x_k}$$

be a differential operator of first order in which the a_k are real numbers, with $a_n = 1$, and let $l_\tau^p(D)$ denote the p.o. with symbol

$$l_\tau^p(\xi) = l_0^p((|\xi''| + \tau)\omega'', \xi_{n-1}, \xi_n),$$

where $p \geqslant 1$ is an integer. We consider the following mixed boundary conditions for (13.31):

$$l_\tau^p(D)u|_{\mathbf{R}^{n-1}_+} = g_1(x'), \qquad u(x)|_{\mathbf{R}^{n-1}_-} = g_2(x'). \tag{13.36}$$

Here

$$\hat{b}_1(\xi') = il_0^p\Big((|\xi''| + \tau)\omega'', \xi_{n-1}, -i\sqrt{\xi_{n-1}^2 + (|\xi''| + \tau)^2}\Big),$$
$$m_1 = p, \quad \hat{b}_2(\xi') = i, m_2 = 0.$$

We note that $\hat{b}_1(\xi') \neq 0$ since $a_n \neq 0$.

Let us find the factorization index of the symbol

$$b_\tau(\xi') = l_0\Big((|\xi''| + \tau)\omega'', \xi_{n-1}, -i\sqrt{\xi_{n-1}^2 + (|\xi''| + \tau)^2}\,\Big).$$

We fix $\xi'' \neq 0$. Then, as ξ_{n-1} varies from $+\infty$ to $-\infty$, the function

$$\frac{b_\tau(\xi')}{\sqrt{\xi_{n-1}^2 + (|\xi''| + \tau)^2}} = -i + \frac{\sum_{k=1}^{n-2} a_k(\xi_k/|\xi''|)(|\xi''| + \tau) + a_{n-1}\xi_{n-1}}{\sqrt{\xi_{n-1}^2 + (|\xi''| + \tau)^2}}$$

describes a curve in the complex plane with initial point $b_0(0, +1) = -i + a_{n-1}$ and terminal point $b_0(0, -1) = -i - a_{n-1}$, so that the increment in the argument of this function is equal to $-2 \operatorname{arc\,tg} a_{n-1}$. Thus, by (6.51), the factorization index $\kappa^{(1)}$ of $b_\tau(\xi')$ is given by

$$\kappa^{(1)} = \frac{1}{2} - \frac{1}{\pi} \operatorname{arc\,tg} a_{n-1}. \tag{13.37}$$

Consequently, the factorization index of the symbol $\hat{b}_1\hat{b}_2^{-1} = b_\tau^p(\xi')$ is equal to

$$p\left(\frac{1}{2} - \frac{1}{\pi}\operatorname{arc\,tg} a_{n-1}\right)$$

and hence inequalities (13.23) take the form

$$s > p + 1/2, \qquad p\kappa^{(1)} < s + r < p\kappa^{(1)} + 1. \tag{13.38}$$

We note that if $p = 1$ and $a_{n-1} < 0$, then (13.38) will be satisfied when $r = 0$ and $3/2 < s < 3/2 - \pi^{-1} \operatorname{arc\,tg} a_{n-1}$. But if $a_{n-1} > 0$, then we must choose $r < 0$.

EXAMPLE 13.3. *Oblique derivative discontinuity*. Consider the following mixed boundary conditions for equation (13.31):

$$l_{\tau 1}(D)u|_{\mathbf{R}_+^{n-1}} = g_1(x'), \qquad l_{\tau 2}(D)u|_{\mathbf{R}_-^{n-1}} = g_2(x'), \tag{13.39}$$

where

$$l_{\tau i}(\xi) = \sum_{k=1}^{n-2} a_k^{(i)} \frac{\xi_k}{|\xi''|}(|\xi''| + \tau) + a_{n-1}^{(i)}\xi_{n-1} + a_n^{(i)}\xi_n,$$

and the $a_k^{(i)}$ are real numbers, $1 \leqslant k \leqslant n$, $1 \leqslant i \leqslant 2$, $a_n^{(1)} = a_n^{(2)} = 1$. Let

$$b_{\tau i}(\xi') = l_{\tau i}\Big(\xi', -i\sqrt{\xi_{n-1}^2 + (|\xi''| + \tau)^2}\,\Big),$$

$i = 1, 2$. Since $a_n^{(1)} \neq 0$ and $a_n^{(2)} \neq 0$, we have $b_{\tau 1}(\xi') \neq 0$ and $b_{\tau 2}(\xi') \neq 0$. Consequently, the Šapiro-Lopatinskiĭ condition is satisfied. By (13.37), the factorization index of the symbol $b_{\tau 1}(\xi')b_{\tau 2}^{-1}(\xi')$ is given by

$$\kappa = \frac{1}{\pi} \operatorname{arc\,tg} a_{n-1}^{(2)} - \frac{1}{\pi} \operatorname{arc\,tg} a_{n-1}^{(1)}. \tag{13.40}$$

Consequently,

$$\kappa_0 = \kappa + m_2 = 1 + \frac{1}{\pi} \operatorname{arc\,tg} a_{n-1}^{(2)} - \frac{1}{\pi} \operatorname{arc\,tg} a_{n-1}^{(1)}.$$

Thus, if s and r satisfy

$$s > 3/2, \qquad \kappa_0 < s + r < \kappa_0 + 1$$

then the mixed problem (13.31), (13.39) has a unique solution belonging to $H_{s,r}(\mathbf{R}_+^n)$ for any

$$g_1(x') \in H_{s+r-3/2}(\mathbf{R}_+^{n-1}), \qquad g_2(x') \in H_{s+r-3/2}(\mathbf{R}_-^{n-1}).$$

6. *Asymptotic behavior of the solution of a mixed problem near the hyperline* $x_n = x_{n-1} = 0$. Suppose $u(x', x_n) \in H_{s,r}(\mathbf{R}^n_+)$ is the solution of the mixed problem (13.1′), (13.2′), with the numbers s and r satisfying (13.23), and suppose $lf \in S(\mathbf{R}^n)$ and $lg_i \in S(\mathbf{R}^{n-1})$, $i = 1, 2$. According to (13.9) and (13.29), $u(x', x_n)$ has the following form for $x_n > 0$:

$$u(x', x_n) = F^{-1}\frac{\widetilde{lf}(\xi)}{\hat{L}(\xi)} + iF_{\xi'}^{-1}\left[e^{-ix_n\hat{\lambda}_2(\xi')}\hat{b}_2^{-1}\hat{b}_+^{-1}(\Pi^+\tilde{\varphi}_1(\xi') + \Pi^-\tilde{\varphi}_2(\xi'))\right], \tag{13.41}$$

where

$$\tilde{\varphi}_1(\xi') = \hat{b}_-^{-1}(\xi')\big(\widetilde{lg_1}(\xi') - \tilde{f}_1(\xi')\big), \qquad \tilde{\varphi}_2(\xi') = \hat{b}_+(\xi')\big(\widetilde{lg_2}(\xi') - \tilde{f}_2(\xi')\big), \tag{13.42}$$

and the functions $\tilde{f}_1(\xi')$ and $\tilde{f}_2(\xi')$ are defined by (13.13). Since $lf \in S(\mathbf{R}^n)$ and $lg_i(x') \in S(\mathbf{R}^{n-1})$, the functions $\tilde{\varphi}_i(\xi')$ satisfy

$$|\tilde{\varphi}_i(\xi')| \leqslant C_N(1 + |\xi'|)^{-N}, \qquad \forall N,\ i = 1, 2. \tag{13.43}$$

From (13.41) it follows that $u(x', x_n) \in C^\infty$ for $x_n > 0$, with $F^{-1}\dfrac{\widetilde{lf}(\xi)}{\hat{L}(\xi)} \in C^\infty(\overline{\mathbf{R}}^n_+)$.

The function $\hat{\lambda}_2(\xi')$ clearly admits the following expansion for $\xi_{n-1} > 0$:

$$\hat{\lambda}_2(\xi'', \xi_{n-1}) = \xi_{n-1}\hat{\lambda}_2\left(\frac{\xi''}{\xi_{n-1}}, +1\right) = \sum_{k=0}^{N} \hat{\mu}_{1k}(\xi'')\xi_{n-1}^{1-k} + \hat{R}_{N1}(\xi'', \xi_{n-1}), \tag{13.44}$$

where $\deg_{\xi''} \mu_{1k}(\xi'') = k$, $\hat{\mu}_{1k}(\xi'') = \mu_{1k}((1 + |\xi''|)\omega'')$, $\mu_{10} = \lambda_2(0, +1)$, $\operatorname{Im} \mu_{10} < 0$, and

$$|\hat{R}_{N1}(\xi'', \xi_{n-1})| \leqslant C_N\frac{(1 + |\xi''|)^{N+1}}{|\xi_{n-1}|^N} \tag{13.45}$$

for $\xi_{n-1} > 0$. The following analogous expansion holds for $\xi_{n-1} < 0$:

$$\hat{\lambda}_2(\xi'', \xi_{n-1}) = |\xi_{n-1}|\hat{\lambda}_2\left(\frac{\xi''}{|\xi_{n-1}|}, -1\right) = \sum_{k=0}^{N} \hat{\mu}_{2k}(\xi'')\xi_{n-1}^{1-k} + \hat{R}_{N2}(\xi'', \xi_{n-1}), \tag{13.46}$$

in which $\mu_{20} = -\lambda_2(0, -1)$, $\operatorname{Im} \mu_{20} > 0$, and $\hat{R}_{N2}(\xi'', \xi_{n-1})$ satisfies the estimate of form (13.45) for $\xi_{n-1} < 0$.

Let

$$z_1 = -(x_{n-1} + x_n\mu_{10}), \qquad z_2 = x_{n-1} + x_n\mu_{20} \tag{13.47}$$

and suppose $T > 0$ is such that

$$\sum_{k=1}^{N} |\hat{\mu}_{1k}(\xi'')|\xi_{n-1}^{-k} + \frac{|R_{N1}|}{\xi_{n-1}} \leqslant \frac{1}{2}|\operatorname{Im} \mu_{10}|$$

for $\xi_{n-1}/(1 + |\xi''|) \geqslant T$. Then (13.44) implies that for $\xi_{n-1} \geqslant T(1 + |\xi''|)$

$$e^{-ix_n\hat{\lambda}_2(\xi') - ix_{n-1}\xi_{n-1}} = e^{iz_1\xi_{n-1} - ix_n\hat{\mu}_{11}(\xi'')} \sum_{k=0}^{N} \sum_{p=0}^{k} x_n^p \hat{\gamma}_{1pk}(\xi'')\xi_{n-1}^{-k} + R_{N3}(x, \xi'), \tag{13.48}$$

where $\deg \gamma_{1pk}(\xi'') = p + k$, $\hat{\gamma}_{100} = 1$, $\hat{\gamma}_{10k}(\xi'') = 0$ for $k \geqslant 1$, and $\hat{R}_{N3}$ satisfies the following estimate for $\xi_{n-1} \geqslant T(1 + |\xi''|)$ and $0 \leqslant x_n \leqslant 1$:

$$\left| \frac{\partial^r}{\partial x_n^r} \hat{R}_{N3}^{(1)} \right| \leqslant C_{Nr} \frac{(1 + |\xi''|)^{2(N+1)+r}}{|\xi_n|^{N+1}}, \qquad \forall r \geqslant 0. \tag{13.49}$$

Analogously, for $\xi_{n-1} < - T(1 + |\xi''|)$, where T is sufficiently large, we have

$$e^{-ix_n\hat{\lambda}_2(\xi') - ix_n\xi_{n-1}} = e^{-iz_2\xi_{n-1} - ix_n\hat{\mu}_{21}(\xi'')} \sum_{k=0}^{N} \sum_{p=0}^{k} x_n^p \hat{\gamma}_{2pk}(\xi'')\xi_{n-1}^{-k} + \hat{R}_{N4}(x_n, \xi'), \tag{13.50}$$

where $\hat{\gamma}_{200} = 1$, $\hat{\gamma}_{20k} = 0$ for $k \geqslant 1$, $\deg_{\xi''} \gamma_{2pk} = p + k$ and $\hat{R}_{N4}$ satisfies the estimate of form (13.49) for $\xi_{n-1} < - T(1 + |\xi''|)$.

Clearly, $\hat{b}_2^{-1}(\xi'', \xi_{n-1})$ admits expansions of form (13.44) and (13.46) for $\xi_{n-1} > 0$ and $\xi_{n-1} < 0$ respectively (cf. (10.5)). We also expand $\hat{b}_+^{-1}(\xi'', \xi_{n-1})$ by (9.2) and $\Pi^+\tilde{\varphi}_1$, $\Pi^-\tilde{\varphi}_2$ by (5.36).

In this way, taking (13.48) and (13.50) into account, we get

$$\begin{aligned} u(x', x_n) = {} & w(x', x_n) + w_1(x', x_n) + iF_{\xi''}^{-1}\Bigg[\frac{1}{2\pi}\int_{T(1+|\xi''|)}^{\infty}\Bigg(e^{iz_1\xi_{n-1} - ix_n\mu_{11}(\xi'')} \\ & \times \sum_{k=1}^{N} \sum_{p=0}^{k-1} \sum_{r=0}^{k-1} x_n^p d_{1kpr}(\xi'')\xi_{n-1}^{-\kappa_0 - k} \ln^r \xi_{n-1} + R_{N5}(x, \xi'', \xi_{n-1})\Bigg) d\xi_{n-1} \\ & + \frac{1}{2\pi}\int_{-\infty}^{-T(1+|\xi''|)}\Bigg(e^{-iz_2\xi_{n-1} - ix_n\hat{\mu}_{21}(\xi'')} \sum_{k=1}^{N} \sum_{p=0}^{k-1} \sum_{r=0}^{k-1} x_n^p d_{2kpr}(\xi'')|\xi_{n-1}|^{-\kappa_0 - k} e^{-(\kappa+1)\pi i} \\ & \times \ln^r|\xi_{n-1}| + R_{N6}(x, \xi'', \xi_{n-1})\Bigg) d\xi_{n-1}\Bigg], \end{aligned} \tag{13.51}$$

where

$$w_1(x', x_n) = F_{\xi''}^{-1} \frac{i}{2\pi} \int_{-T(1+|\xi''|)}^{T(1+|\xi''|)} e^{-x_n\hat{\lambda}_2(\xi'') - ix_n\xi_{n-1}}$$
$$\times \hat{b}_2^{-1}\hat{b}_+^{-1}(\Pi^+\tilde{\varphi}_1 + \Pi^-\tilde{\varphi}_2(\xi'))d\xi_{n-1} \in C^\infty(\overline{\mathbf{R}_+^n}), \tag{13.52}$$

$$d_{1100}(\xi'') = b_2^{-1}(0, +1)\Pi'\left[\hat{b}_-^{-1}(\xi')(\widetilde{lg_1} - \tilde{f}_1) - \hat{b}_+(\xi')(\widetilde{lg_2} - \tilde{f}_2)\right], \tag{13.53}$$

$$d_{2100}(\xi'') = b_2^{-1}(0, -1)\Pi'\left[\hat{b}_-^{-1}(\xi')(\widetilde{lg_1} - \tilde{f}_1) - \hat{b}_+(\xi')(\widetilde{lg_2} - \tilde{f}_2)\right]. \tag{13.53'}$$

The functions R_{N5} and R_{N6} in (13.51) satisfy

$$|D_x^p R_{N5}| + |D_x^p R_{N6}|$$
$$\leqslant C_{pNM}(1 + |\xi''|)^{-M}(1 + |\xi_{n-1}|)^{-\operatorname{Re}\kappa_0 - N - 1 + \varepsilon + |p|}, \qquad \forall M, \forall \varepsilon > 0. \tag{13.54}$$

The functions $d_{ikpr}(\xi'')$ satisfy $|d_{ikpr}(\xi'')| \leqslant C_M(1 + |\xi''|)^{-M}$ and can be determined in explicit form by formulas analogous to (13.53) and (13.53′). By (13.54), the function

$$F_{\xi''}^{-1}\int_{T(1+|\xi''|)}^{\infty} R_{N5}(x, \xi'', \xi_{n-1})d\xi_{n-1} + F_{\xi''}^{-1}\int_{-\infty}^{-T(1+|\xi''|)} R_{N6}(x, \xi'', \xi_{n-1})d\xi_{n-1}$$

can be continuously differentiated $\operatorname{Re}\kappa_0 + N - 1$ times in $\overline{\mathbf{R}_+^n}$.

Let

$$J^{(1)}_{\alpha+i\beta,p} = \int_{T(1+|\xi''|)}^{\infty} \xi_{n-1}^{\alpha+i\beta} \ln^p \xi_{n-1} e^{iz_1\xi_{n-1}}\, d\xi_{n-1}, \tag{13.55}$$

$$J^{(2)}_{\alpha+i\beta,p} = \int_{-\infty}^{-T(1+|\xi''|)} |\xi_{n-1}|^{\alpha+i\beta} \ln^p|\xi_{n-1}| e^{-iz_2\xi_{n-1}}\, d\xi_{n-1}$$
$$= \int_{T(1+|\xi''|)}^{\infty} \xi_{n-1}^{\alpha+i\beta} \ln^p \xi_{n-1} e^{iz_2\xi_{n-1}}\, d\xi_{n-1}. \tag{13.55'}$$

We note that the integrals (13.55) and (13.55′) differ from

$$F_{\xi_{n-1}}(\theta(\xi_{n-1})\xi_{n-1}^{\alpha+i\beta}\, ln^p\, \xi_{n-1})$$

(see Example 2.3) by a function that is analytic in a neighborhood of the origin.

Suppose $\kappa_0 \neq 0, +1, +2, \ldots$. Then, calculating the integrals of form (13.55), (13.55′) in (13.51) and expanding the functions $e^{-ix_n\mu_{11}}$ and $e^{-ix_n\mu_{21}}$, we get

$$u(x', x_n) = b_2^{-1}(0, +1)d_0(x'')z_1^{\kappa_0} - e^{-\kappa\pi i}b_2^{-1}(0, -1)d_0(x'')z_2^{\kappa_0}$$
$$+ \sum_{i=1}^{2}\sum_{r=0}^{N}\sum_{p+k=1}^{N} d^{(1)}_{ipkr}(x'')x_n^p z_i^{\kappa_0+k}\ln^r z_i + u_N(x', x_n), \quad \forall N, x_n > 0, \tag{13.56}$$

where $u_N \in C^{N+[\operatorname{Re}\kappa_0]}(\overline{\mathbf{R}^n_+})$ the functions $d^{(1)}_{ipkr}(x'')$ belong to $C^\infty(\mathbf{R}^{n-1})$ and can be determined in terms of lg_i and $f_i(x')$, $i = 1, 2$, and

$$d_0(x'') = \frac{i}{2\pi} e^{-i(\pi/2)\kappa_0}\Gamma(-\kappa_0) F^{-1}_{\xi''}\Pi'\left[\hat{b}^{-1}_-(\xi')\left(\widetilde{lg_1} - \tilde{f}_1\right) - \hat{b}_+(\xi')\left(\widetilde{lg_2} - \tilde{f}_2\right)\right]. \tag{13.57}$$

When κ_0 is a nonnegative integer, the functions $z_1^{\kappa_0}$ and $z_2^{\kappa_0}$ in the first two summands of (13.56) should be replaced by $z_1^{\kappa_0} \ln z_1$ and $z_2^{\kappa_0} \ln z_2$ respectively. It thus follows from (13.56) that $u(x', x_n) \in C^\infty$ for $x_n + |x_{n-1}| > 0$ and that in a neighborhood of the hyperline $x_n = x_{n-1} = 0$ the solution $u(x', x_n)$ has a smoothness analogous to r^{κ_0} or $r^{\kappa_0} \ln r$, where $r = \sqrt{x_n^2 + x_{n-1}^2}$.

CHAPTER IV

SYSTEMS OF ELLIPTIC PSEUDODIFFERENTIAL EQUATIONS IN A HALFSPACE

§14. Conditions for the normal solvability of a pseudodifferential operator on a halfline

1. Let $A(\xi) = \|a_{ij}(\xi)\|_{i,j=1}^{p}$ be a matrix consisting of symbols of class $O_{\alpha+i\beta}^{\infty}$. Such a matrix is said to be *elliptic* if $\det A(\xi) \neq 0$ for $\xi \neq 0$. We put, as in Chapters II and III, $\hat{A}(\xi', \xi_n) = A((1 + |\xi'|)\omega, \xi_n)$, $\omega = \xi'/|\xi'|$, and consider the following system of pseudodifferential equations (ψ d.e.'s) in $\mathbf{R}_+^n$ with symbol $\hat{A}(\xi', \xi_n)$:

$$p\hat{A}u_+ = f, \tag{14.1}$$

where $f(x) = (f_1(x), \ldots, f_p(x))$, $u_+ = (u_1^+(x), \ldots, u_p^+(x))$, with $f_k(x) \in H_{s-\alpha}(\mathbf{R}_+^n)$, $u_k^+(x) \in H_s^+(\mathbf{R}^n)$, $1 \leqslant k \leqslant p$. For brevity, the spaces of the vector valued functions $f(x)$ and $u_+(x)$ will also be denoted by $H_{s-\alpha}(\mathbf{R}_+^n)$ and $H_s^+(\mathbf{R}^n)$.

We will show that the system of ψ d.e.'s (14.1) is equivalent to a system of ψ d.e.'s of the form

$$p\hat{A}_0 v_+ = g, \tag{14.2}$$

where $v_+ \in H_0^+(\mathbf{R}^n)$, $g \in H_0(\mathbf{R}_+^n)$ and $\deg_\xi A_0(\xi', \xi_n) = 0$.

Let $u_- = lf - \hat{A}u_+$, where lf is an arbitrary extension of f onto $\mathbf{R}^n$ belonging to $H_{s-\alpha}(\mathbf{R}^n)$. Then $u_- \in H_{s-\alpha}^-(\mathbf{R}^n)$ and the system (14.1) is equivalent to

$$\hat{A}u_+ + u_- = lf \tag{14.3}$$

with unknowns $u_+ \in H_s^+(\mathbf{R}^n)$ and $u_- \in H_{s-\alpha}^-(\mathbf{R}^n)$.

Let $\hat{\Lambda}_\pm^s$ denote the p.o.'s with symbols $\hat{\Lambda}_\pm^s(\xi) = (\xi_n \pm i|\xi'| \pm i)^s$ (see Example 4.4), and let $v_+ = \hat{\Lambda}_+^s u_+$, $g = p\hat{\Lambda}_-^{s-\alpha-i\beta} lf$. By Theorem 4.4 and Lemma

4.6, $v_+ \in H_0^+(\mathbf{R}^n)$ and $g \in H_0(\mathbf{R}^n_+)$, with g not depending on the choice of the extension lf. Applying the p.o. $\hat{\Lambda}_-^{s-\alpha-i\beta}$ to (14.3) and making the substitution $u_+ = \hat{\Lambda}_+^{-s} v_+$, we obtain the following system of equations, which is equivalent to (14.3):

$$\hat{A}_0 v_+ + \hat{\Lambda}_-^{s-\alpha-i\beta} u_- = \hat{\Lambda}_-^{s-\alpha-i\beta} lf, \tag{14.4}$$

where $\hat{A}_0$ is the p.o. with symbol $\hat{A}_0(\xi) = \hat{A}(\xi)\hat{\Lambda}_-^{s-\alpha-i\beta}(\xi)\hat{\Lambda}_+^{-s}(\xi)$, and $\deg_\xi A_0(\xi) = 0$. Since $\hat{\Lambda}_-^{s-\alpha-i\beta} u_- \in H_0^-(\mathbf{R}^n)$ (see Example 4.4) and hence $p\hat{\Lambda}_-^{s-\alpha-i\beta} u_- = 0$, it follows that (14.4) is equivalent to (14.2), with $g = p\hat{\Lambda}_-^{s-\alpha-i\beta} lf$.

We denote by $g_+(x) \in H_0^+(\mathbf{R}^n)$ the function that is equal to $g(x)$ for $x_n > 0$ and vanishes for $x_n < 0$. The system of ψ d.e.'s in a halfspace (14.2) is clearly equivalent to the following system of ψ d.e.'s in $\mathbf{R}^n$:

$$\theta^+ \hat{A}_0 v_+ = g_+, \tag{14.5}$$

which is obtained from (14.2) by extending the left and right sides by zero for $x_n < 0$. In (14.5), θ^+ denotes the operator of multiplication by the function $\theta(x_n)$ (see §5).

2. A continuous linear operator A acting from a Banach space B_1 into a Banach space B_2 is called a *Fredholm* (*normally solvable*) *operator* if

1) the homogeneous equation $Au = 0$ has a finite number of linearly independent solutions in B_1, i.e. $\dim \ker A < \infty$, and

2) the inhomogeneous equation $Au = f$ is solvable under the fulfilment of a finite number of conditions of the form $(\varphi_r, f) = 0$, $1 \leqslant r \leqslant l$, where the φ_r are continuous linear functionals on B_2.

Clearly, 2) is equivalent to the following two conditions (see [65]):

2_1) $\dim \ker A^* < \infty$, where A^* is the adjoint of A, and

2_2) the range of A is closed in B_2, i.e. $\operatorname{Im} A = \overline{\operatorname{Im} A} \subset B_2$.

The difference $\kappa = \dim \ker A - \dim \ker A^*$ is called the *index* of A. As is well known (see, for example, [65]), the index of a Fredholm operator does not vary under continuous (in the uniform operator topology) deformations of it in the class of Fredholm operators.

A classical example of a Fredholm operator is an operator of the form $I + T$, where I is the identity operator in a Banach space B and T is a compact operator in B (here $B_1 = B_2 = B$). In particular, if $B = L_2(\mathbf{R}^n)$ and $Tu = \int_{-\infty}^{\infty} T(x,y)u(y)dy$ is an integral operator with kernel $T(x,y) \in L_2(\mathbf{R}^{2n})$ (and hence a Hilbert-Schmidt operator), then $A = I + T$ is a Fredholm operator.

An operator A is normally solvable if there exists a continuous operator R from B_2 into B_1 such that

$$RA = I_1 + T_1, \tag{14.6}$$

$$AR = I_2 + T_2, \tag{14.6'}$$

where I_1 (I_2) is the identity operator in B_1 (B_2) and T_1 (T_2) is a compact operator in B_1 (B_2). In fact, (14.6) and the Fredholm-Riesz theorems (see [83]) imply condition 1) (as well as condition 2_2)) for the normal solvability of A, while (14.6′) implies condition 2).

An operator R satisfying (14.6) and (14.6′) is called a *regularizer* of the operator A.

3. Let us fix $\omega = \xi'/|\xi'|$ and consider on the halfline $\mathbf{R}^1_+$ the system of ψ d.e.'s

$$pA(\omega, D_n)u_+(x_n) = f(x_n), \tag{14.7}$$

where

$$A(\omega, D_n)u_+(x_n) = \frac{1}{2\pi}\int_{-\infty}^{\infty} A(\omega, \xi_n)\tilde{u}_+(\xi_n)e^{ix_n\xi_n}\,d\xi_n$$

is the p.o. in $\mathbf{R}^1$ with symbol $A(\omega, \xi_n)$, $f(x_n) \in H_{s-\alpha}(\mathbf{R}^1_+)$ and $u_+(x_n) \in H_s^+(\mathbf{R}^1)$.

As will be shown in §16, the study of the system of ψ d.e.'s in $\mathbf{R}^n_+$ of form (14.1) is based on an investigation of the family of ψ d.e.'s $pA(\omega, D_n)u_+(x_n)$ in $\mathbf{R}^1_+$ depending on the parameter $\omega = \xi'/|\xi'|$. Therefore the rest of this section will be devoted to an investigation of (14.7) or the equivalent system

$$\theta^+A_0(\omega, D_n)v_+(x_n) = g_+(x_n), \tag{14.8}$$

where $g_+(x_n) \in H_0^+(\mathbf{R}^1)$, $v_+(x_n) \in H_0^+(\mathbf{R}^1)$, the parameter ω is fixed and

$$A_0(\omega, \xi_n) = A(\omega, \xi_n)(\xi_n - i)^{s-\alpha-i\beta}(\xi_n + i)^{-s}. \tag{14.8'}$$

For brevity, the class of functions of form (14.8′), where $A(\xi', \xi_n) \in O^\infty_{\alpha+i\beta}$, will be denoted by O_0.

Since $\deg_\xi A_0(\xi', \xi_n) = 0$, we have

$$A_0(\omega, \xi_n) = A_0\left(\frac{\omega}{|\xi_n|}, \frac{\xi_n}{|\xi_n|}\right).$$

If $\xi_n \geqslant 0$ and $\xi_n \to +\infty$, then

$$A_0(\omega, \xi_n) = A_0\left(\frac{\omega}{|\xi_n|}, +1\right) \to A_0(0, +1) = a_+.$$

On the other hand, if $\xi_n < 0$ and $\xi_n \to -\infty$, then

$$A_0(\omega, \xi_n) = A_0\left(\frac{\omega}{|\xi_n|}, -1\right) \to A_0(0, -1) = a_-.$$

We note that $\det a_\pm \neq 0$ inasmuch as $A_0(\xi', \xi_n)$ is an elliptic matrix.

4. We first consider the case

$$a_+ = a_-. \tag{14.9}$$

LEMMA 14.1. *Under the fulfilment of condition* (14.9) *the elliptic system* (14.8) *defines a Fredholm operator in* $H_0^+(\mathbf{R}^1)$.

PROOF. By Lemma 5.2, the Fourier transform with respect to x_n of (14.8), is

$$\Pi^+ A_0(\omega, \xi_n)\tilde{v}_+(\xi_n) = \tilde{g}_+(\xi_n), \tag{14.10}$$

where the operator Π^+ acts on a function $\tilde{u}(\xi_n) \in S(\mathbf{R}^1)$ according to the formula

$$\Pi^+ \tilde{u}(\xi_n) = \frac{i}{2\pi}\int_{-\infty}^{\infty} \frac{\tilde{u}(\eta_n)}{\xi_n - \eta_n + i0}\, d\eta_n,$$

and is defined at a function $\tilde{u}(\xi_n) \in \tilde{H}_0(\mathbf{R}^1)$ by continuity (see Theorem 5.1).

We wll show that the operator

$$\tilde{R}\tilde{g}_+ = \Pi^+ A_0^{-1}(\omega, \xi_n)\tilde{g}_+(\xi_n), \tag{14.11}$$

which is bounded in $\tilde{H}_0^+(\mathbf{R}^1)$ by virtue of Theorem 5.1 and Lemma 4.4, is a regularizer of the operator $\tilde{\mathfrak{A}}$ defined by (14.10). To prove (14.6′) we note that

$$\begin{aligned}\tilde{\mathfrak{A}}\tilde{R}\tilde{g}_+ &= \Pi^+ A_0(\omega, \xi_n)\Pi^+ A_0^{-1}(\omega, \eta_n)\tilde{g}_+(\eta_n)\\ &= \Pi^+ A_0(A_0^{-1}\Pi^+\tilde{g}_+ + C\tilde{g}_+) = \tilde{g}_+ + \Pi^+ A_0 C\tilde{g}_+,\end{aligned} \tag{14.12}$$

where

$$C\tilde{g}_+ = \Pi^+ A_0^{-1}\tilde{g}_+ - A_0^{-1}\Pi^+\tilde{g}_+ = \frac{i}{2\pi}\int_{-\infty}^{\infty} C(\omega, \xi_n, \eta_n)\tilde{g}_+(\eta_n)d\eta_n, \tag{14.13}$$

$$C(\omega, \xi_n, \eta_n) = \frac{A_0^{-1}(\omega, \eta_n) - A_0^{-1}(\omega, \xi_n)}{\xi_n - \eta_n}. \tag{14.14}$$

It will be seen that C is a Hilbert-Schmidt operator. Since $A(\xi', \xi_n) \in O_{\alpha+i\beta}^{\infty}$, the following estimates hold (cf. the proof of Lemma 6.1):

$$|A_0^{-1}(\omega, \xi_n) - a_+^{-1}| \leqslant C(1 + |\xi_n|)^{-1}, \tag{14.15}$$

$$\left|\frac{\partial}{\partial \xi_n} A_0^{-1}(\omega, \xi_n)\right| \leqslant C(1 + |\xi_n|)^{-2}. \tag{14.15′}$$

But these estimates imply

$$|C(\omega, \xi_n, \eta_n)| \leqslant C(1 + |\xi_n|)^{-1}(1 + |\eta_n|)^{-1}. \tag{14.16}$$

For suppose $|\xi_n - \eta_n| < 1/2(1 + |\xi_n|)$. Then $1 + |\eta_n| \leqslant 1 + |\xi_n| + |\xi_n - \eta_n| < 3/2(1 + |\xi_n|)$, and hence (14.15′) and Lagrange's formula for the remainder in Taylor's theorem imply

$$\begin{aligned}|C(\omega, \xi_n, \eta_n)| &\leqslant \left|\frac{\partial}{\partial \xi_n} A_0^{-1}(\omega, \xi_n + \theta(\eta_n - \xi_n))\right|\\ &\leqslant \frac{C_1}{(1 + |\xi_n|)^2} \leqslant \frac{C_2}{(1 + |\xi_n|)(1 + |\eta_n|)}.\end{aligned}$$

Suppose now $|\xi_n - \eta_n| \geqslant 1/2(1 + |\xi_n|)$. Then $1 + |\eta_n| \leqslant 3|\xi_n - \eta_n|$, and hence (14.15) implies

$$|C(\omega, \xi_n, \eta_n)| = \frac{1}{|\xi_n - \eta_n|}\left(|A_0^{-1}(\omega, \eta_n) - a_+| + |A_0^{-1}(\omega, \xi_n) - a_+|\right)$$
$$\leqslant \frac{C}{(1 + |\xi_n|)(1 + |\eta_n|)}.$$

Thus (14.16) holds for any ξ_n and η_n, which means that C is a Hilbert-Schmidt operator and hence that $\Pi^+ A_0 \tilde{C}$ is a compact operator (inasmuch as it is the product of a bounded operator and a compact operator). Equality (14.6′) is proved; (14.6) is proved analogously.

We note that R is called a *left* (*right*) *regularizer* if (14.6) ((14.6′)) is valid.

Let k denote the dimension of the kernel of $\mathfrak{A}$, let l denote the dimension of its cokernel, i.e. the dimension of the kernel of $\mathfrak{A}^*$, and let $\kappa = k - l$ denote the index of $\mathfrak{A}$.

LEMMA 14.2. *Suppose the conditions of Lemma* 14.1 *are satisfied. Then*

$$\kappa = \frac{1}{2\pi}\Delta \arg \det A_0(\omega, \xi_n)\Big|_{\xi_n = +\infty}^{\xi_n = -\infty}, \tag{14.17}$$

i.e. $2\pi\kappa$ *is equal to the increment in the argument of* $\det A_0(\omega, \xi_n)$ *as* ξ_n *varies from* $+\infty$ *to* $-\infty$.

PROOF. As was noted above, the index of the Fredholm operator $\mathfrak{A}$ does not vary under continuous deformations of the operator in the class of Fredholm operators. By virtue of (14.9), the matrix $A_0(\omega, \xi_n)$ can be regarded, after the change of variables $z = (\xi_n - i)/(\xi_n + i)$ as a continuous matrix valued function on the circle $|z| = 1$ with values in the group of nonsingular matrices of order p. We denote the space of such functions by G. As is well known (see, for example, [89]), every such matrix valued function is homotopic in G to a diagonal matrix valued function of the form $A_0^{(1)} = \|a_k^{(1)}\delta_{jk}\|_{j,k=1}^p$, where $a_1^{(1)}$ is a continuous function on the circle $|z| = 1$ and $a_k^{(1)} = 1$ for $2 \leqslant k \leqslant p$. It can be assumed without loss of generality that the homotopy is realized in the class of infinitely differentiable matrices of class O_0. The index of $\mathfrak{A}$ is equal to the index of the operator $\tilde{\mathfrak{A}}_1 = \Pi^+ A_0^{(1)}$ and hence to the index of the scalar operator $\Pi^+ a_1^{(1)}$. By (14.9), the increment in the argument of the determinant is an integral multiple of 2π and hence does not vary under a deformation. In addition, $\det A_0^{(1)} = a_1^{(1)}$, so that

$$\Delta \arg \det A_0|_{\xi_n = +\infty}^{\xi_n = -\infty} = \Delta \arg \det A_0^{(1)}|_{\xi_n = +\infty}^{\xi_n = -\infty}.$$

Thus it remains to establish (14.17) for $\Pi^+ a_1^{(1)}$.

According to the results of §7, when $s = \alpha + i\beta = 0$, the index of $\Pi^+ a_1^{(1)}$ is equal to the index of the homogeneous factorization of the symbol $a_1^{(1)}(\omega, \xi_n)$, which proves (14.17) by virtue of (6.51).

REMARK 14.1. In the scalar case (see §7) either k or l must vanish, so that the index of $\Pi^+ a_1^{(1)}$ determines the dimensions of its kernel and cokernel. In the matrix case, however, the index of $\mathcal{Q}$ does not yield complete information on the dimensions of the kernel and cokernel.

5. Let us now consider the general case

$$a_+ \neq a_-. \tag{14.18}$$

We will find conditions under which $\tilde{\mathcal{Q}} = \Pi^+ A_0$ is a Fredholm operator in $\tilde{H}_0^+(\mathbf{R}^1)$.

Since $A_0(\xi', \xi_n)$ is elliptic, the matrix $a_+^{-1} a_-$ is nonsingular. Let $a_1, \ldots, a_p$ be the eigenvalues of $a_+^{-1} a_-$, with repetitions according to their multiplicities. It will be assumed that

$$\textit{none of the eigenvalues } a_j,\ 1 \leqslant j \leqslant p, \textit{ are negative real numbers.} \tag{14.19}$$

We define the logarithm of the matrix $a_+^{-1} a_-$ by means of the formula (see [47])

$$\ln a_+^{-1} a_- = \frac{i}{2\pi} \int_\Gamma \ln z \left(a_+^{-1} a_- - zI\right)^{-1} dz, \tag{14.20}$$

where I is the identity matrix and Γ is a closed contour that lies in the complex z plane with a cut along the negative real axis and encloses all of the eigenvalues a_j, $1 \leqslant j \leqslant p$. The branch of $\ln z$ in (14.20) that is chosen is real on the positive real axis.

Let

$$b = \frac{1}{2\pi i} \ln a_+^{-1} a_-, \qquad b_\pm(\xi_n) = \exp\{\ln(\xi_n \pm i) b\}. \tag{14.21}$$

Here we have also chosen the branch of the logarithm that is real on the positive real axis. Let b_j denote the eigenvalues of the matrix $b = (1/2\pi i) \ln a_+^{-1} a_-$. Then (14.19) and (14.20) imply

$$-1/2 < \operatorname{Re} b_j < 1/2, \qquad 1 \leqslant j \leqslant p. \tag{14.22}$$

We note that under an arbitrary choice of $b = (1/2\pi i) \ln a_+^{-1} a_-$ condition (14.19) is equivalent to

$$\operatorname{Re} b_j - 1/2 \in \mathbf{R} \setminus \mathbf{Z}, \qquad 1 \leqslant j \leqslant p. \tag{14.23}$$

Let us show that the matrix $a_+b_-^{-1}(\xi_n)b_+(\xi_n)$ has the same limits for $\xi_n \to \pm\infty$ as the matrix $A_0(\omega, \xi_n)$. We have

$$a_+b_-^{-1}(\xi_n)b_+(\xi_n) = a_+ \exp\left\{\ln\frac{\xi_n + i}{\xi_n - i}\, b\right\}.$$

Therefore, as $\xi_n \to +\infty$,

$$\ln\frac{\xi_n + i}{\xi_n - i} \to 0,$$

and hence $a_+b_-^{-1}b_+ \to a_+ = A_0(0, +1)$. Further, as $\xi_n \to -\infty$,

$$\ln\frac{\xi_n + i}{\xi_n - i} = \arg(\xi_n + i) - \arg(\xi_n - i) \to 2\pi i$$

and hence $a_+b_-^{-1}b_+ \to a_+e^{2\pi i b} = a_+a_+^{-1}a_- = a_-$.

Let

$$P_\pm\tilde{u}_+ = b_\pm^{-1}\Pi^+b_\pm\tilde{u}_+. \tag{14.21$'$}$$

LEMMA 14.3. *If condition* (14.22) *is satisfied, then the operators* $P_\pm$ *are bounded in* $\tilde{H}_0(\mathbf{R}^1) = L_2(\mathbf{R}^1)$.

PROOF. Let b_0 be the Jordan normal form of b and let N be the nonsingular matrix that transforms b into b_0:

$$b = N^{-1}b_0N. \tag{14.24}$$

Then $b_\pm(\xi_n) = \exp\{\ln(\xi_n \pm i)N^{-1}b_0N\} = N^{-1}b_0^\pm(\xi_n)N$, where $b_0^\pm(\xi_n) = \exp\{\ln(\xi_n \pm i)b_0\}$.

When b_0 is a diagonal matrix, the matrices $b_0^\pm(\xi_n)$ are also diagonal and have the form $b_0^\pm(\xi_n) = \|(\xi_n \pm i)^{b_j}\delta_{jk}\|_{j,k=1}^p$. In the general case, to each r_k-dimensional Jordan cell of b_0 there corresponds a triangular r_k-dimensional cell $\|b_{pl}^{(j)}(\xi_n \pm i)\|_{p,l=1}^{r_k}$, where (see [47])

$$b_{pl}^{(j)}(\xi_n \pm i) = (\xi_n \pm i)^{b_j}\ln^{l-p}(\xi_n \pm i) \quad \text{for } 1 \leqslant p \leqslant l \leqslant r_k,$$
$$b_{pl}^{(j)}(\xi_n \pm i) = 0 \quad \text{for } p > l. \tag{14.25}$$

Since the matrix N does not depend on ξ_n and hence commutes with Π^+, $P_\pm\tilde{u}_+ = b_\pm^{-1}\Pi^+b_\pm\tilde{u}_+ = N^{-1}b_0^\pm\Pi^+(b_0^\pm)^{-1}N\tilde{u}_+$. Consequently, Lemma 14.3 will be proved if we establish the estimate

$$\left\|b_0^\pm\Pi^+\left(b_0^\pm\right)^{-1}\tilde{v}_+\right\|_0 \leqslant C\|\tilde{v}_+\|_0, \qquad \forall\tilde{v}_+(\xi_n) \in \tilde{H}_0(\mathbf{R}^1). \tag{14.26}$$

We have

$$b_0^\pm\Pi^+\left(b_0^\pm\right)^{-1}\tilde{v}_+ = \frac{i}{2\pi}\int_{-\infty}^{\infty}\frac{\exp\left\{\ln\dfrac{\xi_n \pm i}{\eta_n \pm i}\, b_0\right\}\tilde{v}_+(\eta_n)\,d\eta_n}{\xi_n - \eta_n + i0}.$$

Therefore (14.26) follows from Lemma 5.3 by virtue of (14.25).

LEMMA 14.4. *Suppose a matrix $B(\xi)$ satisfies the estimates (cf.* (14.15) *and* (14.15′))

$$|B(\xi_n) - I| < C(1 + |\xi_n|)^{-1}, \qquad \left|\frac{\partial B(\xi_n)}{\partial \xi_n}\right| \leqslant C(1 + |\xi_n|)^{-2} \tag{14.27}$$

and suppose (14.22) *is satisfied. Then the operator $C = P_+B - BP_+$, i.e. the commutator of P_+ and B, is a Hilbert-Schmidt operator.*

PROOF. We have

$$C\tilde{u}_+ = \frac{i}{2\pi}\int_{-\infty}^{\infty} C(\xi_n, \eta_n)\tilde{u}_+(\eta_n)d\eta_n, \tag{14.28}$$

where

$$C(\xi_n, \eta_n) = \frac{b_+^{-1}(\xi_n)b_+(\eta_n)B(\eta_n) - B(\xi_n)b_+^{-1}(\xi_n)b_+(\eta_n)}{\xi_n - \eta_n}. \tag{14.29}$$

We will prove that

$$|C(\xi_n, \eta_n)| \leqslant C(1 + |\xi_n|)^{-1/2-\delta_0}(1 + |\eta_n|)^{-1/2-\delta_0}, \tag{14.30}$$

where $0 < \delta_0 < 1/2 - \max_{1 \leqslant j \leqslant p}|\mathrm{Re}\, b_j|$.

Suppose $|\xi_n - \eta_n| < \frac{1}{2}(1 + |\xi_n|)$. Then (14.27) and Lagrange's formula for the remainder in Taylor's theorem imply

$$|B(\xi_n) - B(\eta_n)| \leqslant \frac{C|\xi_n - \eta_n|}{(1 + |\xi_n|)^2}.$$

Analogously,

$$|b_+^{-1}(\xi_n)b_+(\eta_n) - I| = \left|\exp\left\{\ln\left(1 - \frac{\xi_n - \eta_n}{\xi_n + i}\right)b\right\} - I\right| \leqslant C\frac{|\xi_n - \eta_n|}{1 + |\xi_n|}.$$

Therefore

$$\begin{aligned}|C(\xi_n, \eta_n)| &\leqslant \frac{|B(\xi_n) - B(\eta_n)|}{|\xi_n - \eta_n|} \\ &\quad + \frac{|b_+^{-1}(\xi_n)b_+(\eta_n) - I|(|B(\eta_n) - I| + |B(\xi_n) - I|)}{|\xi_n - \eta_n|} \\ &\leqslant \frac{C}{(1 + |\xi_n|)^2} + \frac{C}{(1 + |\xi_n|)(1 + |\eta_n|)} \leqslant \frac{C}{(1 + |\xi_n|)(1 + |\eta_n|)},\end{aligned} \tag{14.31}$$

since $1 + |\eta_n| \leqslant 3/2(1 + |\xi_n|)$.

Now suppose $|\xi_n - \eta_n| \geqslant 1/2(1 + |\xi_n|)$. Then $1 + |\eta_n| \leqslant 3|\xi_n - \eta_n|$, and (14.25) implies $|b_+^{-1}(\xi_n)| \leqslant C(1 + |\xi_n|)^{1/2-\delta_0}$, $|b_+(\eta_n)| \leqslant C(1 + |\eta_n|)^{1/2-\delta_0}$, so that

$$|C(\xi_n, \eta_n)| \leqslant \frac{|b_+^{-1}(\xi_n)|\,|b_+(\eta_n)|(|B(\xi_n) - I| + |B(\eta_n) - I|)}{|\xi_n - \eta_n|}$$
$$\leqslant C(1 + |\xi_n|)^{-1/2-\delta_0}(1 + |\eta_n|)^{-1/2-\delta_0}. \tag{14.32}$$

Thus (14.30) is valid for any ξ_n and η_n, and hence C is a Hilbert-Schmidt operator.

THEOREM 14.1. *Suppose $A_0(\omega, \xi_n)$ is an elliptic matrix and either* (14.19) *or the equivalent condition* (14.23) *is satisfied. Then $\Pi^+ A_0$ is a Fredholm operator in $\tilde{H}_0^+(\mathbf{R}^1)$.*

PROOF. The operators P_+ and P_- satisfy

$$P_+A_0^{-1}a_+ = A_0^{-1}a_+P_- + C_1, \tag{14.33}$$

where C_1 is a Hilbert-Schmidt operator. In fact, since the matrices $a_+b_-^{-1}(\xi_n)b_+(\xi_n)$ and $A_0(\omega, \xi_n)$ have the same limits for $\xi_n \to \pm\infty$, the matrix $B(\xi_n) = A_0^{-1}a_+b_-^{-1}(\xi_n)b_+(\xi_n)$ satisfies (14.27). Consequently, Lemma 14.4 implies

$$\begin{aligned} P_+A_0^{-1}a_+ &= P_+Bb_+^{-1}b_- = BP_+b_+^{-1}b_- + Cb_+^{-1}b_- \\ &= A_0^{-1}a_+b_-^{-1}b_+P_+b_+^{-1}b_- + C_1, \end{aligned} \tag{14.34}$$

where $C_1 = Cb_+^{-1}b_-$. Since $b_+P_+b_+^{-1} = \Pi^+$ and $b_+^{-1}(\xi_n)b_-(\xi_n)$ is a bounded matrix, (14.33) follows from (14.34).

We will show that the operator $\tilde{R}\tilde{g}_+ = P_+A_0^{-1}a_+P_-a_+^{-1}\tilde{g}_+$ is a regularizer of Π^+A_0. We have $\Pi^+A_0\tilde{R}\tilde{g}_+ = \Pi^+A_0P_+A_0^{-1}a_+P_-a_+^{-1}\tilde{g}_+$. From (14.33) and the obvious equality $P_-^2 = P_-$ we get

$$\Pi^+A_0\tilde{R}\tilde{g}_+ = \Pi^+A_0\big(A_0^{-1}a_+P_- + C_1\big)P_-a_+^{-1}\tilde{g}_+ = \Pi^+a_+P_-a_+^{-1}\tilde{g}_+ + T_1\tilde{g}_+, \tag{14.35}$$

where $T_1 = \Pi^+A_0C_1P_-a_+^{-1}$ is a compact operator in $\tilde{H}_0^+(\mathbf{R}^1)$. Since a_+ is a constant matrix, $\Pi^+a_+ = a_+\Pi^+$. Further, $\Pi^+P_-\tilde{h} = \Pi^+b_-^{-1}\Pi^+b_-\tilde{h} = \Pi^+b_-^{-1}(b_-\tilde{h} - \Pi^-b_-\tilde{h}) = \Pi^+\tilde{h}$ for any function $\tilde{h} \in \tilde{H}_0(\mathbf{R}^1)$, since $b_-^{-1}\Pi^-b_-\tilde{h} \in \tilde{H}_0^-(\mathbf{R}^1)$ and hence $\Pi^+b_-^{-1}\Pi^-b_-\tilde{h} = 0$. Thus

$$\Pi^+A_0\tilde{R}\tilde{g}_+ = \tilde{g}_+ + T_1\tilde{g}_+, \qquad \forall \tilde{g}_+ \in \tilde{H}_0^+(\mathbf{R}^1). \tag{14.36}$$

Analogously, $\tilde{R}\Pi_+A_0\tilde{u}_+ = P_+A_0^{-1}a_+P_-a_+^{-1}\Pi^+A_0\tilde{u}_+$. We note that

$$P_-a_+^{-1}\Pi^+\tilde{h} = b_-^{-1}\Pi^+b_-a_+^{-1}(\tilde{h} - \Pi^-\tilde{h}) = P_-a_+^{-1}\tilde{h}$$

for any $\tilde{h} \in \tilde{H}_0(\mathbf{R}^1)$, since $b_- a_+^{-1} \Pi^- \tilde{h} \in \tilde{H}^-_{-1/2+\delta_0}(\mathbf{R}^1)$ (see Theorems 4.3 and 5.1) and hence $\Pi^+ b_- a_+^{-1} \Pi^- \tilde{h} = 0$. Further, $P_+ \tilde{u}_+ = \tilde{u}_+$ for any $\tilde{u}_+(x) \in \tilde{H}_0^+(\mathbf{R}^1)$ since, by Theorem 4.3, $b_+ \tilde{u}_+ \in \tilde{H}^+_{-1/2-\delta_0}(\mathbf{R}^1)$ and, by Lemma 5.4, $b_+ \tilde{u}_+ = \Pi^+ b_+ \tilde{u}_+$. Thus, using (14.33), we get

$$\begin{aligned} \tilde{R}\Pi^+ A_0 \tilde{u}_+ &= P_+ A_0^{-1} a_+ P_- a_+^{-1} A_0 \tilde{u}_+ = P_+ (P_+ A_0^{-1} a_+ - C_1) a_+^{-1} A_0 \tilde{u}_+ \\ &= P_+ \tilde{u}_+ + T_2 \tilde{u}_+ = \tilde{u}_+ + T_2 \tilde{u}_+, \end{aligned} \tag{14.37}$$

where $T_2 = -P^+ C_1 a_+^{-1} A_0$ is a compact operator in $\tilde{H}_0^+(\mathbf{R}^1)$. It follows from (14.36) and (14.37) that $\tilde{R}$ is a regularizer of $\Pi^+ A_0$, which is consequently a Fredholm operator.

6. Let us return now from equation (14.10) to (14.7) and determine conditions for the normal solvability of the operator $pA(\omega, D_n)$ in (14.7). The system (14.7) is equivalent to (14.8), with

$$A_0(\omega, \xi_n) = A^{(0)}(\omega, \xi_n) \left(\frac{\xi_n - i}{\xi_n + i} \right)^s,$$

where

$$A^{(0)}(\xi', \xi_n) = A(\xi', \xi_n)(\xi_n - i|\xi'|)^{-\alpha - i\beta}.$$

Consequently, $a_+ = A_0(0, +1) = A^{(0)}(0, +1)$ and $a_- = A_0(0, -1) = A^{(0)}(0, -1)e^{-2\pi i s}$. Let $a_j^{(0)}$, $1 \leqslant j \leqslant p$, be the eigenvalues of the matrix $A^{(0)}(0, -1)[A^{(0)}(0, +1)]^{-1}$. Then (14.23) is equivalent to the condition

$$s - \operatorname{Re} \frac{1}{2\pi i} \ln a_j^{(0)} - \frac{1}{2} \in \mathbf{R} \setminus \mathbf{Z}. \tag{14.38}$$

We thus obtain from Theorem 14.1

THEOREM 14.2. *Suppose $A(\xi', \xi_n) \in O^\infty_{\alpha + i\beta}$ is an elliptic matrix and s is a real number satisfying* (14.38). *Then $pA(\omega, D_n)$ is a Fredholm operator from $H_s^+(\mathbf{R}^1)$ into $H_{s-\alpha}(\mathbf{R}^1_+)$ for any ω with $|\omega| = 1$.*

REMARK 14.2. Let Z_0 denote the set of real numbers s for which (14.38) is not satisfied. It can be shown that when $s \in Z_0$, the operator $pA(\omega, D_n)$ does not have a closed range in $H_{s-\alpha}(\mathbf{R}^1_+)$, although its kernel and cokernel are still of finite dimension. Consequently, $pA(\omega, D_n)$ is not a Fredholm operator for $s \in Z_0$ (cf. the scalar case in Chapter II).

We now determine the dependence of the kernel and cokernel of $pA(\omega, D_n)$ on s.

LEMMA 14.5. *There exists a number $s_+ = s_+(\omega)$ such that* $\dim \ker pA(\omega, D_n) = 0$ *for $s > s_+$.*

PROOF. Let N_s denote the kernel of $pA(\omega, D_n)$ in $H_s^+(\mathbf{R}^1)$. The spaces N_s are finite dimensional. Since $H_{s_1}^+ \subset H_{s_2}^+$ for $s_1 > s_2$, it follows that $N_{s_1} \subset N_{s_2}$. Consequently, there are only two possibilities: a) there exists an s_+ such that $N_s = 0$ for $s > s_+$, or b) $\bigcap_s N_s = N_\infty \neq 0$ is not empty. We will prove Lemma 14.5 by excluding the second possibility.

Suppose there exists an element $u_0^+(x_n) \in H_\infty^+(\mathbf{R}^1)$ that does not identically vanish and satisfies the condition $pA(\omega, D_n)u_0^+ = 0$. Since the operators D_n^k and $pA(\omega, D_n)$ commute, we then have $pA(\omega, D_n)D_n^k u_0^+ = 0$, $\forall k \geqslant 0$, which implies that $D_n^k u_0^+ \in N_\infty$ for any k, with the functions $D_n^k u_0^+$, $k = 0, 1, \ldots,$ being linearly independent. But this contradicts the fact that N_∞ is finite dimensional. Therefore $N_\infty = 0$.

LEMMA 14.6. *There exists an* $s_- = s_-(\omega)$ *such that* $\dim \operatorname{coker} pA(\omega, D_n) = 0$ *for* $s < s_-$.

PROOF. We first find the form of the adjoint of $pA(\omega, D_n)$. By definition, the adjoint $(pA(\omega, D_n))^*$ acts from $H_{s-\alpha}^*(\mathbf{R}_+^1) = H_{\alpha-s}^+(\mathbf{R}^1)$ into $(H_s^+(\mathbf{R}^1))^* = H_{-s}(\mathbf{R}_+^1)$ (see §4.6), and

$$(pA(\omega, D_n)u_+, v_+) = (u_+, (pA(\omega, D_n))^* v_+), \qquad \forall u_+ \in H_s^+(\mathbf{R}^1),$$
$$\forall v_+ \in H_{\alpha-s}^+(\mathbf{R}^1). \tag{14.39}$$

By §4.6 (see (4.63)),

$$\begin{aligned}(pA(\omega, D_n)u_+, v_+) &= (A(\omega, D_n)u_+, v_+)\\ &= \frac{1}{2\pi}\int_{-\infty}^{\infty} A(\omega, \xi_n)\tilde{u}_+(\xi_n)\cdot\tilde{v}_+(\xi_n)\,d\xi_n,\end{aligned} \tag{14.40}$$

where $f(\xi_u)\cdot g(\xi_n)$ denotes the scalar product in the p-dimensional complex space $\mathbf{C}^p$. From (14.40) we get that

$$\begin{aligned}(pA(\omega, D_n)u_+, v_+) &= \frac{1}{2\pi}\int_{-\infty}^{\infty} \tilde{u}_+(\xi_n)\cdot A^*(\omega, \xi_n)\tilde{v}_+(\xi_n)\,d\xi_n\\ &= (u_+, A^*(\omega, D_n)v_+) = (u_+, pA^*(\omega, D_n)v_+),\end{aligned} \tag{14.41}$$

where $A^*(\omega, \xi_n)$ is the adjoint of $A(\omega, \xi_n)$. By (14.39) and (14.41), $(pA(\omega, D_n))^* = pA^*(\omega, D_n)$. The operator $pA^*(\omega, D_n)$ is a bounded operator from $H_{\alpha-s}^+(\mathbf{R}^1)$ into $H_{-s}(\mathbf{R}_+^1)$ and, by Theorem 14.2, is a Fredholm operator for $s \notin Z_0$. By Lemma 14.5, there exists an s_1^+ such that $\dim \ker pA^*(\omega, D_n) = 0$ for $\alpha - s > s_1^+$. Thus

$$\dim \ker pA^*(\omega, D_n) = \dim \operatorname{coker} pA(\omega, D_n) = 0$$

for $s < \alpha - s_1^+ = s_-$.

§15. An algebra of operators on a halfline containing the pseudodifferential operators

1. As was shown in §14, $\theta^+A_0^{-1}(\omega, D_n)$ is a regularizer of an operator $\theta^+A_0(\omega, D_n)$ if condition (14.9) is satisfied. But if $a_+ \neq a_-$, then $\theta^+A(\omega, D_n)$ does not have a regularizer of the same class. In this section we widen the class of operators so that it contains regularizers of the operators $\theta^+A_0(\omega, D_n)$.

Let $M(t)$ be a matrix belonging to C^∞ for $t > 0$ and satisfying

$$\left| t^k \frac{d^k M(t)}{dt^k} \right| \leqslant \begin{cases} C_\varepsilon t^{-\delta-\varepsilon} & \text{for } 0 < t < 1, 0 \leqslant k < \infty, \\ C_\varepsilon t^{-1+\delta+\varepsilon} & \text{for } 1 \leqslant t < \infty, 0 \leqslant k < \infty, \end{cases} \tag{15.1}$$

where, for a given function, $\delta \in (0, 1/2)$ is a fixed number and ε is any positive number. We denote by $\mathfrak{M}$ the class of integral operators M of the form

$$\hat{M}u_+ = \int_0^\infty M\left(\frac{t}{\tau}\right) \frac{u_+(\tau)}{\tau}\, d\tau, \qquad \forall u_+(\tau) \in C_0^\infty(\mathbf{R}^1_+), \tag{15.2}$$

where $M(t)$ satisfies (15.1). Applying the Mellin transform to (15.2), changing the order of integration and making the change of variables $t = t'\tau$, we get

$$Mu_+ = \int_0^\infty t^{s-1}\left(\int_0^\infty M\left(\frac{t}{\tau}\right)\frac{u_+(\tau)}{\tau}\, d\tau\right) dt = \hat{M}(s)\hat{u}_+(s), \tag{15.2$'$}$$

where $\hat{M}(s) = \int_0^\infty M(t)t^{s-1}dt$, which will be called the *symbol* of the operator M, is the Mellin transform of the matrix $M(t)$. By Lemma 2.3, $\hat{M}(s)$ is analytic in the strip $\delta < \operatorname{Re} s < 1 - \delta$, and

$$|s^m \hat{M}(s)| \leqslant C_{\varepsilon m}, \quad \delta + \varepsilon \leqslant \operatorname{Re} s \leqslant 1 - \delta - \varepsilon, \quad \forall \varepsilon > 0, \quad \forall m \geqslant 0. \tag{15.3}$$

Conversely, if $\hat{M}(s)$ is analytic in the strip $\delta < \operatorname{Re} s < 1 - \delta$ and satisfies (15.3), then $M(t)$ belongs to C^∞ for $t > 0$ and satisfies (15.1)

Consider the class of operators on the halfline $\mathbf{R}^1_+$ that can be represented in the form

$$\theta^+A_0(\omega, D_n)u_+ + \theta^+\chi(x_n)Mu_+ + \theta^+Tu_+, \quad \forall u_+ \in C_0^\infty(\mathbf{R}^1_+), \tag{15.4}$$

where $A_0(\omega, D_n)$ is the p.o. in $\mathbf{R}^1$ with symbol $A(\omega, \xi_n)$, $\chi(t)$ is a fixed element of $C_0^\infty(\mathbf{R}^1)$ such that $\chi(t) \equiv 1$ for $|t| \leqslant 1$, M is an integral operator of class $\mathfrak{M}$ and T is a compact operator in $H_0(\mathbf{R}^1) = L_2(\mathbf{R}^1)$. It will be shown in this section that the operators of form (15.4) form an algebra, i.e. their sums and products can also be represented in the form (15.4).

EXAMPLE 15.1. Let us find the form of the operators

$$\Pi^{\pm}u_+ = \pm\frac{i}{2\pi}\int_0^\infty \frac{u_+(y_n)dy_n}{x_n \pm i0 - y_n}$$

and

$$M_0u_+ = \frac{i}{2\pi}\int_0^\infty \frac{u_+(y_n)dy_n}{x_n + y_n}, \qquad x_n > 0,$$

after an application of the Mellin transform. From Example 2.1 (see (2.27)) we have

$$\int_0^\infty \frac{t^{s-1}dt}{t-a} = 2\pi i\frac{e^{(s-1)\ln a}}{1-e^{2\pi is}},$$

where a is not a positive number and $\ln a = \ln|a| + i\arg a$, $0 < \arg a < 2\pi$. Consequently

$$\pm\frac{i}{2\pi}\int_0^\infty \frac{t^{s-1}dt}{t \pm i\varepsilon - 1} = \mp\frac{e^{(s-1)\ln(1\mp i\varepsilon)}}{1-e^{2\pi is}}, \qquad \varepsilon > 0, \tag{15.5}$$

$$\frac{i}{2\pi}\int_0^\infty \frac{t^{s-1}dt}{t+1} = -\frac{e^{(s-1)\ln(-1)}}{1-e^{2\pi is}} = \frac{e^{\pi is}}{1-e^{2\pi is}}. \tag{15.6}$$

Therefore, analogously to (15.2′), for $0 < \operatorname{Re} s < 1$

$$\widehat{\Pi^+u_+} = \lim_{\varepsilon\to 0}\int_0^\infty x_n^{s-1}\left(+\frac{i}{2\pi}\int_0^\infty \frac{u_+(y_n)dy_n}{x_n + i\varepsilon - y_n}\right)dx_n = -\frac{e^{2\pi is}}{1-e^{2\pi is}}\hat{u}_+(s), \tag{15.7}$$

$$\widehat{\Pi^-u_+} = \lim_{\varepsilon\to 0}\int_0^\infty x_n^{s-1}\left(-\frac{i}{2\pi}\int_0^\infty \frac{u_+(y_n)dy_n}{x_n - i\varepsilon - y_n}\right)dx_n = \frac{1}{1-e^{2\pi is}}\hat{u}_+(s), \tag{15.8}$$

$$\widehat{M_0u_+} = \frac{e^{i\pi s}}{1-e^{2\pi is}}\hat{u}_+(s). \tag{15.9}$$

We note that $e^{2\pi is}/(1-e^{2\pi is})$ and $1/(1-e^{2\pi is})$ are analytic in the strip $0 < \operatorname{Re} s < 1$ and bounded for $\varepsilon < \operatorname{Re} s < 1-\varepsilon$, $\varepsilon > 0$, with

$$\frac{e^{2\pi is}}{1-e^{2\pi is}} = o\left(\frac{1}{|s|^k}\right), \qquad \frac{1}{1-e^{2\pi is}} = 1 + o\left(\frac{1}{|s|^k}\right) \tag{15.10}$$

for $\operatorname{Im} s \to +\infty$, $\varepsilon < \operatorname{Re} s < 1-\varepsilon$, $\forall k > 0$, and

$$\frac{e^{2\pi is}}{1-e^{2\pi is}} = -1 + o\left(\frac{1}{|s|^k}\right), \quad \frac{1}{1-e^{2\pi is}} = o\left(\frac{1}{|s|^k}\right) \tag{15.10′}$$

for $\operatorname{Im} s \to -\infty$, $\varepsilon < \operatorname{Re} s < 1-\varepsilon$, $\forall k > 0$.

From Lemma 2.3 and relations (15.7), (15.8), (15.10) and (15.10′) we obtain

LEMMA 15.1. *Suppose M_1 and M_2 are operators of class $\mathfrak{M}$. Then the operators M_1M_2, $\Pi^{\pm}M_1$, $M_1\Pi^{\pm}$ are also of class $\mathfrak{M}$ and have the symbols*

$$\hat{M}_1(s)\hat{M}_2(s), \qquad \frac{1}{1-e^{2\pi is}}\hat{M}_1(s), \qquad -\frac{e^{2\pi is}}{1-e^{2\pi is}}\hat{M}_1(s)$$

respectively.

2. Let us find an expression for a p.o. $A_0(\omega, D_n)$ in $\mathbf{R}^1$ in the form of an integral operator of convolution type (cf. §3). We will use the notation $a_0 = F_{\xi_n}^{-1}A_0(\omega, \xi_n)$.

LEMMA 15.2. *Suppose $A_0(\omega, \xi_n) \in O_0$. Then $a_0 \in C^\infty$ for $x_n \neq 0$ and has the following properties*:

$$|x_n^k a_0(x_n)| \leqslant C_k, \qquad \forall k \geqslant 0, \quad |x_n| \geqslant 1, \tag{15.11}$$

$$\begin{aligned} A_0(\omega, D_n)u_+ &= a_0 * u_+ = a_+\Pi^- u_+ + a_-\Pi^+ u_+ + b*u_+ \\ &= \frac{a_+ + a_-}{2}u_+ + (a_- - a_+)\Pi^0 u_+ + b*u_+, \end{aligned} \tag{15.12}$$

where $b(x_n) \in L_2(\mathbf{R}^1)$, $a_+ = A_0(0, +1)$, $a_- = A_0(0, -1)$, and

$$\Pi^0 u_+ = \frac{i}{2\pi}\,\text{p.v.}\int_0^\infty \frac{u_+(y_n)dy_n}{x_n - y_n}.$$

PROOF. Since $A_0(\omega, \xi_n) \in O_0$, we have

$$\left|\frac{\partial^k A_0(\omega, \xi_n)}{\partial \xi_n^k}\right| \leqslant C_k(1 + |\xi_n|)^{-k}, \qquad \forall k \geqslant 0, \tag{15.13}$$

from which it follows that $a_0(x_n)$ belongs to C^∞ for $x_n \neq 0$ and has property (15.11).

We now put $B(\omega, \xi_n) = A_0(\omega, \xi_n) - a_+\theta(\xi_n) - a_-(1 - \theta(\xi_n))$. Then $|B(\omega, \xi_n)| \leqslant C(1 + |\xi_n|)^{-1}$ (cf. (14.15)), which implies $b(x_n) = F_{\xi_n}^{-1}B(\omega, \xi_n) \in L_2(\mathbf{R}^1)$. Analogously to (5.15), we have

$$F_{\xi_n}^{-1}\theta(\xi_n)e^{-\varepsilon\xi_n} = \frac{1}{2\pi}\int_0^\infty e^{-\varepsilon\xi_n - ix_n\xi_n}\,d\xi_n = \frac{1}{2\pi}\frac{1}{ix_n + \varepsilon} = -\frac{i}{2\pi}\frac{1}{x_n - i\varepsilon},$$

and hence (cf. the proof of Lemma 5.2)

$$F_{\xi_n}^{-1}\theta(\xi_n) = -\frac{i}{2\pi}\frac{1}{x_n - i0}. \tag{15.14}$$

Analogously,

$$F_{\xi_n}^{-1}(1 - \theta(\xi_n)) = \frac{i}{2\pi}\frac{1}{x_n + i0}, \tag{15.15}$$

and the decomposition (15.12) follows.

3. Suppose $A_1(\omega, \xi_n)$ and $A_2(\omega, \xi_n)$ are of class O_0, and that $a_1^\pm = A_1(0, \pm 1)$ and $a_2^\pm = A_2(0, \pm 1)$. Whereas the product of two p.o.'s in $\mathbf{R}_+^1$ is not a p.o., it is shown in the following lemma that the product of the operators $\theta^+A_1(\omega, D_n)$ and $\theta^+A_2(\omega, D_n)$ can be represented in the form (15.4).

LEMMA 15.3. *The following representation holds*:

$$\theta^+A_1(\omega, D_n)\theta^+A_2(\omega, D_n)u_+ = \theta^+A_3(\omega, D_n)u_+ + \theta^+\chi(x_n)M_{12}u_+ + \theta^+T_1u_+,$$
$$\forall u_+ \in C_0^\infty(\mathbf{R}^1_+), \tag{15.16}$$

where $A_3(\omega, D_n)$ is the p.o. in $\mathbf{R}^1$ with symbol $A_3(\omega, \xi_n) = A_1(\omega, \xi_n)A_2(\omega, \xi_n)$, M_{12} is the integral operator of class $\mathfrak{M}$ with symbol

$$\hat{M}_{12}(s) = (a_1^+ - a_1^-)(a_2^+ - a_2^-)\frac{e^{2\pi i s}}{(1 - e^{2\pi i s})^2}, \tag{15.17}$$

and T is a compact operator in $H_0(\mathbf{R}^1)$.

PROOF. We have

$$\theta^+A_1(\omega, D_n)\theta^+A_2(\omega, D_n)u_+ = \theta^+A_1(\omega, D_n)A_2(\omega, D_n)u_+ - \theta^+A_1(\omega, D_n)\theta^-A_2(\omega, D_n)u_+, \tag{15.18}$$

where θ^- is the operator of multiplication by $(1 - \theta(x_n)) = \theta(-x_n)$. Since $\theta(x_n)\theta(-x_n) = 0$, it follows by virtue of (15.12) that

$$\theta^+A_1(\omega, D_n)\theta^-A_2(\omega, D_n)u_+ = \theta^+\left[(a_1^- - a_1^+)\Pi^0 + B_1\right]\theta^-\left[(a_2^- - a_2^+)\Pi^0 + B_2\right]u_+, \tag{15.19}$$

where $B_iu_+ = b_i * u_+$, $b_i(x_n) \in L_2(\mathbf{R}^1)$, $i = 1, 2$.

The integral operator $\chi(x_n)B_1v = \int_{-\infty}^{\infty}\chi(x_n)b_1(x_n - y_n)v(y_n)dy_n$ is clearly a Hilbert-Schmidt operator. By (15.11),

$$|(1 - \chi(x_n))a_1(x_n - t)| \leqslant C_N(x_n + |t|)^{-N},$$

for $x_n \geqslant 1$, $t \leqslant 0$ and all integral N. Consequently,

$$(1 - \chi(x_n))\theta^+A_1(\omega, D_n)\theta^-$$

is also a Hilbert-Schmidt operator. Thus

$$\begin{aligned}\theta^+A_1(\omega, D_n)\theta^-A_2(\omega, D_n)u_+ &= \chi(x_n)\theta^+A_1(\omega, D_n)\theta^-A_2(\omega, D_n)u_+ \\ &\quad + (1 - \chi(x_n))\theta^+A_1\theta^-A_2u_+ \\ &= \chi(x_n)\theta^+(a_1^- - a_1^+) \\ &\quad \times\Pi^0\theta^-\left[(a_2^- - a_2^+)\Pi^0 + B_2\right]u_+ + T_1u_+,\end{aligned} \tag{15.20}$$

where T_1 is a compact operator in $H_0(\mathbf{R}^1)$. Here we have

$$\theta^+\Pi^0\theta^-\Pi^0u_+ = \theta(x_n)\left(\frac{i}{2\pi}\right)^2\int_{-\infty}^0\frac{1}{x_n - y_n}\left(\theta(-y_n)\int_0^\infty\frac{u_+(z_n)dz_n}{y_n - z_n}\right)dy_n. \tag{15.21}$$

Making the change of variables $y_n = -t$ in (15.21), we get

$$\theta^+\Pi^0\theta^-\Pi^0 u_+ = -M_0^2 u_+, \quad \text{where } M_0 u_+ = \frac{i}{2\pi}\int_0^\infty \frac{u_+(y_n)dy_n}{x_n + y_n}$$

(see Example 15.1). Consequently, taking into account the fact that the commutator of $\chi(x_n)$ and Π^0 and the product $\chi(x_n)B_2$ are Hilbert-Schmidt operators, we get

$$\theta^+A_1(\omega, D_n)\theta^-A_2(\omega, D_n)u_+ = -\chi(x_n)(a_1^- - a_1^+)(a_2^- - a_2^+)M_0^2u_+ + T_2u_+, \tag{15.22}$$

where T_2 is a compact operator. It follows from (15.18) and (15.22) by virtue of Lemma 15.1 that the representation (15.16) is valid.

LEMMA 15.4. *Suppose $A_0(\omega, \xi_n) \in O_0$ and M is an integral operator of class $\mathfrak{M}$. Then*

$$\theta^+A_0(\omega, D_n)\theta^+\chi(x_n)Mu_+ = \theta^+\chi(x_n)M_1u_+ + \theta^+T_1u_+, \tag{15.23}$$

$$\theta^+\chi(x_n)M\theta^+A_0(\omega, D_n)u_+ = \theta^+\chi(x_n)M_2u_+ + \theta^+T_2u_+, \tag{15.24}$$

where M_1 and M_2 are the integral operators of class $\mathfrak{M}$ with symbols

$$\hat{M}_1(s) = \left(a_+\frac{1}{1-e^{2\pi is}} - a_-\frac{e^{2\pi is}}{1-e^{2\pi is}}\right)\hat{M}(s), \tag{15.25}$$

$$\hat{M}_2(s) = \hat{M}(s)\left(a_+\frac{1}{1-e^{2\pi is}} - a_-\frac{e^{2\pi is}}{1-e^{2\pi is}}\right), \tag{15.26}$$

and T_1 and T_2 are compact operators.

PROOF. By (15.12),

$$\theta^+A_0(\omega, D_n)\theta^+\chi(x_n)Mu_+ = \theta^+(a_+\Pi^- + a_-\Pi^+ + B)\theta^+\chi(x_n)Mu_+, \tag{15.27}$$

where $Bv = b*v$. We note that $B\chi(x_n)$ and the commutators of the operators $\Pi^\pm$ and $\chi(x_n)$ are Hilbert-Schmidt operators. Consequently, permuting $\Pi^\pm$ and $\chi(x_n)$ in (15.27) and applying Lemma 15.1, we get (15.23).

Analogously, by Lemma 15.1, we have

$$\begin{aligned}\theta^+\chi(x_n)M\theta^+A_0(\omega, D_n)u_+ &= \theta^+\chi(x_n)M\theta^+(a_+\Pi^- + a_-\Pi^+ + B)u_+ \\ &= \theta^+\chi(x_n)M_2u_+ + \theta^+\chi(x_n)M\theta^+Bu_+,\end{aligned} \tag{15.28}$$

where M_2 is the operator of class $\mathfrak{M}$ with symbol (15.26). Here we note that the commutator $\theta^+(\chi(x_n)M - M\chi(x_n))$ has the form

$$\int_0^\infty \frac{(\chi(x_n) - \chi(y_n))M(x_n/y_n)}{y_n}u_+(y_n)dy_n, \qquad x_n > 0. \tag{15.29}$$

But the estimate

$$\left|M\left(\frac{x_n}{y_n}\right)\right| \leqslant \frac{C_\varepsilon y_n^{1-\delta-\varepsilon}}{x_n^{\delta+\varepsilon}(x_n+y_n)^{1-2\delta-2\varepsilon}},$$

which follows from (15.1), implies (15.29) is a Hilbert-Schmidt operator, and hence $\theta^+\chi(x_n)M\theta^+B\theta^+$ is a Hilbert-Schmidt operator. Thus, as a consequence of (15.29), we have

LEMMA 15.5. *Suppose M_1 and M_2 are operators of class $\mathfrak{M}$. Then*

$$\chi M_1 \chi M_2 u_+ = \chi M_3 u_+ + \theta^+ T u_+,$$

where M_3 is the operator of class $\mathfrak{M}$ with symbol $\hat{M}_3(s) = \hat{M}_1(s)\hat{M}_2(s)$ and T is a compact operator in $L_2(\mathbf{R}^1)$.

4. By the *symbol* of an operator of form (15.4) is meant the pair of matrices

$$A_0(\omega, \xi_n), \qquad -\infty \leqslant \xi_n \leqslant +\infty,$$

$$a_+\frac{1}{1-e^{2\pi i s}} - a_-\frac{e^{2\pi i s}}{1-e^{2\pi i s}} + \hat{M}(s), \qquad \operatorname{Re} s = 1/2, \tag{15.30}$$

where $\hat{M}(s)$ is the symbol of the operator M and $a_\pm = A_0(0, \pm 1) = \lim_{\xi_n \to \pm\infty} A_0(\omega, \xi_n)$. Lemmas 15.3–15.5 imply

THEOREM 15.1. *Suppose $\mathcal{Q}_1 = \theta^+A_1(\omega, D_n) + \theta^+\chi(x_n)M_1 + \theta^+T_1$ and $\mathcal{Q}_2 = \theta^+A_2(\omega, D_n) + \theta^+\chi(x_n)M_2 + \theta^+T_2$ are two operators of form* (15.4) *with symbols*

$$A_1(\omega, \xi_n), \qquad a_1^+\frac{1}{1-e^{2\pi i s}} - a_1^-\frac{e^{2\pi i s}}{1-e^{2\pi i s}} + \hat{M}_1(s)$$

and

$$A_2(\omega, \xi_n), \qquad a_2^+\frac{1}{1-e^{2\pi i s}} - a_1^-\frac{e^{2\pi i s}}{1-e^{2\pi i s}} + \hat{M}_2(s),$$

where $a_1^\pm = A_1(0, \pm 1)$, $a_2^\pm = A_2(0, \pm 1)$. Then their product $\mathcal{Q}_3$ can also be represented in the form (15.4):

$$\mathcal{Q}_3 = \mathcal{Q}_1\mathcal{Q}_2 = \theta^+A_3(\omega, D_n) + \theta^+\chi(x_n)M_3 + \theta^+T_3, \tag{15.31}$$

and has the symbol

$$A_3(\omega, \xi_n), \qquad a_3^+\frac{1}{1-e^{2\pi i s}} - a_3^-\frac{e^{2\pi i s}}{1-e^{2\pi i s}} + \hat{M}_3(s),$$

which is equal to the product of the symbols of $\mathcal{Q}_1$ and $\mathcal{Q}_2$, i.e.

$$A_3(\omega, \xi_n) = A_1(\omega, \xi_n)A_2(\omega, \xi_n), \tag{15.32}$$

$$a_3^+ \frac{1}{1 - e^{2\pi i s}} - a_3^- \frac{e^{2\pi i s}}{1 - e^{2\pi i s}} + \hat{M}_3(s)$$
$$= \left(a_1^+ \frac{1}{1 - e^{2\pi i s}} - a_1^- \frac{e^{2\pi i s}}{1 - e^{2\pi i s}} + \hat{M}_1(s) \right)$$
$$\cdot \left(a_2^+ \frac{1}{1 - e^{2\pi i s}} - a_2^- \frac{e^{2\pi i s}}{1 - e^{2\pi i s}} + \hat{M}_2(s) \right). \qquad (15.33)$$

In fact, by Lemmas 15.3–15.5, the operator $\mathcal{A}_3 = \mathcal{A}_1 \mathcal{A}_2$ is representable in the form (15.31), with $A_3(\omega, \xi_n) = A_1(\omega, \xi_n) A_2(\omega, \xi_n)$, and

$$\hat{M}_3(s) = (a_1^+ - a_1^-)(a_2^+ - a_2^-) \frac{e^{2\pi i s}}{(1 - e^{2\pi i s})^2}$$
$$+ \hat{M}_1(s) \left(a_2^+ \frac{1}{1 - e^{2\pi i s}} - a_2^- \frac{e^{2\pi i s}}{1 - e^{2\pi i s}} \right)$$
$$+ \left(a_1^+ \frac{1}{1 - e^{2\pi i s}} - a_1^- \frac{e^{2\pi i s}}{1 - e^{2\pi i s}} \right) \hat{M}_2(s) + \hat{M}_1(s) \hat{M}_2(s). \qquad (15.34)$$

Since $a_3^\pm = a_1^\pm a_2^\pm$ and

$$\left(\frac{a_1^+}{1 - e^{2\pi i s}} - \frac{a_1^- e^{2\pi i s}}{1 - e^{2\pi i s}} \right) \left(\frac{a_2^+}{1 - e^{2\pi i s}} - \frac{a_2^- e^{2\pi i s}}{1 - e^{2\pi i s}} \right)$$
$$= \frac{a_1^+ a_2^+}{1 - e^{2\pi i s}} - \frac{a_1^- a_2^- e^{2\pi i s}}{1 - e^{2\pi i s}} + (a_1^+ - a_1^-)(a_2^+ - a_2^-) \frac{e^{2\pi i s}}{(1 - e^{2\pi i s})^2},$$

the validity of (15.33) follows from (15.34).

Theorem 15.1 permits one to find conditions for the normal solvability of an operator $\mathcal{A}$ of form (15.4).

COROLLARY 15.1. *An operator* $\mathcal{A}$ *of form* (15.4) *is a Fredholm operator if the determinants of its symbol* (15.30) *are different from zero, i.e. if*

$$\det A_0(\omega, \xi_n) \neq 0, \qquad -\infty \leqslant \xi_n \leqslant +\infty,$$
$$\det \left(\frac{a_+}{1 - e^{2\pi i s}} - \frac{a_- e^{2\pi i s}}{1 - e^{2\pi i s}} + \hat{M}(s) \right) \neq 0, \quad \operatorname{Re} s = 1/2. \qquad (15.35)$$

For there then exists the operator R of form (15.4) whose symbol is the inverse of the symbol of $\mathcal{A}$, viz.

$$Ru_+ = \theta^+ A_0^{-1}(\omega, D_n) u_+ + \theta^+ \chi(x_n) M_1 u_+, \qquad (15.36)$$

where

$$\hat{M}_1(s) = \left(\frac{a_+}{1-e^{2\pi i s}} - \frac{a_- e^{2\pi i s}}{1-e^{2\pi i s}} + \hat{M}(s)\right)^{-1} - \left(\frac{a_+^{-1}}{1-e^{2\pi i s}} - \frac{a_-^{-1} e^{2\pi i s}}{1-e^{2\pi i s}}\right). \tag{15.37}$$

(We note that the operator M_1 is of class $\mathfrak{M}$ by virtue of (15.10) and (15.10′).) It therefore follows from Theorem 15.1 that R is a regularizer of $\mathcal{A}$ and hence that $\mathcal{A}$ is a Fredholm operator.

In particular (when $M = T = 0$), an operator of the form $\mathcal{A} = \theta^+ A_0(\omega, D_n)$, is a Fredholm operator if

$$\det A_0(\omega, \xi_n) \neq 0, \quad -\infty \leqslant \xi_n \leqslant +\infty, \tag{15.38}$$

$$\det\left(\frac{a_+}{1-e^{2\pi i s}} - \frac{a_- e^{2\pi i s}}{1-e^{2\pi i s}}\right) \neq 0, \quad \operatorname{Re} s = 1/2. \tag{15.38′}$$

Here (15.38′) can be given a more convenient form. Since (15.38) implies $\det a_+ \neq 0$ and $e^{2\pi i s} = -e^{-2\pi\tau}$, $1 - e^{2\pi i s} = 1 + e^{-2\pi\tau}$ for $s = 1/2 + i\tau$, $-\infty < \tau < +\infty$, condition (15.38′) is equivalent to

$$\det\left(e^{2\pi\tau} I + a_+^{-1} a_-\right) \neq 0 \quad \text{for } -\infty < \tau < +\infty. \tag{15.39}$$

When τ ranges over the real axis, $e^{2\pi\tau}$ ranges over the positive real axis. Thus (15.38′) means that the matrix $a_+^{-1}a_-$ cannot have negative real eigenvalues and hence is equivalent to condition (14.19).

Remark 15.1. It can be shown that (15.35) holds for any Fredholm operator of form (15.4).

5. Let us find the index of a Fredholm operator $\mathcal{A}$ of form (15.4).

Theorem 15.2. *The index of an operator $\mathcal{A}$ satisfying* (15.35) *is given by*

$$\operatorname{ind} \mathcal{A} = \frac{1}{2\pi} \Delta \arg \det A_0(\omega, \xi_n)\Big|_{\xi_n=+\infty}^{\xi_n=-\infty}$$

$$+ \frac{1}{2\pi} \Delta \arg \det\left(\frac{a_+}{1-e^{2\pi i s}} - \frac{a_- e^{2\pi i s}}{1-e^{2\pi i s}} + \hat{M}(s)\right)\Bigg|_{s=1/2-i\infty}^{s=1/2+i\infty} \tag{15.40}$$

where the right side is an integer since $A_0(\omega, \xi_n)$ for $\xi_n = -\infty$ ($\xi_n = +\infty$) is equal to

$$N(s) = \frac{a_+}{1-e^{2\pi i s}} - \frac{a_- e^{2\pi i s}}{1-e^{2\pi i s}} + \hat{M}(s)$$

for $s = 1/2 - i\infty$ ($s = 1/2 + i\infty$) (see (15.10) *and* (15.10′)).

PROOF. We first continuously deform the operator $\mathfrak{A}$ with preservation of condition (15.35) into an operator of form (15.4) for which

$$a_+ = a_- = I. \tag{15.41}$$

Since the left and right sides of (15.40) do not vary under such a deformation, it suffices to prove (15.40) under the fulfilment of condition (15.41). We note that in this case each of the two summands in the right side of (15.40) is an integer.

Under the fulfilment of condition (15.41) both the matrix $A_0(\omega, \xi_n)$ and the matrix $N(s) = I + \hat{M}(s)$ can be continuously deformed into a diagonal matrix all of whose diagonal elements except one are equal to unity. The homotopy in this connection can be realized so that the operators remain in the form (15.4) and condition (15.35) is preserved. Consequently, the system of equations $\mathfrak{A}u_+ = f$ can be continuously deformed into a diagonal system of the form

$$\begin{aligned} \theta^+ a_1(\omega, D_n)u_1^+ &= f_1^+, \\ u_2^+ + \theta^+\chi m_2 u_2^+ &= f_2^+, \\ u_3^+ &= f_3^+, \\ \cdots\cdots\cdots\cdots & \\ u_p^+ &= f_p^+, \end{aligned} \tag{15.42}$$

where $a_1(\omega, \xi_n)$ is a scalar symbol of class O_0, $a_1(0, \pm 1) = 1$, and m_2 is a scalar operator of class $\mathfrak{M}$. We note that in the case $p = 1$, the system obtained by adding an artificial equation $u_2^+ = f_2^+$ to the equation $\mathfrak{A}u_1^+ = f_1^+$ (which clearly does not change the index) can be continuously deformed into a system of form (15.42).

Clearly, the index of system (15.42) is equal to the sum of the indices of the first and second equations of the system, so that it suffices to prove (15.40) separately for each of the first two equations. The validity of (15.40) for the first equation of (15.42) is a consequence of Lemma 14.2, while its validity for the second equation is proved as follows.

Let $\chi_1(x_n)$ denote the function that is equal to 1 for $x_n \in [0, 1]$ and vanishes for $x_n \notin [0, 1]$. Since estimates (15.1) imply that $\theta^+\chi m_2 u_2^+$ differs from $\chi_1 m_2 u_2^+$ by a compact operator in $H_0(\mathbf{R}^1)$, the second equation is clearly equivalent to the following equation on [0, 1]:

$$v_2^+ + \chi_1 m_2 v_2^+ = g_2^+, \tag{15.43}$$

where $v_2^+ = u_2^+$, $g_2^+ = f_2^+ - \chi_1 m_2(1 - \chi_1)f_2^+$, for $x_n \in [0, 1]$.

To find the index of equation (15.43), we make the change of variables $x_n = e^{-t}$, $0 \leqslant t < \infty$, and let $w_2(t) = v_2(e^{-t})e^{-\frac{t}{2}}$. Then (15.43) goes over into the integral equation of convolution type on a halfline

$$w_2(t) + \int_0^\infty e^{-(t-\tau)/2} m_2(e^{-(t-\tau)}) w_2(\tau)\, d\tau = h_2(t), \tag{15.44}$$

whose form is analogous to (14.8) and whose symbol is equal to

$$1 + \tilde{m}_2(\xi_n) = 1 + \int_{-\infty}^{\infty} e^{-\frac{t}{2}} m_2(e^{-t}) e^{it\xi_n}\, dt. \tag{15.45}$$

Making the change of variables $e^{-t} = x_n$ in (15.45), we get

$$\tilde{m}_2(\xi_n) = \int_0^\infty m_2(x_n) x_n^{-1/2 - i\xi_n} dx_n = \hat{m}_2\left(\frac{1}{2} - i\xi_n\right),$$

where $\hat{m}_2(s)$ is the Mellin transform of $m_2(x_n)$. Thus, by Lemma 14.2, the index of (15.43) is equal to

$$\frac{1}{2\pi} \Delta \arg(1 + \tilde{m}_2(\xi_n)) \Big|_{\xi_n = +\infty}^{\xi_n = -\infty} = \frac{1}{2\pi} \Delta \arg(1 + \hat{m}_2(s)) \Big|_{s = 1/2 - i\infty}^{s = 1/2 + i\infty}.$$

Q.E.D.

Let us derive a formula for the index of the operator $\mathcal{Q}_0 = \theta^+ A_0(\omega, D_n)$. By (15.40),

$$\operatorname{ind} \mathcal{Q}_0 = \frac{1}{2\pi} \Delta \arg \det A_0(\omega, \xi_n) \Big|_{\xi_n = +\infty}^{\xi_n = -\infty}$$

$$+ \frac{1}{2\pi} \Delta \arg \det \left(\frac{a_+}{1 - e^{2\pi i s}} - \frac{a_- e^{2\pi i s}}{1 - e^{2\pi i s}} \right) \Big|_{s = 1/2 - i\infty}^{s = 1/2 + i\infty}. \tag{15.46}$$

If $s = 1/2 + i\tau$, then $e^{2\pi i s} = -e^{-2\pi\tau}$. Let

$$t = \frac{1}{1 - e^{2\pi i s}} = \frac{1}{1 + e^{-2\pi\tau}}.$$

When τ ranges over the real axis, t ranges over $[0, 1]$. Consequently, (15.46) can be written in the form

$$\operatorname{ind} \mathcal{Q}_0 = \frac{1}{2\pi} \Delta \arg \det A_0(\omega, \xi_n) \Big|_{\xi_n = +\infty}^{\xi_n = -\infty} + \frac{1}{2\pi} \Delta \arg \det [t a_+ + (1 - t) a_-] \Big|_{t=0}^{t=1}. \tag{15.47}$$

This can be given the following geometric interpretation (see [52]). When ξ_n varies from $+\infty$ to $-\infty$, the matrices $A_0(\omega, \xi_n)$ describe a curve, which is generally open, in the space of nonsingular matrices of order p. We join the matrices $a_+ = A_0(0, +1)$ and $a = A_0(0, -1)$ by the line segment $ta_+ + (1 - t)a_-$ in the space of matrices (by the line segment joining the points a_+ and a_- in the complex plane, when $A_0(\omega, \xi_n)$ is a scalar symbol). But one of the conditions for the normal solvability of $\mathcal{Q}_0$ is that the matrices $ta_+ + (1 - t)a_-$ be nonsingular for $t \in [0, 1]$. As a result, we obtain a closed curve in the space of nonsingular matrices. The increment in the argument of the determinant around this curve is equal to 2π times the index of $\mathcal{Q}_0$.

REMARK 15.2. Theorem 14.1 and Corollary 15.1 provide two methods of constructing a regularizer of the operator $\mathcal{Q}_0 = \theta^+ A_0(\omega, D_n)$. Here we indicate a third method.

Let $b_\pm(\xi_n)$ be the same matrices as in §14.5 and suppose (14.22) is satisfied. Then Lemma 14.3 implies that the operator $b_+^{-1}(\xi_n)\Pi^+ b_-(\xi_n)a_+^{-1}$, which is the inverse of $\Pi^+ a_+ b_-^{-1}(\xi_n)b_+(\xi_n)$, is bounded in $\tilde{H}_0^+(\mathbf{R}^1)$. We will show that the following operator is a regularizer of $\theta^+ A_0(\omega, D_n)$:

$$R_0 g_+ = \chi(x_n)F_{\xi_n}^{-1} b_+^{-1}(\xi_n)\Pi^+ b_-(\xi_n)a_+^{-1}\tilde{g}_+(\xi_n) + (1-\chi(x_n))\theta^+ F_{\xi_n}^{-1}A_0^{-1}(\omega, \xi_n)\tilde{g}_+(\xi_n). \tag{15.48}$$

(Note that $(1-\chi(x_n))\theta(x_n) \in C^\infty(\mathbf{R}^1)$.)

We have

$$A_0(\omega, \xi_n) = a_+ b_-^{-1}(\xi_n)b_+(\xi_n) + B(\omega, \xi_n), \quad \text{where } |B(\omega, \xi_n)| \leqslant C(1+|\xi_n|)^{-1}.$$

As was noted in the proof of Lemma 15.3, the operators $\chi(x_n)B(\omega, D_n)$, $B(\omega, D_n)\chi(x_n)$ and the commutators of $\chi(x_n)$, $A_0(\omega, D_n)$ and $\chi(x_n)$, $a_+ b_-^{-1}(D_n)b_+(D_n)$ are Hilbert-Schmidt operators. Consequently,

$$\begin{aligned}\theta^+ A_0(\omega, D_n)R_0 g_+ &= \theta^+\big(a_+ b_-^{-1}(D_n)b_+(D_n) + B(\omega, D_n)\big)R_0 g_+\\ &= \chi(x_n)g_+ + \theta^+ T_1 g_+\\ &\quad + (1-\chi(x_n))\theta^+ A_0(\omega, D_n)F_{\xi_n}^{-1}A_0^{-1}(\omega, \xi_n)\tilde{g}_+(\xi_n) + \theta^+ T_2 g_+\\ &= g_2 + \theta^+ T_3 g_+,\end{aligned} \tag{15.49}$$

where the T_i $(i = 1, 2, 3)$ are compact operators in $H_0(\mathbf{R}^1)$. Analogously,

$$R_0\theta^+ A_0(\omega, D_n)u_+ = u_+ + \theta^+ T_4 u_+, \tag{15.50}$$

where T_4 is a compact operator.

§16. Boundary value problems for systems of elliptic pseudodifferential equations in a halfspace

1. *Theorem on the unique solvability of a boundary value problem.* Suppose $\hat{A}(D)$ is a p.o. with symbol $\hat{A}(\xi) = \|\hat{a}_{ij}(\xi)\|_{i,j=1}^p$, such that $A(\xi) \in O_{\alpha+i\beta}^\infty$ and the ellipticity condition

$$\det A(\xi) \neq 0 \quad \text{for } \xi \neq 0 \tag{16.1}$$

holds, and suppose s is a real number satisfying (14.38). Then, by Theorem 14.2, $pA(\omega, D_n)$ is a Fredholm operator from $H_s^+(\mathbf{R}^1)$ into $H_{s-\alpha}(\mathbf{R}_+^1)$ for any ω on the unit sphere S^{n-2} in $\mathbf{R}^{n-1}$.

Let $n_+(\omega)$ and $n_-(\omega)$ denote the dimensions of the kernel and cokernel of $pA(\omega, D_n)$. As is well known (see Corollary 16.1 below), the functions $n_+(\omega)$

and $n_-(\omega)$ are bounded on S^{n-2}: $\max_\omega n_+(\omega) < \infty$, $\max_\omega n_-(\omega) < \infty$. By analogy with the scalar case (see Chapter II), for the correct formulation of a boundary value problem in $\mathbf{R}^n_+$ we add at least $\max_\omega n_-(\omega)$ potentials in the equation $p\hat{A}(D)u_+ = f$ in order to eliminate the cokernel of $pA(\omega, D_n)$, and assign at least $\max_\omega n_+(\omega)$ boundary conditions on the hyperplane $x_n = 0$ in order to eliminate the kernel of $pA(\omega, D_n)$.

Let m_+ and m_- be arbitrary integers such that $m_+ \geqslant \max_\omega n_+(\omega)$, $m_- \geqslant \max_\omega n_-(\omega)$ and $m_+ - m_- = n_+(\omega) - n_-(\omega) = m$, where m is the index of $pA(\omega, D_n)$. We note that m does not depend on ω for $n > 2$ since S^{n-2} is connected and $pA(\omega, D_n)$ continuously depends on ω in the uniform operator topology. It will be assumed that, in addition, $m = n_+(\omega) - n_-(\omega)$ does not depend on ω for $n = 2$.

Suppose $B_j(\xi', \xi_n) \in O'_{\beta_j}$, $C_k(\xi', \xi_n) \in O'_{\gamma_k}$, $1 \leqslant j \leqslant m_+$, $1 \leqslant k \leqslant m_-$, with $B_j(\xi)$ and $C_k(\xi)$ being respectively row and column vectors of dimension p and

$$\operatorname{Re}\beta_j < s - 1/2, \qquad \operatorname{Re}\gamma_k < \alpha - s - 1/2, \tag{16.2}$$

and suppose $E_{jk}(\xi')$ is a homogeneous function of degree $\beta_{jk} = \beta_j + \gamma_k - \alpha - i\beta + 1$ that is continuous for $\xi' \neq 0$.

We consider the following boundary value problem in $\mathbf{R}^n_+$:

$$p\hat{A}u_+ + \sum_{k=1}^{m_-} p\hat{C}_k(v_k(x') \times \delta(x_n)) = f(x), \tag{16.3}$$

$$p'\hat{B}_j u_+ + \sum_{k=1}^{m_-} \hat{E}_{jk}(D')v_k(x') = g_j(x'), \qquad 1 \leqslant j \leqslant m_+, \tag{16.4}$$

where $\hat{A}$, $\hat{C}_k$ and $\hat{B}_j$ are the p.o.'s in $\mathbf{R}^n$ with symbols

$$\hat{A}(\xi', \xi_n), \qquad C_k((1 + |\xi'|)\omega, \xi_n), \qquad B_j((1 + |\xi'|)\omega, \xi_n)$$

and $\hat{E}_{jk}(D')$ is the p.o. in $\mathbf{R}^{n-1}$ with symbol $E_{jk}((1 + |\xi'|)\omega)$.

Let

$$\mathcal{H}_s^{(1)}(\mathbf{R}^n) = H_s^+(\mathbf{R}^n) \times \prod_{k=1}^{m_-} H_{s-\alpha+\operatorname{Re}\gamma_k+1/2}(\mathbf{R}^{n-1}),$$

$$\mathcal{H}_s^{(2)}(\mathbf{R}^n) = H_{s-\alpha}(\mathbf{R}^n_+) \times \prod_{j=1}^{m_+} H_{s-\operatorname{Re}\beta_j-1/2}(\mathbf{R}^{n-1}),$$

and let $\mathfrak{A}$ be the operator corresponding to (16.3) and (16.4). It follows from Lemmas 4.4, 8.1, Theorem 4.2 and (16.2) that $\mathfrak{A}$ is a bounded operator from $\mathcal{H}_s^{(1)}(\mathbf{R}^n)$ into $\mathcal{H}_s^{(2)}(\mathbf{R}^n)$.

We will find conditions for the unique solvability of the boundary value problem (16.3), (16.4).

We associate with problem (16.3), (16.4) the following family of boundary value problems on a halfline that depend on the parameter $\omega \in S^{n-2}$:

$$pA(\omega, D_n)v_+(x_n) + \sum_{k=1}^{m_-} \rho_k pC_k(\omega, D_n)\delta(x_n) = g(x_n), \tag{16.5}$$

$$p'B_j(\omega, D_n)v_+(x_n) + \sum_{k=1}^{m_-} E_{jk}(\omega)\rho_k = h_j, \qquad 1 \leqslant j \leqslant m_+, \tag{16.6}$$

where the ρ_k and h_j are complex numbers, $v_+(x_n) \in H_s^+(\mathbf{R}^1)$ and $g(x_n) \in H_{s-\alpha}(\mathbf{R}^1_+)$.

Let $\mathcal{H}_s^{(1)}(\mathbf{R}^1) = H_s^+(\mathbf{R}^1) \times \mathbf{C}^{m_-}$ and $\mathcal{H}_s^{(2)}(\mathbf{R}^1) = H_{s-\alpha}(\mathbf{R}^1_+) \times \mathbf{C}^{m_+}$, where $\mathbf{C}^k$ is a k-dimensional vector space, and let $\mathfrak{A}(\omega)$ be the operator corresponding to the system of equations (16.5), (16.6). We note that $\mathfrak{A}(\omega)$ is a bounded operator from $\mathcal{H}_s^{(1)}(\mathbf{R}^1)$ into $\mathcal{H}_s^{(2)}(\mathbf{R}^1)$ that continuously depends on $\omega \in S^{n-2}$ in the uniform operator topology.

Let us investigate in more detail the relation between $\mathfrak{A}$ and $\mathfrak{A}(\omega)$. The Fourier transform with respect to x' of problem (16.3), (16.4) is

$$p\hat{A}(\xi', D_n)\tilde{w}_+(\xi', x_n) + \sum_{k=1}^{m_-} p\hat{C}_k(\xi', D_n)(\tilde{v}_k(\xi') \times \delta(x_n)) = \tilde{h}(\xi', x_n), \tag{16.7}$$

$$p'\hat{B}_j(\xi', D_n)\tilde{w}_+(\xi', x_n) + \sum_{k=1}^{m_-} \hat{E}_{jk}(\xi')\tilde{v}_k(\xi') = \tilde{g}_j(\xi'), \qquad 1 \leqslant j \leqslant m_+, \tag{16.8}$$

where $\tilde{w}_+(\xi', x_n) = F_{x'}u_+(x', x_n)$ and $\tilde{h}(\xi', x_n) = F_{x'}f(x', x_n)$. Making the change of variables $x_n = y_n/(1 + |\xi'|)$ in (16.7), (16.8) and setting $\xi_n = (1 + |\xi'|)\eta_n$, we get

$$\begin{aligned}\tilde{w}_+\left(\xi', \frac{y_n}{1 + |\xi'|}\right) &= \frac{1}{2\pi}\int_{-\infty}^{\infty} \tilde{u}_+(\xi', \xi_n)e^{-i\frac{y_n}{1+|\xi'|}\xi_n}\,d\xi_n \\ &= F_{\eta_n}^{-1}\big[(1 + |\xi'|)\tilde{u}_+(\xi', (1 + |\xi'|)\eta_n)\big].\end{aligned}$$

Analogously,

$$\begin{aligned}\hat{A}(\xi', D_n)\tilde{w}_+(\xi', x_n) &= F_{\xi_n}^{-1}\hat{A}(\xi', \xi_n)\tilde{u}_+(\xi', \xi_n) \\ &= \frac{1}{2\pi}\int_{-\infty}^{\infty} \hat{A}(\xi', \xi_n)\tilde{u}_+(\xi', \xi_n)e^{-i\frac{y_n}{1+|\xi'|}\xi_n}\,d\xi_n \\ &= (1 + |\xi'|)^{\alpha+i\beta}F_{\eta_n}^{-1}A(\omega, \eta_n)(1 + |\xi'|)\tilde{u}_+(\xi', (1 + |\xi'|)\eta_n) \\ &= (1 + |\xi'|)^{\alpha+i\beta}A(\omega, D_n)\tilde{w}_+\left(\xi', \frac{y_n}{1 + |\xi'|}\right).\end{aligned}$$

Consequently, from (16.7) and (16.8) we get

$$pA(\omega, D_n)\tilde{w}_+\left(\xi', \frac{y_n}{1+|\xi'|}\right) + \sum_{k=1}^{m_-} pC_k(\omega, D_n)\delta(x_n)\tilde{\rho}_k(\xi') = \tilde{g}\left(\xi', \frac{y_n}{1+|\xi'|}\right), \tag{16.9}$$

$$p'B_j(\omega, D_n)\tilde{w}_+\left(\xi', \frac{y_n}{1+|\xi'|}\right) + \sum_{k=1}^{m_-} E_{jk}(\omega)\tilde{\rho}_k(\xi') = \tilde{h}_j(\xi'), \quad 1 \leqslant j \leqslant m_+, \tag{16.10}$$

where

$$\tilde{\rho}_k(\xi') = \tilde{v}_k(\xi')(1+|\xi'|)^{\gamma_k - \alpha - i\beta + 1},$$
$$\tilde{g}\left(\xi', \frac{y_n}{1+|\xi'|}\right) = (1+|\xi'|)^{-\alpha - i\beta}\tilde{h}\left(\xi', \frac{y_n}{1+|\xi'|}\right),$$
$$\tilde{h}_j(\xi') = (1+|\xi'|)^{-\beta_j}\tilde{g}_j(\xi').$$

We note that $\tilde{\rho}_k(\xi') \in \tilde{H}_{s-1/2}(\mathbf{R}^{n-1})$ and $\tilde{h}_j(\xi') \in \tilde{H}_{s-1/2}(\mathbf{R}^{n-1})$, $1 \leqslant k \leqslant m_-$, $1 \leqslant j \leqslant m_+$.

Since

$$\int_{-\infty}^{\infty}(1+|\xi'|+|\xi_n|)^{2s}|\tilde{u}_+(\xi', \xi_n)|^2\, d\xi' d\xi_n$$
$$= \int_{-\infty}^{\infty}\left(\int_{-\infty}^{\infty}(1+|\eta_n|)^{2s}|(1+|\xi'|)\tilde{u}_+(\xi', (1+|\xi'|)\eta_n)|^2 d\eta_n\right)(1+|\xi'|)^{2s-1}d\xi', \tag{16.11}$$

the function

$$\tilde{w}_+\left(\xi', \frac{y_n}{1+|\xi'|}\right) = F_{\eta_n}^{-1}(1+|\xi'|)\tilde{u}_+(\xi', (1+|\xi'|)\eta_n)$$

can be regarded as a member of the Sobolev-Slobodeckiĭ space $H_{s-1/2}(\mathbf{R}^{n-1}, H_s^+(\mathbf{R}^1))$ of order $s - 1/2$ consisting of those functions that are defined in $\mathbf{R}^{n-1}$, take their values in $H_s^+(\mathbf{R}^1)$ for almost all $\xi' \in \mathbf{R}^{n-1}$ and have a finite norm (16.11). Further, the first of the two vectors

$$\tilde{V}(\xi') = \left(\tilde{w}_+\left(\xi', \frac{y_n}{1+|\xi'|}\right), \tilde{\rho}_1(\xi'), \ldots, \tilde{\rho}_{m_-}(\xi')\right),$$
$$\tilde{\Phi}(\xi') = \left(\tilde{g}\left(\xi', \frac{y_n}{1+|\xi'|}\right), \tilde{h}_1(\xi'), \ldots, \tilde{h}_{m_+}(\xi')\right),$$

belongs to the Sobolev-Slobodeckiĭ space $\mathcal{H}_s^{(3)}(\mathbf{R}^n) = H_{s-1/2}(\mathbf{R}^{n-1}, \mathcal{H}_s^{(1)}(\mathbf{R}^1))$ of order $s - 1/2$ consisting of the functions in $\mathbf{R}^{n-1}$ with values belonging to

$\mathcal{H}_s^{(1)}(\mathbf{R}^1)$ for almost all $\xi' \in \mathbf{R}^{n-1}$, while the second vector $\tilde{\tilde{\Phi}}(\xi')$ belongs to the space $\mathcal{H}_s^{(4)}(\mathbf{R}^n) = H_{s-1/2}(\mathbf{R}^{n-1}, \mathcal{H}_s^{(2)}(\mathbf{R}^1))$ of functions of $\xi' \in \mathbf{R}^{n-1}$ with values in $\mathcal{H}_s^{(2)}(\mathbf{R}^1)$. Therefore, letting $\mathfrak{A}^{(1)}$ denote the operator corresponding to system (16.9), (16.10), we can write that system in the form

$$\mathfrak{A}^{(1)}\tilde{V} = \tilde{\tilde{\Phi}}. \tag{16.12}$$

System (16.9), (16.10) is equivalent to (16.3), (16.4) and hence is uniquely solvable if and only if (16.3), (16.4) is.

We will assume

$$\textit{system } (16.5), (16.6) \textit{ is uniquely solvable for any } \omega \in S^{n-2}. \tag{16.13}$$

THEOREM 16.1. *Suppose conditions* (16.1), (14.38) *and* (16.13) *are satisfied. Then the operator* $\mathfrak{A}$ *bijectively and continuously maps* $\mathcal{H}_s^{(1)}(\mathbf{R}^n)$ *onto* $\mathcal{H}_s^{(2)}(\mathbf{R}^n)$.

PROOF. Let $R(\omega)$ denote the inverse of the operator $\mathfrak{A}(\omega)$. Since $\mathfrak{A}(\omega)$ continuously depends on ω in the uniform operator topology and is invertible for any $\omega \in S^{n-2}$, it follows that $R(\omega)$ also continuously depends on ω in the uniform operator topology for the space of bounded operators from $\mathcal{H}_s^{(2)}(\mathbf{R}^1)$ into $\mathcal{H}_s^{(1)}(\mathbf{R}^1)$. In particular, the norm of $R(\omega)$ is bounded on S^{n-2}.

For a fixed $\xi' \in \mathbf{R}^{n-1}$ the operator $\mathfrak{A}^{(1)}$ takes a vector $\tilde{V}(\xi') \in \mathcal{H}_s^{(1)}(\mathbf{R}^1)$ into the vector $\mathfrak{A}(\omega)\tilde{V}(\xi') \in \mathcal{H}_s^{(2)}(\mathbf{R}^1)$, where $\omega = \xi'/|\xi'|$.

Let $\tilde{\tilde{\Phi}}(\xi')$ denote an arbitrary vector of $\mathcal{H}_s^{(4)}(\mathbf{R}^n)$ and let $R^{(1)}$ denote the operator acting from $\mathcal{H}_s^{(1)}(\mathbf{R}^n)$ into $\mathcal{H}_s^{(3)}(\mathbf{R}^n)$ according to the following rule: for a fixed $\xi' \in \mathbf{R}^{n-1}$ it takes $\tilde{\tilde{\Phi}}(\xi')$ into the vector $R(\omega)\tilde{\tilde{\Phi}}(\xi') \in \mathcal{H}_s^{(1)}(\mathbf{R}^1)$. Since $R(\omega)$ is continuous in ω, we have $R(\omega)\tilde{\tilde{\Phi}}(\xi') \in \mathcal{H}_s^{(3)}(\mathbf{R}^n)$, with

$$\left|\left|\left| R(\omega)\tilde{\tilde{\Phi}}(\xi') \right|\right|\right|_1 \leqslant C \left|\left|\left| \tilde{\tilde{\Phi}}(\xi') \right|\right|\right|_2, \tag{16.14}$$

where $|||\cdot|||_1$ and $|||\cdot|||_2$ denote the norms in $\mathcal{H}_s^{(3)}(\mathbf{R}^n)$ and $\mathcal{H}_s^{(4)}(\mathbf{R}^n)$ respectively. Consequently, $R^{(1)}$ is a bounded operator from $\mathcal{H}_s^{(4)}(\mathbf{R}^n)$ into $\mathcal{H}_s^{(3)}(\mathbf{R}^n)$, with, clearly,

$$\mathfrak{A}^{(1)}R^{(1)} = I_4, \qquad R^{(1)}\mathfrak{A}^{(1)} = I_3,$$

where I_3 (I_4) is the identity operator in $\mathcal{H}_s^{(3)}(\mathbf{R}^n)$ ($\mathcal{H}_s^{(4)}(\mathbf{R}^n)$). Thus, under the fulfilment of the conditions of Theorem 16.1, systems (16.9), (16.10) and (16.3), (16.4) are uniquely solvable.

It follows from Theorem 16.1 that the operator $\mathfrak{A}$ has a bounded inverse R from $\mathcal{H}_s^{(2)}(\mathbf{R}^n)$ into $\mathcal{H}_s^{(1)}(\mathbf{R}^n)$, i.e. there exists a bounded solution R of the equations

$$R\mathfrak{A} = I_1, \qquad \mathfrak{A}R = I_2, \tag{16.15}$$

where I_1 (I_2) is the identity operator in $\mathcal{H}_s^{(1)}(\mathbf{R}^n)$ ($\mathcal{H}_s^{(2)}(\mathbf{R}^n)$).

REMARK 16.1. The operators $\mathfrak{A}^{(1)}$ and $R^{(1)}$ can be regarded as pseudodifferential operators in $\mathbf{R}^{n-1}$ with operator valued symbols $\mathfrak{A}(\omega)$ and $R(\omega)$.

REMARK 16.2. Suppose s_0 satisfies (14.38). Then there clearly exists an interval (s_-, s_+) containing s_0 such that (14.38) continues to be satisfied by all $s \in (s_-, s_+)$; hence Theorem 16.1 remains valid for any $s \in (s_-, s_+)$ under the assumption that $\operatorname{Re} \beta_j < s_- - 1/2$, $\operatorname{Re} \gamma_k < a - s_+ - 1/2$, $1 < j < m_+$, $1 < k < m_-$, with the norm of the inverse R of $\mathfrak{A}$ being bounded with respect to $s \in [s_- + \varepsilon, s_+ - \varepsilon]$, $\varepsilon > 0$, by virtue of the interpolation theorem (see [62]).

2. *Conditions for the existence of a uniquely solvable boundary value problem.* Let us determine when condition (16.13) is satisfied. It can be assumed without loss of generality that $s = \alpha + i\beta = 0$ (cf. §14.1).

We require some elementary facts from the theory of Fredholm operators (see [7]). Suppose $A(\omega)$ is for each $\omega \in S^{n-2}$ a Fredholm operator acting from a Banach space B_1 into a Banach space B_2 and continuously depending on ω in the uniform operator topology. Let $n_+(\omega)$ ($n_-(\omega)$) denote the dimension of the kernel (cokernel) of $A(\omega)$.

LEMMA 16.1. *There exist vectors $g_k \in B_2$, $1 \leqslant k \leqslant m_-$, such that the operator*

$$\mathcal{A}(\omega)(u, c_1, \dots, c_{m_-}) = A(\omega)u + \sum_{k=1}^{m_-} c_k g_k, \tag{16.16}$$

acting from $B_1 \times \mathbf{C}^{m_-}$ into B_2, does not have a cokernel for any $\omega \in S^{n-2}$.

PROOF. We fix an arbitrary $\omega_0 \in S^{n-2}$. Since the dimension of the cokernel of $A(\omega_0)$ is equal to $n_-(\omega_0)$, there exist vectors $g_{k0} \in B_2$, $1 \leqslant k \leqslant n_-(\omega_0)$ such that the range of the operator

$$\mathcal{A}_0(\omega)\left(u, c_1^{(0)}, \dots, c_{n_-(\omega_0)}^{(0)}\right) = A(\omega)u + \sum_{k=1}^{n_-(\omega_0)} c_k^{(0)} g_{k0}$$

coincides with B_2 for $\omega = \omega_0$. Let $R_0(\omega_0)$ denote a bounded right inverse of $A_0(\omega_0)$, so that $\mathcal{A}_0(\omega_0)R_0(\omega_0) = I$, where I is the identity operator in B_2. We have

$$\mathcal{A}_0(\omega)R_0(\omega_0) = I + B(\omega), \quad \text{where } B(\omega) = (\mathcal{A}_0(\omega) - \mathcal{A}_0(\omega_0))R_0(\omega_0).$$

Since $\|B(\omega)\| < 1$ for $|\omega - \omega_0| < \varepsilon$, where $\varepsilon > 0$ is sufficiently small, the operator $R_0(\omega) = R_0(\omega_0)(I + B(\omega))^{-1}$ is a right inverse of $A_0(\omega)$ for $|\omega - \omega_0| < \varepsilon$. Consequently, $A_0(\omega)$ does not have a cokernel for $|\omega - \omega_0| < \varepsilon$.

We choose from the covering of S^{n-2} obtained in this way a finite subcovering $U_j = U(\omega_j)$, $1 \leqslant j \leqslant N$. Let g_{kj}, $1 \leqslant k \leqslant n_-(\omega_j)$, be vectors in B_2 such that the operator $A(\omega)u + \sum_{k=1}^{n_-(\omega_j)} c_k^{(j)} g_{kj}$ does not have a cokernel for

$\omega \in U_j$. Then the operator

$$A(\omega)u + \sum_{j=1}^{N} \sum_{k=1}^{n_-(\omega_j)} c_k^j g_{kj},$$

acting from $B_1 \times C^{m_-}$ into B_2, where $m_- = \sum_{j=1}^N n_-(\omega_j)$, does not have a cokernel for any $\omega \in S^{n-2}$.

Applying Lemma 16.1 to both $A(\omega)$ and $A^*(\omega)$, we get

COROLLARY 16.1. *The inequalities*

$$\max_\omega n_-(\omega) \leqslant m_- < \infty, \qquad \max_\omega n_+(\omega) < \infty$$

hold.

With the use of Lemma 16.1, we can define symbols $C_k(\xi', \xi_n) \in O'_{\gamma_k}$, $1 \leqslant k \leqslant m_-$, $m_- \geqslant \max_\omega n_-(\omega)$, such that the system

$$pA(\omega, D_n)v_+ + \sum_{k=1}^{m_-} \rho_k p C_k(\omega, D_n)\delta(x_n) = g(x_n) \qquad (16.17)$$

has a solution $(v_+(x_n), \rho_1, \dots, \rho_{m_-}) \in H_0^+(\mathbf{R}^1) \times \mathbf{C}^{m_-}$ for any $\omega \in S^{n-2}$ and $g(x_n) \in H_0(\mathbf{R}^1_+)$. In fact, by Lemma 16.1, we can choose functions $g_k(x_n) \in H_0(\mathbf{R}^1)$ so that the system

$$pA(\omega, D_n)v_+ + \sum_{k=1}^{m_-} \rho_k p g_k(x_n) = g(x_n) \qquad (16.18)$$

is solvable for any $\omega \in S^{n-2}$ and $g(x_n) \in H_0(\mathbf{R}^1_+)$. But the functions $g_k(x_n)$ can be approximated by functions $g_{k0}(x_n) \in S(\mathbf{R}^1)$ so closely in the norm of $H_0(\mathbf{R}^1)$ that the system of equations obtained upon replacing $g_k(x_n)$ by $g_{k0}(x_n)$ in (16.18) is still solvable. We now put

$$C_k(\xi', \xi_n) = |\xi'|^{\gamma_k} C_{k0}\left(\frac{\xi_n}{|\xi'|}\right),$$

where $C_{k0}(\xi_n) = F^{-1}g_{k0}(x_n)$. Then $C_k(\xi', \xi_n) \in O'_{\gamma_k}$ and the operator defined by the left side of (16.17) when $C_k(\omega, \xi_n) = C_{k0}(\xi_n)$ does not have a cokernel for any $\omega \in S^{n-2}$.

Let us now determine when there exist boundary operators $\hat{B}_j(D)$ and $\hat{E}_{jk}(D')$ such that system (16.5), (16.6) is uniquely solvable.

LEMMA 16.2. *Suppose system* (16.17) *has a solution* $(v_+(x_n), \rho_1, \dots, \rho_{m_-}) \in H_0^+(\mathbf{R}^1) \times \mathbf{C}^{m_-}$ *for any* $\omega \in S^{n-2}$ *and* $g(x_n) \in H_0(\mathbf{R}^1_+)$. *Then there exist boundary conditions*

$$p'B_j(\omega, D_n)v_+ + \sum_{k=1}^{m_-} E_{jk}(\omega)\rho_k = h_j, \qquad 1 \leqslant j \leqslant m_+, \qquad (16.19)$$

ensuring the unique solvability of system (16.17), (16.19) *for any* $\omega \in S^{n-2}$ *if and only if the solutions of the homogeneous system* (16.17) *have a basis that is continuous on* S^{n-2}.

PROOF. Suppose $w_k = (v_+^{(k)}, \rho_1^{(k)}, \ldots, \rho_{m_-}^{(k)}) \in H_0^+(\mathbf{R}^1) \times \mathbf{C}^{m_+}$ is a solution of system (16.17), (16.19) corresponding to the right sides $g = 0$, $h_j = \delta_{jk}$, $1 \leqslant k \leqslant m_+$. By assumption, this solution is unique. The vectors w_k continuously depend on ω since the inverse operator $R(\omega)$ is continuous in ω. Clearly, $w_1, \ldots, w_{m_+}$ are linearly independent and for any ω form a basis in the kernel of the operator $\mathcal{Q}(\omega)$ defined by the left side of (16.17)

Conversely, suppose the vectors $w_k = (v_+^k, \rho_1^{(k)}, \ldots, \rho_{m_+}^{(k)})$, $1 \leqslant k \leqslant m_+$, form a basis in the kernel of $\mathcal{Q}(\omega)$ that is continuous in $\omega \in S^{n-2}$, and let $E_{jk}(\omega) = \overline{\rho_k^{(j)}}$, $B_{j0}(\omega, \xi_n) = \overline{F_{x_n} v_+^{(k)}(\omega, x_n)}$. We add to (16.17) the boundary conditions

$$p' B_{j0}(\omega, D_n) v_+ + \sum_{k=1}^{m_-} E_{jk}(\omega)\rho_k = h_j, \qquad 1 \leqslant j \leqslant m_+. \tag{16.20}$$

By taking into account the fact that

$$p' B_{j0}(\omega, D_n) v_+ = \frac{1}{2\pi}\int_{-\infty}^{\infty} \tilde{v}_+(\omega, \xi_n) \cdot \overline{\tilde{v}_+^{(j)}(\omega, \xi_n)}\, d\xi_n = (v_+, v_+^{(j)}),$$

we can write (16.20) in the form

$$\{w, w_j\} = h_j, \qquad 1 \leqslant j \leqslant m_+,$$

where $\{w, w_j\} = (v_+, v_+^{(j)}) + \sum_{k=1}^{m_-} \rho_k \overline{\rho_k^{(j)}}$ is the scalar product in $H_0^+(\mathbf{R}^1) \times \mathbf{C}^{m_-}$. Consequently, the boundary value problem (16.17), (16.20) is uniquely solvable for any $\omega \in S^{n-2}$. But the symbols $B_{j0}(\omega, \xi_n)$ can be approximated arbitrarily closely by symbols $B_j(\omega, \xi_n)$ that are continuous in ω and belong to $S(\mathbf{R}^1)$ with respect to ξ_n, so that (16.17), (16.20) will remain uniquely solvable if the $B_{j0}(\omega, \xi_n)$ are replaced by the $B_j(\omega, \xi_n)$.

LEMMA 16.3. *Suppose a Fredholm operator* $\mathcal{Q}(\omega)$ *does not have a cokernel for any* $\omega \in S^{n-2}$. *Then the kernel of* $\mathcal{Q}(\omega)$ *has a basis that is locally continuous in* ω.

PROOF. Suppose $e_0 \in \ker \mathcal{Q}(\omega_0)$. By Lemma 16.1, the operator $\mathcal{Q}(\omega)$ has a right inverse $R(\omega)$ for $|\omega - \omega_0| < \varepsilon$ that continuously depends on ω. Let $e(\omega) = e_0 - R(\omega)\mathcal{Q}(\omega)e_0$. Then $\mathcal{Q}(\omega)e(\omega) = 0$, with $e(\omega_0) = e_0$.

Suppose the vectors e_j, $1 \leqslant j \leqslant n_+(\omega_0)$, form a basis in $\ker \mathcal{Q}(\omega_0)$. As was just shown, there exist functions $e_j(\omega)$, $1 \leqslant j \leqslant n_+(\omega_0)$, belonging to $\ker \mathcal{Q}(\omega)$ for $|\omega - \omega_0| < \varepsilon$ and such that $e_j(\omega_0) = e_j$. Consequently, for $|\omega - \omega_0|$ sufficiently small the functions $e_j(\omega)$, $1 \leqslant j \leqslant n_+(\omega_0)$, are linearly independent and form a basis in $\ker \mathcal{Q}(\omega)$, since $n_+(\omega) = n_+(\omega_0)$.

3. *Index of a family of Fredholm operators.* By Lemma 16.2, the fulfilment of (16.13) is connected with the existence of a continuous (on S^{n-2}) basis for the kernel of the operator $\mathcal{Q}(\omega)$ defined by system (16.17). In order to determine when such a basis exists, we require some information from K-theory (see [7]).

Let M be a compact k-dimensional C^∞-manifold (in this section we confine ourselves to the case $M = S^{n-2}$), let $\{U_j\}$, $1 \leqslant j \leqslant N$, denote a finite covering of M and let a nonsingular $m \times m$ matrix valued function $c_{ij}(x)$ of class C^∞ be defined in each $U_i \cap U_j$. It is assumed that the matrices c_{ij} satisfy the following conditions:

1) $c_{ij}(x) = c_{ji}^{-1}(x)$ for $x \in U_i \cap U_j$,

2) $c_{ij}(x)c_{jp}(x) = c_{ip}(x)$ for $x \in U_i \cap U_j \cap U_p$.

Let E denote the quotient space obtained from the union

$$\bigcup_{j=1}^{N} (U_j \times \mathbf{C}^m \times \{j\})$$

by identifying two points (x, ξ, i) and (y, η, j) when $x = y$ and $\xi = c_{ij}(x)\eta$. We call the set E together with the system $\{U_j, c_{ij}\}$, $1 \leqslant i \leqslant N$, $1 \leqslant j \leqslant N$, a *vector bundle over M with fiber $\mathbf{C}^m$ and base space M.* The matrices c_{ij} are called *transition matrices.*

Let E' be another vector bundle over M with fiber $\mathbf{C}^m$ defined by the same covering $\{U_j\}$ but different transition matrices $c'_{ij}(x)$. The bundles E and E' are said to be *isomorphic* if there is defined in each U_j a nonsingular matrix valued function $c_j(x)$ of class C^∞ such that $c'_{ij}(x) = c_i c_{ij} c_j^{-1}$ for $x \in U_i \cap U_j$.

When the bundles E and E' are defined by different coverings $\{U_j\}$, $1 \leqslant j \leqslant N$, and $\{U'_i\}$, $1 \leqslant i \leqslant N'$, we go over in each bundle to the same finer covering $\{V_p\}$, $1 \leqslant p \leqslant N''$, formed by all possible intersections $U_j \cap U'_i$; in this connection, in E (E') the transition matrix for $V_i \cap V_j$ is set equal to I or to the transition matrix for $U_i \cap U_j$ ($U'_i \cap U'_j$), when U_i and U_j (U'_i and U'_j) are different neighborhoods containing V_i and V_j respectively. Then the original bundles E and E' are said to be *isomorphic* if the new bundles defined by the covering $\{V_p\}$ are isomorphic.

A bundle that is isomorphic to the product bundle $M \times \mathbf{C}^m$ is said to be *trivial.* The set of all isomorphism classes of vector bundles over M is denoted by $\mathrm{Vect}(M)$. The trivial bundle with fiber $\mathbf{C}^m$ is denoted by $\underline{m}$.

Let E and E' be vector bundles over M with fibers $\mathbf{C}^{m_1}$ and $\mathbf{C}^{m_2}$. It can be assumed without loss of generality that E and E' are defined by the same covering $\{U_j\}$. By the *direct sum* $E \oplus E'$ of E and E' is meant the vector bundle over M with fiber $\mathbf{C}^{m_1+m_2}$, defined by the covering $\{U_j\}$ and the transition matrices

$$c''_{ij} = \left\| \begin{matrix} c_{ij} & 0 \\ 0 & c'_{ij} \end{matrix} \right\|.$$

Let $E - F$ denote the formal difference of two vector bundles over M with fibers $\mathbf{C}^{m_1}$ and $\mathbf{C}^{m_2}$. Two differences $E - F$ and $E' - F'$ are said to be *equivalent* if there exists a vector bundle Q such that $E \oplus F' \oplus Q \approx E' \oplus F \oplus Q$, where $\approx$ means that the bundles in question are isomorphic. The set of equivalence classes of formal differences is denoted by $K(M)$. The set $K(M)$ is an Abelian group if one puts $(E - F) + (E' - F') = E \oplus E' - F \oplus F'$ and $-(E - F) = F - E$.

Each bundle E defines an element of $K(M)$ of the form $E - O$, where O is the bundle with fiber equal to zero. We denote this element by $[E]$. Clearly, an arbitrary element of $K(M)$ can be written in the form $[E] - [F]$, where E and F are vector bundles.

Let $A(\omega)$ be a family of Fredholm operators continuously depending on $\omega \in S^{n-2}$. By Lemma 16.1, there exist m_- continuous functions $g_k(\omega)$, $1 \leqslant k \leqslant m_-$, on S^{n-2} such that the operator

$$\mathcal{A}(\omega)w = A(\omega)u + \sum_{k=1}^{m_-} c_k g_k(\omega), \qquad w = (u, c_1, \ldots, c_{m_-})$$

does not have a cokernel. In this case $\ker \mathcal{A}(\omega)$ is a vector bundle over S^{n-2}. In fact, by Lemma 16.3, a finite covering $\{U_j\}$, $1 \leqslant j \leqslant N$, can be found such that $\ker \mathcal{A}(\omega)$ has a continuous basis in each U_j. Then $\ker \mathcal{A}(\omega)$ becomes a vector bundle over S^{n-2} if the transition matrix c_{ij} for $U_i \cap U_j$ is the matrix mapping the ith basis onto the jth basis.

The element $[\ker \mathcal{A}(\omega)] - [\underline{m}_-]$, of $K(S^{n-2})$ is called the *index* of the family $A(\omega)$ and denoted by $k(A(\omega))$. We will show that $k(A(\omega))$ does not depend on the choice of the functions $g_k(\omega)$ used in the construction of the operator $\mathcal{A}(\omega)$ mapping $B_1 \times \mathbf{C}^{m_-}$ onto B_2.

Let $\mathcal{A}'(\omega)w' = A(\omega)u + \sum_{k=1}^{m_1} c'_k g'_k(\omega)$ be an operator mapping $B_1 \times \mathbf{C}^{m_1}$ onto B_2 for all $\omega \in S^{n-2}$, where $w' = (u, c'_1, \ldots, c'_{m_1})$, and let $k'(A(\omega)) = [\ker \mathcal{A}'(\omega)] - [\underline{m}_1]$. We will prove that $k'(A(\omega)) = k(A(\omega))$ by examining the kernel of the following operator, which maps $B_1 \times \mathbf{C}^{m_- + m_1}$ onto B_2:

$$\mathcal{A}''(\omega)w'' = A(\omega)u + \sum_{k=1}^{m_-} c_k g_k(\omega) + \sum_{k=1}^{m_1} c'_k g'_k(\omega),$$

where $w'' = (u, c_1, \ldots, c_{m_-}, c'_1, \ldots, c'_{m_1})$. By Lemma 16.1, the operators $\mathcal{A}(\omega)$ and $\mathcal{A}'(\omega)$ have continuous (in ω) right inverses $R(\omega)$ and $R'(\omega)$ (in the proof of Lemma 16.1 a right inverse was constructed only locally but could have been constructed globally by means of a partition of unity on S^{n-2}). Thus, writing the operator $\mathcal{A}''(\omega)$ in the form

$$\mathcal{A}''(\omega)w'' = \mathcal{A}(\omega)w + \sum_{k=1}^{m_1} c'_k g'_k(\omega),$$

we get that the general solution of the equation $\mathcal{A}''(\omega)w'' = 0$ can be represented in the form

$$w'' = \left(w_0 - \sum_{k=1}^{m_1} R(\omega)g'_k(\omega)c'_k, c'_1, \ldots, c'_{m_1} \right),$$

where w_0 is an arbitrary solution of the equation $\mathcal{A}(\omega)w = 0$ and $c'_1, \ldots, c'_{m_1}$ are arbitrary numbers. It follows that the bundle ker $\mathcal{A}''(\omega)$ is isomorphic to the direct sum of the bundle ker $\mathcal{A}(\omega)$ and the trivial bundle $\underline{m}_1$: ker $\mathcal{A}''(\omega) \approx$ ker $\mathcal{A}(\omega) \oplus \underline{m}_1$. Analogously, writing $\mathcal{A}''(\omega)$ in the form

$$\mathcal{A}'(\omega)w' + \sum_{k=1}^{m_-} c_k g_k(\omega),$$

we get that ker $\mathcal{A}''(\omega) \approx$ ker $\mathcal{A}'(\omega) \oplus \underline{m}_-$. Consequently, [ker $\mathcal{A}(\omega)$] $-$ $[\underline{m}_-]$ $=$ [ker $\mathcal{A}'(\omega)$] $-$ $[\underline{m}_1]$, i.e. $k(A(\omega)) = k'(A(\omega))$.

THEOREM 16.2. *There exist for sufficiently large m_+ and m_- ($m_+ - m_- = m$ is the index of the operator $pA(\omega, D_n)$ for a fixed ω) p.o.'s $\hat{B}_j(D)$, $\hat{E}_{jk}(D')$, $\hat{C}_k(D)$, $1 \leqslant j \leqslant m_+$, $1 \leqslant k \leqslant m_-$, such that* (16.13) *is satisfied if and only if the index of the family of Fredholm operators $pA(\omega, D_n)$ is equal to $[\underline{m}_+] - [\underline{m}_-] = \operatorname{sgn} m[\underline{m}]$.*

PROOF. Suppose (16.13) is satisfied. Then Lemma 16.2 implies that ker $\mathcal{A}(\omega)$ is a trivial bundle and hence that the index of the family $pA(\omega, D_n)$ is equal to $[\underline{m}_+] - [\underline{m}_-] = \operatorname{sgn} m[\underline{m}]$.

Conversely, suppose $k(pA(\omega, D_n)) = \operatorname{sgn} m[\underline{m}]$. By Lemma 16.2, there exist m_- operators of potential type such that the operator $\mathcal{A}(\omega)$ defined by the left side of (16.17) effects a mapping onto $H_0(\mathbf{R}^1_+)$ for all ω. Then [ker $\mathcal{A}(\omega)$] $= \operatorname{sgn} m[\underline{m}] + [\underline{m}_-] = [\underline{m}_+]$. By definition, the equality [ker $\mathcal{A}(\omega)$] $= [\underline{m}_+]$ means that there exists a vector bundle F such that

$$\ker \mathcal{A}(\omega) \oplus F \approx \underline{m}_+ \oplus F. \tag{16.21}$$

As is well known (see [7]), for any vector bundle F there exists a vector bundle F' such that $F \oplus F' \approx \underline{m}_1$. Consequently, from (16.21) we get

$$\ker \mathcal{A}(\omega) \oplus F \oplus F' \approx \underline{m}_+ \oplus \underline{m}_1 \approx \underline{m}_2.$$

Thus, if one adds m_1 arbitrary operators of potential type to the operator $\mathcal{A}(\omega)$, the kernel of the operator

$$\mathcal{A}''(\omega)w'' = \mathcal{A}(\omega)w + \sum_{k=1}^{m_1} c'_k g'_k(\omega)$$

will be a trivial bundle and hence Lemma 16.2 will imply the existence of p.o.'s $\hat{B}_j(D)$ and $\hat{E}_{jk}(D')$ such that (16.13) is satisfied.

The following theorem determines when the index of the family of operators $pA(\omega, D_n)$ is equal to sgn $m[\underline{m}]$. Let I_N be the identity matrix in $\mathbf{C}^N$.

THEOREM 16.3. *In order for the index of the family of operators $pA(\omega, D_n)$ to be equal to* sgn $m[\underline{m}]$, *it is necessary and sufficient that for sufficiently large N the matrix*

$$\left\| \begin{matrix} A(\xi', \xi_n) & 0 \\ 0 & I_N \end{matrix} \right\|$$

be homotopic to the matrix

$$\left\| \begin{matrix} \left(\dfrac{\xi_n + i|\xi'|}{\xi_n - i|\xi'|} \right)^m I_1 & 0 \\ 0 & I_{N+p-1} \end{matrix} \right\|$$

in the class of elliptic matrices belonging to O_0 and satisfying (14.22).

PROOF. *Sufficiency*. We note that the index of a family of operators does not vary under a continuous deformation in the class of Fredholm operators. For suppose $A(\omega, t)$ is a homotopy for a family $A(\omega)$, i.e. $A(\omega, 0) = A(\omega)$. We fix an arbitrary point $t_0 \in [0, 1]$. Then Lemma 16.1 implies the existence of vectors g_k, $1 \leqslant k \leqslant m_-$, such that the operator

$$\mathcal{A}(\omega, t)w = A(\omega, t)u + \sum_{k=1}^{m_-} c_k g_k$$

does not have a cokernel for any $\omega \in S^{n-2}$ and sufficiently small $|t - t_0|$. If $|t - t_0|$ is sufficiently small, it follows by virtue of Lemma 16.3 that the vector bundle ker $\mathcal{A}(\omega, t)$ is isomorphic to ker $\mathcal{A}(\omega, t_0)$. Thus $k(A(\omega, t)) = [\ker \mathcal{A}(\omega, t)] - [\underline{m}_-] = k(A(\omega, t_0))$ for $|t - t_0| < \varepsilon$, i.e. $k(A(\omega, t))$ does not depend on t.

Clearly, the indices of the families $pA(\omega, D_n)$ and

$$p\left\| \begin{matrix} A(\omega, D_n) & 0 \\ 0 & I_N \end{matrix} \right\|$$

are the same. If (14.22) is satisfied, then an elliptic matrix homotopy remains within the class of Fredholm operators in $H_0^+(\mathbf{R}^1)$. Consequently, the index of the family

$$p\begin{Vmatrix} A(\omega, D_n) & 0 \\ 0 & I_N \end{Vmatrix}$$

is equal to the index of the family

$$p\begin{Vmatrix} (D_n + i)^m (D_n - i)^{-m} I_1 & 0 \\ 0 & I_{N+p-1} \end{Vmatrix}.$$

But the latter index is clearly equal to sgn $m[\underline{m}]$ (see §7). Thus $k(pA(\omega, D_n))$ $=$ sgn $m[\underline{m}]$.

Necessity. We will continuously deform the matrix $A(\xi', \xi_n)$ into simpler forms. We have

$$A(\xi', \xi_n) = A(0, +1)\exp\left\{ \ln \frac{\Lambda_+(\xi)}{\Lambda_-(\xi)} b \right\} A_1(\xi', \xi_n),$$

where b is the same matrix as in (14.21), $\Lambda_\pm(\xi) = \xi_n \pm i|\xi'|$, and the matrix $A_1(\xi', \xi_n) \in O_0$ and satisfies the condition $A_1(0, \pm 1) = I$. Therefore the matrix

$$A^{(t)} = A(0, +1)\exp\left\{ t \ln \frac{\Lambda_+(\xi)}{\Lambda_-(\xi)} b \right\} A_1(\xi', \xi_n)$$

effects a homotopy between the matrices $A(\xi',\xi_n)$ and $A_2(\xi', \xi_n) = A(0, +1)A_1(\xi', \xi_n)$, with (14.22) being satisfied for any $t \in [0, 1]$. The matrix $A_2(\xi', \xi_n) \in O_0$, and

$$A_2(0, +1) = A_2(0, -1). \tag{16.22}$$

The next homotopy will be realized in the class of elliptic matrices satisfying (16.22). Let $z = \Lambda_+(\xi)/\Lambda_-(\xi)$. Then $|z| = 1$ and

$$\xi_n = -i|\xi'|\frac{1 + z}{1 - z}.$$

By virtue of (16.22), the matrix

$$A_2\left(\xi', -i|\xi'|\frac{1 + z}{1 - z}\right)$$

can be expanded in a uniformly convergent Fourier series

$$\sum_{k=-\infty}^{\infty} a_k(\omega) z^k, \qquad \omega = \frac{\xi'}{|\xi'|}.$$

Thus, if M is sufficiently large, the matrix

$$A_3(\xi', \xi_n) = \sum_{k=-M}^{M} a_k(\omega)z^k\left(\sum_{k=-M}^{M} a_k(\omega)\right)^{-1} A_2(0, +1) = \sum_{k=-M}^{M} b_k(\omega)z^k$$

is homotopic to $A_2(\xi', \xi_n)$, with $A_3(0, +1) = A_3(0, -1)$.

We write the matrix $A_3(\xi', \xi_n)$ in the form $A_3(\xi', \xi_n) = z^{-M}p(\xi', \xi_n)$, where $p(\xi', \xi_n) = \Sigma_{k=0}^{2M} p_k(\omega)z^k$ is a polynomial matrix in z. The matrix

$$\begin{Vmatrix} p & 0 \\ 0 & I_{2Mp} \end{Vmatrix}$$

is homotopic to the linear matrix in z

$$a(\omega)z + b(\omega) = \begin{Vmatrix} p_0 & p_1 \cdots p_{2M} \\ -zI_p & I_p 0 \ldots 0 \\ 0 & -zI_p I_p 0 \ldots 0 \\ \cdots & \cdots \\ 0 & \ldots 0 - zI_p I_p \end{Vmatrix},$$

in view of the fact that the matrix

$$\begin{Vmatrix} p & 0 \\ 0 & I_{2Mp} \end{Vmatrix}$$

can be obtained from the matrix $az + b$ by elementary transformations that can be converted into a homotopy by introducing the parameter $t \in [0, 1]$ when adding one row (or column) to another. Further, since $a(\omega) + b(\omega) = a_0$ is homotopic to $I_{(2M+1)p}$ we have

$$a(\omega)z + b(\omega) = a_0(\xi_n + c(\omega)|\xi'|)\Lambda_-^{-1}.$$

Thus the matrix

$$\begin{Vmatrix} A_3 & 0 \\ 0 & I_{2Mp} \end{Vmatrix}$$

is homotopic to the matrix

$$A_4(\xi', \xi_n) = A_5(\xi_n)\Lambda_-^{-1}(\xi_n I_{(2M+1)p} + c(\omega)|\xi'|),$$

where

$$A_5 = \begin{Vmatrix} z^{-M}I_p & 0 \\ 0 & I_{2Mp} \end{Vmatrix}.$$

Since $\xi_n I_{(2M+1)p} + c(\omega)|\xi'|$ is an elliptic matrix, the eigenvalues of the matrix $c(\omega)$ do not lie on the real axis for any $\omega \in S^{n-2}$. Let

$$P_{\pm}(\omega) = \frac{i}{2\pi}\int_{\Gamma_{\pm}} (c(\omega) - \lambda)^{-1} d\lambda,$$

where Γ_+ (Γ_-) is a contour lying in the upper (lower) halfplane that encloses all of the eigenvalues in the upper (lower) halfplane of the matrix $c(\omega)$ for all $\omega \in S^{n-2}$. Then $P_+(\omega) + P_-(\omega) = I_{(2M+1)p}$. For each $\omega \in S^{n-2}$ the matrix $P_+(\omega)$ ($P_-(\omega)$) defines a projection of $\mathbf{C}^{(2M+1)p}$ onto the invariant subspace $M_+(\omega)$ ($M_-(\omega)$) of $c(\omega)$ corresponding to the eigenvalues of $c(\omega)$ in the upper (lower) halfplane. We have

$$\begin{aligned}\Lambda_-^{-1}(\xi)(\xi_n I_{(2M+1)p} &+ c(\omega))\\ &= \Lambda_-^{-1}\big[(\xi_n P_+(\omega) + c(\omega)P_+(\omega)) + (\xi_n P_-(\omega) + c(\omega)P_-(\omega))\big].\end{aligned}$$

The matrix

$$(\xi_n + i\tau)P_+(\omega) + c(\omega)P_+(\omega)((\xi_n + i\tau)P_-(\omega) + c(\omega)P_-(\omega))$$

isomorphically maps $M_+(\omega)$ ($M_-(\omega)$) onto itself for any $\tau \geqslant 0$ ($\tau \leqslant 0$). Consequently, since $M_+(\omega) \oplus M_-(\omega) = C^{(2M+1)p}$, for any $\omega \in S^{n-2}$, the matrix

$$\begin{aligned}A_6^{(N)} = (\xi_n - i|\xi'| - iN|\xi'|)^{-1}\big[&(\xi_n + iN|\xi'|)P_+(\omega)\\ &+ c(\omega)P_+(\omega) + (\xi_n - iN|\xi'|)P_-(\omega) + c(\omega)P_-(\omega)\big]\end{aligned}$$

is nonsingular for any $N \geqslant 0$. Letting N tend to $+\infty$, we get that the matrix $\Lambda_-^{-1}(\xi)(\xi_n I_{(2M+1)p} + c(\omega))$ is homotopic to the matrix

$$\frac{\Lambda_+}{\Lambda_-} P_+(\omega) + P_-(\omega).$$

Thus the matrix $A_4(\xi', \xi_n)$ is homotopic to the matrix

$$A_6(\xi', \xi_n) = A_5\left(\frac{\Lambda_+}{\Lambda_-} P_+(\omega) + P_-(\omega)\right).$$

We continuously deform the matrix

$$\begin{Vmatrix} A_5(z) & 0 \\ 0 & 1 \end{Vmatrix},$$

which is infinitely differentiable on the circle $|z| = 1$, into the matrix

$$\begin{Vmatrix} I_{(2M+1)p-1} & 0 \\ 0 & z^{-Mp}I_1 \end{Vmatrix}$$

(cf. the proof of Lemma 14.2). Then the matrix

$$A_7(\xi', \xi_n) = \begin{Vmatrix} A_6 & 0 \\ 0 & 1 \end{Vmatrix}$$

is homotopic to the matrix

$$A_8(\xi', \xi_n) = \begin{Vmatrix} zP_+(\omega) + P_-(\omega) & 0 \\ 0 & z^{-Mp}I_1 \end{Vmatrix}.$$

with the index of the family of operators $pA_8(\omega, D_n)$ being equal to sgn $m[\underline{m}]$.

On the other hand, clearly,

$$k(pA_8(\omega, D_n)) = k\left(p\left(\frac{\Lambda_+}{\Lambda_-}P_+(\omega) + P_-(\omega)\right)\right) + k\left(p\frac{\Lambda_-^{Mp}}{\Lambda_+^{Mp}}\right),$$

with $k(\Lambda_-^{Mp}\Lambda_+^{-Mp}) = -[\underline{Mp}]$. Consequently,

$$k(\Lambda_+\Lambda_-^{-1}P_+(\omega) + P_-(\omega)) = [\,\underline{Mp + m}\,].$$

The cokernel of the operator $p(\Lambda_+\Lambda_-^{-1}P_+(\omega) + P_-(\omega))$ is equal to zero, since $(\Lambda_+(\xi)/\Lambda_-(\xi))P_+(\omega) + P_-(\omega)$ is a "minus" symbol, while the kernel of the operator $p(\Lambda_+\Lambda_-^{-1}P_+ + P_-)$ is equivalent to $M_+(\omega)$ since the general solution of the system $(\Lambda_+\Lambda_-^{-1}P_+(\omega) + P_-(\omega))\tilde{u}_+ = \tilde{u}_-$, $\tilde{u}_\pm \in \tilde{H}_0^\pm(\mathbf{R}^1)$, has the form $\tilde{u}_+ = \Lambda_+^{-1}\rho_+$, $\tilde{u}_- = \Lambda_-^{-1}\rho_+$, where $\rho_+ \in M_+(\omega)$. Thus $[M_+(\omega)] = [\underline{Mp + m}]$. Also $[M_-(\omega)] = [\underline{(M + 1)p - m}]$ since $[M_-(\omega)] = [\underline{(2M + 1)p}] - [M_+(\omega)]$. If now the number M has been chosen sufficiently large, $M_+(\omega)$ and $M_-(\omega)$ will be trivial bundles, so that a basis that is continuous on S^{n-2} can be constructed in them. Let $Q(\omega)$ be a matrix whose columns are the vectors of this basis. Then $P_\pm(\omega) = Q^{-1}(\omega)P_\pm^0 Q(\omega)$, where

$$P_+^0 = \begin{Vmatrix} I_{Mp+m} & 0 \\ 0 & 0 \end{Vmatrix}, \qquad P_-^0 = \begin{Vmatrix} 0 & 0 \\ 0 & I_{(M+1)p-m} \end{Vmatrix}.$$

Consequently,

$$A_8 = \begin{Vmatrix} Q^{-1}(zP_+^0 + P_-^0)Q & 0 \\ 0 & z^{-Mp}I_1 \end{Vmatrix}.$$

To complete the proof of Theorem 16.3 we need the following lemma (see [89]).

LEMMA 16.4. *Any continuous nonsingular $N \times N$ matrix $Q(\omega)$ on S^k is homotopic in the class of nonsingular matrices on S^k to a matrix of the form*

$$\begin{Vmatrix} I_{N-p} & 0 \\ 0 & Q_0(\omega) \end{Vmatrix},$$

in which $Q_0(\omega)$ is a $p \times p$ matrix, with $p \leqslant (k + 1)/2$.

(We note that we have already used this lemma for $k = 1$.)

By Lemma 16.4, $Q(\omega)$ is homotopic to the matrix

$$Q_1(\omega) = \begin{Vmatrix} I_{Mp+m} & 0 \\ 0 & Q_2(\omega) \end{Vmatrix}.$$

Consequently, $A_8(\xi', \xi_n)$ is homotopic to the matrix

$$A_9(\xi', \xi_n) = \begin{Vmatrix} zI_{Mp+m} & 0 & 0 \\ 0 & I_{(M+1)p-m} & 0 \\ 0 & 0 & z^{-Mp}I_1 \end{Vmatrix},$$

since $Q_1^{-1}P_{\pm}^0 Q_1 = P_{\pm}^0$.

Since the matrix A_9 is infinitely differentiable on the circle $|z| = 1$, it follows from Lemma 16.4 for $k = 1$ that A_9 is homotopic to the matrix

$$\begin{Vmatrix} \left(\frac{\Lambda_+}{\Lambda_-}\right)^m I_1 & 0 \\ 0 & I_{(2M+1)p-1} \end{Vmatrix},$$

Q.E.D.

REMARK 16.3. According to the Bott periodicity theorem (see [7]), when n is odd, any elliptic matrix

$$\begin{Vmatrix} A & 0 \\ 0 & I_N \end{Vmatrix},$$

is homotopic to

$$\begin{Vmatrix} \left(\frac{\Lambda_+}{\Lambda_-}\right)^m I_1 & 0 \\ 0 & I_{N-1+p} \end{Vmatrix}$$

if N is sufficiently large. When n is even, this is no longer the case. Consequently, when n is even, there exist elliptic systems of p.o.'s for which there does not exist a uniquely solvable boundary value problem of form (16.3), (16.4).

§17. Construction of the solution of a boundary value problem on a halfline and in a halfspace

1. In the present section we carry out a more detailed investigation of the form of the solution of the boundary value problems (16.3), (16.4) and (16.5), (16.6). This investigation is needed for the construction of a parametrix for these problems. In the scalar case the solution of a boundary value problem for an elliptic p.o. in $\mathbf{R}_+^n$ was found in explicit form by means of a factorization of the scalar symbol. A factorization also exists for fixed ξ' in the case of an elliptic matrix, but the factors are generally discontinuous in ξ'. Therefore factoring does not play as fundamental a role in the case of systems of p.o.'s as in the scalar case. Also, the factorization of a matrix is not as efficient as the factorization of a scalar symbol.

2. We will distinguish a class of elliptic matrices whose factorization factors smoothly depend on parameters. As a preliminary, we note the following lemma on integrals of Cauchy type, which is a generalization of Lemma 6.1.

LEMMA 17.1. *Suppose $b(\omega, \eta_n) \in C^\infty$ with respect to (ω, η_0) and*

$$|D^p_\omega D^k_{\eta_n} b(\omega, \eta_n)| \leqslant C_{pk}(1 + |\eta_n|)^{-\gamma-k}, \qquad 0 < \gamma < 1, \qquad k \geqslant 0, |p| \geqslant 0. \tag{17.1}$$

Then the function

$$B_+(\omega, \xi_n + i\tau) = \frac{i}{2\pi}\int_{-\infty}^{\infty} \frac{b(\omega, \eta_n)}{\xi_n + i\tau - \eta_n}\, d\eta_n$$

is analytic with respect to $\xi_n + i\tau$ for $\tau > 0$, is infinitely differentiable with respect to $(\omega, \xi_n + i\tau)$ for $\tau \geqslant 0$ and satisfies the estimates

$$|D^p_\omega D^k_{\zeta_n} B_+(\omega, \zeta_n)| \leqslant C_{pk}(1 + |\xi_n| + |\tau|)^{-\gamma-k}, \qquad \zeta_n = \xi_n + i\tau. \tag{17.2}$$

PROOF. Since $|D^p_\omega b(\omega, \eta_n)| \leqslant C_p(1 + |\eta_n|)^{-\gamma}$ and

$$\left| D^p_\omega \frac{\partial b(\omega, \eta_n)}{\partial \eta_n} \right| \leqslant C_p(1 + |\eta_n|)^{-\gamma-1},$$

in completely the same way as in the proof of Lemma 6.1 we get that the function $B_+(\omega, \xi_n + i\tau)$ is continuous in $(\omega, \xi_n + i\tau)$ for $\tau \geqslant 0$ and satisfies (17.2) for $p = k = 0$. Further, for $\tau > 0$ the function $B_+(\omega, \xi_n + i\tau)$ is clearly analytic with respect to $\xi_n + i\tau$, with

$$D^p_\omega D^k_{\zeta_n} B_+(\omega, \xi_n + i\tau) = \frac{i}{2\pi}\int_{-\infty}^{\infty} \frac{D^p_\omega D^k_{\eta_n} b(\omega, \eta_n)}{\xi_n + i\tau - \eta_n}\, d\eta_n, \qquad \zeta_n = \xi_n + i\tau, \tag{17.3}$$

for arbitrary p and k. Expanding (17.3) by the expansion formula for an integral of Cauchy type (see (5.36)), we get

$$D^p_\omega D^k_{\zeta_n} B_+(\omega, \xi_n + i\tau) = \sum_{r=0}^{k} \frac{i\Pi'(\eta_n + i)^{r-1} D^p_\omega D^k_{\eta_n} b(\omega, \eta_n)}{(\xi_n + i\tau + i)^r} + \frac{i}{2\pi(\xi_n + i\tau + i)^k} \int \frac{(\eta_n + i)^k D^p_\omega D^k_{\eta_n} b(\omega, \eta_n)}{\xi_n + i\tau - \eta_n}\, d\eta_n, \tag{17.4}$$

where

$$\Pi'(\eta_n + i)^{r-1} D_\omega^p D_{\eta_n}^k b(\omega, \eta_n) = \frac{1}{2\pi}\int_{-\infty}^{\infty} (\eta_n + i)^{r-1} D_\omega^p D_{\eta_n}^k b(\omega, \eta_n) d\eta_n. \tag{17.5}$$

Since $r \leqslant k$, we find upon integrating by parts in (17.5) that

$$\Pi'(\eta_n + i)^{r-1} D_\omega^p D_{\eta_n}^k(\omega, \eta_n) = 0.$$

On the other hand, the function $(\eta_n + i)^k D_\omega^p D_{\eta_n}^k b(\omega, \eta_n)$ satisfies estimates of form (17.1). Consequently, as in the proof of Lemma 6.1, we get that the function

$$(\xi_n + i\tau + i)^k D_\omega^p D_{\xi_n}^k B_+(\omega, \xi_n + i\tau)$$

is continuous in $(\omega, \xi_n + i\tau)$ for $\tau \geqslant 0$, and

$$|(\xi_n + i\tau + i)^k D_\omega^p D_{\xi_n}^k B_+(\omega, \xi_n + i\tau)| \leqslant C_{pk}(1 + |\xi_n| + |\tau|)^{-\gamma}.$$

Lemma 17.1 is proved.

COROLLARY 17.1. *Suppose* $P_+ = b_+^{-1}(\xi_n)\Pi^+ b_+(\eta_n)$ *is the operator defined by* (14.21′) *and the function* $g(\omega, \eta_n)$ *satisfies* (17.1) *for* $\gamma > 1/2 + \delta_0$, *where*

$$0 < \delta_0 < 1/2 - \max_{1 \leqslant j \leqslant p} |\operatorname{Re} b_j|$$

(*the* b_j *are the eigenvalues of the matrix* b (*see* (14.21)). *Then*

$$|D_\omega^p D_{\zeta_n}^k P_+ g| \leqslant C_{pk}(1 + |\xi_n| + |\tau|)^{-1/2-\delta_0-k}, \qquad \forall p, \forall k, \zeta_n = \xi_n + i\tau, \tau \geqslant 0. \tag{17.6}$$

In fact, by (14.25) (cf. the proof of Lemma 14.3) it suffices to establish (17.6) for integrals of the form

$$\frac{i}{2\pi}\int_{-\infty}^{\infty}\left(\frac{\xi_n + i\tau + i}{\eta_n + i}\right)^{b_j} \ln^q\left(\frac{\xi_n + i\tau + i}{\eta_n + i}\right)\frac{g(\omega, \eta_n)}{\xi_n + i\tau - \eta_n} d\eta_n.$$

Since $0 < 1/2 + \delta_0 + \operatorname{Re} b_j < 1$ for any integer $j \in [1, p]$, (17.6) follows from Lemma 17.1.

An analogous assertion is valid for the operator $P_- = b_-^{-1}(\xi_n)\Pi^+ b_-(\eta_n)$.

We also need the following addition to Lemma 14.4.

LEMMA 17.2. *Suppose* $B(\omega, \eta_n) \in O_0$ *and* $B(0, \pm 1) = I$. *Then* $C = P_+B - BP_+$ *is an integral operator whose kernel* $C(\omega, \xi_n, \eta_n)$ *is infinitely differentiable with respect to* (ω, ξ_n, η_n) *and satisfies*

$$|D_\omega^p D_{\xi_n}^k C(\omega, \xi_n, \eta_n)| \leqslant C_{pk}(1 + |\xi_n|)^{-1/2-\delta_0-k}(1 + |\eta_n|)^{-1/2-\delta_0} \tag{17.7}$$

for any p *and* k.

In fact, since $B(\omega, \eta_n) \in O_0$ and $B(0, \pm 1) = I$, we have

$$|B(\omega, \eta_n) - I| \leqslant C(1 + |\eta_n|)^{-1}, \qquad |D_\omega^p D_{\eta_n}^k B(\omega, \eta_n)| \leqslant C(1 + |\eta_n|)^{-k-1}$$

for $k \geqslant 0$, $|p| \geqslant 0$. It follows from (14.29) that $C(\omega, \xi_n, \eta_n) \in C^\infty$ with respect to the totality of variables, while (17.7) for $k = 0$ follows from Lemma 14.4. The case $k > 0$ is proved analogously.

We can now prove the main lemma of this subsection.

LEMMA 17.3. *Suppose $A_0(\omega, \eta_n) \in O_0$ is an elliptic matrix satisfying* (14.22) *and the operator $\theta^+ A_0(\omega, D_n)$ is invertible in $H_0^+(\mathbf{R}^1)$ for all $\omega \in S^{n-2}$. Then the matrix $A_0(\omega, \eta_0)$ admits the factorization*

$$A_0(\omega, \eta_n) = A_-(\omega, \eta_n) A_+(\omega, \eta_n), \tag{17.8}$$

where

$$A_+^{-1}(\omega, \eta_n) = b_+^{-1}(\eta_n) + c_+(\omega, \eta_n), \qquad A_-(\omega, \eta_n) = a_+ b_-^{-1}(\eta_n) + c_-(\omega, \eta_n), \tag{17.9}$$

the matrices $b_\pm(\eta_n)$ are the same as in (14.21), *the matrix $c_+(\omega, \eta_n + i\tau)$ is analytic with respect to $\eta_n + i\tau$ for $\tau > 0$, is infinitely differentiable with respect to $(\omega, \eta_n + i\tau)$ for $\tau \geqslant 0$ and satisfies*

$$|D_\omega^p D_{\zeta_n}^k c_+(\omega, \eta_n + i\tau)| \leqslant \frac{C_{pk}}{(1 + |\eta_n| + |\tau|)^{1/2 + \delta_0 + k}},$$

$$\tau \geqslant 0, \qquad \zeta_n = \xi_n + i\tau, \qquad k \geqslant 0, \qquad |p| \geqslant 0, \tag{17.10}$$

and, analogously, the matrix $c_-(\omega, \eta_n + i\tau)$ is analytic with respect to $\xi_n + i\tau$ for $\tau < 0$, is of class C^∞ with respect to $(\omega, \eta_n + i\tau)$ for $\tau \leqslant 0$ and satisfies (17.10) *for $\tau \leqslant 0$. In addition,*

$$\det A_+(\omega, \eta_n + i\tau) \neq 0 \text{ for } \tau \geqslant 0, \qquad \det A_-(\omega, \eta_n + i\tau) \neq 0 \text{ for } \tau \leqslant 0. \tag{17.11}$$

PROOF. The matrix $A_0(\omega, \eta_0)$ can be represented in the form

$$A_0(\omega, \eta_n) = a_+ b_-^{-1}(\eta_n) b_+(\eta_n) + a_0(\omega, \eta_n), \tag{17.12}$$

where

$$a_0(\omega, \eta_n) = O\left(\frac{1}{1 + |\eta_n|}\right).$$

Let us find matrices $c_+(\omega, \eta_n)$ and $c_-(\omega, \eta_n)$ such that

$$A_0(\omega, \eta_n)\big(b_+^{-1}(\eta_n) + c_+(\omega, \eta_n)\big) = a_+ b_-^{-1}(\eta_n) + c_-(\omega, \eta_n). \tag{17.13}$$

By (17.12),

$$A_0(\omega, \eta_n) c_+(\omega, \eta_n) = c_-(\omega, \eta_n) - a_0(\omega, \eta_n) b_+^{-1}(\eta_n). \tag{17.14}$$

Since

$$a_0(\omega, \eta_n)b_+^{-1}(\eta_n) = O\left(\frac{1}{(1+|\eta_n|)^{1/2+\delta_0}}\right)$$

and consequently belongs to $H_0(\mathbf{R}^1)$, we have $\Pi^\pm a_0(\omega, \eta_n)b_+^{-1}(\eta_n) \in H_0^\pm(\mathbf{R}^1)$ (see §5). Applying the operator Π^+ to (17.14), we get

$$\Pi^+ A_0(\omega, \eta_n)c_+(\omega, \eta_n) = -\Pi^+ a_0(\omega, \eta_n)b_+^{-1}(\eta_n). \tag{17.15}$$

By assumption, (17.15) and the equivalent equation (17.14) are uniquely solvable. Consequently, there exist unique matrices $c_+(\omega, \eta_n) \in H_0^+(\mathbf{R}^1)$ and $c_-(\omega, \eta_n) \in H_0^-(\mathbf{R}^1)$ satisfying (17.14). Since the matrix $\Pi^+ a_0(\omega, \eta_n)b_+^{-1}(\eta_n)$ is clearly infinitely differentiable with respect to ω in the norm of $H_0^+(\mathbf{R}^1)$ and the inverse $R(\omega)$ of $\Pi^+ A_0(\omega, \eta_n)$ also belongs to C^∞ with respect to ω, it follows that the $c_\pm(\omega, \eta_n)$ are infinitely differentiable matrices with respect to ω with values in $H_0^\pm(\mathbf{R}^1)$.

Let us show that the matrix $c_+(\omega, \eta_n)$ has the properties indicated in Lemma 17.3. Applying the left regularizer $\tilde{R} = P_+ A_0^{-1}(\omega, \eta_n)a_+ P_- a_+^{-1}$, constructed in §14 to (17.15), we get, analogously to (14.37),

$$c_+(\omega, \eta_n) = P_+ C b_+^{-1} b_- a_+^{-1} A_0 c_+ - P_+ A_0^{-1} a_+ P_- a_+^{-1} a_0 b_+^{-1}, \tag{17.16}$$

where $C = P_+ A_0^{-1} a_+ b_-^{-1} b_+ - A_0^{-1} a_+ b_-^{-1} b_+ P_+$, by Lemma 17.2, is an integral operator with an infinitely differentiable kernel, satisfying (17.7). By Corollary 17.1, the matrix $P_+ A_0^{-1} a_+ P_- a_+^{-1} a_0 b_+^{-1}$ belongs to C^∞ with respect to $(\omega, \xi_n + i\tau)$ for $\tau \geqslant 0$ and satisfies the estimates of form (17.10). Since $a_1 = b_+^{-1} b_- a_+^{-1} A_0 c_+$ belongs to C^∞ with respect to ω and takes its values in $H_0(\mathbf{R}^1)$ with respect to η_n, the matrix Ca_1 is infinitely differentiable with respect to (ω, ξ_n) and satisfies the estimates of form (17.1) for $\gamma > 1/2 + \delta_0$. Consequently, by Corollary 17.1, $P_+ Ca_1 \in C^\infty$ with respect to $(\omega, \xi_n + i\tau)$ for $\tau \geqslant 0$ and satisfies (17.10).

It follows from (17.14) that $c_-(\omega, \eta_n)$ also has the properties indicated in Lemma 17.3.

Let us now show that the first of relations (17.11) is satisfied. Suppose the contrary:

$$\det\left(b_+^{-1}(\xi_{n0} + i\tau_0) + c_+(\omega_0, \xi_{n0} + i\tau_0)\right) = 0$$

for some ω_0 and $\xi_{n0} + i\tau_0$ with $\tau_0 \geqslant 0$. Then there exist numbers $\alpha_1, \ldots, \alpha_p$, not all of which are equal to zero, such that the linear combination

$$\sum_{k=1}^{p} \alpha_k s_k^+(\omega_0, \xi_n + i\tau)$$

of columns of the matrix $b_+^{-1}(\omega_0, \xi_n + i\tau) + c_+(\omega_0, \xi_n + i\tau)$ vanishes when $\xi_n + i\tau = \xi_{n0} + i\tau_0$. Further, the function

$$s_+(\omega_0, \xi_n + i\tau) = \sum_{k=1}^{p} \frac{\alpha_k s_k^+(\omega_0, \xi_n + i\tau)}{(\xi_n + i\tau) - (\xi_{n0} + i\tau_0)}.$$

belongs to $H_0^+(\mathbf{R}^1)$, and, by (17.13),

$$A_0(\omega_0, \xi_n)s_+(\omega_0, \xi_n) = s_-(\omega_0, \xi_n), \tag{17.17}$$

where

$$s_-(\omega_0, \xi_n) = \sum_{k=1}^{p} \frac{\alpha_k s_k^-(\omega_0, \xi_n)}{\xi_n - (\xi_{n0} + i\tau_0)},$$

with $s_1^-, \dots, s_p^-$ being the columns of the matrix $a_+ b_-^{-1}(\xi_n) + c_-(\omega_0, \xi_n)$. Since $\tau_0 \geqslant 0$, we have $s_-(\omega_0, \xi_n) \in \tilde{H}_0^-(\mathbf{R}^1)$. But then (17.17) contradicts the assumption concerning the absence of a kernel for the operator $\Pi^+ A_0(\omega, \xi_n)$. Consequently,

$$\det\big(b_+^{-1}(\xi_n + i\tau) + c_+(\omega, \xi_n + i\tau)\big) \neq 0 \qquad \text{for } \tau \geqslant 0, \quad \omega \in S^{n-2}.$$

An analogous contradiction can be obtained if it is assumed that

$$\det\big(a_+ b_-^{-1}(\xi_n^{(0)} + i\tau^{(0)}) + c_-(\omega^{(0)}, \xi_n^{(0)} + i\tau^{(0)})\big) = 0$$

for $\tau^{(0)} \leqslant 0$.

THEOREM 17.1. *Suppose $A(\xi', \xi_n) \in O_{\alpha+i\beta}^{\infty}$ and the matrix*

$$A_0(\xi', \xi_n) = (\xi_n - i|\xi'|)^{-\alpha-i\beta+s}(\xi_n + i|\xi'|)^{-s}A(\xi', \xi_n)$$

satisfies the conditions of Lemma 17.3. Then the matrix $A(\xi', \xi_n)$ admits the factorization

$$A(\xi', \xi_n) = A_-(\xi', \xi_n)A_+(\xi', \xi_n), \tag{17.18}$$

in which

$$A_+(\xi', \xi_n) = \Lambda_+^s(\xi)b_+(\xi', \xi_n)\left(I + c_+^{(1)}\left(\frac{\xi'}{|\xi'|}, \frac{\xi_n}{|\xi'|}\right)\right), \tag{17.19}$$

$$A_-(\xi', \xi_n) = \left(I + c_-^{(1)}\left(\frac{\xi'}{|\xi'|}, \frac{\xi_n}{|\xi'|}\right)\right)a_+ b_-^{-1}(\xi', \xi_n)\Lambda_-^{\alpha+i\beta-s}(\xi), \tag{17.20}$$

where

$$\Lambda_\pm(\xi) = \xi_n \pm i|\xi'|, \quad b_\pm(\xi', \xi_n) = \exp\{\ln \Lambda_\pm(\xi)b\},$$

$$a_+ = A(0, +1), \quad b = \frac{1}{2\pi i}\ln(A^{-1}(0, +1)A(0, -1)e^{\pi i(\alpha+i\beta-2s)}),$$

and the matrix

$$c_+^{(1)}\left(\frac{\xi'}{|\xi'|}, \frac{\xi_n}{|\xi'|}\right) \qquad \left(c_-^{(1)}\left(\frac{\xi'}{|\xi'|}, \frac{\xi_n}{|\xi'|}\right)\right)$$

is analytic with respect to $\xi_n + i\tau$ *for* $\tau > 0$ $(\tau < 0)$, *is infinitely differentiable with respect to* $(\xi', \xi_n + i\tau)$ *for* $\xi' \neq 0, \tau \geqslant 0$ $(\tau \leqslant 0)$, *and satisfies the estimates*

$$\left| D_{\xi'}^p D_{\xi_n}^k c_{\pm}^{(1)}\left(\frac{\xi'}{|\xi'|}, \frac{\xi_n + i\tau}{|\xi'|}\right)\right| \leqslant \frac{C_{pk}|\xi'|^{2\delta_0 - |p|}}{(|\xi'| + |\xi_n| + |\tau|)^{2\delta_0 + k}}, \qquad k \geqslant 0, |p| \geqslant 0, \tag{17.21}$$

with

$$\det\left(I + c_+^{(1)}\left(\frac{\xi'}{|\xi'|}, \frac{\xi_n + i\tau}{|\xi'|}\right)\right) \neq 0 \quad \text{for } |\xi'| + |\xi_n| + \tau > 0, \tau \geqslant 0,$$

$$\det\left(I + c_-^{(1)}\left(\frac{\xi'}{|\xi'|}, \frac{\xi_n + i\tau}{|\xi'|}\right)\right) \neq 0 \quad \text{for } |\xi'| + |\xi_n| + |\tau| > 0, \tau \leqslant 0.$$

PROOF. Let $\omega = \dfrac{\xi'}{|\xi'|}$ and $\xi_n = |\xi'|\eta_n$. Then $A_0(\xi', \xi_n) = A_0(\omega, \eta_n)$. By Lemma 17.3,

$$\begin{aligned} A_0(\omega, \eta_n) &= (a_+ b_-^{-1}(\eta_n) + c_-(\omega, \eta_n))(b_+^{-1}(\eta_n) + c_+(\omega, \eta_n))^{-1} \\ &= (I + c_-(\omega, \eta_n) b_-(\eta_n) a_+^{-1}) \\ &\quad \times a_+ b_-^{-1}(\eta_n) b_+(\eta_n)(I + c_+(\omega, \eta_n) b_+(\eta_n))^{-1}. \end{aligned} \tag{17.22}$$

Let

$c_-^{(1)}(\omega, \eta_n) = c_-(\omega, \eta_n) b_-(\eta_n) a_+^{-1}$, $c_+^{(1)}(\omega, \eta_n) = (I + c_+(\omega, \eta_n) b_+(\eta_n))^{-1} - I$.

Then, returning from the variables (ω, η_n) to the variables (ξ', ξ_n) and taking into account the fact that $b_-^{-1}(\eta_n) b_+(\eta_n) = b_-^{-1}(\xi', \xi_n) b_+(\xi', \xi_n)$, we get

$$\begin{aligned} A_0(\xi', \xi_n) &= \left(I + c_-^{(1)}\left(\frac{\xi'}{|\xi'|}, \frac{\xi_n}{|\xi'|}\right)\right) \\ &\quad \times a_+ b_-^{-1}(\xi', \xi_n) b_+(\xi', \xi_n)\left(1 + c_+^{(1)}\left(\frac{\xi'}{|\xi'|}, \frac{\xi_n}{|\xi'|}\right)\right). \end{aligned}$$

Inequalities (17.21) follow from (17.10) and the obvious estimates

$$|D_{\xi_n}^k b_\pm(\xi_n + i\tau)| \leqslant c_k(1 + |\xi_n| + |\xi_n| + |\tau|)^{1/2 - \delta_0 - k}, \qquad k \geqslant 0.$$

REMARK 17.1. If the conditions of Theorem 17.1 are satisfied, then the system of equations $p\hat{A}(D)u_+ = f$ has a unique solution $u_+ \in H_s^+(\mathbf{R}^n)$ for any $f \in H_{s-\alpha}(\mathbf{R}_+^n)$. Theorem 17.1 permits one to give an explicit formula for the Fourier transform of this solution (cf. §7, Theorem 7.1):

$$\begin{aligned} \tilde{u}_+(\xi) &= \hat{\Lambda}_+^{-s}(\xi)(I + \hat{c}_+^{(1)})^{-1}\hat{b}_+^{-1}(\xi', \xi_n) \\ &\quad \cdot \Pi^+ \hat{b}_-(\xi', \xi_n) a_+^{-1}(I + \hat{c}_-^{(1)})^{-1}\hat{\Lambda}_-^{s-\alpha-i\beta} lf. \end{aligned} \tag{17.23}$$

As in the proof of Theorem 7.1, it can be verified that the function $u_+ = F^{-1}\tilde{u}_+(\xi)$, where $\tilde{u}_+(\xi)$ is given by (17.23), is actually a solution of the system $p\hat{A}(D)u_+ = f$. The estimate $\|u_+\|_s < C\|f\|^+_{s-\alpha}$ follows from Lemma 5.3 (cf. the proof of Lemma 14.3).

EXAMPLE 17.1 [62]. We will cite an example of an operator $pA(\omega, D_n)$ that is uniquely solvable for any $\omega \in S^{n-2}$.

A matrix $A(\xi', \xi_n) \in O^\infty_\alpha$ is said to be *strongly elliptic* if the matrix

$$A_R(\xi', \xi_n) = \frac{A(\xi', \xi_n) + A^*(\xi', \xi_n)}{2}$$

is positive definite for $|\xi'| + |\xi_n| > 0$, i.e. $A_R(\xi', \xi_n)v \cdot v > |\xi|^\alpha |v|^2$ for any vector $v \in \mathbf{C}^p$.

Let us show that the system of equations

$$p\hat{A}(D)u_+ = f \tag{17.24}$$

has a unique solution $u_+ \in H^+_{\alpha/2}(\mathbf{R}^n)$ for any $f \in H_{-\alpha/2}(\mathbf{R}^n_+)$. Since $(H^+_{\alpha/2})^* = H_{-\alpha/2}(\mathbf{R}^n_+)$ (see §4.6), we get upon taking the Fourier transform with respect to (x', x_n) of the scalar product of (17.24) and $u_+(x', x_n)$ that

$$(\hat{A}(\xi)\tilde{u}_+, \tilde{u}_+) = (lf, \tilde{u}_+), \tag{17.25}$$

where $lf \in H_{-\alpha/2}(\mathbf{R}^n)$ is an arbitrary extension of $f \in H_{-\alpha/2}(\mathbf{R}^n_+)$. On the other hand,

$$\operatorname{Re}(\hat{A}(\xi)\tilde{u}_+, \tilde{u}_+) = (\hat{A}_R(\xi)\tilde{u}_+, \tilde{u}_+) > C\|u_+\|^2_{\alpha/2}.$$

It consequently follows from (17.25) by virtue of the Cauchy-Schwarz-Bunjakovskiĭ inequality (4.4) that

$$\|u_+\|_{\alpha/2} < C\|f\|^+_{-\alpha/2}. \tag{17.26}$$

The adjoint system of (17.24) has the form

$$p\hat{A}^*(D)v_+ = g, \tag{17.27}$$

where $\hat{A}^*(D)$ is the p.o. with symbol $A^*(\xi)$, $g \in H_{-\alpha/2}(\mathbf{R}^n_+)$ and $v_+ \in H^+_{\alpha/2}(\mathbf{R}^n)$. Since the matrix $\hat{A}^*(\xi)$ is also strongly elliptic, the following estimate, which is analogous to (17.24), holds for any solution $v_+ \in H^+_{\alpha/2}(\mathbf{R}^n)$ of (17.27):

$$\|v_+\|_{\alpha/2} < C\|g\|^+_{-\alpha/2}. \tag{17.28}$$

We denote by $R_{A^*} \subset H_{-\alpha/2}(\mathbf{R}^n_+)$ the range of the operator $p\hat{A}^*(D)$. The functional (lf, v_+) is continuous on R_{A^*} by virtue of (17.28). Consequently, according to the Riesz representation theorem, there exists a function $u_+ \in H^+_{\alpha/2}(\mathbf{R}^n)$ such that

$$(lf, v_+) = (u_+, \hat{A}^*(D)v_+). \tag{17.29}$$

Since v_+ is any function belonging to $H^+_{\alpha/2}(\mathbf{R}^n)$, it follows from (17.29) that u_+ satisfies (17.24). By (17.26), this solution is unique.

The unique solvability for any $\omega \in S^{n-2}$ of the system $pA(\omega, D)u_+(x_n) = f(x_n)$, where $f \in H_{-\alpha/2}(\mathbf{R}^1_+)$ and $u_+ \in H^+_{\alpha/2}$, is proved analogously.

REMARK 17.2. It follows from Theorem 17.1 that a strongly elliptic matrix $A(\xi', \xi_n)$ admits the factorization

$$A(\xi', \xi_n) = \left(I + c_-^{(1)}\left(\frac{\xi'}{|\xi'|}, \frac{\xi_n}{|\xi'|}\right)\right)a_+ b_-^{-1}(\xi', \xi_n)\Lambda_-^{\alpha/2}(\xi)$$
$$\times \Lambda_+^{\alpha/2}(\xi) b_+(\xi', \xi_n)\left(I + c_+^{(1)}\left(\frac{\xi'}{|\xi'|}, \frac{\xi_n}{|\xi'|}\right)\right). \tag{17.30}$$

EXAMPLE 17.2. Suppose a matrix $A(\xi) \in O_0^\infty$ is close to the identity matrix:

$$|A(\xi) - I| < \varepsilon. \tag{17.31}$$

Then $A(\xi)$ is clearly a strongly elliptic matrix and hence admits a factorization of form (17.30) for $\alpha = 0$. We note that the factorization factors of $A(\xi)$ can be found by the method of successive approximations if ε is sufficiently small. For let $a_0(\omega, \eta_n) = A(\omega, \eta_n) - a_+b_-^{-1}(\eta_n)b_+(\eta_n)$, and suppose $|a_0(\omega, \eta_n)| < \varepsilon_1$, where ε_1 is sufficiently small. Then (17.15) can be written in the form

$$\Pi^+a_+b_-^{-1}b_+c_+ + \Pi^+a_0c_+ = -\Pi^+a_0b_+^{-1}. \tag{17.32}$$

Applying the operator $b_+^{-1}\Pi^+b_-a_+^{-1}$ to (17.32), we get

$$c_+ + P_1c_+ = c_0, \tag{17.33}$$

where $c_0 = -b_+^{-1}\Pi^+b_-a_+^{-1}\Pi^+a_0b_+^{-1}$ and $P_1 = b_+^{-1}\Pi^+b_-a_+^{-1}\Pi^+a_0$. Since $\|P_1c_+\|_0 \leqslant C\varepsilon_1\|c_+\|_0$, it follows that

$$c_+ = c_0 - P_1c_0 + P_1^2c_0 - P_1^3c_0 + \dots \tag{17.34}$$

if $C\varepsilon_1 < 1$.

3. Let us now consider the general case.

THEOREM 17.2. *Suppose $A(\xi', \xi_n) \in O_{\alpha+i\beta}^\infty$ is an elliptic matrix and s is a real number satisfying* (14.38). *Then there exist (nonunique) matrices $Q_-(\xi', \xi_n)$, $Q_+(\xi', \xi_n)$ and an integer N such that*

$$A(\xi', \xi_n) = \Lambda_-^{\alpha+i\beta-s+N}(\xi)Q_-(\xi)Q_+(\xi)P_{2N}^{-1}(\xi)\Lambda_+^{s+N}(\xi), \tag{17.35}$$

$$Q_-(\xi) = (I + q_N^-(\xi))a_+b_-^{-1}(\xi), \tag{17.36}$$

$$Q_+(\xi) = b_+(\xi)(I + q_N^+(\xi)), \tag{17.37}$$

where the matrices a_+ and $b_\pm(\xi)$ are the same as in (17.19) *and* (17.20), *the matrices $q_N^\pm(\xi)$ and $I + q_N^\pm(\xi)$ have the same properties as the matrices $c_\pm^{(1)}$, $I + c_\pm^{(1)}$ in* (17.19) *and* (17.20), *and the matrix $P_{2N}(\xi)$ is a homogeneous function of degree $2N$, a polynomial in ξ_n and such that*

$$\det P_{2N}(\xi) \neq 0, \qquad P_{2N}(0, \pm 1) = I. \tag{17.38}$$

PROOF. We have

$$A(\xi', \xi_n) = \Lambda_-^{\alpha+i\beta-s}(\xi)\Lambda_+^s(\xi)a_+b_-^{-1}(\xi)b_+A_1(\xi), \tag{17.39}$$

where $A_1(\xi) \in O_0$ is an elliptic matrix such that $A_1(0, +1) = A_1(0, -1) = I$.

Let $z = (\xi_n + i|\xi'|)/(\xi_n - i|\xi'|)$. Then

$$\xi_n = i|\xi'|\frac{z+1}{z-1}.$$

Since $A_1(0, +1) = A_1(0, -1)$, the matrix

$$A_1\left(\xi', i|\xi'|\frac{z+1}{z-1}\right) = A_1\left(\omega, i\frac{z+1}{z-1}\right)$$

can be expanded in a uniformly convergent Fourier series (cf. §16):

$$A_1\left(\omega, i\frac{z+1}{z-1}\right) = \sum_{k=-\infty}^{\infty} a_k(\omega) z^k = \sum_{k=-\infty}^{\infty} a_k(\omega)\left(\frac{\Lambda_+}{\Lambda_-}\right)^k.$$

By taking a sufficiently large segment of the Fourier series, we can get

$$A_1(\xi', \xi_n) = B_N(\xi) R_N(\xi), \tag{17.40}$$

where $|B_N(\xi) - I| < \varepsilon$, with ε being sufficiently small, and the matrix $R_N(\xi', \xi_n)$ is a homogeneous function of degree zero, a rational function in ξ_n and such that $R_N(0, \pm 1) = I$ and $\det R_N(\xi', \xi_n) \neq 0$ for $\xi \neq 0$.

From Theorem 17.1 we have the factorization (see Example 17.2)

$$a_+ b_-^{-1}(\xi) b_+(\xi) B_N(\xi) = (I + q_N^-(\xi)) a_+ b_-^{-1}(\xi) b_+(\xi)(I + q_N^+(\xi)); \tag{17.41}$$

while $R_N^{-1}(\xi)$ is representable in the form

$$R_N^{-1}(\xi) = \Lambda_+^{-N}(\xi)\Lambda_-^{-N}(\xi) P_{2N}(\xi), \tag{17.42}$$

where $P_{2N}(\xi)$ is a polynomial matrix in ξ_n. Thus the factorization (17.35) follows from (17.39), (17.41) and (17.42).

Consider the operator

$$Rg = \hat{\Lambda}_+^{-s-2N}(D)\hat{P}_{2N}(D)\hat{Q}_+^{-1}(D)\theta^+\hat{Q}_-^{-1}(D)\hat{\Lambda}_-^{s-\alpha-i\beta}(D)lg,$$
$$\forall g \in H_{s-\alpha}(\mathbf{R}_+^n), \tag{17.43}$$

where $\hat{Q}_\pm(D)$ are the p.o.'s with symbols (17.36) and (17.37).

LEMMA 17.4. *The operator R is a bounded operator from $H_{s-\alpha}(\mathbf{R}_+^n)$ into $H_s^+(\mathbf{R}^n)$ whose kernel is equal to zero.*

PROOF. The boundedness of R follows from Lemma 5.3 and Theorem 4.3 (cf. the proof of Lemma 14.3).

Suppose $Rg = 0$. Then $\theta^+\hat{Q}_-^{-1}(D)\hat{\Lambda}_-^{s-\alpha-i\beta}(D)lg = 0$. Consequently, $\hat{Q}_-^{-1}(D)\hat{\Lambda}_-^{s-\alpha-i\beta}lg = v_-$, where $v_- \in H_\delta^-(\mathbf{R}^n)$ for $|\delta| < 1/2$. By Theorem 4.3, $lg(\xi) = \hat{Q}_-(\xi)\hat{\Lambda}_-^{-s+\alpha+i\beta}(\xi)\tilde{v}_-(\xi)$ is a "minus" function. Thus $g = plg = 0$.

We note that Rg does not depend on the choice of the extension lg.

REMARK 17.3. The representation (17.35) can be used to obtain more constructive proofs of some of the results of §§14 and 16.

Consider the following bounded operator from $H_{s-\alpha-N}(\mathbf{R}_+^1)$ into $H_{s-N}^+(\mathbf{R}^1)$:

$$R_N g = \Lambda_+^{-s-N}(D_n) P_{2N}(\omega, D_n) Q_+^{-1}(\omega, D_n)$$
$$\cdot\theta^+ Q_-^{-1}(\omega, D_n)\Lambda_-^{s-\alpha-i\beta-N}(D_n) lg(x_n), \tag{17.44}$$

where $\Lambda_\pm(\eta_n) = \eta_n \pm i$. By (17.35),

$$pA(\omega, D_n) R_N g = pQ_-\Lambda_-^{-s+\alpha+i\beta+N}\theta^+ Q_-^{-1}\Lambda_-^{s-\alpha-i\beta-N} lg = g. \tag{17.45}$$

Thus $pA(\omega, D_n)$ does not have a cokernel in $H_{s-\alpha-N}(\mathbf{R}^1_+)$ for any $\omega \in S^{n-2}$, so that we have obtained another proof of Lemma 14.6.

Now let

$$R^{(0)}g = \Lambda_+^{-s-2N}(D_n)P_{2N}(\omega, D_n)Q_+^{-1}(\omega, D_n) \cdot \theta^+ Q_-^{-1}\Lambda_+^N\Lambda_-^{s-N-\alpha-i\beta}(D_n)lg(x_n). \tag{17.46}$$

Then

$$pA(\omega, D_n)R^{(0)}g = p\Lambda_-^{\alpha+i\beta-s+N}Q_-\Lambda_+^{-N}\theta^+\Lambda_+^N Q_-^{-1}\Lambda_-^{s-N-\alpha-i\beta}lg.$$

Applying the expansion formula for an integral of Cauchy type, we get

$$\Lambda_+^{-N}(\eta_n)\Pi^+\Lambda_+^N(\eta_n)\tilde{h}(\eta_n) = \Pi^+\tilde{h} - \sum_{k=1}^{N} \Lambda_+^{-k}i\Pi'\Lambda_+^{k-1}\tilde{h}, \tag{17.47}$$

where

$$\tilde{h} = Q_-^{-1}(\omega, \eta_n)\Lambda_-^{s-\alpha-i\beta-N}(\eta_n)lg(\eta_n).$$

Consequently,

$$pA(\omega, D_n)R^{(0)}g = g - \sum_{k=1}^{N} ip\Lambda_-^{\alpha+i\beta-s+N}Q_-\Lambda_+^{-k}(p'\Lambda_+^{k-1}Q_-^{-1}\Lambda_-^{s-\alpha-i\beta-N}lg \times \delta(x_n)). \tag{17.48}$$

Thus the operator

$$\mathcal{A}(\omega)(u_+, \rho_1, \dots, \rho_N) = pA(\omega, D_n)u_+ + \sum_{k=1}^{N} ip\Lambda_-^{\alpha+i\beta-s+N}Q_-\Lambda_+^{-k}(\rho_k \times \delta(x_n)) \tag{17.49}$$

does not have a cokernel in $H_{s-\alpha}(\mathbf{R}^1_+)$ for any $\omega \in S^{n-2}$, inasmuch as the equation

$$\mathcal{A}(\omega)(u_+, \rho_1, \dots, \rho_N) = g$$

has the solution $u_+ = R^{(0)}g$, $\rho_k = p'\Lambda_+^{k-1}Q_-^{-1}\Lambda_-^{s-\alpha-i\beta-N}lg$ for any $g \in H_{s-\alpha}(\mathbf{R}^1_+)$, so that we have obtained a new proof of Lemma 16.1.

The operator (17.43) can be used to reduce (i) the boundary value problem (16.3), (16.4) to a system of pseudodifferential equations in $\mathbf{R}^{n-1}$ and (ii) the boundary value problem (16.5), (16.6) to a system of linear algebraic equations. By (17.35) and the relation $\theta^+w = w - \theta^-w$,

$$\begin{aligned} Rp\hat{A}(D)u_+ &= \hat{\Lambda}_+^{-s-2N}\hat{P}_{2N}(D)\hat{Q}_+^{-1}\theta^+\hat{Q}_+\hat{P}_{2N}^{-1}\hat{\Lambda}_+^{s+N}\hat{\Lambda}_-^N u_+ \\ &= \hat{\Lambda}_-^N\hat{\Lambda}_+^{-N}u_+ - \hat{\Lambda}_+^{-s-2N}\hat{P}_{2N}\hat{Q}_+^{-1}\theta^-w, \end{aligned} \tag{17.50}$$

where $w = \hat{Q}_+\hat{P}_{2N}^{-1}\hat{\Lambda}_+^{s+N}\hat{\Lambda}_-^N u_+$.

Let us show that

$$\theta^-w = \sum_{k=1}^{2N} \theta^-\hat{Q}_{k1}(D)(p'\hat{Q}_{k2}(D)\hat{\Lambda}_+^s u_+ \times \delta(x_n)). \tag{17.51}$$

We have

$$\theta^{-}w=\theta^{-}\hat{Q}_{+}\theta^{-}\hat{P}_{2N}^{-1}\hat{\Lambda}_{+}^{s+N}\hat{\Lambda}_{-}^{N}u_{+}, \tag{17.52}$$

since $\hat{Q}_{+}(D)$ is a "plus" operator. The operator $\hat{\Lambda}_{+}^{N}\hat{\Lambda}_{-}^{N}$ is a differential operator with respect to x_n. Consequently, by Leibniz's formula (cf. the expansion formula for an integral of Cauchy type),

$$\hat{\Lambda}_{+}^{N}\hat{\Lambda}_{-}^{N}\theta^{-}P_{2N}^{-1}\Lambda_{+}^{s}u_{+}=\theta^{-}\hat{P}_{2N}^{-1}\hat{\Lambda}_{+}^{N+s}\hat{\Lambda}_{-}^{N}u_{+}+\sum_{k=1}^{2N}\hat{q}_{k1}(D)\left(p'D_n^{k-1}\hat{P}_{2N}^{-1}\hat{\Lambda}_{+}^{s}u_{+}\times\delta\right), \tag{17.53}$$

where the $\hat{q}_{k1}(D)$ are differential operators with respect to x_n. Analogously,

$$\hat{P}_{2N}\theta^{-}\hat{P}_{2N}^{-1}\hat{\Lambda}_{+}^{s}u_{+}=\theta^{-}\hat{\Lambda}_{+}^{s}u_{+}+\sum_{k=1}^{2N}\hat{q}_{k2}(D)\left(p'D_n^{k-1}\hat{P}_{2N}^{-1}\hat{\Lambda}_{+}^{s}u_{+}\times\delta\right). \tag{17.54}$$

Since $\theta^{-}\hat{\Lambda}_{+}^{s}u_{+}=0$, the elimination of the left sides of (17.53) and (17.54) leads to the relation

$$\theta^{-1}\hat{P}_{2N}^{-1}\hat{\Lambda}_{+}^{N+s}\hat{\Lambda}_{-}^{N}u_{+}=\sum_{k=1}^{2N}\left(\hat{\Lambda}_{+}^{N}\hat{\Lambda}_{-}^{N}\hat{P}_{2N}^{-1}\hat{q}_{k2}(D)-\hat{q}_{k1}(D)\right)\left(p'D_n^{k-1}P_{2N}^{-1}\hat{\Lambda}_{+}^{s}u_{+}\times\delta\right), \tag{17.55}$$

from which (17.51) follows upon putting

$$\hat{Q}_{k2}=D_n^{k-1}\hat{P}_{2N}^{-1},\qquad \hat{Q}_{k1}=\hat{Q}_{+}\left(\hat{\Lambda}_{+}^{N}\hat{\Lambda}_{-}^{N}\hat{P}_{2N}^{-1}\hat{q}_{k2}-\hat{q}_{k1}\right).$$

We note that the left side and hence the right side of (17.55) belong to $H_0(\mathbf{R}^n)$. Therefore, if in $\tilde{\Lambda}_{+}^{N}\hat{\Lambda}_{-}^{N}\hat{P}_{2N}^{-1}\hat{q}_{k2}$ the differential operator with respect to x_n is separated out, then it must be cancelled by $\hat{q}_{k1}(D)$.

Applying the operator $\hat{\Lambda}_{+}^{s}R$ to (16.3) and taking (17.50) and (17.51) into account, we get

$$\hat{\Lambda}_{-}^{N}\hat{\Lambda}_{+}^{s-N}u_{+}=\sum_{k=1}^{2N}\hat{C}_{k1}(D)(w_k\times\delta)-\sum_{k=1}^{m_-}\hat{C}_{k2}(D)(v_k\times\delta)+\hat{\Lambda}_{+}^{s}Rf, \tag{17.56}$$

where

$$C_{k1}(\xi)=\Lambda_{+}^{-2N}(\xi)P_{2N}Q_{+}^{-1}(\xi)\Pi^{-}Q_{k1}(\xi),\qquad 1\leqslant k\leqslant 2N, \tag{17.57}$$

$$C_{k2}(\xi)=\Lambda_{+}^{-2N}(\xi)P_{2N}(\xi)Q_{+}^{-1}\Pi^{+}Q_{-}^{-1}(\xi)\Lambda_{-}^{s-\alpha-i\beta}(\xi)C_k(\xi),\qquad 1\leqslant k\leqslant m_-, \tag{17.57'}$$

$$w_k(x')=p'\hat{Q}_{k2}(D)\hat{\Lambda}_{+}^{s}u_{+},\qquad 1\leqslant k\leqslant 2N. \tag{17.58}$$

We note that system (17.56) is equivalent to (16.3), since $\ker\hat{\Lambda}_{+}^{s}R=0$.

Let $\Phi_+(x)$ denote the right side of (17.56), which belongs to $H_0^+(\mathbf{R}^n)$. The equation $\hat{\Lambda}_-^N \hat{\Lambda}_+^{s-N} u_+ = \Phi_+$ has a solution $u_+ \in H_s^+(\mathbf{R}^n)$ if and only if

$$\theta^- \hat{\Lambda}_+^N \hat{\Lambda}_-^{-N} \Phi_+ = 0, \tag{17.59}$$

in which case

$$u_+ = \hat{\Lambda}_+^{-s+N} \hat{\Lambda}_-^{-N} \Phi_+ = \hat{\Lambda}_+^{-s} \theta^+ \hat{\Lambda}_+^N \hat{\Lambda}_-^{-N} \Phi_+.$$

A condition equivalent to (17.59) is

$$p' \hat{\Lambda}_-^{-k} \Phi_+ = 0, \qquad 1 \leqslant k \leqslant N. \tag{17.60}$$

In fact, since $\Pi^- \hat{\Lambda}_-^{-N}(\xi) \hat{\Lambda}_+^N \tilde{\Phi}_+(\xi) = 0$, the function

$$\hat{\Lambda}_-^{-N}(\xi', \xi_n + i\tau) \hat{\Lambda}_+^N(\xi', \xi_n + i\tau) \tilde{\Phi}_+(\xi', \xi_n + i\tau)$$

is analytic for $\tau > 0$. This is possible if and only if

$$\frac{\partial^{k-1}}{\partial \zeta_n^{k-1}} \tilde{\Phi}_+(\xi', \zeta_n)\Big|_{\zeta_n = i(|\xi'|+1)} = 0, \qquad 1 \leqslant k \leqslant N.$$

But

$$\frac{\partial^{k-1}}{\partial \zeta_n^{k-1}} \tilde{\Phi}_+(\xi', \zeta_n) = \frac{i}{2\pi} (-1)^{k-1} (k-1)! \int_{-\infty}^{\infty} \frac{\tilde{\Phi}_+(\xi', \eta_n)}{(\zeta_n - \eta_n)^k} \, d\eta_n, \qquad \operatorname{Im} \zeta_n > 0. \tag{17.60'}$$

Thus (17.59) and (17.60) are equivalent.

It follows from (17.60) that (17.56) cannot have a solution $u_+ \in H_s^+(\mathbf{R}^n)$ unless

$$\sum_{k=1}^{2N} \hat{b}_{jk}^{(1)}(D') w_k(x') - \sum_{k=1}^{m_-} \hat{b}_{jk}^{(2)}(D') v_k(x') = f_j(x'), \qquad 1 \leqslant j \leqslant N, \tag{17.61}$$

where

$$b_{jk}^{(1)}(\xi') = \Pi' \Lambda_-^{-j} C_{k1}(\xi), \qquad b_{jk}^{(2)}(\xi') = \Pi' \Lambda_-^{-j} C_{k2}(\xi). \tag{17.62}$$

Further, writing (17.56) in the form

$$u_+ = \sum_{k=1}^{2N} \hat{\Lambda}_+^{N-s} \hat{\Lambda}_-^{-N} \hat{C}_{k1}(w_k \times \delta) - \sum_{k=1}^{m_-} \hat{\Lambda}_+^{N-s} \hat{\Lambda}_-^{-N} \hat{C}_{k2}(v_k \times \delta) + \hat{\Lambda}_+^N \hat{\Lambda}_-^{-N} R f \tag{17.63}$$

and substituting this form in the boundary conditions (16.4), we get

$$\sum_{k=1}^{2N} \hat{b}_{jk}^{(3)}(D') w_k(x') + \sum_{k=1}^{m_-} \hat{b}_{jk}^{(4)}(D') v_k(x') = g_j(x') - p' \hat{B}_j \hat{\Lambda}_+^N \hat{\Lambda}_-^{-N} R f, \tag{17.64}$$

where

$$b_{jk}^{(3)}(\xi') = \Pi' B_j(\xi)\Lambda_+^{N-s}\Lambda_-^{-N}C_{k1}(\xi),$$
$$b_{jk}^{(4)}(\xi') = E_{jk}(\xi') - \Pi' B_j\Lambda_+^{N-s}\Lambda_-^{-N}C_{k2}. \tag{17.65}$$

Thus the boundary value problem (16.3), (16.4) is equivalent to the system (17.58), (17.61), (17.63), (17.64).

To solve this system, we apply the operator $p'\hat{Q}_{j2}(D)\hat{\Lambda}_+^s$ to (17.63). In this way, letting

$$b_{jk}^{(5)}(\xi') = \Pi' Q_{j2}\Lambda_+^N\Lambda_-^{-N}C_{k1}, \qquad b_{jk}^{(6)}(\xi') = \Pi' Q_{j2}\Lambda_+^N\Lambda_-^{-N}C_{k2}(\xi), \tag{17.66}$$

we get

$$w_j(x') = \sum_{k=1}^{2N} \hat{b}_{jk}^{(5)}(D')w_k(x') - \sum_{k=1}^{m_-} \hat{b}_{jk}^{(6)}(D')v_k(x') + p'\hat{Q}_{j2}\hat{\Lambda}_+^{N+s}\hat{\Lambda}_-^{-N}Rf. \tag{17.67}$$

Consequently, the boundary value problem (16.3), (16.4) is equivalent to the system (17.61), (17.64), (17.67) of $3N + m_+$ pseudodifferential equations in $\mathbf{R}^{n-1}$ for the unknowns $v_k(x')$, $1 \leqslant k \leqslant m_-$, and $w_k(x')$, $1 \leqslant k \leqslant 2N$. For suppose $v_k^{(0)}(x')$, $1 \leqslant k \leqslant m_-$, $w_p^{(0)}(x')$, $1 \leqslant p \leqslant 2N$, is a solution of system (17.61), (17.64), (17.67). We determine $u_+^{(0)}$ in terms of the $v_k^{(0)}$ and $w_p^{(0)}$ by means of (17.63). If we now again apply the operator $p'\hat{Q}_{j2}\hat{\Lambda}_+^s$ to (17.63), we will get by virtue of (17.67) that

$$w_j^{(0)}(x') = p'\hat{Q}_{j2}\hat{\Lambda}_+^s u_+^{(0)}, \qquad 1 \leqslant j \leqslant 2N.$$

Thus $u_+^{(0)}(x)$, $v_k^{(0)}(x')$, $1 \leqslant k \leqslant m_-$, is a solution of problem (16.3), (16.4).

Suppose condition (16.13) is satisfied. Then system (17.61), (17.64), (17.67) is elliptic, i.e. its symbol $M(\xi')$ has maximal rank $2N + m_-$ for all $\xi' \neq 0$, which implies the existence of a left inverse $L(\xi')$ of the matrix $M(\xi')$. The unknowns $w_k(x')$, $1 \leqslant k \leqslant 2N$ and $v_j(x')$, $1 \leqslant j \leqslant m_-$, can therefore be found by applying the p.o. with symbol $L(\xi')$ to system (17.61), (17.64), (17.67), after which $u_+(x)$ can be determined by (17.63). Thus, under the fulfilment of condition (16.13), we have obtained an explicit expression for the unique solution of the boundary value problem (16.3), (16.4).

When carried out for the boundary value problem (16.5), (16.6), the preceding constructions convert system (17.61), (17.64), (17.67) into a system of linear algebraic equations. The system of form (17.63) for fixed ω is a system of integral equations with degenerate kernel. The above method is actually a repetition of the usual procedure for solving an integral equation with degenerate kernel.

Remark 17.4. Besides the boundary value problem (16.3), (16.4) one can consider the more general boundary value problem

$$p\hat{A}u_+ + \sum_{k=1}^{m_0} p\hat{G}_{k1}(D)(p'\hat{G}_{k2}u_+ \times \delta(x_n)) + \sum_{k=1}^{m_-} p\hat{C}_k(v_k(x')) \times \delta(x_n) = f(x), \quad (17.68)$$

$$p\hat{B}_j u_+ + \sum_{k=1}^{m_-} \hat{E}_{jk}(D')v_k(x') = g_j(x'), \qquad 1 < j < m_+, \quad (17.69)$$

where $\hat{A}$, $\hat{B}_j$, $\hat{C}_k$ and $\hat{E}_{jk}$ are the same as in (16.3) and (16.4), while $\hat{G}_{k1}(\xi)$ and $\hat{G}_{k2}(\xi)$ are respectively p-dimensional column, row vectors of classes $D_{\alpha+i\beta+1-\delta_k}$ (see §10) and O'_{δ_k}, with $\operatorname{Re}\delta_k < s - 1/2$, and m_0 is a positive integer. This problem is of interest since summands of the form $p\hat{G}_{k1}(p'\hat{G}_{k2}u_+ \times \delta)$ show up in the analysis of problem (16.3), (16.4) (see (17.56) or (17.63)) as well as in the analysis of an ordinary boundary value problem for an elliptic differential equation with differential boundary conditions.

Analogously to (16.5), (16.6), we associate with problem (17.68), (17.69) a family of boundary value problems on a halfline. Let $\mathfrak{A}_1(\omega)$ denote the bounded operator from $\mathcal{H}_s^{(1)}(\mathbf{R}^1)$ into $\mathcal{H}_s^{(2)}(\mathbf{R}^1)$ defined by this family. We note that the difference between $\mathfrak{A}_1(\omega)$ and the operator $\mathfrak{A}(\omega)$ defined by (16.5), (16.6) is the operator

$$K(\omega)v_+(x_n) = \sum_{k=1}^{m_0} (p'G_{k2}(\omega, D_n)v_+(x_n))pG_{k1}(\omega, D_n)\delta$$

with a finite-dimensional range. Under the assumption that $\mathfrak{A}_1(\omega)$ is invertible for all $\omega \in S^{n-2}$, it can be proved in exactly the same way as in §16 that problem (17.68), (17.69) is uniquely solvable in the same function spaces as (16.3), (16.4). It is also evident that the construction of the inverse operator in §17.3 goes over almost without change to the case of problem (17.68), (17.69).

We note that when the index m of the operator $\mathfrak{A}(\omega)$ is nonpositive (nonnegative) it is possible to choose the operator $K(\omega)$ so that

$$\dim\ker(\mathfrak{A}(\omega) + K(\omega)) = 0 \qquad (\dim\operatorname{coker}(\mathfrak{A}(\omega) + K(\omega)) = 0)$$

for all $\omega \in S^{n-2}$.

CHAPTER V

PSEUDODIFFERENTIAL OPERATORS WITH VARIABLE SYMBOLS

§18. Theorems on the boundedness and composition of pseudodifferential operators

1. *The symbol class* S_α^t. If $\tilde{A}'(\eta, \xi)$ is a measurable function in the Lebesgue sense that satisfies the estimates

$$(1+|\eta|)^N|\tilde{A}'(\eta, \xi)| \leqslant C_N(1+|\xi|)^\alpha, \qquad \forall N, \tag{18.1}$$

then

$$A'(x, \xi) = F_\eta^{-1}\tilde{A}'(\eta, \xi) = \frac{1}{(2\pi)^n}\int_{-\infty}^{\infty} \tilde{A}'(\eta, \xi)e^{-i(x,\eta)}\,d\eta, \tag{18.2}$$

is an infinitely differentiable function with respect to x such that $|D_x^pA'(x, \xi)| \leqslant C_p(1+|\xi|)^\alpha$ and $D_x^pA'(x, \xi) \to 0$ as $|x| \to \infty$ for any p. We denote by $S_\alpha^0 = S_\alpha^0(\mathbf{R}^n)$ the class of functions of the form

$$A(x, \xi) = A(\infty, \xi) + A'(x, \xi), \tag{18.3}$$

where $A(\infty, \xi)$ is a measurable function satisfying the estimate $|A(\infty, \xi)| \leqslant C(1+|\xi|)^\alpha$ (cf. §3), and $A'(x, \xi) = F_\eta^{-1}\tilde{A}'(\eta, \xi)$, with $\tilde{A}'(\eta, \xi)$ satisfying estimates (18.1).

Let $t = m + \gamma$, where m is a nonnegative integer and $\gamma \in (0, 1]$. A function $A(x, \xi)$ is said to be of class $S_\alpha^t = S_\alpha^t(\mathbf{R}^n)$ if $D_\xi^kA(x, \xi)$ is an absolutely continuous function of class $S_{\alpha-|k|}^0$ for $0 \leqslant |k| \leqslant m$ and $D_\xi^kA(x, \xi) \in S_{\alpha-t}^0$ for $|k| = m+1$. The intersection $\bigcap_{m=0}^\infty S_\alpha^m$ is denoted by S_α^∞.

By the *pseudodifferential operator* (p.o.) with *symbol* $A(x, \xi) \in S_\alpha^t$, $t \geqslant 0$, is meant the operator A defined on the functions $u(x) \in S(\mathbf{R}^n)$ by the formula (cf. (3.2))

$$Au = \frac{1}{(2\pi)^n} \int_{-\infty}^{\infty} A(x, \xi)\tilde{u}(\xi)e^{-i(x,\xi)}\, d\xi. \tag{18.4}$$

If $A(x, \xi) = \sum_{|k|=0}^m a_k(x)\xi^k$ is a polynomial in ξ, then (18.4) implies $Au = \sum_{|k|=0}^m a_k(x)D^k u(x)$ (cf. (3.3)), i.e. A is a differential operator with variable coefficients. By analogy with differential operators, we will sometimes denote the p.o. with symbol $A(x, \xi)$ by $A(x, D)$.

The symbol classes $O_{\alpha+i\beta}^\infty$ *and* $\hat{O}_{\alpha+i\beta}^\infty$. The class $O_{\alpha+i\beta}^\infty$ consists of those functions $A_0(x, \xi)$ that are infinitely differentiable with respect to x and ξ ($\xi \neq 0$), homogeneous of degree $\alpha + i\beta$ with respect to ξ, and independent of x for $|x| \geqslant N$: $A_0(x, \xi) = A_0(\infty, \xi)$ for $|x| \geqslant N$.

A function $A(x, \xi)$ is said to be of class $\hat{O}_{\alpha+i\beta}^\infty$ if $A(x, \xi) \in C^\infty(\mathbf{R}^n \times \mathbf{R}^n)$, $A(x, \xi) = A(\infty, \xi)$ for $|x| \geqslant N$ and there exists a function $A_0(x, \xi) \in O_{\alpha+i\beta}^\infty$ such that $A(x, \xi) - A_0(x, \xi)(1 - \chi(\xi)) \in S_{\alpha-1}^\infty$, where $\chi(\xi) \in C_0^\infty(\mathbf{R}^n)$ and $\chi(\xi) = 1$ for $|\xi| \leqslant 1$. Clearly, $\hat{O}_{\alpha+i\beta}^\infty \subset S_\alpha^\infty$.

The main class of p.o.'s considered in the book are the p.o.'s with symbols belonging to $\hat{O}_{\alpha+i\beta}^\infty$. But since the factorization factors of a symbol of class $O_{\alpha+i\beta}^\infty$ are not infinitely differentiable with respect to ξ, we must also consider symbols of class S_α^t with a finite smoothness in ξ.

We note that, as in §3, it is possible to define p.o.'s with symbols $A_0(x, \xi) \in O_{\alpha+i\beta}^\infty$, where a p.o. with symbol from $O_{\alpha+i\beta}^\infty$ differs from a p.o. with symbol $A(x, \xi) = (1 - \chi(\xi))A_0(x, \xi) \in \hat{O}_{\alpha+i\beta}^\infty$ by an integral operator with an infinitely differentiable kernel.

2. *Boundedness of pseudodifferential operators in* $H_s(\mathbf{R}^n)$.

THEOREM 18.1. *Suppose* $A(x, \xi) \in S_\alpha^0$. *Then the p.o.* $A(x, D)$ *is a bounded operator from* $H_s(\mathbf{R}^n)$ *into* $H_{s-\alpha}(\mathbf{R}^n)$ *for any* s:

$$\|Au\|_{s-\alpha} \leqslant K_s\|u\|_s. \tag{18.5}$$

PROOF. We have

$$A(x, D)u = A(\infty, D)u + A'(x, D)u \tag{18.6}$$

for $u(x) \in S(\mathbf{R}^n)$. Clearly, since $A(\infty, D)u = F^{-1}A(\infty, \xi)\tilde{u}(\xi)$,

$$\|A(\infty, D)u\|_{s-\alpha} \leqslant K^{(0)}\|u\|_s, \tag{18.7}$$

where

$$K^{(0)} = \sup_\xi |A(\infty, \xi)|(1 + |\xi|)^{-\alpha}. \tag{18.8}$$

The Fourier transform of the function

$$v(x) = A'(x, D)u = \frac{1}{(2\pi)^n}\int_{-\infty}^{\infty} A'(x, \xi)\tilde{u}(\xi)e^{-i(x,\xi)}\, d\xi$$

has the form

$$\tilde{v}(\eta) = \frac{1}{(2\pi)^n}\int_{-\infty}^{\infty} \tilde{A}'(\eta - \xi, \xi)\tilde{u}(\xi)\, d\xi. \tag{18.9}$$

By (18.1),

$$|\tilde{v}(\eta)| \leqslant \frac{1}{(2\pi)^n}\int_{-\infty}^{\infty} \frac{C_N(1 + |\xi|)^\alpha|\tilde{u}(\xi)|d\xi}{(1 + |\eta - \xi|)^N}. \tag{18.10}$$

On the other hand, for any real t

$$(1 + |\eta|)^t \leqslant (1 + |\xi|)^t(1 + |\xi - \eta|)^{|t|} \tag{18.11}$$

(see (1.7)). Consequently, multiplying (18.10) by $(1 + |\eta|)^{s-\alpha}$, we get

$$(1 + |\eta|)^{s-\alpha}|\tilde{v}(\eta)| \leqslant \frac{1}{(2\pi)^n}\int_{-\infty}^{\infty} \frac{C_N(1 + |\xi|)^s|\tilde{u}(\xi)|d\xi}{(1 + |\eta - \xi|)^{N-|s-\alpha|}}. \tag{18.12}$$

To complete the proof of Theorem 18.1, we need the following simple lemma.

LEMMA 18.1. *Suppose*

$$f(\eta) = \int_{-\infty}^{\infty} B(\eta - \xi)g(\xi)\, d\xi, \tag{18.13}$$

where $g(\xi) \in S(\mathbf{R}^n)$ *and* $\int_{-\infty}^{\infty}|B(\eta)|d\eta = B_0 < \infty$. *Then*

$$\|f\|_0 \leqslant B_0\|g\|_0. \tag{18.14}$$

In fact, the Fourier transform of (18.13) is

$$\tilde{f}(x) = \tilde{B}(x)\tilde{g}(x), \tag{18.15}$$

where $\tilde{B}(x) = \int_{-\infty}^{\infty} B(\eta)e^{i(x,\eta)}\, d\eta$. Since

$$|\tilde{B}(x)| \leqslant \int_{-\infty}^{\infty}|B(\eta)|d\eta = B_0,$$

estimate (18.14) follows from (18.15).

Applying Lemma 18.1 to (18.12) after setting

$$N = |s - \alpha| + n + 1,$$

$$B(\eta - \xi) = \frac{C_N}{(2\pi)^n}\frac{1}{(1 + |\eta - \xi|)^{n+1}}, \quad g(\xi) = (1 + |\xi|)^s|\tilde{u}(\xi)|.$$

we get

$$\|v\|_{s-\alpha} \leqslant C'_N \|u\|_s. \tag{18.16}$$

Thus the validity of (18.5) for all $u(x) \in S(\mathbf{R}^n)$ follows from (18.7) and (18.16). Since $S(\mathbf{R}^n)$ is dense in $H_s(\mathbf{R}^n)$ (see Theorem 4.1), the operator $A(x, D)$ can be extended by continuity to a bounded operator from $H_s(\mathbf{R}^n)$ into $H_{s-\alpha}(\mathbf{R}^n)$, with (18.5) remaining in force for any $u \in H_s(\mathbf{R}^n)$.

REMARK 18.1. An operator is said to be of *order* α if it is a bounded operator from $H_s(\mathbf{R}^n)$ into $H_{s-\alpha}(\mathbf{R}^n)$ for all s. By Theorem 18.1, a p.o. $A(x, D)$ with symbol $A(x, \xi) \in S_\alpha^0$ is of order α. In the sequel we will need an estimate of the norm of such an operator $A(x, D)$ in terms of its symbol $A(x, \xi)$. By (18.8) and (18.12), we have

$$\|A\|_{(s)} \leqslant \sup_\xi \frac{|A(\infty, \xi)|}{(1+|\xi|)^\alpha} + C_0 \sup_{\xi,\eta} \frac{(1+|\eta|)^N |\tilde{A}'(\eta, \xi)|}{(1+|\xi|)^\alpha}, \tag{18.17}$$

where $N = n + 1 + |s - \alpha|$ and C_0 does not depend on $A(x, \xi)$.

3. *Composition of pseudodifferential operators.*

THEOREM 18.2. *Suppose* $A(x, \xi) \in S_\alpha^0$ *and* $B(x, \xi) \in S_\beta^0$. *Then* $A(x, D)B(x, D)$ *is a pseudodifferential operator* $C(x, D)$ *with symbol* $C(x, \xi) \in S_{\alpha+\beta}^0$ *defined by*

$$C(x, \xi) = A(x, \xi)B(\infty, \xi) + \frac{1}{(2\pi)^n} \int_{-\infty}^{\infty} A(x, \xi + \zeta)\tilde{B}'(\zeta, \xi) e^{-i(x,\zeta)}\, d\zeta. \tag{18.18}$$

PROOF. Since $B(x, D) = B(\infty, D) + B'(x, D)$, we have

$$A(x, D)B(x, D) = A(x, D)B(\infty, D) + A(x, D)B'(x, D). \tag{18.19}$$

Clearly, $A(x, D)B(\infty, D)$ is the p.o. with symbol $A(x, \xi)B(\infty, \xi) \in S_{\alpha+\beta}^0$:

$$A(x, D)B(\infty, D)u(x) = \frac{1}{(2\pi)^n} \int_{-\infty}^{\infty} A(x, \xi)B(\infty, \xi)\tilde{u}(\xi) e^{-i(x,\xi)}\, d\xi \tag{18.20}$$

for all $u \in S(\mathbf{R}^n)$.

Analogously to (18.9), the Fourier transform of the function $v(x) = B'(x, D)u(x)$ has the form

$$\tilde{v}(\eta) = \frac{1}{(2\pi)^n} \int_{-\infty}^{\infty} \tilde{B}'(\eta - \xi, \xi)\tilde{u}(\xi)\, d\xi, \tag{18.21}$$

where $\tilde{B}'(\eta, \xi) = F_x B'(x, \xi)$. Consequently,

$$A(x, D)B'(x, D)u = \frac{1}{(2\pi)^n} \int_{-\infty}^{\infty} A(x, \eta)\tilde{v}(\eta)e^{-i(x,\eta)}\, d\eta$$
$$= \frac{1}{(2\pi)^{2n}} \int_{-\infty}^{\infty} A(x, \eta)\tilde{B}'(\eta - \xi, \xi)e^{-i(x,\eta)}\tilde{u}(\xi)\, d\xi d\eta. \tag{18.22}$$

Replacing η by $\zeta = \eta - \xi$ in (18.22), we get

$$A(x, D)B'(x, D)u = \frac{1}{(2\pi)^{2n}} \int_{-\infty}^{\infty} A(x, \xi + \zeta)\tilde{B}'(\zeta, \xi)e^{-i(x,\xi+\zeta)}\tilde{u}(\xi)\, d\xi d\zeta. \tag{18.23}$$

Let

$$C_1(x, \xi) = \frac{1}{(2\pi)^n} \int_{-\infty}^{\infty} A(x, \xi + \zeta)\tilde{B}'(\zeta, \xi)e^{-i(x,\zeta)}\, d\zeta. \tag{18.24}$$

Then (18.23) implies that $A(x, D)B'(x, D)$ is the p.o. with symbol $C_1(x, \xi)$:

$$A(x, D)B'(x, D)u = \frac{1}{(2\pi)^n} \int_{-\infty}^{\infty} C_1(x, \xi)e^{-i(x,\xi)}\tilde{u}(\xi)\, d\xi. \tag{18.25}$$

Let us show that $C_1(x, \xi) \in S^0_{\alpha+\beta}$. We have $C_1(x, \xi) = C_2(x, \xi) + C_3(x, \xi)$, where

$$C_2(x, \xi) = \frac{1}{(2\pi)^n} \int_{-\infty}^{\infty} A(\infty, \xi + \zeta)\tilde{B}'(\zeta, \xi)e^{-i(x,\zeta)}\, d\zeta, \tag{18.26}$$

$$C_3(x, \xi) = \frac{1}{(2\pi)^n} \int_{-\infty}^{\infty} A'(x, \xi + \zeta)\tilde{B}'(\zeta, \xi)e^{-i(x,\zeta)}\, d\zeta. \tag{18.27}$$

The Fourier transforms of (18.26) and (18.27) are

$$\tilde{C}_2(\eta, \xi) = A(\infty, \xi + \eta)\tilde{B}'(\eta, \xi), \tag{18.28}$$

$$\tilde{C}_3(\eta, \xi) = \frac{1}{(2\pi)^n} \int_{-\infty}^{\infty} \tilde{A}'(\eta - \zeta, \xi + \zeta)\tilde{B}'(\zeta, \xi)\, d\zeta. \tag{18.29}$$

Since $A(x, \xi) \in S^0_\alpha$ and $B(x, \xi) \in S^0_\beta$, the following estimates hold for arbitrary N:

$$|\tilde{B}'(\eta, \xi)| \leqslant C_N(1 + |\eta|)^{-2N}(1 + |\xi|)^\beta, \tag{18.30}$$

$$|\tilde{A}'(\eta - \zeta, \xi + \zeta)| \leqslant C_N(1 + |\zeta - \eta|)^{-N}(1 + |\xi - \zeta|)^\alpha, \tag{18.30$'$}$$

$$|A(\infty, \xi + \eta)| \leqslant C(1 + |\xi + \eta|)^\alpha. \tag{18.30$''$}$$

On the other hand, by (18.11),

$$(1+|\xi+\eta|)^{\alpha} \leqslant (1+|\xi|)^{\alpha}(1+|\eta|)^{|\alpha|},$$
$$(1+|\eta-\zeta|)^{-N} \leqslant (1+|\eta|)^{-N}(1+|\zeta|)^{N}. \qquad (18.31)$$

Consequently, for arbitrary N

$$|\tilde{C}_2(\eta,\xi)| \leqslant \frac{C_N(1+|\xi+\eta|)^{\alpha}(1+|\xi|)^{\beta}}{(1+|\eta|)^{2N}} \leqslant C_N\frac{(1+|\xi|)^{\alpha+\beta}}{(1+|\eta|)^{2N-|\alpha|}}, \qquad (18.32)$$

so that $C_2(x,\xi) \in S^0_{\alpha+\beta}$, with $C_2(\infty,\xi)=0$. Analogously, for arbitrary N

$$|\tilde{C}_3(\eta,\xi)| \leqslant C_N\int_{-\infty}^{\infty}\frac{(1+|\xi+\zeta|)^{\alpha}(1+|\xi|)^{\beta}}{(1+|\eta-\zeta|)^{N}(1+|\zeta|)^{2N}}\,d\zeta$$
$$\leqslant C_N\int_{-\infty}^{\infty}\frac{(1+|\xi|)^{\alpha+\beta}(1+|\zeta|)^{|\alpha|+N}}{(1+|\eta|)^{N}(1+|\zeta|)^{2N}}\,d\zeta \leqslant C_N\frac{(1+|\xi|)^{\alpha+\beta}}{(1+|\eta|)^{N}}, \qquad (18.32')$$

and hence also $C_3(x,\xi)\in S^0_{\alpha+\beta}$, with $C_3(\infty,\xi)=0$. Theorem 18.2 is proved.

By imposing additional smoothness conditions on the symbol $A(x,\xi)$, one can extract from (18.18) the principal part of the symbol of the operator $C(x,D)=A(x,D)B(x,D)$. As a preliminary, we prove the following lemma.

LEMMA 18.2. *Suppose $A'(x,\xi)\in S^{m+\gamma}_{\alpha}$, where m is a nonnegative integer, $\gamma\in(0,1]$ and $A'(\infty,\xi)=0$. Then the remainder term in the Taylor expansion of $\tilde{A}'(\eta,\xi+\zeta)$ about $\zeta=0$, viz.*

$$\tilde{A}'(\eta,\xi+\zeta)=\sum_{|k|=0}^{m}\frac{1}{k!}\frac{\partial^k\tilde{A}'(\eta,\xi)}{\partial\xi^k}\zeta^k+\tilde{R}'_{m+1}(\eta,\xi,\zeta) \qquad (18.33)$$

satisfies the estimates

$$(1+|\eta|)^{N}|\tilde{R}'_{m+1}(\eta,\xi,\zeta)| \leqslant C_N(1+|\xi|)^{\alpha-m-\gamma}|\zeta|^{m+|\alpha|+1}, \qquad \forall N. \qquad (18.34)$$

PROOF. Suppose first that $|\zeta|<(1+|\xi|)/2$. We make use of the expression in integral form[1]

$$\tilde{R}'_{m+1}(\eta,\xi,\zeta)=\sum_{|k|=m+1}\frac{(m+1)}{k!}\int_0^1\frac{\partial^k\tilde{A}'(\eta,\xi+t\zeta)}{\partial\xi^k}(1-t)^m\,dt\,\zeta^k \qquad (18.35)$$

for the remainder term $R'_{m+1}(\eta,\xi,\zeta)$. From (18.35) and (18.1) we get

$$(1+|\eta|)^{N}|\tilde{R}'_{m+1}(\eta,\xi,\zeta)| \leqslant C_N\int_0^1(1+|\xi+t\zeta|)^{\alpha-m-\gamma}\,dt\,|\zeta|^{m+1}. \qquad (18.36)$$

---

[1]If the functions $\partial^k\tilde{A}'(\eta,\xi)$, $|k|=m+1$, are continuous, the remainder term can be taken in Lagrange's form.

But $(1 + |\xi + t\zeta|)^{\alpha-m-\gamma} \leqslant C(1 + |\xi|)^{\alpha-m-\gamma}$ for $|\zeta| < (1 + |\xi|)/2$, and hence estimates (18.34) are proved for this case.

Suppose now that $|\zeta| \geqslant (1 + |\xi|)/2$. From (18.33) we get

$$(1 + |\eta|)^N |\tilde{R}'_{m+1}(\eta, \xi, \zeta)|$$
$$\leqslant \sum_{|k|=0}^{m} \frac{(1 + |\eta|)^N}{k!} \left| \frac{\partial^k \tilde{A}'(\eta, \xi)}{\partial \xi^k} \right| |\zeta|^{|k|} + (1 + |\eta|)^N |\tilde{A}'(\eta, \xi + \zeta)|. \tag{18.37}$$

Since $(1 + |\xi + \zeta|)^\alpha \leqslant (1 + |\xi|)^\alpha (1 + |\zeta|)^{|\alpha|}$ (see (18.11)), it follows for $|\zeta| \geqslant (1 + |\xi|)/2$ that

$$\begin{aligned}(1 + |\eta|)^N |\tilde{R}'_{m+1}(\eta, \xi, \zeta)| &\leqslant C_N \sum_{|k|=0}^{m} (1 + |\xi|)^{\alpha-|k|} |\zeta|^{|k|} + C_N (1 + |\xi|)^\alpha (1 + |\zeta|)^{|\alpha|} \\ &\leqslant C_N (1 + |\xi|)^{\alpha-m-\gamma} |\zeta|^{|\alpha|+m+\gamma} \\ &\leqslant C_N (1 + |\xi|)^{\alpha-m-\gamma} |\zeta|^{|\alpha|+m+1}.\end{aligned} \tag{18.38}$$

Lemma 18.2 is proved.

The following assertion is proved analogously.

LEMMA 18.2′. *Suppose $A(\infty, \xi) \in S_\alpha^{m+\gamma}$. Then the remainder term*

$$R_{m+1}(\xi, \zeta) = A(\infty, \xi + \zeta) - \sum_{|k|=0}^{m} \frac{1}{k!} \frac{\partial^k A(\infty, \xi)}{\partial \xi^k} \zeta^k \tag{18.39}$$

satisfies the estimate

$$|R_{m+1}(\xi, \zeta)| \leqslant C(1 + |\xi|)^{\alpha-m-\gamma} |\zeta|^{m+1+|\alpha|}. \tag{18.40}$$

THEOREM 18.3. *Suppose $A(x, \xi) \in S_\alpha^{m+\gamma}$, where m is a nonnegative integer and $\gamma \in (0, 1]$, and suppose $B(x, \xi) \in S_\beta^0$. Then the symbol $C(x, \xi)$ of the operator $C(x, D) = A(x, D)B(x, D)$ admits the expansion*

$$C(x, \xi) = \sum_{|k|=0}^{m} \frac{1}{k!} \frac{\partial^k A(x, \xi)}{\partial \xi^k} D_x^k B(x, \xi) + C^{(m+1)}(x, \xi), \tag{18.41}$$

where $C^{(m+1)}(x, \xi) \in S^0_{\alpha+\beta-m-\gamma}$, with $C^{(m+1)}(\infty, \xi) = 0$.

PROOF. Replacing the symbol $A(x, \xi + \zeta)$ in (18.24) by its Taylor expansion about $\zeta = 0$, we obtain the expansion

$$\begin{aligned}C_1(x, \xi) = &\sum_{|k|=0}^{m} \frac{1}{k!} \frac{\partial^k A(x, \xi)}{\partial \xi^k} \frac{1}{(2\pi)^n} \int_{-\infty}^{\infty} \zeta^k \tilde{B}'(\zeta, \xi) e^{-i(x,\zeta)}\, d\zeta \\ &+ C^{(m+1)}(x, \xi),\end{aligned} \tag{18.42}$$

in which

$$C^{(m+1)}(x, \xi) = \frac{1}{(2\pi)^n} \int_{-\infty}^{\infty} (R_{m+1}(\xi, \zeta) + R'_{m+1}(x, \xi, \zeta)) \tilde{B}'(\zeta, \xi) e^{-i(x,\zeta)} \, d\zeta, \tag{18.43}$$

where $R_{m+1}(\xi, \zeta)$ is defined by (18.39) and $R'_{m+1}(x, \xi, \zeta) = F_\eta^{-1} \tilde{R}'_{m+1}(\eta, \xi, \zeta)$, with the function $\tilde{R}'_{m+1}(\eta, \xi, \zeta)$ being the same as in (18.33). The Fourier transform with respect to x of $C^{(m+1)}(x, \xi)$ has the form

$$\tilde{C}^{(m+1)}(\eta, \xi) = R_{m+1}(\xi, \eta) \tilde{B}'(\eta, \xi) + \frac{1}{(2\pi)^n} \int_{-\infty}^{\infty} \tilde{R}'_{m+1}(\eta - \zeta, \xi, \zeta) \tilde{B}'(\zeta, \xi) \, d\zeta. \tag{18.44}$$

From (18.40) and (18.34) we get (cf. estimates (18.32′))

$$(1 + |\eta|)^N |\tilde{C}^{(m+1)}(\eta, \xi)| \leqslant C_N (1 + |\xi|)^{\alpha + \beta - m - \gamma}, \qquad \forall N. \tag{18.45}$$

Consequently, $C^{(m+1)}(x, \xi) \in S^0_{\alpha+\beta-m-\gamma}$ and $C^{(m+1)}(\infty, \xi) = 0$. Since

$$\frac{1}{(2\pi)^n} \int_{-\infty}^{\infty} \zeta^k \tilde{B}'(\zeta, \xi) e^{-i(x,\zeta)} \, d\zeta = D_x^k B'(x, \xi),$$

the expansion (18.41) follows from (18.19) and (18.42).

We note that

$$\frac{\partial^k A(x, \xi)}{\partial \xi^k} D_x^k B(x, \xi) \in S^0_{\alpha+\beta-|k|}.$$

COROLLARY 18.1. *Suppose $A(x, \xi) \in S^{t_1}_\alpha$, $B(x, \xi) \in S^{t_2}_\beta$, $t_1 > 0$, $t_2 > 0$. Then the commutator $T = A(x, D)B(x, D) - B(x, D)A(x, D)$ is a pseudodifferential operator with symbol $T(x, \xi) \in S^0_{\alpha+\beta-t_0}$, where $t_0 = \min(1, t_1, t_2)$.*

In fact, applying Theorem 18.3 for $m = 0$, we get

$$A(x, D)B(x, D) = C_0(x, D) + T_1(x, D),$$
$$B(x, D)A(x, D) = C_0(x, D) + T_2(x, D), \tag{18.46}$$

where $C_0(x, D)$ is the p.o. with symbol $A(x, \xi)B(x, \xi)$, $T_1(x, \xi) \in S^0_{\alpha+\beta-t_{10}}$, $T_2(x, \xi) \in S^0_{\alpha+\beta-t_{20}}$, $t_{10} = \min(1, t_1)$, $t_{20} = \min(1, t_2)$. Consequently

$$T(x, \xi) = T_1(x, \xi) - T_2(x, \xi) \in S^0_{\alpha+\beta-t_0}.$$

4. *Compactness of pseudodifferential operators.*

THEOREM 18.4. *Suppose $T'(x, \xi) \in S^0_{-\delta}$, $\delta > 0$, and $T'(\infty, \xi) = 0$. Then the p.o. $T'(x, D)$ is compact in $H_s(\mathbf{R}^n)$ for any s.*

PROOF. Let $\chi(\xi) \in C_0^\infty(\mathbf{R}^n)$, $\chi(\xi) = 1$ for $|\xi| \leqslant 1$, and let $T'_\varepsilon(x, \xi) = T'(x, \xi)\chi(\varepsilon\xi)$. We will show that the norm of $T'(x, D) - T'_\varepsilon(x, D)$ in $H_s(\mathbf{R}^n)$

tends to zero as $\varepsilon \to 0$. By (18.17),

$$\|T'(x, D) - T'_\varepsilon(x, D)\|_{(s)} \leqslant C_0 \sup_{\xi,\eta} (1 + |\eta|)^{n+1+|s|} |\tilde{T}'(\eta, \xi)||1 - \chi(\varepsilon\xi)|. \tag{18.47}$$

Since $T'(x, \xi) \in S^0_{-\delta}$, we have

$$(1 + |\eta|)^{n+1+|s|} |\tilde{T}'(\eta, \xi)| \leqslant C(1 + |\xi|)^{-\delta}.$$

Consequently, as $\varepsilon \to 0$,

$$\|T'(x, D) - T'_\varepsilon(x, D)\|_{(s)} \leqslant C \sup_\xi (1 + |\xi|)^{-\delta} |1 - \chi(\varepsilon\xi)| \to 0. \tag{18.48}$$

We will now prove that $T'_\varepsilon(x, D)$ is compact in $H_s(\mathbf{R}^n)$. It will then follow from (18.48) that $T'(x, D)$ is also compact. Let $v(x) = T'_\varepsilon(x, D)u$. We have

$$\tilde{v}(\eta) = \int_{-\infty}^{\infty} \tilde{T}'_\varepsilon(\eta - \xi, \xi)\tilde{u}(\xi)\, d\xi.$$

For any N the kernel $\tilde{T}'_\varepsilon(\eta - \xi, \xi)$ of the integral operator $\tilde{T}'_\varepsilon$ defined by this relation satisfies

$$|\tilde{T}'_\varepsilon(\eta - \xi, \xi)| \leqslant C_{\varepsilon,N}(1 + |\xi|)^{-N}(1 + |\eta|)^{-N}.$$

But the integral operator $\tilde{T}'_\varepsilon$ acting in $\tilde{H}_s(\mathbf{R}^n)$ is isomorphic to the integral operator acting in $\tilde{H}_0(\mathbf{R}^n)$ with kernel $(1 + |\eta|)^s \tilde{T}'_\varepsilon(\eta - \xi, \xi)(1 + |\xi|)^{-s}$. Since the latter is therefore a Hilbert-Schmidt operator, the operator $\tilde{T}'_\varepsilon$ is compact.

5. *Elliptic p.o.'s in* $\mathbf{R}^n$. An operator $A(x, D) = \|a_{ij}(x, D)\|^p_{i,j=1}$ that is a matrix of p.o.'s with symbols $a_{ij}(x, \xi) \in S^t_\alpha$, $t > 0$, is said to be *elliptic* if $|\det A(x, \xi)| \geqslant C(1 + |\xi|)^{\alpha p - \delta}$ outside some ball $|x|^2 + |\xi|^2 < N^2$, where $0 \leqslant \delta < \min(t, 1)$. If, in particular, $A_0(x, \xi) = \|a_{ij}(x, \xi)\|^p_{i,j=1} \in O^\infty_{\alpha + i\beta}$ and $\det A_0(x, \xi) \neq 0$ for $\xi \neq 0$, then the operator $A(x, \xi) = A_0(x, \xi)(1 - \chi(\xi))$, is clearly elliptic.

THEOREM 18.5. *An elliptic p.o. $A(x, D)$ with symbol $A(x, \xi) \in S^t_\alpha$, $t > 0$, is a Fredholm operator acting from $H_s(\mathbf{R}^n)$ into $H_{s-\alpha}(\mathbf{R}^n)$ for any s.*

PROOF. Let $R(x, D)$ denote the p.o. with symbol

$$R(x, \xi) = \left(1 - \chi\left(\frac{x}{N}\right)\chi\left(\frac{\xi}{N}\right)\right)A^{-1}(x, \xi).$$

Then $R(x, \xi) \in S^t_{-\alpha+\delta}$, with $R(\infty, \xi) = A^{-1}(\infty, \xi)$. By Theorem 18.3,

$$A(x, D)R(x, D) = C_0(x, D) + T'(x, D), \tag{18.49}$$

where $C_0(x, D)$ is the p.o. with symbol $C_0(x, \xi) = A(x, \xi)R(x, \xi)$, and $T'(x, \xi) \in S^0_{-t_0}$, $t_0 > 0$, with $T'(\infty, \xi) = 0$. Since $C_0(x, \xi) - I \in S^t_{-N}$, for any

N, where I is the identity operator, it follows from (18.49) that

$$A(x, D)R(x, D) = I + T_1'(x, D), \tag{18.50}$$

where $T_1'(x, \xi) \in S^0_{-t_0}$, $t_0 > 0$, and $T_1'(\infty, \xi) = 0$. But then, by Theorem 18.4, the operator $T_1'(x, D)$ is compact in $H_{s-\alpha}(\mathbf{R}^n)$, and hence $R(x, D)$ is a right regularizer of $A(x, D)$ (see §14). Analogously, by Theorem 18.3,

$$R(x, D)A(x, D) = I + T_2'(x, D), \tag{18.51}$$

where $T_2'(x, \xi) \in S^0_{-t_0}$, $T_2'(\infty, \xi) = 0$, and hence $R(x, D)$ is also a left regularizer of $A(x, D)$. Q.E.D.

§19. Change of variables formula for pseudodifferential operators

1. *Generalized pseudodifferential operators.* A pseudodifferential operator $A(x, D)$ can be written in the form

$$A(x, D)u = \frac{1}{(2\pi)^n}\int_{-\infty}^{\infty} A(x, \xi)e^{-i(x-y,\xi)}u(y)\, dyd\xi, \tag{19.1}$$

in which the integral exists as an iterated integral: one integrates first with respect to y, and then with respect to ξ. We will introduce a formally wider class of operators than (19.1), the definition of which involves locally integrable functions $\tilde{\tilde{a}}_1'(\eta, \zeta, \xi)$ in $\mathbf{R}^{3n}$ satisfying the estimates

$$(1 + |\eta|)^{N_1}(1 + |\zeta|)^{N_2}|\tilde{\tilde{a}}_1'(\eta, \zeta, \xi)| \leqslant C_{N_1N_2}(1 + |\xi|)^{\alpha}, \qquad \forall N_1, \quad \forall N_2. \tag{19.2}$$

Let $S^0_\alpha(\mathbf{R}^n \times \mathbf{R}^n)$ denote the class of functions $a(x, y, \xi)$ that are representable in the form

$$a(x, y, \xi) = a(x, \infty, \xi) + a'(x, y, \xi), \tag{19.3}$$

in which $a(x, \infty, \xi) \in S^0_\alpha(\mathbf{R}^n)$ and

$$a'(x, y, \xi) = a'(\infty, y, \xi) + a_1'(x, y, \xi), \tag{19.3'}$$

where $a'(\infty, y, \xi) \in S^0_\alpha(\mathbf{R}^n)$, with $a'(\infty, \infty, \xi) = 0$, and

$$a_1'(x, y, \xi) = F_\eta^{-1}F_\zeta^{-1}\tilde{\tilde{a}}_1'(\eta, \zeta, \xi),$$

with $\tilde{\tilde{a}}_1'(\eta, \zeta, \xi)$ satisfying (19.2). Let $t = m + \gamma$, where m is a nonnegative integer and $\gamma \in (0, 1]$. A function $a(x, y, \xi)$ is said to be of class $S^t_\alpha(\mathbf{R}^n \times \mathbf{R}^n)$ if

$$D_\xi^k a(x, y, \xi) \in S^0_{\alpha-|k|}(\mathbf{R}^n \times \mathbf{R}^n) \qquad \text{for } 0 \leqslant |k| \leqslant m,$$

$$D_\xi^k a(x, y, \xi) \in S^0_{\alpha-m-\gamma}(\mathbf{R}^n \times \mathbf{R}^n) \qquad \text{for } |k| = m + 1.$$

Consider an operator of the form

$$Au = \frac{1}{(2\pi)^n}\int_{-\infty}^{\infty} a(x, y, \xi)u(y)e^{-i(x-y,\xi)}\, dyd\xi, \qquad \forall u \in S(\mathbf{R}^n), \tag{19.4}$$

where $a(x, y, \xi) \in S^t_\alpha(\mathbf{R}^n \times \mathbf{R}^n)$ and, as in (19.1), the integration is carried out first with respect to y and then with respect to ξ. A pseudodifferential operator of form (19.1) or (18.4) is a special case of an operator of form (19.4), when $a(x, y, \xi)$ does not depend on y. An operator of form (19.4) is called a *generalized pseudodifferential operator* (g.p.o.).

In the following lemma we prove that a g.p.o. can be transformed into an ordinary p.o. of form (18.4). In doing so, we will prove that the integral (19.4) converges as an iterated integral.

LEMMA 19.1. *A generalized pseudodifferential operator A with symbol $a(x, y, \xi) \in S^0_\alpha(\mathbf{R}^n \times \mathbf{R}^n)$ can be written in the form of a p.o. of form* (18.4) *with symbol $A(x, \xi) \in S^0_\alpha(\mathbf{R}^n)$, defined by*

$$A(x, \xi) = a(x, \infty, \xi) + \frac{1}{(2\pi)^n}\int_{-\infty}^{\infty} \tilde{a}'(x, \zeta, \zeta + \xi)e^{-i(x,\zeta)}\, d\zeta, \qquad (19.5)$$

where

$$a(x, y, \xi) = a(x, \infty, \xi) + a'(x, y, \xi) \qquad \textit{and} \qquad \tilde{a}'(x, \zeta, \xi) = F_y a'(x, y, \xi).$$

PROOF. By a property of the Fourier transform (cf. (2.4)),

$$\int_{-\infty}^{\infty} a'(x, y, \xi)u(y)e^{i(y,\xi)}\, dy = \frac{1}{(2\pi)^n}\int_{-\infty}^{\infty} \tilde{a}'(x, \xi - \eta, \xi)\tilde{u}(\eta)\, d\eta, \quad (19.6)$$

where (19.3′) implies

$$\tilde{a}'(x, \xi - \eta, \xi) = \tilde{a}'(\infty, \xi - \eta, \xi) + \tilde{a}'_1(x, \xi - \eta, \xi). \qquad (19.7)$$

Since $\tilde{a}'(x, \zeta, \xi)$ decreases with respect to ζ more rapidly than any power, the integral (19.6) rapidly decreases with respect to ξ. Consequently, (19.4) converges as an iterated integral.

From (19.4), (19.3) and (19.6) we have

$$\begin{aligned} Au = {} & \frac{1}{(2\pi)^n}\int_{-\infty}^{\infty} a(x, \infty, \xi)\tilde{u}(\xi)e^{-i(x,\xi)}\, d\xi \\ & + \frac{1}{(2\pi)^{2n}}\int_{-\infty}^{\infty} \tilde{a}'(x, \xi - \eta, \xi)\tilde{u}(\eta)e^{-i(x,\xi)}\, d\xi d\eta. \end{aligned} \qquad (19.8)$$

We transform the second integral here by replacing ξ by $\zeta + \eta$:

$$\begin{aligned} & \frac{1}{(2\pi)^{2n}}\int_{-\infty}^{\infty} \tilde{a}'(x, \xi - \eta, \xi)\tilde{u}(\eta)e^{-i(x,\xi)}\, d\xi d\eta \\ & = \frac{1}{(2\pi)^n}\int_{-\infty}^{\infty} A'(x, \eta)\tilde{u}(\eta)e^{-i(x,\eta)}\, d\eta, \end{aligned} \qquad (19.9)$$

where

$$A'(x, \eta) = \frac{1}{(2\pi)^n} \int_{-\infty}^{\infty} \tilde{a}'(x, \zeta, \zeta + \eta) e^{-i(x,\zeta)} \, d\zeta. \tag{19.10}$$

Thus (19.5) is proved.

To prove that $A(x, \xi) = a(x, \infty, \xi) + A'(x, \xi) \in S_\alpha^0(\mathbf{R}^n)$ it suffices to show that $A'(x, \xi) \in S_\alpha^0(\mathbf{R}^n)$ and $A'(\infty, \xi) = 0$. The Fourier transform with respect to x of (19.10) is

$$\tilde{A}'(\xi, \eta) = \tilde{a}'(\infty, \xi, \xi + \eta) + \frac{1}{(2\pi)^n} \int_{-\infty}^{\infty} \tilde{\tilde{a}}_1'(\xi - \zeta, \zeta, \zeta + \eta) \, d\eta. \tag{19.11}$$

Analogously to (18.32′), inequalities (18.1), (19.2) and (18.31) imply

$$|\tilde{A}'(\xi, \eta)| \leqslant C_N^{(1)} \frac{(1 + |\xi + \eta|)^\alpha}{(1 + |\xi|)^N} + C_N^{(2)} \int_{-\infty}^{\infty} \frac{(1 + |\zeta + \eta|)^\alpha \, d\zeta}{(1 + |\xi - \zeta|)^N (1 + |\zeta|)^{2N}}$$

$$\leqslant C_N^{(1)} \frac{(1 + |\eta|)^\alpha}{(1 + |\xi|)^{N - |\alpha|}} + C_N^{(2)} \int_{-\infty}^{\infty} \frac{(1 + |\eta|)^\alpha (1 + |\zeta|)^{N + |\alpha|}}{(1 + |\xi|)^N (1 + |\zeta|)^{2N}} \, d\zeta \leqslant \frac{C_N^{(3)} (1 + |\eta|)^\alpha}{(1 + |\xi|)^{N - |\alpha|}}. \tag{19.12}$$

Since N is arbitrarily large, it consequently follows from (19.12) that

$$A'(x, \eta) = F_\xi^{-1} \tilde{A}'(\xi, \eta) \in S_\alpha^0(\mathbf{R}^n) \qquad \text{and} \qquad A'(\infty, \eta) = 0.$$

Lemma 19.1 is proved.

If $a(x, y, \xi) \in S_\alpha^{m+\gamma}(\mathbf{R}^n \times \mathbf{R}^n)$, where m is a nonnegative integer and $\gamma \in (0, 1]$, then the principal part of the symbol (19.5) can be extracted analogously to Theorem 18.3.

LEMMA 19.2. *Suppose $a(x, y, \xi) \in S_\alpha^{m+\gamma}(\mathbf{R}^n \times \mathbf{R}^n)$. Then the symbol $A(x, \xi)$ defined by (19.5) admits the expansion*

$$A(x, \xi) = a(x, \infty, \xi) + \sum_{|k|=0}^{m} \frac{1}{k!} D_y^k \frac{\partial^k}{\partial \xi^k} \tilde{a}'(x, y, \xi)\big|_{y=x} + C_{m+1}'(x, \xi), \tag{19.13}$$

where $C_{m+1}'(x, \xi) \in S_{\alpha - m - \gamma}^0$.

PROOF. The Taylor expansion of $\tilde{a}'(x, \zeta, \xi + z)$ about $z = 0$ is

$$\tilde{a}'(x, \zeta, \xi + z) = \sum_{|k|=0}^{m} \frac{1}{k!} \frac{\partial^k \tilde{a}'(x, \zeta, \xi)}{\partial \xi^k} z^k + R_{m+1}'(x, \zeta, \xi, z). \tag{19.14}$$

Let us show that the symbol

$$C_{m+1}'(x, \xi) = \frac{1}{(2\pi)^n} \int_{-\infty}^{\infty} R_{m+1}'(x, \zeta, \xi, \zeta) e^{-i(x,\zeta)} \, d\zeta$$

belongs to $S^0_{\alpha-m-\gamma}(\mathbf{R}^n)$. By (19.7) we have

$$R'_{m+1}(x, \zeta, \xi, z) = R'_{m+1}(\infty, \zeta, \xi, z) + R'_{m+1,1}(x, \zeta, \xi, z),$$

where

$$R'_{m+1}(\infty, \zeta, \xi, z) = \tilde{a}'(\infty, \zeta, z + \xi) - \sum_{|k|=0}^{m} \frac{1}{k!} \frac{\partial^k \tilde{a}'(\infty, \zeta, \xi)}{\partial \xi^k} z^k, \tag{19.15}$$

$$R'_{m+1,1}(x, \zeta, \xi, z) = \tilde{a}'_1(x, \zeta, z + \xi) - \sum_{|k|=0}^{m} \frac{1}{k!} \frac{\partial^k \tilde{a}'_1(x, \zeta, \xi)}{\partial \xi^k} z^k. \tag{19.15'}$$

Consequently, the Fourier transform with respect to x of $C'_{m+1}(x, \xi)$ is

$$\tilde{C}'_{m+1}(\eta, \xi) = \tilde{R}'_{m+1}(\infty, \eta, \xi, \eta) + \frac{1}{(2\pi)^n} \int_{-\infty}^{\infty} \tilde{R}'_{m+1,1}(\eta - \zeta, \zeta, \xi, \zeta)\, d\zeta. \tag{19.16}$$

By Lemma 18.2,

$$|\tilde{R}'_{m+1}(\infty, \eta, \xi, \eta)| \leqslant C_N \frac{(1 + |\xi|)^{\alpha - m - \gamma} |\eta|^{m + |\alpha| + 1}}{(1 + |\eta|)^N} \leqslant \frac{C_N (1 + |\xi|)^{\alpha - m - \gamma}}{(1 + |\eta|)^{N - m - |\alpha| - 1}}. \tag{19.17}$$

Analogously,

$$|\tilde{R}'_{m+1,1}(\eta - \zeta, \zeta, \xi, \zeta)| \leqslant \frac{C_N (1 + |\xi|)^{\alpha - m - \gamma} |\zeta|^{m + 1 + |\alpha|}}{(1 + |\eta - \zeta|)^N (1 + |\zeta|)^{2N}} \leqslant \frac{C_N (1 + |\xi|)^{\alpha - m - \gamma}}{(1 + |\eta|)^N (1 + |\zeta|)^{N - m - |\alpha| - 1}}. \tag{19.18}$$

It follows from (19.17) and (19.18) that $C'_{m+1}(x, \xi) \in S^0_{\alpha - m - \gamma}$ and $C'_{m+1}(\infty, \xi) = 0$.

Substituting (19.14) for $z = \zeta$ in (19.5) and making use of the relation

$$\frac{1}{(2\pi)^n} \int_{-\infty}^{\infty} \frac{\partial^k \tilde{a}'(x, \zeta, \xi)}{\partial \xi^k} \zeta^k e^{-i(x,\zeta)}\, d\zeta = D_y^k \frac{\partial^k a'(x, y, \xi)}{\partial \xi^k}\bigg|_{y=x}$$

(see (2.2)), we obtain (19.13). Lemma 19.2 is proved.

2. *Adjoint operator*. Suppose A is a p.o. with symbol $A(x, \xi) \in S^0_\alpha(\mathbf{R}^n)$. We denote by A^* the *adjoint* of A, i.e. the operator satisfying

$$(Au, v) = (u, A^*v), \qquad \forall u \in S(\mathbf{R}^n), v \in S(\mathbf{R}^n), \tag{19.19}$$

where $(u, v) = \int_{-\infty}^{\infty} u(x)\overline{v(x)}\, dx$ is the scalar product in $H_0(\mathbf{R}^n)$. By (19.1),

$$(Au, v) = \int_{-\infty}^{\infty} \frac{1}{(2\pi)^n}\left(\int_{-\infty}^{\infty} A(x, \xi) e^{-i(x-y,\xi)} u(y)\, dy d\xi\right) \overline{v(x)}\, dx$$

$$= \int_{-\infty}^{\infty} u(y)\left(\frac{1}{(2\pi)^n}\int_{-\infty}^{\infty} A(x, \xi) e^{-i(x-y,\xi)}\, \overline{v(x)}\, dx d\xi\right) dy. \quad (19.20)$$

Thus, changing the notation in (19.20),

$$A^* v = \frac{1}{(2\pi)^n}\int_{-\infty}^{\infty} \overline{A(y, \xi)}\, e^{-i(x-y,\xi)} v(y)\, dy d\xi, \qquad \forall v \in S(\mathbf{R}^n). \quad (19.21)$$

Consequently, A^* has the form (19.4). By Lemma 19.1, we have

THEOREM 19.1. *Suppose $A(x, D)$ is a p.o. with symbol $A(x, \xi) = A(\infty, \xi) + A'(x, \xi) \in S_\alpha^0$. Then the adjoint operator A^* is also a p.o. with symbol $A^*(x, \xi) \in S_\alpha^0(\mathbf{R}^n)$, defined by*

$$A^*(x, \xi) = \overline{A(\infty, \xi)} + \frac{1}{(2\pi)^n}\int_{-\infty}^{\infty} \overline{\tilde{A}'(-\zeta, \xi + \zeta)}\, e^{-i(x,\zeta)}\, d\zeta, \quad (19.22)$$

where $\tilde{A}'(\zeta, \xi) = F_y A'(y, \xi)$.

In fact, (19.22) follows from (19.5) if one puts

$$a(x, y, \xi) = \overline{A(y, \xi)} = \overline{A(\infty, \xi)} + \overline{A'(y, \xi)}$$

and takes into account the fact that $F_y \overline{A'(y, \xi)} = \overline{\tilde{A}'(-\zeta, \xi)}$.

THEOREM 19.2. *Suppose A is a p.o. with symbol $A(x, \xi) \in S_\alpha^{m+\gamma}$, where m is a nonnegative integer and $\gamma \in (0, 1]$. Then the symbol $A^*(x, \xi)$ of the p.o. A^* admits the expansion*

$$A^*(x, \xi) = \overline{A(\infty, \xi)} + \sum_{|k|=0}^{m} \frac{1}{k!} D_x^k \frac{\partial^k}{\partial \xi^k} \overline{A'(x, \xi)} + A'_{m+1}(x, \xi), \quad (19.23)$$

where $A'_{m+1}(x, \xi) \in S^0_{\alpha-m-\gamma}$.

Theorem 19.2 follows from Lemma 19.2 if one puts

$$a(x, y, \xi) = \overline{A(y, \xi)} = \overline{A(\infty, \xi)} + \overline{A'(y, \xi)}\,.$$

3. *Estimate of a p.o. to within operators of smaller order.*

THEOREM 19.3. *Let $A(x, D)$ be a p.o. with symbol $A(x, \xi) \in S_\alpha^t$, $t > 0$, such that $A(x, \xi) = A(\infty, \xi)$ for large $|x|$ and let*

$$M_0 = \overline{\lim_{\xi \to \infty}} \sup_{x \in \mathbf{R}^n} \frac{|A(x, \xi)|}{(1 + |\xi|)^\alpha}.$$

Then for any $\varepsilon > 0$

$$\|Au\|_{s-\alpha}^2 \leqslant (M_0 + \varepsilon)^2 \|u\|_s^2 + C_{\varepsilon,s}\|u\|_{s-1/2}^2. \tag{19.24}$$

PROOF. By definition of the superior limit, for any $\varepsilon > 0$ there exists an N such that

$$\sup_{x\in\mathbf{R}^n} \frac{|A(x, \xi)|}{(1 + |\xi|)^\alpha} \leqslant M_0 + \frac{\varepsilon}{2}$$

for $|\xi| \geqslant N$. Suppose $\chi(\xi) \in C_0^\infty(\mathbf{R}^n)$, $\chi(\xi) = 1$ for $|\xi| \leqslant 1$ and $0 \leqslant \chi(\xi) \leqslant 1$ for the remaining values of ξ. Let

$$A_1(x, \xi) = A(x, \xi)\left(1 - \chi\left(\frac{\xi}{N}\right)\right) \quad \text{and} \quad A_2(x, \xi) = A(x, \xi)\chi\left(\frac{\xi}{N}\right).$$

Since $A_2(x, D)$ is of order $-\infty$, the estimate

$$\|A_2(x, D)u\|_{s-\alpha}^2 \leqslant C_s\|u\|_{s-1/2}^2 \tag{19.25}$$

is obviously valid. Consequently, it suffices to prove (19.24) for $A_1(x, D)$.

We note that

$$|A_1(x, \xi)| \leqslant \left(M_0 + \frac{\varepsilon}{2}\right)(1 + |\xi|)^\alpha \tag{19.26}$$

for all $x, \xi \in \mathbf{R}^n$. On the other hand, we have

$$\|A_1 u\|_{s-\alpha}^2 = (\Lambda_0^{s-\alpha}A_1 u, \Lambda_0^{s-\alpha}A_1 u) = \left(A_1^*\Lambda_0^{2(s-\alpha)}A_1 u, u\right), \tag{19.27}$$

where Λ_0 is the p.o. with symbol $1 + |\xi| \in S_1^1$. Let

$$B(x, \xi) = \left((M_0 + \varepsilon)^2(1 + |\xi|)^{2s} - |A_1(x, \xi)|^2(1 + |\xi|)^{2(s-\alpha)}\right)^{1/2}. \tag{19.28}$$

By virtue of (19.26) and the fact that $B(x, \xi) = B(\infty, \xi)$ for large $|x|$, we have $B(x, \xi) \in S_s^{t_0}$, where $t_0 = \min(t, 1)$. It follows from Theorems 18.2, 18.3, 19.1 and 19.2 that

$$(M_0 + \varepsilon)^2\Lambda_0^{2s}u - A_1^*\Lambda_0^{2(s-\alpha)}A_1 u = B^*Bu + Tu, \qquad \forall u \in S(\mathbf{R}^n), \tag{19.29}$$

where B^* is the adjoint of $B(x, D)$ and $T(x, D)$ is a p.o. with symbol $T(x, \xi) \in S_{2s-t_0}^0$. Consequently, the scalar product of (19.29) and $u(x)$ is

$$\|A_1 u\|_{s-\alpha}^2 = (M_0 + \varepsilon)^2\|u\|_s^2 - \|Bu\|_0^2 + (Tu, u). \tag{19.30}$$

Since $\|Bu\|_0^2 \geqslant 0$ and $|(Tu, u)| \leqslant \|Tu\|_{-s+t_0/2}\|u\|_{s-t_0/2} \leqslant C\|u\|_{s-t_0/2}^2$, it follows from (19.30) that

$$\|A_1 u\|_{s-\alpha}^2 \leqslant (M_0 + \varepsilon)^2\|u\|_s^2 + C\|u\|_{s-t_0/2}^2. \tag{19.31}$$

Thus, if $t_0 = 1$, (19.24) is proved. If, on the other hand, $t_0 < 1$, (19.24) is proved by making use of the following inequality for $s_1 = s - t_0/2$, $s_2 = s$ and $s_3 = s - 1/2$:

$$\|u\|_{s_1}^2 < \varepsilon\|u\|_{s_2}^2 + C_\varepsilon\|u\|_{s_3}^2, \tag{19.32}$$

which is obviously valid for any ε when $s_3 < s_1 < s_2$.

In Chapter VI we will frequently use the following theorem.

THEOREM 19.4. *Suppose $A(x, D)$ is a p.o. with symbol $A(x, \xi) \in \hat{O}^\infty_{\alpha+i\beta}$. Then for an arbitrary point $x_0 \in \mathbf{R}^n$ and arbitrary $\varepsilon > 0$ there exists a neighborhood $U_0 \in \mathbf{R}^n$ such that for any $\varphi(x) \in C_0^\infty(U_0)$ the operator $\varphi(x)A(x, D)$ admits a decomposition of the form*

$$\varphi(x)A(x, D) = \varphi(x)(A(x_0, D) + K(x, D) + T(x, D)), \tag{19.33}$$

in which $A(x_0, D)$ is the p.o. with symbol $A(x_0, \xi)$ "frozen" at the point x_0, $K(x, D)$ is a p.o. with norm at most ε:

$$\|K(x, D)u\|_{s-\alpha} \leqslant \varepsilon\|u\|_s, \tag{19.34}$$

and $T(x, D)$ is a p.o. of order at most $\alpha - 1$.

PROOF. We choose the neighborhood U_0 so that

$$\max_{\xi\in\mathbf{R}^n} |A(x, \xi) - A(x_0, \xi)|(1 + |\xi|)^{-\alpha} \leqslant \frac{\varepsilon}{4} \qquad \text{for } x \in \overline{U}_0. \tag{19.35}$$

Let $\varphi(x) \in C_0^\infty(U_0)$, let $\psi(x)$ denote a function belonging to $C_0^\infty(U_0)$ such that $\varphi(x)\psi(x) \equiv \varphi(x)$, $|\psi(x)| \leqslant 1$, and let $B(x, \xi) = \psi(x)(A(x, \xi) - A(x_0, \xi))$. Then

$$\varphi(x)A(x, \xi) = \varphi(x)A(x_0, \xi) + \varphi(x)B(x, \xi).$$

By (19.35) and Theorem 19.3,

$$\|B(x, D)u\|_{s-\alpha} \leqslant \frac{\varepsilon}{2}\|u\|_s + C_\varepsilon\|u\|_{s-1/2}. \tag{19.36}$$

Let

$$K(x, \xi) = B(x, \xi)\left(1 - \chi\left(\frac{\xi}{N}\right)\right) \qquad \text{and} \qquad T(x, \xi) = B(x, \xi)\chi\left(\frac{\xi}{N}\right),$$

where $\chi(\xi/N)$ is the same function as in the proof of Theorem 19.3. Then $T(x, D)$ is of order $-\infty$ and, by (19.36),

$$\|K(x, D)u\|_{s-\alpha} \leqslant \frac{\varepsilon}{2}\|(1 - \chi(D/N))u\|_s + C_\varepsilon\|(1 - \chi(D/N))u\|_{s-1/2}. \tag{19.37}$$

Since $0 \leqslant 1 - \chi(\frac{\varepsilon}{N}) \leqslant 1$, we have

$$\left\| \left(1 - \chi\left(\frac{D}{N}\right)\right)u \right\|_s \leqslant \|u\|_s.$$

Further, by definition of the norm in $H_{s-1/2}(\mathbf{R}^n)$, we have

$$\left\| \left(1 - \chi\left(\frac{D}{N}\right)\right)u \right\|_{s-1/2}^2 = \int_{-\infty}^{\infty} \frac{(1 - \chi(\xi/N))^2 |\tilde{u}(\xi)|^2}{1 + |\xi|} (1 + |\xi|)^{2s}\, d\xi \leqslant \frac{1}{1+N} \|u\|_s^2,$$

since

$$\frac{(1 - \chi(\xi/N))^2}{1 + |\xi|} \leqslant \frac{1}{1+N}.$$

Consequently,

$$\|K(x, D)u\|_{s-\alpha} \leqslant \varepsilon \|u\|_s$$

if N is sufficiently large. Theorem 19.4 is proved.

4. *Pseudolocal property*. In this and the following subsections we will confine ourselves to p.o.'s with a symbol of class $S_\alpha^\infty(\mathbf{R}^n)$.

Let A denote the following g.p.o. with symbol $a(x, y, \xi) \in S_\alpha^\infty(\mathbf{R}^n \times \mathbf{R}^n)$:

$$Au = \frac{1}{(2\pi)^n} \int_{-\infty}^{\infty} a(x, y, \xi) e^{-i(x-y,\xi)} u(y)\, dy d\xi, \tag{19.38}$$

and let A_ε denote the g.p.o.

$$A_\varepsilon u = \frac{1}{(2\pi)^n} \int_{-\infty}^{\infty} a(x, y, \xi) \chi(\varepsilon\xi) e^{-i(x-y,\xi)} u(y)\, dy d\xi, \tag{19.39}$$

where $\chi(\xi)$ is the same function as in subsection 3. Clearly, A_ε is an integral operator with an infinitely differentiable kernel

$$E_\varepsilon(x, y) = \frac{1}{(2\pi)^n} \int_{-\infty}^{\infty} a(x, y, \xi) \chi(\varepsilon\xi) e^{-i(x-y,\xi)}\, d\xi. \tag{19.40}$$

We note that $A_\varepsilon u \to Au$ as $\varepsilon \to 0$ for each $x \in \mathbf{R}^n$, since (19.39) converges as an iterated integral.

By the *kernel* of the operator A is meant the generalized function $E_0 \in S'(\mathbf{R}^n \times \mathbf{R}^n)$ such that $(E_0, v(x)\overline{u(y)}) = (Au, v)$ for all $u, v \in S(\mathbf{R}^n)$. It follows from (19.38) that E_0 is the generalized function that takes an arbitrary function $w(x, y) \in S(\mathbf{R}^n \times \mathbf{R}^n)$ into the number

$$(E_0, w) = \frac{1}{(2\pi)^n} \int_{-\infty}^{\infty} a(x, y, \xi) e^{-i(x-y,\xi)} \overline{w(x, y)}\, dx dy d\xi. \tag{19.41}$$

Like (19.38), the integral (19.41) exists if one integrates first with respect to x and y and then with respect to ξ.

Clearly, $(E_\varepsilon(x, y), w(x, y)) \to (E_0, w)$ as $\varepsilon \to 0$ for any $w(x, y) \in S(\mathbf{R}^n \times \mathbf{R}^n)$; that is, $E_\varepsilon \to E_0$ in $S'(\mathbf{R}^n \times \mathbf{R}^n)$ (see §1). Let us show that E_0 is an infinitely differentiable function off the diagonal $x = y$ in $\mathbf{R}^n \times \mathbf{R}^n$.

THEOREM 19.5. *For an arbitrary $\delta > 0$ the function*

$$E_0^{(\delta)} = \left(1 - \chi\left(\frac{x - y}{\delta}\right)\right) E_0$$

belongs to $C^\infty(\mathbf{R}^n \times \mathbf{R}^n)$.

PROOF. Let

$$E_\varepsilon^{(\delta)} = \left(1 - \chi\left(\frac{x - y}{\delta}\right)\right) E_\varepsilon(x, y).$$

Since

$$(x_k - y_k) e^{-i(x-y,\xi)} = i \frac{\partial}{\partial \xi_k} e^{-i(x-y,\xi)},$$

we have

$$E_\varepsilon^{(\delta)}(x, y) = \frac{1}{(2\pi)^n} \int_{-\infty}^{\infty} \left(1 - \chi\left(\frac{x - y}{\delta}\right)\right) \chi(\varepsilon\xi) \frac{a(x, y, \xi)}{|x - y|^{2N}} \Delta_\xi^N e^{-i(x-y,\xi)}\, d\xi, \tag{19.42}$$

where $\Delta_\xi = -\Sigma_{k=1}^n \partial^2/\partial\xi_k^2$. We note that

$$\frac{(1 - \chi((x - y)/\delta))}{|x - y|^{2N}} \in C^\infty(\mathbf{R}^n \times \mathbf{R}^n),$$

since $1 - \chi(\frac{x - y}{\delta}) = 0$ for $|x - y| \leqslant \delta$. Integrating by parts in (19.42), we get

$$E_\varepsilon^{(\delta)}(x, y) = \int_{-\infty}^{\infty} \Delta_\xi^N(\chi(\varepsilon\xi) b_N(x, y, \xi)) e^{-i(x-y,\xi)}\, d\xi, \qquad \forall N, \tag{19.43}$$

where

$$b_N(x, y, \xi) = \frac{1}{(2\pi)^n} \frac{(1 - \chi((x - y)/\delta))}{|x - y|^{2N}} a(x, y, \xi).$$

Let us show that $E_\varepsilon^{(\delta)}(x, y)$ converges to $E_0^{(\delta)}$ uniformly in x and y as $\varepsilon \to 0$, and that

$$E_0^{(\delta)} = \int_{-\infty}^{\infty} \Delta_\xi^N b_N(x, y, \xi) e^{-i(x-y,\xi)}\, d\xi. \tag{19.44}$$

The function $\partial^k \chi(\varepsilon\xi)/\partial\xi^k$ differs from zero for $1/\varepsilon \leqslant |\xi| \leqslant C_0/\varepsilon$ if $|k| > 0$. Since $a(x, y, \xi) \in S_\alpha^\infty(\mathbf{R}^n \times \mathbf{R}^n)$, we have

$$\left|\Delta_\xi^N(\chi(\varepsilon\xi) b_N(x, y, \xi)) - \chi \Delta_\xi^N b_N(x, y, \xi)\right| \leqslant C \sum_{k=1}^{2N} \varepsilon^k (1 + |\xi|)^{-n-1+k}$$

for $N > \dfrac{\alpha + n + 1}{2}$. Consequently,

$$\left|\int_{-\infty}^{\infty}\left(\Delta_\xi^N(\chi(\varepsilon\xi)b_N(x, y, \xi)) - \chi\Delta_\xi^N b_N(x, y, \xi)\right)e^{-i(x-y,\xi)}\,d\xi\right|$$
$$\leqslant C\sum_{k=1}^{2N}\varepsilon^k\int_{1/\varepsilon}^{C_0/\varepsilon}(1 + |\xi|)^{-n-1+k}|\xi|^{n-1}d|\xi| \to 0 \quad \text{as } \varepsilon \to 0. \tag{19.45}$$

Thus

$$E_\varepsilon^{(\delta)}(x, y) \to \int_{-\infty}^{\infty}\Delta_\xi^N b_N(x, y, \xi)e^{-i(x-y,\xi)}\,d\xi$$

uniformly in x and y. Since, on the other hand, $E_\varepsilon^{(\delta)} \to E_0^{(\delta)}$ in $S'(\mathbf{R}^n \times \mathbf{R}^n)$, equality (19.44) is valid.

Let M be an arbitrary positive integer, and take $N > (\alpha + n + 1 + M)/2$. Since $a(x, y, \xi) \in S_\alpha^\infty(\mathbf{R}^n \times \mathbf{R}^n)$, we have

$$\left|D_x^p D_y^q \Delta_\xi^N b_N(x, y, \xi)\right| \leqslant C_{pqN}(1 + |\xi|)^{-n-1-M}, \qquad \forall p,\ \forall q.$$

Consequently, the integral (19.44) is M times continuously differentiable with respect to x and y. Since M is arbitrary, it follows that $E_0^{(\delta)} \in C^\infty(\mathbf{R}^n \times \mathbf{R}^n)$.

COROLLARY 19.1 (pseudolocal property). *Suppose $\varphi(x) \in C_0^\infty(\mathbf{R}^n)$, $\psi(x) \in C_0^\infty(\mathbf{R}^n)$, with the supports of $\varphi(x)$ and $\psi(x)$ not intersecting. Suppose A is a g.p.o. with symbol $a(x, y, \xi) \in S_\alpha^\infty(\mathbf{R}^n \times \mathbf{R}^n)$. Then $\psi A\varphi$ is an integral operator with an infinitely differentiable kernel.*

In fact, since $\operatorname{supp}\varphi \cap \operatorname{supp}\psi = \varnothing$, there exists a $\delta > 0$ such that the symbol $\psi(x)a(x, y, \xi)\varphi(y) = 0$ for $|x - y| \leqslant \delta$. Consequently, by Theorem 19.5, $\psi A\varphi$ is an integral operator with kernel of class $C^\infty(\mathbf{R}^n \times \mathbf{R}^n)$.

REMARK 19.1. A symbol

$$\chi\left(\frac{x-y}{\delta}\right)a(x, y, \xi), \quad \text{where } a(x, y, \xi) \in S_\alpha^\infty(\mathbf{R}^n \times \mathbf{R}^n),$$

is generally not of class $S_\alpha^\infty(\mathbf{R}^n \times \mathbf{R}^n)$. In fact, $a(x, y, \xi)$ has the representation

$$a(x, y, \xi) = a_0(\xi) + a_1(x, y, \xi),$$

where $a_1(x, y, \xi) \in S_0^\infty(\mathbf{R}^n \times \mathbf{R}^n)$ and $a_1(x, y, \xi) \to 0$ as $|x| + |y| \to \infty$. It is easily verified that

$$\chi\left(\frac{x-y}{\delta}\right)a_1(x, y, \xi) \in S_\alpha^\infty(\mathbf{R}^n \times \mathbf{R}^n),$$

while

$$\chi\left(\frac{x-y}{\delta}\right)a_0(\xi) \notin S_\alpha^\infty(\mathbf{R}^n \times \mathbf{R}^n).$$

Let us show that Lemmas 19.1 and 19.2 remain valid for a g.p.o. with symbol of the form $\chi((x-y)/\delta)a(x,y,\xi)$. In view of the above representation, it suffices to consider a g.p.o. with symbol of the form $\chi((x-y)/\delta)a_0(\xi)$, where $a_0(\xi) \in S_\alpha^\infty$.

Repeating the proof of Lemma 19.1, we get that the g.p.o. with symbol $\chi((x-y)/\delta)a_0(\xi)$ can be transformed into the p.o. of form (18.4) with symbol $A_0(\eta)$ defined by

$$A_0(\eta) = \frac{1}{(2\pi)^n}\int_{-\infty}^{\infty} \delta^n \tilde{\chi}(-\delta\zeta)a_0(\zeta+\eta)\,d\zeta, \tag{19.46}$$

where $\tilde{\chi}(\eta) = F_y\chi(y)$. Since $\tilde{\chi}(\delta\zeta) \in S(\mathbf{R}^n)$, we have $A_0(\eta) \in S_\alpha^\infty(\mathbf{R}^n)$. We note that (19.46) coincides with (19.10) if in (19.10) one puts

$$a'(x,y,\xi) = \chi\left(\frac{x-y}{\delta}\right)a_0(\xi).$$

In regard to Lemma 19.2, since $\chi(x) = 1$ for $|x| \leqslant 1$, we have $\chi(0) = 1$ and $\partial^k\chi(0)/\partial x^k = 0$ for $|k| > 0$. Consequently,

$$\frac{1}{(2\pi)^n}\int_{-\infty}^{\infty} \delta^n\tilde{\chi}(-\delta\zeta)\,d\zeta = \chi(0) = 1,$$

$$\frac{1}{(2\pi)^n}\int_{-\infty}^{\infty} \delta^n\zeta^k\tilde{\chi}(-\delta\zeta)\,d\zeta = 0 \quad \text{for } |k| > 0.$$

Thus, expanding $a_0(\zeta+\eta)$ about $\zeta = 0$ by Taylor's formula and using Lemma 18.2, we get

$$A_0(\eta) = a_0(\eta) + C_{N+1}(\eta), \qquad \forall N, \tag{19.47}$$

where $C_{N+1}(\eta) \in S_{\alpha-N-1}^\infty(\mathbf{R}^n)$. It follows from (19.47) that (19.13) remains valid for any m, with the remainder term $C'_{m+1}(x,\xi)$ belonging to $S_{\alpha-m-1}^\infty$, for a symbol of form

$$\chi\left(\frac{x-y}{\delta}\right)a(x,y,\xi).$$

5. *Change of variables formula*. Let A be a p.o. with symbol $A(x,\xi) \in S_\alpha^\infty(\mathbf{R}^n)$, let $x = s(\check{x})$ denote an infinitely differentiable bijection from $\mathbf{R}^n$ onto itself that is equal to the identity map for $|x| \geqslant N$, let $\check{u}(\check{x}) = u(s(\check{x}))$ and let B denote the operator A written in the new coordinates, i.e. $(\bar{B}\check{u})(\check{x}) = (Au)(s(\check{x}))$. We will show that B can be represented in the form of a sum of a p.o. and an integral operator with an infinitely differentiable kernel.

We have

$$Au = A_0u + T_\infty u,$$

where

$$A_0u = \frac{1}{(2\pi)^n}\int_{-\infty}^{\infty} A(x,\xi)\left(\frac{x-y}{\delta}\right)e^{-i(x-y,\xi)}u(y)\,dyd\xi, \tag{19.48}$$

$$T_\infty u = \frac{1}{(2\pi)^n}\int_{-\infty}^{\infty} A(x,\xi)\left(1-\chi\left(\frac{x-y}{\delta}\right)\right)e^{-i(x-y,\xi)}u(y)\,dyd\xi, \tag{19.49}$$

with $\chi(x) \in C_0^\infty(\mathbf{R}^n)$ being the same function as in subsection 3. The number δ in (19.48) and (19.49) is assumed to be sufficiently small and will be chosen below. By Theorem 19.5, T_∞ is an integral operator with kernel of class $C^\infty(\mathbf{R}^n \times \mathbf{R}^n)$.

We make the change of variables $x = s(\check{x}), y = s(\check{y})$ in (19.48) and (19.49), and let $\check{u}(\check{y}) = u(s(\check{y}))$, $(B_0\check{u})(\check{x}) = (A_0u)(s(\check{x}))$ and $(\check{T}_\infty\check{u})(\check{x}) = (T_\infty u)(s(\check{x}))$. Clearly, $\check{T}_\infty$ is, as before, an integral operator with an infinitely differentiable kernel, while B_0 takes the form

$$B_0\check{u} = \frac{1}{(2\pi)^n}\int_{-\infty}^{\infty} A(s(\check{x}), \xi)\chi\left(\frac{s(\check{x}) - s(\check{y})}{\delta}\right)$$

$$\times e^{-i(s(\check{x})-s(\check{y}),\xi)}\check{u}(\check{y})\left|\frac{Ds(\check{y})}{D\check{y}}\right| d\check{y}\, d\xi, \tag{19.50}$$

where $Ds(\check{y})/D\check{y}$ is the Jacobian matrix of the transformation $y = s(\check{y})$ and $|Ds(\check{y})/D\check{y}|$ is its determinant. It is assumed that

$$\left|\frac{Ds(\check{y})}{D\check{y}}\right| \neq 0 \quad \text{for any } \check{y} \in \mathbf{R}^n. \tag{19.51}$$

By Taylor's theorem,

$$s(\check{x}) - s(\check{y}) = \int_0^1 \frac{d}{dt} s(\check{y} + t(\check{x} - \check{y}))\, dt = H(\check{x}, \check{y})(\check{x} - \check{y}), \tag{19.52}$$

where $H(\check{x}, \check{y})$ is a matrix of class $C^\infty(\mathbf{R}^n \times \mathbf{R}^n)$. We note that $H(\check{y}, \check{y}) = Ds(\check{y})/D\check{y}$. Let $|H(\check{x}, \check{y})|$ denote the determinant of $H(\check{x}, \check{y})$. Since $|Ds(\check{y})/D\check{y}| \neq 0$, we have $|H(\check{x}, \check{y})| \neq 0$ if $|\check{x} - \check{y}|$ is sufficiently small. Suppose the number δ in (19.50) has been chosen so small that

$$|H(\check{x}, \check{y})| \neq 0 \quad \text{for } |s(\check{x}) - s(\check{y})| \leqslant \delta. \tag{19.53}$$

We have

$$(s(\check{x}) - s(\check{y}), \xi) = (H(\check{x}, \check{y})(\check{x} - \check{y}), \xi) = (\check{x} - \check{y}, H^*(\check{x}, \check{y}), \xi), \tag{19.54}$$

where $H^*(\check{x}, \check{y})$ is the adjoint of the matrix $H(\check{x}, \check{y})$. Substituting (19.54) in (19.50) and making the change of variables

$$\eta = H^*(\check{x}, \check{y})\xi, \tag{19.55}$$

we get

$$B_0\check{u} = \frac{1}{(2\pi)^n}\int_{-\infty}^{\infty} A(s(\check{x}), (H^*(\check{x}, \check{y}))^{-1}\eta)\chi\left(\frac{s(\check{x}) - s(\check{y})}{\delta}\right)$$

$$\times e^{-i(\check{x}-\check{y},\eta)}\left|\frac{Ds(\check{y})}{D\check{y}}\right||H(\check{x}, \check{y})|^{-1}\check{u}(\check{y})\, d\check{y}\, d\eta. \tag{19.56}$$

To justify this change of variables it is necessary, as in the preceding subsection, to introduce the factor $\chi(\varepsilon\xi)$ in the integrand of (19.50), make the change of variables (19.55) and then pass to the limit for $\varepsilon \to 0$. The limits for $\varepsilon \to 0$ exist since both of the integrals (19.50) and (19.55) converge as iterated integrals.

From (19.56) we see that B_0 is the g.p.o. with symbol

$$b(\check{x}, \check{y}, \eta) = A\big(s(\check{x}), (H^*(\check{x}, \check{y}))^{-1}\eta\big)\chi\left(\frac{s(\check{x}) - s(\check{y})}{\delta}\right) \times \left|\frac{Ds(\check{y})}{D\check{y}}\right| |H(\check{x}, \check{y})|^{-1}. \tag{19.57}$$

By Lemma 19.1 and Remark 19.1, B_0 can be transformed into the p.o. of form (18.4) with symbol

$$B_0(\check{x}, \eta) = \frac{1}{(2\pi)^n}\int_{-\infty}^{\infty} \tilde{b}(\check{x}, \zeta, \zeta + \eta)e^{-i(\check{x},\zeta)}\, d\zeta, \tag{19.58}$$

where

$$\tilde{b}(\check{x}, \zeta, \eta) = F_{\check{y}}b(\check{x}, \check{y}, \eta) = \int_{-\infty}^{\infty} A\big(s(\check{x}), (H^*)^{-1}\eta\big)\chi\left(\frac{s(\check{x}) - s(\check{y})}{\delta}\right) \times \left|\frac{Ds(\check{y})}{D\check{y}}\right| |H(\check{x}, \check{y})|^{-1}e^{i(\check{y},\zeta)}\, d\check{y}. \tag{19.59}$$

It follows from Lemma 19.2 and Remark 19.1 that the expansion

$$B_0(\check{x}, \eta) = b(\check{x}, \check{x}, \eta) + \sum_{k=1}^{N} \frac{1}{k!} D_y^k \frac{\partial^k b(\check{x}, \check{y}, \eta)}{\partial \eta^k}\bigg|_{\check{y}=\check{x}} + C_{N+1}(\check{x}, \eta), \tag{19.60}$$

where $C_{N+1}(\check{x}, \eta) \in S^{\infty}_{\alpha-N-1}(\mathbf{R}^n)$, holds for any N. Since $H(\check{x}, \check{x}) = Ds(\check{x})/D\check{y}$, when $\check{x} = \check{y}$, the principal part of $B_0(\check{x}, \eta)$ has the form

$$b(\check{x}, \check{x}, \eta) = A\left(s(\check{x}), \left(\left(\frac{Ds(\check{x})}{D\check{x}}\right)^*\right)^{-1}\eta\right). \tag{19.61}$$

Thus B has the representation $B = B_0 + \check{T}_\infty$, where $\check{T}_\infty$ is an integral operator with an infinitely differentiable kernel:

$$\check{T}_\infty \check{u} = \int \check{T}_\infty(\check{x}, \check{y})\check{u}(\check{y})\, d\check{y}.$$

Let us now show that $\check{T}_\infty$ is a p.o. with symbol of class $S^{\infty}_{-\infty}$. Since $\check{x} = s(\check{x})$ for $|\check{x}| \geqslant N$, it follows by virtue of (19.44) that the kernel $\check{T}_\infty(\check{x}, \check{y})$ has the representation

$$\check{T}_\infty(\check{x}, \check{y}) = t_0(\check{x}, \check{x} - \check{y}) + t_1(\check{x}, \check{y}),$$

in which $t_1(\check{x}, \check{y})$ satisfies

$$|D_{\check{x}}^{p} D_{\check{y}}^{k} t_1(\check{x}, \check{y})| \leqslant C_{pkN}(1 + |\check{x}| + |\check{y}|)^{-N}$$

for any N, p and k, while $t_0(\check{x}, z)$ satisfies the estimates analogous to those satisfied by a function of class $S_{-\infty}^{\infty}$. Let $\tilde{T}_\infty(\check{x}, \xi)$ denote the Fourier transform with respect to $\check{y}$ of the complex conjugate of $\check{T}_\infty(\check{x}, \check{y})$:

$$\tilde{T}_\infty(\check{x}, \xi) = F_{\check{y}}\, \overline{t_0(\check{x}, \check{x} - \check{y})} + F_{\check{y}}\, \overline{t_1(\check{x}, \check{y})} = T_0(\check{x}, \xi) e^{i(\check{x}, \xi)} + T_1(\check{x}, \xi).$$

The estimates for $t_0(\check{x}, z)$ and $t_1(\check{x}, \check{y})$ imply that

$$T_0(\check{x}, \xi) \in S_{-\infty}^{\infty}, \ |D_{\check{x}}^{p} D_{\xi}^{k} T_1(\check{x}, \xi)| \leqslant C_{pkN}(1 + |\check{x}| + |\xi|)^{-N}, \quad \forall N, p, k.$$

In this way, applying Parseval's equality, we get

$$\check{T}_\infty \check{u} = \int \check{T}_\infty(\check{x}, \check{y}) \check{u}(\check{y})\, d\check{y} = \frac{1}{(2\pi)^n} \int \overline{\tilde{T}(\check{x}, \xi)}\, \check{u}(\xi)\, d\xi$$

$$= \frac{1}{(2\pi)^n} \int T(\check{x}, \xi) e^{-i(\check{x}, \xi)} \check{u}(\xi)\, d\xi,$$

where $T(\check{x}, \xi) = \overline{T_0(\check{x}, \xi)} + \overline{T_1(\check{x}, \xi)} e^{i(\check{x}, \xi)}$ is a symbol of class $S_{-\infty}^{\infty}$. We have thus proved

THEOREM 19.6. *The operator B obtained from the p.o. A as a result of the change of variables $x = s(\check{x})$ is also a p.o. Its symbol $B(\check{x}, \xi)$ has the form $B(\check{x}, \xi) = B_0(\check{x}, \xi) + T(\check{x}, \xi)$, where the symbol $B_0(\check{x}, \xi)$ is defined by* (19.58) *and the symbol $T(\check{x}, \xi)$ is of class $S_{-\infty}^{\infty}$. Further, $B_0(\check{x}, \eta)$ has the expansion* (19.60) *for any N and its principal part has the form* (19.61).

§20. Factorization of elliptic pseudodifferential operators

1. *Construction of a more precise regularizer for an elliptic p.o. in* $\mathbf{R}^n$. Let $A(x, \xi) = \|a_{ij}(x, \xi)\|_{i,j=1}^{p}$ be an elliptic symbol of class S_α^t, $t > 0$ (see §18.5). In §18 we constructed a p.o. $R_0(x, D)$ with symbol $R_0(x, \xi) \in S_{-\alpha+\delta}^{t}$ such that

$$A(x, D) R_0(x, D) = I + T_1'(x, D), \tag{20.1}$$

$$R_0(x, D) A(x, D) = I + T_2'(x, D), \tag{20.1$'$}$$

where I is the identity operator and $T_i'(x, \xi) \in S_{-t_0}^{0}(\mathbf{R}^n)$, $i = 1, 2$, with $T_i'(\infty, \xi) = 0$, $t_0 > 0$. In this subsection we will construct a more precise regularizer of $A(x, D)$, viz. an operator $R(x, D)$ such that

$$A(x, D) R(x, D) = I + T_\infty^{(1)}(x, D), \tag{20.2}$$

$$R(x, D) A(x, D) = I + T_\infty^{(2)}(x, D), \tag{20.2$'$}$$

where $T_\infty^{(i)}(x, \xi) \in S_{-\infty}^{0}$, $T_\infty^{(i)}(\infty, \xi) = 0$, $i = 1, 2$.

LEMMA 20.1. *Suppose $a'(x, \xi)$ is a matrix whose elements are of class $S^0_{-\gamma}$, with $\gamma \geqslant 0$, $a'(\infty, \xi) = 0$ and*

$$(1+|\eta|)^{n+1}|\tilde{a}'(\eta, \xi)| \leqslant \varepsilon(1+|\xi|)^{-\gamma}, \tag{20.3}$$

where ε is sufficiently small. Then there exists a p.o. $r'(x, D)$ with symbol $r'(x, \xi) \in S^0_{-\gamma}$, $r'(\infty, \xi) = 0$, such that $I + r'(x, D)$ is the inverse of the operator $I + a'(x, D)$.

PROOF. We first construct a p.o. $r'(x, D)$ such that

$$(I + a'(x, D))(I + r'(x, D))u = u, \qquad \forall u(x) \in S(\mathbf{R}^n), \tag{20.4}$$

or, equivalently,

$$r'(x, D)u + a'(x, D)u + a'(x, D)r'(x, D)u = 0. \tag{20.5}$$

It follows directly from the definition of a p.o. that (20.5) will be satisfied if and only if

$$\frac{1}{(2\pi)^n}\int_{-\infty}^{\infty} r'(x, \xi)\tilde{u}(\xi)e^{-i(x,\xi)}\,d\xi + \frac{1}{(2\pi)^n}\int_{-\infty}^{\infty} a'(x, \xi)\tilde{u}(\xi)e^{-i(x,\xi)}\,d\xi$$
$$+ \frac{1}{(2\pi)^{2n}}\int_{-\infty}^{\infty} a'(x, \eta)e^{-i(x,\eta)}\tilde{r}'(\eta - \xi, \xi)\tilde{u}(\xi)\,d\xi d\eta = 0, \tag{20.6}$$

where $\tilde{r}'(\eta, \xi) = F_x r'(x, \xi)$. Since the function $\tilde{u}(\xi) \in S(\mathbf{R}^n)$ is arbitrary, (20.6) is equivalent to

$$r'(x, \xi)e^{-i(x,\xi)} + a'(x, \xi)e^{-i(x,\xi)}$$
$$+ \frac{1}{(2\pi)^n}\int_{-\infty}^{\infty} a'(x, \eta)e^{-i(x,\eta)}\tilde{r}'(\eta - \xi, \xi)\,d\eta = 0. \tag{20.7}$$

Multiplying (20.7) by $e^{i(x,\xi)}$ and making the change of variables $\eta - \xi = \zeta$, we get

$$r'(x, \xi) + \frac{1}{(2\pi)^n}\int_{-\infty}^{\infty} a'(x, \zeta + \xi)e^{-i(x,\zeta)}\tilde{r}'(\zeta, \xi)\,d\zeta = -a'(x, \xi). \tag{20.8}$$

Thus a p.o. $r'(x, D)$ satisfies (20.4) if and only if its symbol $r'(x, \xi)$ satisfies

$$r'(x, \xi) + a'(x, \xi + D_x)r'(x, \xi) = -a'(x, \xi). \tag{20.9}$$

The Fourier transform with respect to x of (20.9) (or (20.8)) is

$$\tilde{r}'(\eta, \xi) + \frac{1}{(2\pi)^n}\int_{-\infty}^{\infty} \tilde{a}'(\eta - \zeta, \zeta + \xi)\tilde{r}'(\zeta, \xi)\,d\zeta = -\tilde{a}'(\eta, \xi), \tag{20.10}$$

where $\tilde{a}'(\eta, \xi) = F_x a'(x, \xi)$. We denote by $\tilde{A}'$ the operator

$$\tilde{A}'\tilde{r}'(\eta, \xi) = \frac{1}{(2\pi)^n}\int_{-\infty}^{\infty} \tilde{a}'(\eta - \zeta, \zeta + \xi)\tilde{r}'(\zeta, \xi)\,d\zeta. \tag{20.11}$$

Multiplying (20.11) by $(1+|\xi|)^\gamma$ and applying (20.3), we get

$$\sup_{\eta,\xi}(1+|\xi|)^\gamma|\tilde{A}'\tilde{r}'(\eta,\xi)| \leqslant \frac{\varepsilon}{(2\pi)^n}\sup_{\eta,\xi}\int_{-\infty}^{\infty}\frac{(1+|\xi|)^\gamma|\tilde{r}'(\zeta,\xi)|\,d\zeta}{(1+|\eta-\zeta|)^{n+1}}$$
$$\leqslant \frac{\varepsilon}{(2\pi)^n}\int_{-\infty}^{\infty}\frac{d\zeta}{(1+|\zeta|)^{n+1}}\sup_{\zeta,\xi}(1+|\xi|)^\gamma|r'(\zeta,\xi)|. \tag{20.12}$$

Thus, if ε is sufficiently small, the norm of $\tilde{A}'$ in the space of bounded functions of (η, ξ) with weight $(1+|\xi|)^\gamma$ is less than unity. Consequently, (20.10) has a unique solution $\tilde{r}'(\eta,\xi)$, which satisfies

$$|\tilde{r}'(\eta,\xi)| \leqslant C(1+|\xi|)^{-\gamma}, \qquad \gamma \geqslant 0, \tag{20.13}$$

and is given by

$$\tilde{r}' = -\tilde{a}' + \tilde{A}'\tilde{a}' - \tilde{A}'^2\tilde{a}' + \ldots + (-1)^n\tilde{A}'^n\tilde{a}' + \ldots \tag{20.14}$$

Let us show that $\tilde{r}'(\eta,\xi)$ satisfies

$$|(1+|\eta|)^N\tilde{r}'(\eta,\xi)| \leqslant C_N(1+|\xi|)^{-\gamma}, \qquad \forall N, \tag{20.15}$$

and hence that $r'(x,\xi) = F_\eta^{-1}\tilde{r}'(\eta,\xi) \in S^0_{-\gamma}$. To this end we multiply (20.10) by η_i, $1 \leqslant i \leqslant n$, and let $\tilde{r}_i'(\eta,\xi) = \eta_i\tilde{r}'(\eta,\xi)$. Then $\tilde{r}_i'(\eta,\xi)$ satisfies

$$\tilde{r}_i'(\eta,\xi) + \frac{1}{(2\pi)^n}\int_{-\infty}^{\infty}\tilde{a}'(\eta-\zeta,\zeta+\xi)\tilde{r}_i'(\zeta,\xi)\,d\zeta$$
$$= -\eta_i\tilde{a}'(\eta,\xi) - \frac{1}{(2\pi)^n}\int_{-\infty}^{\infty}(\eta_i-\zeta_i)\tilde{a}'(\eta-\zeta,\zeta+\xi)\tilde{r}'(\zeta,\xi)\,d\zeta. \tag{20.16}$$

Since $|\tilde{a}'(\eta,\xi) \leqslant C_N(1+|\eta|)^{-N}(1+|\xi|)^{-\gamma}$, $\forall N$, the right side of (20.16) remains bounded with respect to η and ξ after being multiplied by $(1+|\xi|)^\gamma$. Consequently, as in the proof of the solvability of (20.10), we get $(1+|\xi|)^\gamma|\tilde{r}_i'(\eta,\xi)| \leqslant C$, $1 \leqslant i \leqslant n$. Thus $|\tilde{r}'(\eta,\xi)| \leqslant C(1+|\eta|)^{-1}(1+|\xi|)^{-\gamma}$. Analogously, multiplying (20.16) by η_k, $1 \leqslant k \leqslant n$, we get

$$|\tilde{r}'(\eta,\xi) \leqslant C(1+|\eta|)^{-2}(1+|\xi|)^{-\gamma},$$

and so on. In this way, we successively prove (20.15) for any N.

By (20.3), the norm of $a'(x,D)$ in $H_0(\mathbf{R}^n)$ is less than unity. Consequently, the operator $I + a'(x,D)$ is invertible in $H_0(\mathbf{R}^n)$. Therefore $I + r'(x,D)$ is not only a right but also a left inverse of $I + a'(x,D)$. Q.E.D.

We now apply Lemma 20.1 to the construction of a more precise regularizer of the elliptic operator $A(x,D)$. Suppose, as in the preceding sections, that $\chi(\xi) \in C_0^\infty(\mathbf{R}^n)$, $\chi(\xi) = 1$ for $|\xi| \leqslant 1$, and let

$$V_1'(x,\xi) = T_1'(x,\xi)(1-\chi(\delta\xi)), \quad V_2'(x,\xi) = T_1'(x,\xi)\chi(\delta\xi).$$

Then $V_2'(x, \xi) \in S^0_{-\infty}$, $V_2'(\infty, \xi) = 0$. Since $T_1'(x, \xi) \in S^0_{-t_0}$ and $t_0 > 0$, it follows for sufficiently small δ that $\tilde{V}_1'(\eta, \xi)$ satisfies the estimate of form (20.3) in which $\gamma = t_0 - \varepsilon_1$, with ε_1 being arbitrarily small. By Lemma 20.1, the p.o. $I + V_1'(x, D)$ has an inverse $I + r'(x, D)$, with $r'(x, \xi) \in S^0_{-t_0+\varepsilon_1}$ and $r'(\infty, \xi) = 0$. Applying the p.o. $I + r'(x, D)$ to (20.1) from the right, we obtain (20.2), where Theorem 18.2 implies that

$$R(x, D) = R_0(x, D)(I + r'(x, D)) = R_0(x, D) + R_1'(x, D)$$

and

$$T^{(1)}_\infty(x, D) = V_2'(x, D)(I + r'(x, D)).$$

The existence of an operator $R^{(1)}(x, D)$ for which the relation

$$R^{(1)}(x, D)A(x, D) = I + T_\infty(x, D) \tag{20.17}$$

of form (20.2′) holds is proved analogously.

It follows from (20.2) that $R(x, D)$ and $R^{(1)}(x, D)$ differ by a p.o. $T^{(3)}_\infty(x, D)$ with symbol $T^{(3)}_\infty(x, \xi) \in S^0_{-\infty}$. In fact, applying $R(x, D)$ to (20.17) from the right and making use of (20.2), we get

$$R^{(1)}(x, D) - R(x, D) = T^{(3)}_\infty(x, D),$$

where $T^{(3)}_\infty(x, D) = T_\infty(x, D)R(x, D) - R^{(1)}(x, D)T^{(1)}_\infty(x, D)$ and $T^{(3)}_\infty(x, \xi) \in S^0_{-\infty}$. Consequently, the elliptic operator $A(x, D)$ has the more precise (left and right) regularizer $R(x, D)$.

2. *"Plus" and "minus" p.o.'s with variable symbols.* A symbol $A_+(x, \xi) = A_+(\infty, \xi) + A_+'(x, \xi) \in S^t_\alpha$ is called a *"plus" symbol* if for almost all $(\eta, \xi') \in \mathbf{R}^n \times \mathbf{R}^{n-1}$ the functions $A_+(\infty, \xi)$ and $\tilde{A}_+'(\eta, \xi) = F_x A_+'(x, \xi)$ can be analytically continued with respect to $\xi_n + i\tau$ into the halfplane $\tau > 0$, are continuous in the closed halfplane $\tau \geqslant 0$ and satisfy the estimates

$$|A_+(\infty, \xi', \xi_n + i\tau)| \leqslant C(1 + |\xi'| + |\xi_n| + |\tau|)^\alpha, \tag{20.18}$$

$$|\tilde{A}_+'(\eta, \xi', \xi_n + i\tau)| \leqslant C_N(1 + |\eta|)^{-N}(1 + |\xi'| + |\xi_n| + |\tau|)^\alpha \tag{20.18′}$$

for $\tau \geqslant 0$. Analogously, a symbol $A_-(x, \xi) = A_-(\infty, \xi) + A_-'(x, \xi) \in S^t_\alpha$ is called a *"minus" symbol* if for almost all $(\eta, \xi') \in \mathbf{R}^n \times \mathbf{R}^{n-1}$ the functions $A_-(\infty, \xi)$ and $\tilde{A}_-'(\eta, \xi) = F_x A_-'(x, \xi)$ can be analytically continued with respect to $\xi_n + i\tau$ into the halfplane $\tau < 0$, are continuous for $\tau \leqslant 0$ and satisfy the estimates of form (20.18), (20.18′) for $\tau \leqslant 0$.

LEMMA 20.2. *A p.o. $A_+(x, D)$ with "plus" symbol $A_+(x, \xi) \in S^0_\alpha$ is a bounded operator from $H_s^+(\mathbf{R}^n)$ into $H^+_{s-\alpha}(\mathbf{R}^n)$ for any s.*

PROOF. For any function $u_+(x) \in C_0^\infty(\mathbf{R}^n_+)$ we have

$$A_+(x, D)u_+ = \frac{1}{(2\pi)^n}\int_{-\infty}^{\infty} A_+(x, \xi', \xi_n)\tilde{u}_+(\xi', \xi_n)e^{-i(x',\xi')-ix_n\xi_n}\, d\xi_n. \tag{20.19}$$

We will show that $A_+(x, D)u_+ = 0$ for $x_n < 0$, i.e. $\operatorname{supp} A_+(x, D)u_+ \in \overline{\mathbf{R}^n_+}$. Since $A_+(x, D)$ is a bounded operator from $H_s(\mathbf{R}^n)$ into $H_{s-\alpha}(\mathbf{R}^n)$ (see Theorem 18.1), the assertion of the lemma will then follow by virtue of the fact that $C_0^\infty(\mathbf{R}^n_+)$ is dense in $H_s^+(\mathbf{R}^n)$ and $H_s^+(\mathbf{R}^n)$ is a closed subspace of $H_s(\mathbf{R}^n)$ (cf. the proof of Theorem 4.3).

Let $v = A_+(x, D)u_+$. From the definition of a "plus" symbol it follows for any fixed $x \in \mathbf{R}^n_-$ that for almost all $\xi' \in \mathbf{R}^{n-1}$ the symbol $A_+(x, \xi)$ can be analytically continued with respect to $\xi_n + i\tau$ into the halfplane $\tau > 0$, is continuous for $\tau \geqslant 0$ and satisfies

$$|A_+(x, \xi', \xi_n + i\tau)| \leqslant C(1 + |\xi'| + |\xi_n| + |\tau|)^\alpha, \qquad \tau \geqslant 0.$$

Displacing the contour of integration in (20.19) into the halfplane $\tau > 0$ with the use of Cauchy's theorem, we get

$$e^{-x_n\tau}v(x', x_n) = \frac{1}{(2\pi)^n}\int_{-\infty}^{\infty} A(x, \xi', \xi_n + i\tau)\tilde{u}_+(\xi', \xi_n + i\tau)e^{-i(x',\xi') - ix_n\xi_n}\, d\xi'\, d\xi_n. \tag{20.20}$$

By Lemma 2.2, the integral (20.20) admits the estimate

$$|e^{-x_n\tau}v(x', x_n)| \leqslant C_N\int_{-\infty}^{\infty}(1 + |\xi'| + |\xi_n| + |\tau|)^\alpha(1 + |\xi'| + |\xi_n| + |\tau|)^{-N}\, d\xi'\, d\xi_n = C_N^{(1)}, \tag{20.21}$$

from which it follows upon letting $\tau \to +\infty$ that $v = 0$ for $x_n < 0$.

LEMMA 20.3. *Suppose $A_+(x, \xi)$ and $B_+(x, \xi)$ are "plus" symbols of classes S^0_α and S^0_β respectively. Then the symbol $C_+(x, \xi)$ of the product $A_+(x, D)B_+(x, D)$ is a "plus" symbol of class $S^0_{\alpha+\beta}$.*

PROOF. By Theorem 18.2,

$$C_+(x, \xi) = A_+(x, \xi)B_+(\infty, \xi) + C_1(x, \xi),$$

where

$$C_1(x, \xi) = \frac{1}{(2\pi)^n}\int_{-\infty}^{\infty} A_+(x, \xi + \zeta)\tilde{B}'_+(\zeta, \xi)e^{-i(x,\zeta)}\, d\zeta. \tag{20.22}$$

Clearly, $A_+(x, \xi)B_+(\infty, \xi)$ is a "plus" symbol. Furthermore,

$$\tilde{C}_1(\eta, \xi) = A_+(\infty, \xi + \eta)\tilde{B}'_+(\eta, \xi) + \frac{1}{(2\pi)^n}\int_{-\infty}^{\infty}\tilde{A}'_+(\eta - \zeta, \zeta + \xi)\tilde{B}'_+(\zeta, \xi)\, d\zeta, \tag{20.23}$$

which implies the $C_1(x, \xi)$ is a "plus" symbol such that (cf. the proof of Theorem 18.2)

$$|\tilde{C}_1(\eta, \xi', \xi_n + i\tau)| \leqslant C_N(1 + |\eta|)^{-N}(1 + |\xi'| + |\xi_n| + |\tau|)^{\alpha+\beta},$$
$$\tau \geqslant 0, \quad N \geqslant 0. \qquad (20.24)$$

Lemma 20.3 is proved.

Suppose $a'_+(x, \xi) \in S^0_{-\varepsilon}$ is a "plus" symbol, $\varepsilon > 0$, $a'_+(\infty, \xi) = 0$, and let

$$r_+^{(N)}(x, D) = -a'_+(x, D) + (a'_+(x, D))^2 - (a'_+(x, D))^3$$
$$+ \cdots + (-1)^N(a'_+(x, D))^N. \qquad (20.25)$$

By Lemma 20.3, $r_+^{(N)}(x, \xi)$ is a "plus" symbol of class $S^0_{-\varepsilon}$, while the p.o. $r_+^{(N)}(x, D)$ clearly satisfies the relations

$$(I + a'_+(x, D))(I + r_+^{(N)}(x, D)) = I + T^+_{N\varepsilon}(x, D), \qquad (20.26)$$
$$(I + r_+^{(N)}(x, D))(I + a'_+(x, D)) = I + T^+_{N\varepsilon}(x, D), \qquad (20.27)$$

where I is the identity operator and $T^+_{N\varepsilon}(x, \xi)$ is a "plus" symbol of class $S^0_{-(N+1)\varepsilon}$. The following lemma, which is analogous to Lemma 20.1, permits one to construct a more precise regularizer for the "plus" operator $I + a'_+(x, D)$.

LEMMA 20.4. *Suppose $a'_+(x, \xi) \in S^0_{-\varepsilon}$ is a "plus" symbol, $\varepsilon > 0$ and $a'_+(\infty, \xi) = 0$. Then there exists a "plus" symbol $r'_+(x, \xi) \in S^0_{-\varepsilon/2}$ such that*

$$(I + a'_+(x, D))(I + r'_+(x, D)) = I + T^+_\infty(x, D), \qquad (20.28)$$
$$(I + r'_+(x, D))(I + a'_+(x, D)) = I + V^+_\infty(x, D), \qquad (20.29)$$

where $T^+_\infty(x, \xi)$ and $V^+_\infty(x, \xi)$ are "plus" symbols of class $S^0_{-\infty}$.

PROOF. Let $\chi_+(x_n) \in C_0^\infty(\mathbf{R}^1_+)$ and $\tilde{\chi}_+(\xi_n) = F\chi(x_n)$, with $\tilde{\chi}_+(0) = 1$, let $\chi(\xi')$ denote a function of class $C_0^\infty(\mathbf{R}^{n-1})$ that is equal to 1 for $|\xi'| \leqslant 1$, and let

$$a_+^{(1)}(x, \xi) = a'_+(x, \xi)(1 - \chi(\delta\xi')\tilde{\chi}_+(\delta\xi_n)),$$
$$a_+^{(2)}(x, \xi) = a'_+(x, \xi)\chi(\delta\xi')\tilde{\chi}_+(\delta\xi_n).$$

By Lemma 2.2, $\tilde{\chi}_+(\delta(\xi_n + i\tau))$ is an analytic function decreasing more rapidly than any polynomial in the halfplane $\tau \geqslant 0$. Consequently, $a_+^{(2)}(x, \xi)$ is a "plus" symbol of class $S^0_{-\infty}$. For any ε_1 it is possible to choose a δ so small that

$$|\tilde{a}_+^{(1)}(\eta, \xi', \xi_n + i\tau)| \leqslant \varepsilon_1(1 + |\eta|)^{-n-1}(1 + |\xi'| + |\xi_n| + |\tau|)^{-\varepsilon/2}, \quad \tau \geqslant 0.$$
$$(20.30)$$

To find the p.o. $r'_+(x, D)$ such that

$$(I + a_+^{(1)}(x, D))(I + r'_+(x, D)) = I, \tag{20.31}$$

we note that the Fourier transform $\tilde{r}'_+(\eta, \xi)$ with respect to x of its symbol $r'_+(x, \xi)$ must satisfy the equation

$$\tilde{r}'_+(\eta, \xi) + \frac{1}{(2\pi)^n}\int_{-\infty}^{\infty} \tilde{a}_+^{(1)}(\eta - \zeta, \zeta + \xi)\tilde{r}'_+(\zeta, \xi)\, d\zeta = -\tilde{a}_+^{(1)}(\eta, \xi), \tag{20.32}$$

which is analogous to (20.10). As in the proof of Lemma 20.1, we get that (20.32) has a unique solution $\tilde{r}'_+(\eta, \xi)$, which is equal to the sum of a uniformly convergent series of form (20.14).

Consequently, for almost all (η, ξ') the function $\tilde{r}'_+(\eta, \xi)$ is analytic with respect to $\xi_n + i\tau$ for $\tau > 0$ and is continuous for $\tau \geqslant 0$, and

$$|\tilde{r}'_+(\eta, \xi', \xi_n + i\tau)| \leqslant C(1 + |\xi'| + |\xi_n| + |\tau|)^{-\varepsilon/2}, \qquad \tau \geqslant 0.$$

Further, as in the proof of Lemma 20.1, it can be shown that

$$|\tilde{r}'_+(\eta, \xi', \xi_n + i\tau)| \leqslant C_N(1 + |\eta|)^{-N}(1 + |\xi'| + |\xi_n| + |\tau|)^{-\varepsilon/2}$$

for any N. Thus $r'_+(x, \xi)$ is a "plus" symbol of class $S^0_{-\varepsilon/2}$. (20.28) and (20.29) are satisfied by virtue of (20.31) and Lemma 20.3. Lemma 20.4 is proved.

The following lemmas, which are analogous to Lemmas 20.2–20.4, are valid for p.o.'s with "minus" symbols.

LEMMA 20.5. *A p.o. $A_-(x, D)$ with "minus" symbol $A_-(x, \xi) \in S^0_\alpha$ is a bounded operator from $H_s^-(\mathbf{R}^n)$ into $H_{s-\alpha}^-(\mathbf{R}^n)$ for any s. The product $C_-(x, D)$ of two p.o.'s $A_-(x, D)$ and $B_-(x, D)$ with "minus" symbols $A_-(x, \xi) \in S^0_\alpha$ and $B_-(x, \xi) \in S^0_\beta$ is a p.o. with "minus" symbol $C_-(x, \xi) \in S^0_{\alpha+\beta}$.*

LEMMA 20.6. *Suppose $a'_-(x, \xi) \in S^0_{-\varepsilon}$ is a "minus" symbol, $\varepsilon > 0$ and $a'_-(\infty, \xi) = 0$. Then there exists a "minus" symbol $r'_-(x, \xi) \in S^0_{-\varepsilon/2}$ such that*

$$(I + a'_-(x, D))(I + r'_-(x, D)) = I + T^-_\infty(x, D),$$

$$(I + r'_-(x, D))(I + a'_-(x, D)) = I + V^-_\infty(x, D),$$

where $T^-_\infty(x, \xi)$, $V^-_\infty(x, \xi)$ are "minus" symbols of class $S^0_{-\infty}$.

3. *Factorization of an elliptic p.o.*

LEMMA 20.7. *Suppose*

$$q'(x, \xi) \in S^0_{-\gamma}, \qquad \frac{\partial q'(x, \xi)}{\partial \xi_n} \in S^0_{-\gamma-1},$$

where $0 < \gamma < 1$, and $q'(\infty, \xi) = 0$. Then there exist a "plus" symbol $q'_+(x, \xi) \in S^0_{-\gamma+\varepsilon}$ and a "minus" symbol $q'_-(x, \xi) \in S^0_{-\gamma+\varepsilon}$ such that

$$(I + q'(x, D))(I + q'_+(x, D)) = I + q'_-(x, D) + T'_\infty(x, D), \tag{20.33}$$

where $T'_\infty(x, \xi) \in S^0_{-\infty}$, and $\varepsilon > 0$ can be chosen arbitrarily small.

PROOF. As above, let $\chi(\xi) \in C_0^\infty(\mathbf{R}^n)$, $\chi(\xi) = 1$ for $|\xi| \leqslant 1$, and let

$$q_1'(x, \xi) = q'(x, \xi)(1 - \chi(\delta\xi)), \qquad q_2'(x, \xi) = q'(x, \xi)\chi(\delta\xi).$$

Clearly, $q_2'(x, \xi) \in S^0_{-\infty}$. For any ε and ε_1 it is possible to find a δ such that

$$|\tilde{q}_1'(\eta, \xi)|(1 + |\eta|) + \left|\frac{\partial \tilde{q}_1'(\eta, \xi)}{\partial \xi_n}\right|(1 + |\xi|) \leqslant \varepsilon_1(1 + |\eta|)^{-(n+1)}(1 + |\xi|)^{-\gamma+\varepsilon/2}. \tag{20.34}$$

We will select "plus" and "minus" symbols $q'_+(x, \xi)$ and $q'_-(x, \xi)$ so that

$$(I + q_1'(x, D))(I + q'_+(x, D)) = I + q'_-(x, D). \tag{20.35}$$

As in the proof of Lemma 20.1, it follows from (20.35) that (cf. (20.8))

$$q'_+(x, \xi) + \frac{1}{(2\pi)^n}\int_{-\infty}^{\infty} q_1'(x, \zeta + \xi)e^{-i(x,\zeta)}\tilde{q}'_+(\zeta, \xi)\, d\zeta = -q_1'(x, \xi) + q'_-(x, \xi). \tag{20.36}$$

We take the Fourier transform with respect to x of (20.36) and apply the operator $\Pi^+_{\xi_n}$. Since $\Pi^+_{\xi_n}\tilde{q}'_+(\eta, \xi) = \tilde{q}'_+(\eta, \xi)$ and $\Pi^+_{\xi_n}\tilde{q}'_-(\eta, \xi) = 0$, we get

$$\tilde{q}'_+(\eta, \xi) + \frac{1}{(2\pi)^n}\int_{-\infty}^{\infty} \Pi^+_{\xi_n}\tilde{q}_1'(\eta - \zeta, \zeta + \xi)\tilde{q}'_+(\zeta, \xi)\, d\zeta = -\Pi^+_{\xi_n}\tilde{q}_1'(\eta, \xi). \tag{20.37}$$

We will prove that (20.37) has a unique solution if the number ε_1 in (20.34) is sufficiently small.

Let us show that $\Pi^+_{\xi_n}$ is bounded in the space H of functions $\tilde{a}(\eta, \xi)$ in $\mathbf{R}^n \times \mathbf{R}^n$ with finite norm

$$\|\tilde{a}(\eta, \xi)\|^2 = \sup_{\eta, \xi'} \int_{-\infty}^{\infty}\left[|\tilde{a}(\eta, \xi)|^2(1 + |\xi|)^{-1+2\gamma-2\varepsilon}(1 + |\eta|)^2 + \left|\frac{\partial \tilde{a}(\eta, \xi)}{\partial \xi_n}\right|^2(1 + |\xi|)^{1+2\gamma-2\varepsilon}\right] d\xi_n. \tag{20.38}$$

Since $-1 < -1 + 2\gamma - 2\varepsilon < 1$ for sufficiently small ε, it follows by virtue of Lemma 5.3 that

$$\int_{-\infty}^{\infty} |\Pi^+_{\xi_n}\tilde{a}(\eta, \xi)|^2(1 + |\xi|)^{-1+2\gamma-2\varepsilon}\, d\xi_n \leqslant C\int_{-\infty}^{\infty} |\tilde{a}(\eta, \xi)|^2(1 + |\xi|)^{-1+2\gamma-2\varepsilon}\, d\xi_n. \tag{20.39}$$

On the other hand, from the expansion formula for an integral of Cauchy type (see (5.36)) we have

$$\frac{\partial}{\partial \xi_n}\Pi^+_{\xi_n}\tilde{a} = \Pi^+_{\xi_n}\frac{\partial \tilde{a}}{\partial \xi_n} = \hat{\Lambda}^{-1}_+(\xi)i\Pi'\frac{\partial \tilde{a}}{\partial \xi_n} + \hat{\Lambda}^{-1}_+(\xi)\Pi^+_{\xi_n}\hat{\Lambda}_+\frac{\partial \tilde{a}}{\partial \xi_n} = \hat{\Lambda}^{-1}_+(\xi)\Pi^+_{\xi_n}\hat{\Lambda}_+\frac{\partial \tilde{a}}{\partial \xi_n}, \tag{20.40}$$

since

$$\Pi'\frac{\partial \tilde{a}}{\partial \xi_n} = \frac{1}{2\pi}\int_{-\infty}^{\infty}\frac{\partial \tilde{a}}{\partial \xi_n}\,d\xi_n = 0.$$

Consequently, using (20.39) and (20.40), we obtain

$$\int_{-\infty}^{\infty}\left|\frac{\partial}{\partial \xi_n}\Pi^+_{\xi_n}\tilde{a}\right|^2(1+|\xi|)^{1+2\gamma-2\varepsilon}\,d\xi_n \leqslant C\int_{-\infty}^{\infty}\left|\Pi^+_{\xi_n}\hat{\Lambda}_+\frac{\partial \tilde{a}}{\partial \xi_n}\right|^2(1+|\xi|)^{-1+2\gamma-2\varepsilon}\,d\xi_n$$
$$\leqslant C\int_{-\infty}^{\infty}\left|\frac{\partial \tilde{a}}{\partial \xi_n}\right|^2(1+|\xi|)^{1+2\gamma-2\varepsilon}\,d\xi_n, \tag{20.41}$$

which together with (20.39) imply the boundedness of $\Pi^+_{\xi_n}$ in H. Thus $\Pi^+_{\xi_n}\tilde{q}_1'(\eta,\xi) \in H$, since $\tilde{q}_1'(\eta,\xi) \in H$ by virtue of (20.34).

Let us now estimate the norm in H of the operator

$$\tilde{Q}\tilde{a}(\eta,\xi) = \frac{1}{(2\pi)^n}\int_{-\infty}^{\infty}\Pi^+_{\xi_n}\tilde{q}_1'(\eta-\zeta,\zeta+\xi)\tilde{a}(\zeta,\xi)\,d\zeta \tag{20.42}$$

in (20.37). From (20.39) and (20.34) we have

$$\int_{-\infty}^{\infty}|\tilde{Q}\tilde{a}|^2(1+|\xi|)^{-1+2\gamma-2\varepsilon}\,d\xi_n$$
$$\leqslant C\int_{-\infty}^{\infty}\left|\int_{-\infty}^{\infty}|\tilde{q}_1'(\eta-\zeta,\zeta+\xi)|\,|\tilde{a}(\zeta,\xi)|\,d\zeta\right|^2\cdot(1+|\xi|)^{-1+2\gamma-2\varepsilon}\,d\xi_n$$
$$\leqslant C\varepsilon_1^2\int_{-\infty}^{\infty}\left|\int_{-\infty}^{\infty}\frac{|\tilde{a}(\zeta,\xi)|\,d\zeta}{(1+|\eta-\zeta|)^{n+2}}\right|^2(1+|\xi|)^{-1+2\gamma-2\varepsilon}\,d\xi_n. \tag{20.43}$$

On the other hand, the Cauchy-Schwarz-Bunjakovskiĭ inequality and the inequalities of form (18.11) imply

$$\left|\int_{-\infty}^{\infty}\frac{|\tilde{a}(\zeta,\xi)|\,d\zeta}{(1+|\eta-\zeta|)^{n+2}}\right|^2 \leqslant \int_{-\infty}^{\infty}\frac{d\zeta}{(1+|\eta-\zeta|)^{n+1}}\int_{-\infty}^{\infty}\frac{|\tilde{a}(\zeta,\xi)|^2\,d\zeta}{(1+|\eta-\zeta|)^{n+3}}$$
$$\leqslant C\int_{-\infty}^{\infty}\frac{(1+|\zeta|)^2|\tilde{a}(\zeta,\xi)|^2}{(1+|\eta|)^2(1+|\eta-\zeta|)^{n+1}}\,d\zeta. \tag{20.44}$$

Consequently,

$$\int_{-\infty}^{\infty} |\tilde{Q}\tilde{a}|^2 (1+|\eta|)^2 (1+|\xi|)^{-1+2\gamma-2\varepsilon}\, d\xi_n$$

$$\leqslant C\varepsilon_1^2 \int_{-\infty}^{\infty} \frac{d\zeta}{(1+|\eta-\zeta|)^{n+1}} \sup_{\zeta,\xi'} \int |\tilde{a}(\zeta,\xi)|^2 (1+|\zeta|)^2 (1+|\xi|)^{-1+2\gamma-2\varepsilon}\, d\xi_n$$

$$= C_1\varepsilon_1^2 \sup_{\zeta,\xi'} \int_{-\infty}^{\infty} |\tilde{a}(\zeta,\xi)|^2 (1+|\zeta|)^2 (1+|\xi|)^{-1+2\gamma-2\varepsilon}\, d\xi_n. \tag{20.45}$$

Further, differentiating (20.42) with respect to ξ_n and applying (20.41), we get

$$\int_{-\infty}^{\infty} \left| \frac{\partial}{\partial \xi_n} \tilde{Q}\tilde{a} \right|^2 (1+|\xi|)^{1+2\gamma-2\varepsilon}\, d\xi_n \leqslant C(I_1+I_2), \tag{20.46}$$

where

$$I_1 = \int_{-\infty}^{\infty} \left| \int_{-\infty}^{\infty} \left| \frac{\partial \tilde{q}_1'(\eta-\zeta,\zeta+\xi)}{\partial \xi_n} \right| |\tilde{a}(\zeta,\xi)|\, d\zeta \right|^2 (1+|\xi|)^{1+2\gamma-2\varepsilon}\, d\xi_n, \tag{20.47}$$

$$I_2 = \int_{-\infty}^{\infty} \left| \int_{-\infty}^{\infty} |\tilde{q}_1'(\eta-\zeta,\zeta+\xi)| \left| \frac{\partial \tilde{a}(\zeta,\xi)}{\partial \xi_n} \right| d\zeta \right|^2 (1+|\xi|)^{1+2\gamma-2\varepsilon}\, d\xi_n. \tag{20.48}$$

On the other hand, by (20.34) and (18.11),

$$\left| \frac{\partial \tilde{q}_1'(\eta-\zeta,\zeta+\xi)}{\partial \xi_n} \right| \leqslant \frac{\varepsilon_1}{(1+|\eta-\zeta|)^{n+1}(1+|\zeta+\xi|)} \leqslant \frac{\varepsilon_1(1+|\zeta|)}{(1+|\eta-\zeta|)^{n+1}(1+|\xi|)}, \tag{20.49}$$

$$|\tilde{q}_1'(\eta-\zeta,\zeta+\xi)| \leqslant \frac{\varepsilon_1}{(1+|\eta-\zeta|)^{n+1}}. \tag{20.49$'$}$$

Consequently,

$$I_1 \leqslant C\varepsilon_1^2 \sup_{\zeta,\xi'} \int_{-\infty}^{\infty} |\tilde{a}(\zeta,\xi)|^2 (1+|\zeta|)^2 (1+|\xi|)^{-1+2\gamma-2\varepsilon}\, d\xi_n, \tag{20.50}$$

$$I_2 \leqslant C\varepsilon_1^2 \sup_{\zeta,\xi'} \int_{-\infty}^{\infty} \left| \frac{\partial \tilde{a}(\zeta,\xi)}{\partial \xi_n} \right|^2 (1+|\xi|)^{1+2\gamma-2\varepsilon}\, d\xi_n. \tag{20.50$'$}$$

Thus, as a result of (20.45), (20.46), (20.50) and (20.50′),

$$\|\tilde{Q}\tilde{a}\| \leqslant C\varepsilon_1 \|\tilde{a}\|, \tag{20.51}$$

which implies that the norm of $\tilde{Q}$ in H can be made less than unity by taking ε_1 sufficiently small.

Therefore (20.37) has a unique solution $\tilde{q}'_+(\eta, \xi)$, which belongs to H and is given by the series

$$\tilde{q}'_+(\eta, \xi) = -\Pi^+_{\xi_n} \tilde{q}'_1(\eta, \xi) + \tilde{Q}\Pi^+_{\xi_n} \tilde{q}'_1(\eta, \xi) - \tilde{Q}^2\Pi^+_{\xi_n} \tilde{q}'_1(\eta, \xi) + \dots \tag{20.52}$$

Further, multiplying (20.37) by η_i, $1 \leqslant i \leqslant n$, as in the proof of Lemma 20.1, we obtain the estimates

$$\|\eta_i \tilde{q}'_+(\eta, \xi)\| \leqslant C, \qquad 1 \leqslant i \leqslant n. \tag{20.53}$$

Continuing as in the proof of Lemma 20.1, we get that for any N

$$\|(1 + |\eta|)^N \tilde{q}'_+(\eta, \xi)\|^2 = \sup_{\eta, \xi'} (1 + |\eta|)^{2N}$$

$$\times \int_{-\infty}^{\infty} \left[|\tilde{q}'_+|^2 (1 + |\xi|)^{-1+2\gamma-2\varepsilon}(1 + |\eta|)^2 + \left| \frac{\partial \tilde{q}'_+}{\partial \xi_n} \right|^2 (1 + |\xi|)^{1+2\gamma-2\varepsilon} \right] d\xi_n < \infty. \tag{20.54}$$

Let us show that $q'_+(x, \xi) = F^{-1}_\eta \tilde{q}^{-1}_+(\eta, \xi)$ is a "plus" symbol of class $S^0_{-\gamma+\varepsilon}$. By (20.37),

$$\tilde{q}'_+(\eta, \xi) = \Pi^+_{\xi_n} \tilde{q}'_+(\eta, \xi).$$

Consequently, according to the Paley-Wiener theorem (see also Remark 5.1), $\tilde{q}_+(\eta, \xi', \xi_n + i\tau)$ admits for almost all (η, ξ') an analytic continuation with respect to $\xi_n + i\tau$ into the halfplane $\tau > 0$, with (20.54) remaining valid under the replacement of ξ_n by $\xi_n + i\tau$, $\tau \geqslant 0$. In addition, for any $\tau \geqslant 0$

$$\tilde{q}'_+(\eta, \xi', \xi_n + i\tau) = -\int_{\xi_n}^{\infty} \frac{\partial \tilde{q}'_+(\eta, \xi', \zeta_n + i\tau)}{\partial \zeta_n} \, d\zeta_n. \tag{20.55}$$

Estimating the integral (20.55) with the use of the Cauchy-Schwarz-Bunjakovskiĭ inequality, we get

$$|\tilde{q}'_+(\eta, \xi', \xi_n + i\tau)|^2 \leqslant \int_{\xi_n}^{\infty} \left| \frac{\partial \tilde{q}'_+(\eta, \xi', \zeta_n + i\tau)}{\partial \zeta_n} \right|^2 (1 + |\xi'| + |\zeta_n| + \tau)^{1+2\gamma-2\varepsilon} \, d\zeta_n$$

$$\times \int_{\xi_n}^{\infty} (1 + |\xi'| + |\zeta_n| + \tau)^{-1-2\gamma+2\varepsilon} \, d\zeta_n.$$

Consequently,

$$(1 + |\xi'| + |\xi_n| + \tau)^{\gamma-\varepsilon} |\tilde{q}'_+(\eta, \xi', \xi_n + i\tau)| \leqslant C. \tag{20.56}$$

It is shown as in the proof of Lemma 20.1 that $|\eta|^N \tilde{q}'_+(\eta, \xi)$ satisfies (20.56) for any N.

Analogously, applying $\Pi^-_{\xi_n}$ to (20.36), we get that $q'_-(x, \xi) = F^{-1}_\eta \tilde{q}'_-(\eta, \xi)$ is a "minus" symbol of class $S^0_{-\gamma+\varepsilon}$. Lemma 20.7 is proved.

REMARK 20.1. the class of "plus" and "minus" symbols was defined above for all $x \in \mathbf{R}^n$. It is possible to introduce a wider class of symbols that are "plus" for $x \in \mathbf{R}^n_-$ or "minus" for $x \in \mathbf{R}^n_+$.

A symbol $A_+(x, \xi) = A_+(\infty, \xi) + A'_+(x, \xi) \in S^0_\alpha$ is called *"plus" for* $x \in \mathbf{R}^n_-$ if $A_+(\infty, \xi)$ is a "plus" symbol and for any $p \geqslant 0$ and almost all $(\eta, \xi') \in \mathbf{R}^n \times \mathbf{R}^{n-1}$ the function

$$\Pi^-_{\eta_n} \eta_n^p \tilde{A}'_+(\eta, \xi) = F_x \theta(-x_n) D^p_n A'(x, \xi)$$

can be analytically continued with respect to $\xi_n + i\tau$ into the halfplane $\tau > 0$, is continuous for $\tau \geqslant 0$ and satisfies

$$\int_{-\infty}^{\infty} |\Pi^-_{\eta_n} \eta_n^p \tilde{A}'_+(\eta, \xi', \xi_n + i\tau)|^2 \, d\eta_n \leqslant C_{pN}(1 + |\eta'|)^{-N}(1 + |\xi'| + |\xi_n| + |\tau|)^\alpha, \quad \forall N, \tag{20.57}$$

for $\tau \geqslant 0$. Analogously, a symbol $A_-(x, \xi) = A_-(\infty, \xi) + A'_-(x, \xi)$ is called *"minus" for* $x \in \mathbf{R}^n_+$ if $A_-(\infty, \xi)$ is a "minus" symbol and for any $p \geqslant 0$ and almost all (η, ξ') the function

$$\Pi^+_{\eta_n} \eta_n^p \tilde{A}'_-(\eta, \xi) = F_x \theta(x_n) D^p_n A'(x, \xi)$$

can be analytically continued with respect to $\xi_n + i\tau$ into the halfplane $\tau < 0$, is continuous for $\tau \leqslant 0$ and satisfies the estimates of form (20.57) for $\tau \leqslant 0$.

It can be proved that Lemmas 20.2–20.4 remain valid for p.o.'s with symbols that are "plus" for $x \in \mathbf{R}^n_-$, while Lemmas 20.5 and 20.6 remain in force for symbols that are "minus" for $x \in \mathbf{R}^n_+$.

4. Suppose $A(x, \xi)$ is an elliptic symbol of class $\hat{O}^\infty_{\alpha+i\beta}$ (see §18.5), i.e. there exists an elliptic symbol $A_0(x, \xi) \in O^\infty_{\alpha+i\beta}$ such that $A(x, \xi) = A_0(x, \xi)$ for $|\xi| \geqslant N$. For each $x \in \mathbf{R}^n$ the symbol $A_0(x, \xi)$ has the homogeneous factorization

$$A_0(x, \xi) = A_0^-(x, \xi) A_0^+(x, \xi) \tag{20.58}$$

in accordance with (6.44) and (6.45). The factorization index $\kappa(x) = \deg_\xi A_0^+(x, \xi)$ of $A_0(x, \xi)$ belongs to $C^\infty(\mathbf{R}^n)$ and is independent of x for $|x| \geqslant N_1$. The functions $A_0^\pm(x, \xi)$ have the representations

$$A_0^+(x, \xi) = \Lambda_+^{\kappa(x)}(\xi) A_1^+(x, \xi), \tag{20.59}$$

$$A_0^-(x, \xi) = \Lambda_-^{\alpha+i\beta-\kappa(x)} A_1^-(x, \xi), \tag{20.59'}$$

in which the functions $A_1(x, \xi)$ are homogeneous in ξ of degree 0 and, by Lemma 17.1,

$$|D^p_x D^{k_n}_{\xi_n} A_1^\pm(x, \xi', \xi_n + i\tau)| \leqslant C_{pk_n}(|\xi'| + |\xi_n| + |\tau|)^{-k_n}, \quad \forall p, k_n, \tag{20.60}$$

$$|D^p_x D^{k'}_{\xi'} D^{k_n}_{\xi_n} A_1^\pm(x, \xi', \xi_n + i\tau)| \leqslant C_{pk'k_n}(|\xi'| + |\xi_n| + |\tau|)^{-k_n - 1 + \varepsilon} \times |\xi'|^{-\varepsilon - |k'| + 1},$$
$$\forall \varepsilon > 0, \quad |k'| \geqslant 1, \quad k_n \geqslant 0, \quad |p| \geqslant 0. \tag{20.61}$$

Here $\tau \geqslant 0$ in the estimates for A_1^+ and $\tau \leqslant 0$ in the estimates for A_1^-.

We assume that the oscillation of the real part of the factorization index $\kappa(x)$ is less than 1, i.e. there exist numbers κ_+ and κ_- such that

$$\kappa_- < \operatorname{Re}\kappa(x) < \kappa_+, \qquad \kappa_+ - \kappa_- < 1. \tag{20.62}$$

Let

$$R_0^+(x,\xi) = (A_0^+(x,\xi))^{-1}\Lambda_+^{\kappa_+ + 1}(\xi)\hat{\Lambda}_+^{-\kappa_+ - 1},$$

and

$$R_0^-(x,\xi) = (A_0^-(x,\xi))^{-1}\Lambda_-^{\alpha-\kappa_- + 1}\hat{\Lambda}_-^{-\alpha+\kappa_- - 1}.$$

From (20.60) and (20.61) it follows that $R_0^+(x,\xi)$ is a "plus" symbol of class $S^1_{-\kappa_-}$ and $R_0^-(x,\xi)$ is a "minus" symbol of class $S^1_{\kappa_+ - \alpha}$. By Theorems 18.2 and 18.3,

$$R_0^-(x,D)A(x,D)R_0^+(x,D) = I + q'(x,D), \tag{20.63}$$

where $q'(x,\xi) \in S^0_{-\gamma}$, $\partial q'(x,\xi)/\partial\xi_n \in S^0_{-\gamma-1}$, $q'(\infty,\xi) = 0$, and $\gamma = 1 - \kappa_+ + \kappa_-$. Consequently, by Lemma 20.7, there exist a "plus" p.o. $q'_+(x,D)$ and a "minus" p.o. $q'_-(x,D)$ such that

$$R_0^-(x,D)A(x,D)R_0^+(x,D)(I + q'_+(x,D)) = I + q'_-(x,D) + T_\infty(x,D), \tag{20.64}$$

where $q'_\pm(x,\xi) \in S^0_{-\gamma+\varepsilon}$, $0 < \varepsilon < \gamma$, and $T_\infty(x,\xi) \in S^0_{-\infty}$.

Let us show that there exists a "minus" p.o. $A_1^-(x,D)$ satisfying a relation of the form

$$A_1^-(x,D)R_0^-(x,D) = I + T_\infty^-(x,D), \tag{20.65}$$

where $T_\infty^-(x,\xi) \in S^0_{-\infty}$. By Lemma 20.5 and Theorem 18.3, the symbol

$$A_2^-(x,\xi) = A_0^-(x,\xi)\Lambda_-^{\kappa_+ - \alpha + 1}\hat{\Lambda}_-^{-\kappa_+ + \alpha - 1}$$

satisfies a relation of the form

$$A_2^-(x,D)R_0^-(x,D) = I + a'_-(x,D), \tag{20.66}$$

where $a'_-(x,\xi) \in S^0_{-\varepsilon}$ is a "minus" symbol and $\varepsilon > 0$. On the other hand, by Lemma 20.6, there exists a "minus" symbol $r'_-(x,\xi) \in S^0_{-\varepsilon/2}$ such that

$$(I + r'_-(x,D))(I + a'_-(x,D)) = I + V_\infty^-(x,D), \tag{20.67}$$

where $V_\infty^-(x,\xi) \in S^0_{-\infty}$. Consequently, the operator

$$A_1^-(x,D) = (I + r'_-(x,D))A_2^-(x,D)$$

satisfies a relation of form (20.65).

Thus, applying $A_1^-(x,D)$ to (20.64) from the left, we get that there exist a "plus" symbol $R_+(x,\xi) \in S^0_{-\kappa_-}$ and a "minus" symbol $A_-(x,\xi) \in S^0_{\alpha-\kappa_-}$ such that

$$A(x, D)R_+(x, D) = A_-(x, D) + T^1_\infty(x, D), \tag{20.68}$$

where $T^1_\infty(x, \xi) \in S^0_{-\infty}$.

REMARK 20.2. The symbol $A_0(x, \xi)$ can be represented in the form of a product of homogeneous elliptic symbols, one of which is "plus" for $x \in \mathbf{R}^n_-$ and the other of which is "minus" for $x \in \mathbf{R}^n_+$ (see Remark 20.1). Unlike (20.58), this factorization is not unique and is performed as follows.

We factor $A_0(x, \xi)$ in accordance with (6.44) and (6.45) for $x_n = 0$ and arbitrary $x' \in \mathbf{R}^{n-1}$. We next arbitrarily extend the factor $A_0^-(x', 0, \xi)$ into the halfspace $x_n > 0$ with preservation of its homogeneity, the estimates of form (20.60), (20.61), its analyticity with respect to $\xi_n + i\tau$ for $\tau < 0$ and its ellipticity for $\tau \leqslant 0$. We then put

$$A_0^+(x, \xi) = \frac{A_0(x, \xi)}{A_0^-(x, \xi)}$$

for $x_n \geqslant 0$. After this, we analogously extend $A_0^+(x, \xi)$ into the halfspace $x_n < 0$ and put

$$A_0^-(x, \xi) = \frac{A_0(x, \xi)}{A_0^+(x, \xi)}$$

for $x_n \leqslant 0$.

As a result, we obtain a factorization

$$A_0(x, \xi) = A_0^-(x, \xi)A_0^+(x, \xi)$$

in which $A_0^-(x, \xi)$ is "minus" for $x_n > 0$ and $A_0^+(x, \xi)$ is "plus" for $x_n < 0$. In particular, it is possible to choose $A_0^\pm(x, \xi)$ so that the factorization index $\kappa(x)$ does not depend on x_n: $\kappa(x', x_n) = \kappa(x', 0)$.

REMARK 20.3. The factorization in this section of a scalar elliptic p.o. and the extension by analogy of the analysis of §17 to the case of operators with variable symbols for a system of elliptic p.o.'s permit one to construct a parametrix in the case of a boundary value problem for a system of elliptic p.o.'s (see [34]). This parametrix has precisely the same form as the solution of a boundary value problem in a halfspace for p.o.'s with constant symbols. We note that parametrices can be used, in particular, to study nonelliptic boundary value problems for elliptic p.o.'s. For other applications of parametrices see [42], where the asymptotic behavior of the spectral function for an elliptic boundary value problem is studied, and §27 below, where elliptic p.o.'s with a small parameter are considered.

§21. Pseudodifferential operators in a domain and on a manifold

1. *P.o.'s in a domain.* Let G be a bounded domain in $\mathbf{R}^n$ with a smooth boundary Γ. There then exists a finite covering of $\overline{G}$ by open neighborhoods U_j, $1 \leqslant j \leqslant N$, in which, when $U_j \cap \Gamma \neq \varnothing$, it is possible to introduce a coordinate system $x^{(j)} = (x_1^{(j)}, \dots, x_n^{(j)})$ such that the equation for $U_j \cap \Gamma$ has the form $x_n^{(j)} = 0$. A coordinate system in U_j with this property will be called a *local coordinate system* (l.c.s.). When $U_i \cap U_j \cap \Gamma \neq \varnothing$, the passage from the jth local coordinates to the ith local coordinates is denoted by $x^{(i)} = s_{ij}(x^{(j)})$. It is assumed that the Jacobian matrix $Ds_{ij}(x^{(j)})/Dx^{(j)}$ of this transformation is not singular in $U_i \cap U_j$.

We will sometimes use the following *special local coordinate system* (s.l.c.s.). In a neighborhood of Γ we draw the field of normals to Γ and, in each neighborhood U_j that intersects Γ, we introduce a coordinate system $x^{(j)} = (x_1^{(j)}, \ldots, x_n^{(j)})$, in which the coordinate $x_n^{(j)}$ of an arbitrary point x is the distance from x to Γ, taken with the plus sign if the point lies in the interior of the domain, and with the minus sign otherwise, while the remaining coordinates $x_1^{(j)}, \ldots, x_{n-1}^{(j)}$ of x are the coordinates of the point at which the normal form x to Γ intersects Γ. Clearly, for s.l.c.s.'s when $U_i \cap U_j \cap \Gamma \neq \varnothing$, the formulas for passing from the jth special local coordinates to the ith have the form

$$x'_{(i)} = s'_{ij}(x'_{(j)}), \quad x_n^{(i)} = x_n^{(j)}, \tag{21.1}$$

where $x'_{(i)} = (x_1^{(i)}, \ldots, x_{n-1}^{(i)})$.

Let $\mathring{H}_s(G)$ denote the subspace of $H_s(\mathbf{R}^n)$ consisting of functions with support in $\overline{G}$ (for the case $G = \mathbf{R}^n_\pm$ this space was denoted by $H_s^\pm(\mathbf{R}^n)$ in Chapters I–IV). As in §4, it can be verified that $\mathring{H}_s(G)$ is a closed subspace of $H_s(\mathbf{R}^n)$. Also, let $H_s(G)$ denote the space of generalized functions in G admitting an extension onto $\mathbf{R}^n$ that belongs to $H_s(\mathbf{R}^n)$. As in the case $G = \mathbf{R}^n_+$, the norm in $H_s(G)$ is defined by

$$\|f\|_s^+ = \inf \|lf\|_s,$$

where the infimum is taken over all extensions $lf \in H_s(\mathbf{R}^n)$ (see §4). As in §4, it can be proved that $\mathring{H}_s^*(G) = H_{-s}(G)$ and $H_s^*(G) = \mathring{H}_{-s}(G)$.

LEMMA 21.1. *A bounded operator T_0 from $\mathring{H}_s(G)$ into $\mathring{H}_{s+\varepsilon}(G)$, $\varepsilon > 0$, is compact in $\mathring{H}_s(G)$. Analogously, a bounded operator T from $H_s(G)$ into $H_{s+\varepsilon}(G)$, $\varepsilon > 0$, is compact in $H_s(G)$.*

PROOF. Let Λ_ε denote the p.o. with symbol $(1 + |\xi|^2)^{\varepsilon/2}$ and let $\psi_0(x)$ be a function of class $C_0^\infty(\mathbf{R}^n)$ that is identically equal to 1 in a neighborhood of $\overline{G}$. Then $T_0 u_0 = \psi_0(x)\Lambda_{-\varepsilon}\Lambda_\varepsilon T_0 u_0$ for any $u_0 \in \mathring{H}_s(G)$. But $\Lambda_\varepsilon T_0$ is a bounded operator from $\mathring{H}_s(G)$ into $H_s(\mathbf{R}^n)$, while $\psi_0(x)\Lambda_{-\varepsilon}$ is a compact operator in $H_s(\mathbf{R}^n)$ by virtue of Theorem 18.4. Thus T_0 is a compact operator in $\mathring{H}_s(G)$.

Let l_0 denote a fixed extension operator from $H_s(G)$ into $H_s(\mathbf{R}^n)$ (cf. (4.55)) that satisfies

$$\|l_0 g\|_s \leqslant C\|g\|_s^+.$$

Then $Tf = p\psi_0(x)\Lambda_{-\varepsilon}\Lambda_\varepsilon l_0 Tf$ for any $f \in H_s(G)$. Thus, as the product of the bounded operator $\Lambda_\varepsilon l_0 T$ and a compact operator, T is a compact operator in $H_s(G)$.

REMARK 21.1. From the compactness of T (T_0) in $H_s(G)$ ($\mathring{H}_s(G)$) it follows upon going over to the adjoint operator that T (T_0) is also a compact operator in $H_{s+\varepsilon}(G)$ ($\mathring{H}_{s+\varepsilon}(G)$).

Suppose $A(x, \xi) \in \hat{O}^{\infty}_{\alpha+i\beta}$, i.e. there exists a symbol $A_0(x, \xi) \in O^{\infty}_{\alpha+i\beta}$ such that

$$A(x, \xi) - A_0(x, \xi)(1 - \chi(\xi)) \in S^{\infty}_{\alpha-1},$$

where $\chi(\xi) \in C_0^\infty(\mathbf{R}^n)$, $\chi(\xi) = 1$ for $|\xi| \leqslant 1$. We will say that a p.o. $A(x, D)$ with symbol $A(x, \xi) \in \hat{O}^{\infty}_{\alpha+i\beta}$ is a p.o. of class $\hat{O}^{\infty}_{\alpha+i\beta}$, while the symbol $A_0(x, \xi)$ will be called the *principal homogeneous symbol* (p.h.s.) of $A(x, D)$.

By a *p.o. in a domain G* we will mean, as in §7, an operator of the form $pA(x, D)u_0$, where $u_0 \in \mathring{H}_s(G)$ and p is the operator of restriction of a function to G. By Theorem 18.1, $pA(x, D)u_0$ is a bounded operator from $\mathring{H}_s(G)$ into $H_{s-\alpha}(G)$ for any s.

2. *Factorization index of an elliptic symbol in a domain.* A p.o. $A(x, D)$ of class $\hat{O}^{\infty}_{\alpha+i\beta}$ is elliptic (see §18) if its principal homogeneous symbol is elliptic, i.e.

$$\det A_0(x, \xi) \neq 0 \quad \text{for } \xi \neq 0.$$

Let $\varphi(x) \in C_0^\infty(U_j)$ and $\psi(x) \in C_0^\infty(U_j)$, where U_j is an arbitrary neighborhood intersecting Γ, and consider the operator $\psi A(x, D)\varphi$, i.e. a *localization* of $A(x, D)$ to U_j. Let

$$\psi_j(x^{(j)})A_j^{(0)}(x^{(j)}, \xi^{(j)})\varphi_j(x^{(j)})$$

be the p.h.s. of the operator $\psi A(x, D)\varphi$ written in the jth local coordinates. By Theorem 19.6, when $U_i \cap U_j \cap \Gamma \neq \varnothing$, the p.h.s.'s $A_j^{(0)}(x^{(j)}, \xi^{(j)})$ and $A_i^{(0)}(x^{(i)}, \xi^{(i)})$ are connected in $U_i \cap U_j$ by the relation

$$A_j^{(0)}(x^{(j)}, \xi^{(j)}) = A_i^{(0)}\left\{s_{ij}(x^{(j)}), \left[\left(\frac{Ds_{ij}(x^{(j)})}{Dx^{(j)}}\right)'\right]^{-1}\xi^{(j)}\right\}, \tag{21.2}$$

where the prime denotes the transpose.

Since $x_n^{(i)} = 0$ when $x_n^{(j)} = 0$, it follows that in the Jacobian matrix

$$\frac{Ds_{ij}(x^{(j)})}{Dx^{(j)}} = \|b_{kr}\|^n_{k,r=1} \tag{21.3}$$

we have $b_{nr} = 0$ for $r = 1, \ldots, n-1$ and $b_{nn} > 0$ when $x_n^{(j)} = 0$; hence in the matrix

$$\left[\left(\frac{Ds_{ij}(x^{(j)})}{Dx^{(j)}}\right)'\right]^{-1} = \|c_{kr}\|_{k,r=1}^{n}$$

we have $c_{kn} = 0$ for $k = 1, \ldots, n-1$ and $c_{nn} > 0$ when $x_n^{(j)} = 0$. Consequently, when $x_n^{(j)} = 0$, $\xi^{(i)}$ and $\xi^{(j)}$ are connected by the linear relations

$$\xi_k^{(i)} = \sum_{r=1}^{n-1} c_{kr}\xi_r^{(j)}, \quad 1 \leqslant k \leqslant n-1, \qquad \xi_n^{(i)} = \sum_{r=1}^{n} c_{nr}\xi_r^{(j)}, \qquad c_{nn} > 0. \quad (21.4)$$

Let

$$A_j^{(0)}(x'_{(j)}, 0, \xi'_{(j)}, \xi_n^{(j)}) = A_j^-(x'_{(j)}, \xi'_{(j)}, \xi_n^{(j)})A_j^+(x'_{(j)}, \xi'_{(j)}, \xi_n^{(j)}) \quad (21.5)$$

be the homogeneous factorization of $A_j^{(0)}$ with respect to $\xi_n^{(j)}$ when $x_n^{(j)} = 0$, and let $\kappa_j(x'_{(j)})$ be the factorization index of $A_j^{(0)}$. Analogously, let

$$A_i^{(0)}(x'_{(i)}, 0, \xi'_{(i)}, \xi_n^{(i)}) = A_i^-(x'_{(i)}, \xi'_{(i)}, \xi_n^{(i)})A_i^+(x'_{(i)}, \xi'_{(i)}, \xi_n^{(i)}) \quad (21.6)$$

be the homogeneous factorization of $A_i^{(0)}$ with respect to $\xi_n^{(i)}$ when $x_n^{(i)} = 0$, and let $\kappa_i(x'_{(i)})$ be the factorization index of $A_i^{(0)}$.

It is easily shown that

$$\kappa_j(x'_{(j)}) = \kappa_i(x'_{(i)}), \quad (21.7)$$

where $(x'_{(j)}, 0)$ and $(x'_{(i)}, 0)$ are connected by the transformation s_{ij}, i.e. they are the jth and ith coordinates respectively of one and the same point $x' \in \Gamma$. In fact, substituting

$$(x'_{(i)}, 0) = s_{ij}(x'_{(j)}, 0), \quad \xi^{(i)} = \left[\left(\frac{Ds_{ij}(x'_{(j)}, 0)}{Dx^{(j)}}\right)'\right]^{-1}\xi^{(j)}$$

in (21.6) and making use of (21.2) and (21.4), we obtain another expression for the homogeneous factorization of $A_j^{(0)}$, which must coincide with (21.5) by virtue of the uniqueness of a homogeneous factorization (see §6). This implies, in particular, (21.7).

Thus there is uniquely defined on Γ a function $\kappa(x')$ of class $C^\infty(\Gamma)$ that coincides with $\kappa_j(x'_j)$ in the jth l.c.s. for any U_j intersecting Γ. The function $\kappa(x')$ will be called the *factorization index* of the elliptic symbol $A_0(x, \xi)$ in the domain G.

3. *Sobolev-Slobodeckiĭ spaces on a manifold.* Let M be a compact n-dimensional C^∞-manifold without boundary. This means that there is given a finite covering of M by neighborhoods U_j, $1 \leqslant j \leqslant N$, and injections $x^{(j)} = s_j(x)$ of the neighborhoods U_j onto open balls $V_j \subset \mathbf{R}^n$ such that in every nonempty intersection $U_i \cap U_j$ the function $x^{(i)} = s_{ij}(x^{(j)}) = s_i(s_j^{-1}(x^{(j)}))$ is infinitely differentiable and the Jacobian $|Ds_{ij}(x^{(j)})/Dx^{(j)}|$ does not vanish. We will call U_j a *coordinate neighborhood* and $x^{(j)}$ the jth *local coordinate system* (l.c.s.). Another covering of M by neighborhoods U_k', $1 \leqslant k \leqslant N'$, with injections $y^{(k)} = s_k'(y)$ from the U_k' onto open balls $V_k' \subset \mathbf{R}^n$ is said to be *equivalent* to the system $\{U_j, s_j\}$, $1 \leqslant j \leqslant N$, if for every nonempty intersection of the form $U_k' \cap U_j$ the map $s_k's_j^{-1}$ is infinitely differentiable. The boundary Γ of a smooth domain $G \subset \mathbf{R}^n$ (see §21.1) serves as an example of a manifold M of dimension $n - 1$.

Let $C^\infty(M)$ denote the space of functions that are infinitely differentiable on M, i.e. infinitely differentiable in the l.c.s. of each coordinate neighborhood. A sequence v_n is said to converge in $C^\infty(M)$ if it converges uniformly together with all of its derivatives. By a *generalized function* on M is meant a continuous semilinear functional on $C^\infty(M)$. The space of generalized functions on M is denoted by $S'(M)$.

Let $\{\varphi_j(x)\}$ be a partition of unity subordinate to the covering $\{U_j\}$, $1 \leqslant j \leqslant N$. The *Sobolev-Slobodeckiĭ space* $H_s(M)$ is defined as the totality of generalized functions u on M with finite norm

$$\|u\|_s^2 = \sum_{j=1}^{N} \|\varphi_j u\|_s^2, \tag{21.8}$$

in the right side of which $\|\varphi_j u\|_s^2$ is defined in the jth l.c.s. by (4.1). As in the case of $H_s(\mathbf{R}^n)$, it is possible to also define $H_s(M)$ as the completion of $C^\infty(M)$ in the norm (21.8) (cf. Theorem 4.1).

LEMMA 21.2. *Suppose $y = s(x)$ is a diffeomorphism from $\mathbf{R}^n$ into itself such that $y = x$ for $|x| \geqslant N$. Then the map $u(x) \to v(x) = u(s(x))$ from the function space $S(\mathbf{R}^n)$ into itself extends by continuity to a bijection from $H_s(\mathbf{R}^n)$ onto itself for any s.*

PROOF. Suppose $u(x) \in S(\mathbf{R}^n)$. When $s = 0$, we have

$$\|u\|_0^2 = \int_{-\infty}^{\infty} |u(y)|^2\,dy = \int_{-\infty}^{\infty} |u(s(x))|^2 \left|\frac{Ds(x)}{Dx}\right| dx \leqslant C \int_{-\infty}^{\infty} |v(x)|^2\,dx = C\|v\|_0^2,$$

where $|Ds(x)/Dx|$ is the Jacobian of the transformation $y = s(x)$. Since, by assumption, $|Ds(x)/Dx| \neq 0$, we also have the opposite inequality $\|v\|_0 \leqslant C\|u\|_0$.

When $s \neq 0$, we have $\|u\|_s^2 \leqslant C\|\Lambda_s u\|_0^2$, where Λ_s is the p.o. with symbol $\Lambda_s(\xi) = (1 + |\xi|^2)^{s/2}$. On the other hand, letting $u_1 = \Lambda_s u$ and making the change of variables $y = s(x)$, we get $u_1(s(x)) = (\Lambda_s^{(1)} v)(x)$, where $\Lambda_s^{(1)}$ is an operator of order s by virtue of Theorem 19.6, so that $\|u_1\|_0^2 \leqslant C\|v\|_s^2$ by virtue of Theorem 18.1. Thus $\|u\|_s \leqslant C\|v\|_s$. The opposite inequality $\|v\|_s \leqslant C\|u\|_s$ is proved analogously. Since $S(\mathbf{R}^n)$ is dense in $H_s(\mathbf{R}^n)$ (see Theorem 4.1), the assertion of Lemma 21.2 follows.

Let $\{\varphi_j\}$ and $\{\varphi'_k\}$ be partitions of unity that are respectively subordinate to equivalent systems $\{U_j, s_j\}$, $1 \leqslant j \leqslant N$, and $\{U'_k, s'_k\}$, $1 \leqslant k \leqslant N'$. Since

$$\|\varphi_j u\|_s \leqslant \sum_k \|\varphi_j \varphi'_k u\|_s \leqslant c \sum_k \|\varphi'_k u\|_s$$

it follows from Lemma 21.2 that the space $H_s(M)$ is independent of equivalent systems and their partitions of unity.

4. *P.o.'s on a manifold.* An operator A acting from $C^\infty(M)$ into itself is called a pseudodifferential operator of class $\hat{O}^\infty_{\alpha+i\beta}$ if it satisfies the following two conditions:

1) $\psi A \varphi$ is an integral operator on M with an infinitely differentiable kernel whenever $\varphi(x) \in C^\infty(M)$, $\psi(x) \in C^\infty(M)$ and the supports of $\varphi(x)$ and $\psi(x)$ do not intersect.

2) Whenever the supports of $\varphi(x)$ and $\psi(x)$ lie in the same coordinate neighborhood U_j and intersect, the operator $\psi A \varphi$ written in the jth local coordinates has the form

$$\psi_j(x^{(j)}) A_j(x^{(j)}, D^{(j)}) \varphi_j(x^{(j)}), \tag{21.9}$$

where $A_j(x^{(j)}, D^{(j)})$ is the p.o. in the jth l.c.s. with symbol $A_j(x^{(j)}, \xi^{(j)}) \in \hat{O}^\infty_{\alpha+i\beta}$.

5. *Symbol of a p.o. on a manifold.* By the *cotangent bundle* $T^*(M)$ is meant the vector bundle over M (see §16) with fiber $\mathbf{R}^n$ and transition matrices having in the jth l.c.s. the form

$$C_{ij}(x^{(j)}) = \left[\left(Ds_{ij}(x^{(j)})/Dx^{(j)}\right)'\right]^{-1},$$

where $Ds_{ij}(x^{(j)})/Dx^{(j)}$ is the Jacobian matrix of the transformation of the jth l.c.s. into the ith l.c.s. Thus, if $(x^{(j)}, \xi^{(j)})$ are the coordinates of a point in $U_j \times \mathbf{R}^n$ and $(x^{(i)}, \xi^{(i)})$ are the coordinates of a point in $U_i \times \mathbf{R}^n$, then $T^*(M)$ is determined by identifying the points $(x^{(i)}, \xi^{(i)}, i)$, $(x^{(j)}, \xi^{(j)}, j)$ over $U_i \cap U_j$ when

$$x^{(i)} = s_{ij}(x^{(j)}), \qquad \xi^{(i)} = \left[\left(Ds_{ij}(x^{(j)})/Dx^{(j)}\right)'\right]^{-1} \xi^{(j)}.$$

The elements of $T^*(M)$ will be denoted by (x, ξ_x), and ξ_x will be called a *cotangent vector* at the point $x \in M$. The part of $T^*(M)$ consisting of nonnull cotangent vectors is denoted by $T^*_0(M)$.

Let $A_j^{(0)}(x^{(j)}, \xi^{(j)})$ and $A_i^{(0)}(x^{(i)}, \xi^{(i)})$ be the principal homogeneous symbols of the operators $A_j(x^{(j)}, D^{(j)})$ and $A_i(x^{(i)}, D^{(i)})$ (see (21.9)). By Theorem 19.4, we have (cf. (21.2))

$$A_j^{(0)}(x^{(j)}, \xi^{(j)}) = A_i^{(0)}\left[s_{ij}(x^{(j)}), \left[\left(\frac{Ds_{ij}(x^{(j)})}{Dx^{(j)}}\right)'\right]^{-1}\xi^{(j)}\right]$$

in the intersection $U_i \cap U_j$. Consequently, there is uniquely defined on $T_0^*(M)$ a function $A^{(0)}(x, \xi_x)$ that coincides in any set $U_j \times \mathbf{R}^n$ with $A_j^{(0)}(x^{(j)}, \xi^{(j)})$, in the jth l.c.s. Clearly, $A^{(0)}(x, \xi_x) \in C^\infty$ on $T_0^*(M)$ and is homogeneous in ξ_x of degree $\alpha + i\beta$. The function $A^{(0)}(x, \xi_x)$ will be called the *principal homogeneous symbol* (p.h.s.) of the p.o. A on the manifold M.

A p.o. can be constructed on M whose p.h.s. is a given function $A^{(0)}(x, \xi_x)$. For suppose $\varphi_j(x) \in C_0^\infty(U_j)$ is a partition of unity on M subordinate to the covering $\{U_j\}$, $1 \leqslant j \leqslant N$, and suppose $\psi_j(x)\varphi_j(x) \equiv \varphi_j(x)$. We define a p.o. A by means of the formula

$$Au = \sum_{j=1}^{N} \psi_j A_j \varphi_j u, \tag{21.10}$$

where A_j is the p.o. in $\mathbf{R}^n$ with symbol

$$A_j^{(0)}(x_{(j)}, \xi_{(j)})(1 - \chi(\xi_{(j)})) \in \hat{O}^\infty_{\alpha+i\beta}$$

in the jth l.c.s. The principal homogeneous symbol of the operator (21.10) is the function $A^{(0)}(x, \xi_x)$.

Clearly, two p.o.'s having one and the same principal homogeneous symbol differ by a p.o. of smaller order.

6. *Elliptic p.o.'s on a manifold.* A p.o. A of class $\hat{O}^\infty_{\alpha+i\beta}$ on a manifold M is said to be *elliptic* if its principal homogeneous symbol $A^{(0)}(x, \xi_x)$ satisfies the condition

$$\det A^{(0)}(x, \xi_x) \neq 0 \quad \text{for } \xi_x \neq 0,\ x \in M. \tag{21.11}$$

THEOREM 21.1. *An elliptic p.o. of class $\hat{O}^\infty_{\alpha+i\beta}$ on M is a Fredholm operator from $H_s(M)$ into $H_{s-\alpha}(M)$ for any s.*

For the proof of Theorem 21.1 we need the following lemmas, which follow directly from the results of §§18 and 19.

LEMMA 21.3. *A p.o. A of class $\hat{O}^\infty_{\alpha+i\beta}$ is a bounded operator from $H_s(M)$ into $H_{s-\alpha}(M)$ for any s.*

LEMMA 21.4. *Let A and B be p.o.'s on M of classes $\hat{O}^\infty_{\alpha_1+i\beta_1}$ and $\hat{O}^\infty_{\alpha_2+i\beta_2}$ respectively. Then the product $C = AB$ is a p.o. on M of class $\hat{O}^\infty_{\alpha_3+i\beta_3}$, where $\alpha_3 + i\beta_3 = \alpha_1 + \alpha_2 + i(\beta_1 + \beta_2)$, and the principal homogeneous symbol of C is the product of the principal homogeneous symbols of A and B.*

LEMMA 21.5. *A p.o. T on M of class $\hat{O}^{\infty}_{\alpha+i\beta}$, with $\alpha < 0$, is a compact operator in $H_s(M)$ for any s.*

PROOF OF THEOREM 21.1. Let $R^{(0)}(x, \xi_x)$ denote the (generally matrix-valued) function on $T_0^*(M)$ that is equal to $(A^{(0)}(x, \xi_x))^{-1}$. We construct a p.o. R on M with principal homogeneous symbol $R^{(0)}(x, \xi_x)$ by means of the formula of form (21.10). Then Lemmas 21.3 and 21.4 imply that R is a bounded operator from $H_{s-\alpha}(M)$ into $H_s(M)$ such that

$$AR = I + T_1, \quad RA = I + T_2, \tag{21.12}$$

where I is the identity operator and T_1 and T_2 are p.o.'s on M of order at most -1. Consequently, Lemma 21.5 implies that T_1 and T_2 are compact operators in $H_{s-\alpha}(M)$ and $H_s(M)$ respectively. Thus R is a regularizer of A, and hence A is a Fredholm operator.

7. *P.o.'s of potential type* (*coboundary operators*). Let G be a bounded domain in $\mathbf{R}^n$ with a smooth boundary Γ, and let p' denote the operator of restriction of a function $u(x) \in S(\mathbf{R}^n)$ to Γ.

LEMMA 21.6. *The operator p' extends by continuity to a bounded operator from $H_s(\mathbf{R}^n)$ into $H_{s-1/2}(\Gamma)$ for $s > 1/2$.*

PROOF. Suppose $u(x) \in S(\mathbf{R}^n)$ and $\{\varphi_j(x)\}$ is a partition of unity subordinate to a covering $\{U_j\}$. We have

$$p'u = \sum_j{}' p'\varphi_j u,$$

where $\sum_j{}'$ means that the summation is taken over only those U_j that intersect Γ. Going over in U_j to the jth l.c.s. and using Lemmas 21.2 and Theorem 4.2, we get

$$[p'u]_{s-1/2} \leqslant \sum_j{}'[p'\varphi_j u]_{s-1/2} \leqslant C\sum_j{}'\|\varphi_j u\|_s \leqslant C\|u\|_s, \tag{21.13}$$

where $[p'u]_{s-1/2}$ is the norm in $H_{s-1/2}(\Gamma)$. Lemma 21.6 is proved.

A generalized function $\rho \in S'(\Gamma)$ determines a generalized function $\rho \times \delta(\Gamma) \in S'(\mathbf{R}^n)$ in accordance with the formula

$$(\rho \times \delta(\Gamma), \varphi) = [\rho, p'\varphi], \qquad \forall\varphi \in S(\mathbf{R}^n), \tag{21.14}$$

where $[\rho, v]$ is the value of the functional ρ at a function $v \in C^\infty(\Gamma)$. In particular, if $\mu(x')$ is an integrable function on Γ, then the corresponding regular functional $\mu \in S'(\Gamma)$ defined by

$$[\mu, v] = \int_\Gamma \mu(x')\overline{v(x')}\, ds_\Gamma, \qquad \forall v(x') \in C^\infty(\Gamma), \tag{21.15}$$

where ds_Γ is the surface element of Γ, determines a generalized function $\mu \times \delta(\Gamma) \in S'(\mathbf{R}^n)$ in accordance with the formula

$$(\mu \times \delta(\Gamma), \varphi) = \int_\Gamma \mu(x')\,\overline{p'\varphi(x)}\, ds_\Gamma, \qquad \forall \varphi \in S(\mathbf{R}^n). \tag{21.16}$$

By an *operator of potential type* or *coboundary operator* is meant an operator that maps each function $v(x')$ of class $C^\infty(\Gamma)$ into a generalized function of class $S'(\mathbf{R}^n)$ of the form (cf. §7)

$$A(x, D)(v \times \delta(\Gamma)), \tag{21.17}$$

where $A(x, D)$ is a p.o. of class $\hat{O}^\infty_{\alpha+i\beta}$.

LEMMA 21.7. *Suppose $A(x, \xi) \in \hat{O}^\infty_{\alpha+i\beta}$ and $s < -\alpha - 1/2$. Then the operator $A(x, D)(v \times \delta(\Gamma))$ extends by continuity from $C^\infty(\Gamma)$ to a bounded operator from $H_{s+\alpha+1/2}(\Gamma)$ into $H_s(\mathbf{R}^n)$.*

PROOF. By Theorem 18.1,

$$\|A(x, D)(v \times \delta(\Gamma))\|_s \leqslant C\|v \times \delta(\Gamma)\|_{s+\alpha}, \qquad \forall v \in C^\infty(\Gamma). \tag{21.18}$$

It therefore suffices to show that

$$\|v \times \delta(\Gamma)\|_{s+\alpha} \leqslant C[v]_{s+\alpha+1/2}. \tag{21.19}$$

As in the proof of Lemma 21.6, we have

$$v \times \delta(\Gamma) = \sum_j{}' \varphi_j(x)(v \times \delta(\Gamma)). \tag{21.20}$$

In the jth l.c.s. the function $\varphi_j(x)(v \times \delta(\Gamma))$ has the form

$$a_j(x'_{(j)})\varphi_{j0}(x'_{(j)})v_j(x'_{(j)}) \times \delta(x_n^{(j)}),$$

where $a_j(x'_{(j)})$ is a function of class $C^\infty(\mathbf{R}^{n-1})$ and $\varphi_{j0}(x'_{(j)})$ is the restriction of $\varphi_j(x)$ to Γ in the jth l.c.s. But, in view of the fact that $s + \alpha < -1/2$,

$$\|\varphi_{j1}v_j \times \delta(x_n^{(j)})\|_{s+\alpha} \leqslant C[\varphi_{j1}v_j]_{s+\alpha+1/2}, \tag{21.21}$$

where $\varphi_{j1} = a_j\varphi_{j0}$. Thus (21.19) follows from (21.20), (21.21) and Lemma 21.2.

EXAMPLE 21.1. Suppose $\Lambda^{-2}(\xi) = 1/|\xi|^2$, $n > 3$ and $\mu(x') \in C^\infty(\Gamma)$. Then, by (3.15″),

$$\Lambda^{-2}u = \frac{\Gamma((n-2)/2)}{4\pi^{n/2}} \int_\Gamma \frac{u(y)\,dy}{|x-y|^{n-2}}, \qquad n > 3.$$

Applying $\Lambda^{-2}(D)$ to the generalized function with compact support $\mu \times \delta(\Gamma)$ (see (2.60)), we get

$$\Lambda^{-2}(D)(\mu \times \delta(\Gamma)) = \frac{\Gamma((n-2)/2)}{4\pi^{n/2}} \int_\Gamma \frac{\mu(y')\,ds_\Gamma}{|x-y'|^{n-2}}, \qquad n > 3.$$

The function $\Lambda^{-2}(\mu \times \delta(\Gamma))$ is called a *single layer potential*.

CHAPTER VI

BOUNDARY VALUE PROBLEMS FOR ELLIPTIC PSEUDODIFFERENTIAL OPERATORS IN A BOUNDED DOMAIN WITH A SMOOTH BOUNDARY

§22. Normal solvability of the general boundary value problems in a domain with a smooth boundary

1. *Case of a system of p.e.'s.* Let $A(x, \xi) = \|a_{ij}(x, \xi)\|_{i,j=1}^{p}$ be an elliptic symbol in $\mathbf{R}^n$ of class $\hat{O}^{\infty}_{\alpha+i\beta}$. There then exists a principal homogeneous symbol (p.h.s.) $A_0(x, \xi) \in O^{\infty}_{\alpha+i\beta}$ such that

$$\det A_0(x, \xi) \neq 0, \qquad \xi \neq 0, \tag{22.1}$$

and

$$A(x, \xi) - A_0(x, \xi)(1 - \chi(\xi)) \in S^{\infty}_{\alpha-1}.$$

We will consider the p.o. $pA(x, D)u_0$ in a domain $G \subset \mathbf{R}^n$ with a smooth boundary Γ.

Let $A_j^{(0)}(x^{(j)}, \xi^{(j)})$ be the p.h.s. of the operator $A(x, D)$ in U_j written in the jth local coordinates (see §21.1), and let

$$\begin{aligned} a_+^{(j)}(x'_{(j)}) &= A_j^{(0)}(x'_{(j)}, 0, 0, +1), \\ a_-^{(j)}(x'_{(j)}) &= e^{i\pi(\alpha+i\beta)} A_j^{(0)}(x'_{(j)}, 0, 0, -1), \\ a_0^{(j)}(x'_{(j)}) &= \left(a_+^{(j)}(x'_{(j)})\right)^{-1} a_-^{(j)}(x'_{(j)}). \end{aligned} \tag{22.2}$$

In $U_i \cap U_j \cap \Gamma$, by (21.4) and the homogeneity of $A_j^{(0)}(x^{(j)}, \xi^{(j)})$ in $\xi^{(j)}$, we have

$$\begin{aligned} A_j^{(0)}(x'_{(j)}, 0, 0, +1) &= c_{nn}^{\alpha+i\beta} A_i^{(0)}(x'_{(i)}, 0, 0, +1), \\ A_j^{(0)}(x'_{(j)}, 0, 0, -1) &= c_{nn}^{\alpha+i\beta} A_i^{(0)}(x'_{(i)}, 0, 0, -1), \end{aligned}$$

where $(x'_{(i)}, 0) = s_{ij}(x'_{(j)}, 0)$. Consequently, there is uniquely defined on Γ an infinitely differentiable matrix $a_0(x')$ that coincides in U_j with $a_0^{(j)}(x'_{(j)})$ in the jth l.c.s. Let $\gamma_j(x')$, $1 \leqslant j \leqslant p$, denote the eigenvalues of the matrix

$$\gamma(x') = \frac{1}{2\pi i} \ln a_0(x').$$

Since the matrix $\ln a_0(x')$ is not uniquely defined, the numbers $\gamma_j(x')$ are defined to within an integral summand.

It is assumed that there exists a real number s such that

$$s - \operatorname{Re} \gamma_j(x') - 1/2 \in \mathbf{R} \setminus \mathbf{Z}, \qquad 1 \leqslant j \leqslant p, \quad \forall x' \in \Gamma. \tag{22.3}$$

By Theorem 14.2, for any fixed $x'_{(j)}$ and $\xi'_{(j)} \neq 0$, $1 \leqslant j \leqslant N$, the operator $pA_j^{(0)}(x'_{(j)}, 0, \xi'_{(j)}, D_n^{(j)})$ is a Fredholm operator from $H_s^+(\mathbf{R}^1)$ into $H_{s-\alpha}(\mathbf{R}^1_+)$.

We will consider for the p.o. $A(x, D)$ in the domain G the following boundary value problem of form (16.3), (16.4):

$$pA(x, D)u_0 + \sum_{k=1}^{m_-} pC_k(x, D)(v_k(x') \times \delta(\Gamma)) = f(x), \tag{22.4}$$

$$p'B_j(x, D)u_0 + \sum_{k=1}^{m_-} E_{jk}(x', D')v_k(x') = g_j(x'), \qquad 1 \leqslant j \leqslant m_+, \quad x' \in \Gamma, \tag{22.5}$$

where $B_j(x, D)$ and $C_k(x, D)$ are p.o.'s in $\mathbf{R}^n$ whose symbols $B_j(x, \xi)$ and $C_k(x, \xi)$ are respectively p-dimensional row and column vectors of classes $\hat{O}^\infty_{\beta_j}$ and $\hat{O}^\infty_{\gamma_k}$ with

$$\operatorname{Re} \beta_j < s - 1/2, \quad \operatorname{Re} \gamma_k < \alpha - s - 1/2, \qquad 1 \leqslant j \leqslant m_+, \quad 1 \leqslant k \leqslant m_-, \tag{22.6}$$

and $E_{jk}(x', D')$ is a p.o. on the manifold Γ of class $\hat{O}^\infty_{\beta_{jk}}$ with

$$\beta_{jk} = \beta_j + \gamma_k - \alpha - i\beta + 1. \tag{22.6$'$}$$

Further, $f(x) \in H_{s-\alpha}(G)$ and $g_j(x') \in H_{s - \operatorname{Re} \beta_j - 1/2}(\Gamma)$, $1 \leqslant j \leqslant m_+$, are given functions, while $u_0 \in \mathring{H}_s(G)$ and $v_k(x') \in H_{s-\alpha+\operatorname{Re}\gamma_k+1/2}(\Gamma)$, $1 \leqslant k \leqslant m_-$, are the desired functions.

Let

$$\mathcal{H}_s^{(1)}(G) = \mathring{H}_s(G) \times \prod_{k=1}^{m_-} H_{s-\alpha+\operatorname{Re}\gamma_k+1/2}(\Gamma),$$

and

$$\mathcal{H}_s^{(2)}(G) = H_{s-\alpha}(G) \times \prod_{j=1}^{m_+} H_{s-\operatorname{Re}\beta_j-1/2}(\Gamma).$$

By Theorem 18.1 and Lemmas 21.3, 21.6 and 21.7, the operator $\mathfrak{A}$ defined by the left sides of (22.4) and (22.5) is a bounded operator from $\mathcal{H}_s^{(1)}(G)$ into $\mathcal{H}_s^{(2)}(G)$.

Let U_i be an arbitrary neighborhood intersecting Γ, and let $B_{j0}^{(i)}(x^{(i)}, \xi^{(i)})$ and $C_{k0}^{(i)}(x^{(i)}, \xi^{(i)})$ denote the p.h.s.'s of the operators $B_j(x, D)$ and $C_k(x, D)$ in U_i written in the ith local coordinates. The intersection $U_i' = U_i \cap \Gamma$ is a coordinate neighborhood on Γ in which the restriction of the ith l.c.s. to U_i' is a local coordinate system. Let $E_{jk0}^{(i)}(x'_{(i)}, \xi'_{(i)})$ denote the p.h.s. of the p.o. $E_{jk}(x', D')$ in U_i' written in the ith local coordinates. We will say that the Šapiro-Lopatinskiĭ condition is satisfied for the boundary value problem (22.4), (22.5) if for any neighborhood U_i, $1 \leqslant i \leqslant N$, intersecting Γ and any fixed $x'_{(i)} = (x_1^{(i)}, \ldots, x_{n-1}^{(i)})$, $x_n^{(i)} = 0$ and $\xi'_{(i)} \neq 0$ in the ith l.c.s. the boundary value problem in $\mathbf{R}_+^1$ (cf. (16.5) and (16.6))

$$pA_i^{(0)}\big(x'_{(i)}, 0, \xi'_{(i)}, D_n^{(i)}\big)v_+\big(x_n^{(i)}\big) + \sum_{k=1}^{m_-} \rho_k pC_{k0}^{(i)}\big(x'_{(i)}, 0, \xi'_{(i)}, D_n^{(i)}\big)\delta\big(x_n^{(i)}\big) = g\big(x_n^{(i)}\big), \tag{22.7}$$

$$p'B_{j0}^{(i)}\big(x'_{(i)}, 0, \xi'_{(i)}, D_n^{(i)}\big)v_+\big(x_n^{(i)}\big) + \sum_{k=1}^{m_-} E_{jk0}^{(i)}(x'_{(i)}, \xi'_{(i)})\rho_k = h_j, \quad 1 \leqslant j \leqslant m_+, \tag{22.8}$$

has a unique solution $(v_+(x_n^{(i)}), \rho_1, \ldots, \rho_{m_-}) \in H_s^+(\mathbf{R}^1) \times \mathbf{C}^{m_-}$ for arbitrary $(g(x_n^{(i)}), h_1, \ldots, h_{m_+}) \in H_{s-\alpha}(\mathbf{R}_+^1) \times \mathbf{C}^{m_+}$.

THEOREM 22.1. *Suppose conditions* (22.1), (22.3), (22.6), (22.6′) *and the Šapiro-Lopatinskiĭ condition are satisfied. Then the operator* $\mathfrak{A}$ *defined by the left sides of* (22.4) *and* (22.5) *is a Fredholm operator acting from* $\mathcal{H}_s^{(1)}(G)$ *into* $\mathcal{H}_s^{(2)}(G)$.

PROOF. Let x_0 be an arbitrary point on Γ, let $U_i^{(0)}$ denote a sufficiently small neighborhood of x_0 that is entirely contained in some coordinate neighborhood U_i, and let $\varphi_0(x)$ and $\psi_0(x)$ be arbitrary functions of class $C_0^\infty(U_i^{(0)})$. The operator $\psi_0(x)A(x, D)\varphi_0(x)$ has the following form in the ith l.c.s.:

$$\psi_0^{(i)}(x^{(i)})\big(A_i(x^{(i)}, D^{(i)}) + T_i^{(0)}\big)\varphi_0^{(i)}(x^{(i)}),$$

where $\psi_0^{(i)}(x^{(i)})$ and $\varphi_0^{(i)}(x^{(i)})$ are the functions $\psi_0(x)$ and $\varphi_0(x)$ written in the ith local coordinates, $A_i(x^{(i)}, D^{(i)})$ is the p.o. with symbol

$$A_i^{(0)}(x^{(i)}, \xi^{(i)})(1 - \chi(\xi^{(i)})),$$

in which $A_i^{(0)}(x^{(i)}, \xi^{(i)})$ is the p.h.s. of $A(x, D)$ in the ith l.c.s. in U_i, and $T_i^{(0)}$ is of order at most $\alpha - 1$. On the other hand, by Theorem 19.4, the operator $\psi_0^{(i)} A_i(x^{(i)}, D^{(i)})$ admits a decomposition of the form

$$\psi_0^{(i)}(x^{(i)}) A_i(x^{(i)}, D^{(i)}) = \psi_0^{(i)}(x^{(i)}) \big(A_i(x_0^{(i)}, D^{(i)}) + K_i^{(0)}(x^{(i)}, D^{(i)}) + T_i(x^{(i)}, D^{(i)}) \big),$$

where $x_0^{(i)}$ denotes the coordinates of x_0 in the ith l.c.s., the norm of $K_i^{(0)}(x^{(i)}, D^{(i)})$ as an operator from $H_s(\mathbf{R}^n)$ into $H_{s-\alpha}(\mathbf{R}^n)$ is small if $U_i^{(0)}$ is sufficiently small, and T_i is of order at most $\alpha - 1$. Therefore, letting $\hat{A}_i(x_0^{(i)}, D^{(i)})$ denote the p.o. with symbol

$$A_i\left(x_0^{(i)}, \frac{1 + |\xi'_{(i)}|}{|\xi'_{(i)}|} \xi'_{(i)}, \xi_n^{(i)} \right)$$

as in Chapters II–IV, we can represent the operator $\psi_0(x) A(x, D) \varphi_0(x)$ in the ith l.c.s. in the form

$$\psi_0^{(i)}(x^{(i)}) \big(\hat{A}_i(x_0^{(i)}, D^{(i)}) + K_i^{(0)}(x^{(i)}, D^{(i)}) + T_i^{(1)} \big) \varphi_0^{(i)}(x^{(i)}), \tag{22.9}$$

where $T_i^{(1)}$ is of order at most $\alpha - 1$. The operators

$$\psi_0(x) B_j(x, D) \varphi_0(x), \quad \psi_0(x) C_k(x, D) \varphi_0(x) \quad \text{and} \quad \psi_0'(x') E_{jk}(x', D') \varphi_0'(x'),$$

where $\psi_0'(x')$ and $\varphi_0'(x')$ denote $p'\psi_0(x)$ and $p'\varphi_0(x)$, admit analogous decompositions in the ith l.c.s.

Suppose

$$V = \big(u_0(x), v_1(x'), \dots, v_{m_-}(x') \big) \in \mathcal{H}_s^{(1)}$$

and

$$\Phi = \big(f(x), g_1(x'), \dots, g_{m_+}(x') \big) \in \mathcal{H}_s^{(2)}.$$

For brevity we will use the notation

$$\varphi V = \big(\varphi(x) u_0(x), \varphi'(x') v_1(x'), \dots, \varphi'(x') v_{m_-}(x') \big)$$

and

$$\varphi \Phi = \big(\varphi(x) f(x), \varphi'(x') g_1(x'), \dots, \varphi'(x') g_{m_+}(x') \big),$$

where $\varphi(x) \in C_0^\infty(\mathbf{R}^n)$ and $\varphi'(x') = p'\varphi(x)$.

In particular, if $\operatorname{supp} \varphi \cap \Gamma = \varnothing$, then

$$\varphi V = (\varphi(x) u_0(x), 0, \dots, 0), \quad \varphi \Phi = (\varphi(x) f(x), 0, \dots, 0).$$

Let $\psi_0^{(i)}(x^{(i)}) \mathfrak{A}_i \varphi_0^{(i)}(x^{(i)})$ be the expression for the operator $\psi_0 \mathfrak{A} \varphi_0$ in the ith l.c.s. By (22.9), the operator $\psi_0^{(i)} \mathfrak{A}_i \varphi_0^{(i)}$ can be represented in the form

$$\psi_0^{(i)} \mathfrak{A}_i \varphi_0^{(i)} = \psi_0^{(i)} \big(\hat{\mathfrak{A}}_i^{(0)} + K_i + T_i \big) \varphi_0^{(i)}, \tag{22.10}$$

where $\hat{\mathfrak{A}}_i^{(0)}$ is the operator analogous to the one defined by (16.3) and (16.4), K_i is a bounded operator with small norm from $\mathcal{H}_s^{(1)}(\mathbf{R}^n)$ into $\mathcal{H}_s^{(2)}(\mathbf{R}^n)$ and T_i is a bounded operator from $\mathcal{H}_s^{(1)}(\mathbf{R}^n)$ into $\mathcal{H}_{s+1}^{(2)}(\mathbf{R}^n)$ for any s. Under the fulfilment of the conditions of Theorem 22.1 it follows from Theorem 16.1 that $\hat{\mathfrak{A}}_i^{(0)}$ has a bounded inverse $\hat{R}_i^{(0)}$ from $\mathcal{H}_s^{(2)}(\mathbf{R}^n)$ into $\mathcal{H}_s^{(1)}(\mathbf{R}^n)$. If we choose the neighborhood $U_i^{(0)}$ so small that

$$\|K_i\|_{(s)} < \frac{1}{2\|\hat{R}_1^{(0)}\|_{(s)}}, \tag{22.11}$$

then the operator $R_i^{(0)} = \hat{R}_i^{(0)}(I_2 + K_i\hat{R}_i^{(0)})^{-1}$, where I_2 is the identity operator in $\mathcal{H}_s^{(2)}(\mathbf{R}^n)$, is the inverse of $\hat{\mathfrak{A}}_i^{(0)} + K_i$, i.e.

$$(\hat{\mathfrak{A}}_i^{(0)} + K_i)R_i^{(0)} = I_2, \qquad R_i^{(0)}(\hat{\mathfrak{A}}_i^{(0)} + K_i) = I_1, \tag{22.12}$$

where I_1 is the identity operator in $\mathcal{H}_s^{(1)}(\mathbf{R}^n)$.

From the covering of Γ consisting of neighborhoods $U_i^{(0)}$ we choose a finite subcovering and extend it to a finite covering of the closed domain $\overline{G}$. In order not to complicate the notation, we will denote this covering, as before, by U_i. Let $\{\varphi_i(x)\}$ be a partition of unity subordinate to the covering $\{U_i\}$ and let $\psi_i(x)$ be a function of class $C_0^\infty(U_i)$ such that $\psi_i = 1$ in a neighborhood of supp φ_i. By (22.10) and (22.12), for each neighborhood U_i there exists in the ith l.c.s. a bounded operator $R_i^{(0)}$ from $\mathcal{H}_s^{(2)}(\mathbf{R}^n)$ into $\mathcal{H}_s^{(1)}(\mathbf{R}^n)$ such that

$$R_i^{(0)}\mathfrak{A}_i = I_1 + T_{i1}, \qquad \mathfrak{A}_i R_i^{(0)} = I_2 + T_{i2}, \tag{22.13}$$

where $\mathfrak{A}_i$ is the operator $\mathfrak{A}$ in U_i written in the ith local coordinates, T_{i1} is a bounded operator from $\mathcal{H}_{s-1}^{(1)}(\mathbf{R}^n)$ into $\mathcal{H}_s^{(1)}(\mathbf{R}^n)$ and T_{i2} is a bounded operator from $\mathcal{H}_s^{(2)}(\mathbf{R}^n)$ into $\mathcal{H}_{s+1}^{(2)}(\mathbf{R}^n)$. We denote by $\psi_i R_i \varphi_i$ the operator written in the original coordinates in $\mathbf{R}^n$ that corresponds to the operator $\psi_i^{(0)}(x^{(i)})R_i^{(0)}\varphi_i^{(0)}(x^{(i)})$ in the ith l.c.s. In the case of a neighborhood U_i that does not intersect Γ we denote by R_i the p.o. with symbol $A_0^{-1}(x, \xi)(1 - \chi(\xi))$. For such neighborhoods the operator $\mathfrak{A}$ reduces to the p.o. $A(x, D)$ to within an infinitely smoothing operator. It consequently follows from Theorems 18.2 and 18.3 that the relations of form (22.13) also hold in this case.

The operator R defined by

$$R\Phi = \sum_i \psi_i R_i \varphi_i \Phi, \qquad \forall \Phi \in \mathcal{H}_s^{(2)}(G), \tag{22.14}$$

is a bounded operator from $\mathcal{H}_s^{(2)}(G)$ into $\mathcal{H}_s^{(1)}(G)$ by virtue of Lemma 21.2 on the invariance of the Sobolev-Slobodeckiĭ spaces under a change of variables. We will prove that R is a regularizer of $\mathfrak{A}$.

To show that R is a left regularizer of $\mathfrak{A}$, we note that

$$R\mathfrak{A}V = \sum_i \psi_i R_i \varphi_i \mathfrak{A}\psi_i V + T^{(0)}V, \tag{22.15}$$

where

$$T^{(0)} = \sum_i \psi_i R_i \varphi_i \mathfrak{A}(1 - \psi_i).$$

The pseudolocal property (see §19) implies $\varphi_i \mathfrak{A}(1 - \psi_i)$ is of order $-\infty$, and hence $T^{(0)}$ is at least a bounded operator from $\mathcal{H}^{(1)}_{s-1}(G)$ into $\mathcal{H}^{(1)}_s(G)$. Writing $\varphi_i \mathfrak{A} \psi_i$ in the ith local coordinates and permuting $\mathfrak{A}_i$ and $\varphi_i^{(0)}$, we get by virtue of Corollary 18.1 and (22.13) that

$$\psi_i^{(0)} R_i^{(0)} \varphi_i^{(0)} \mathfrak{A}_i \psi_i^{(0)} = \psi_i^{(0)} R_i^{(0)} (\mathfrak{A}_i \varphi_i^{(0)} + T_i^{(1)}) \psi_i^{(0)} = \varphi_i^{(0)} I_1 + \psi_i^{(0)} T_i^{(2)} \psi_i^{(0)}, \tag{22.16}$$

where $T_i^{(2)}$ is a bounded operator from $\mathcal{H}^{(1)}_{s-1}(\mathbf{R}^n)$ into $\mathcal{H}^{(1)}_s(\mathbf{R}^n)$. Thus, by (22.15) and (22.16),

$$R\mathfrak{A}V = V + T^{(3)}V, \qquad \forall V \in \mathcal{H}^{(1)}_s(G), \tag{22.17}$$

where $T^{(3)}$ is a bounded operator from $\mathcal{H}^{(1)}_{s-1}(G)$ into $\mathcal{H}^{(1)}_s(G)$ and hence, according to Remark 21.1, is compact in $\mathcal{H}^{(1)}_s(G)$.

To show that R is a right regularizer of $\mathfrak{A}$, we let $\psi_{i1}(x)$ be a function of class $C_0^\infty(U_i)$ such that $\psi_{i1}\psi_i \equiv \psi_i$, and we note that

$$\mathfrak{A} R \Phi = \sum_i \psi_{i1} \mathfrak{A} \psi_i R_i \varphi_i \Phi + T_0 \Phi, \tag{22.18}$$

where $T_0 = \Sigma_i (1 - \psi_{i1}) \mathfrak{A} \psi_i R_i \varphi_i$ is at least a bounded operator from $\mathcal{H}^{(2)}_s(G)$ into $\mathcal{H}^{(2)}_{s+1}(G)$. Analogously to (22.16), permuting $\psi_i^{(0)}(x^{(i)})$ and $\mathfrak{A}_i$ in the ith l.c.s., we get

$$\psi_{i1}^{(0)}(x^{(i)}) \mathfrak{A}_i \psi_i^{(0)} R_i^{(0)} \varphi_i^{(0)} = (\psi_i^{(0)} \mathfrak{A}_i + T_{1i}) R_i^{(0)} \varphi_i^{(0)} = \varphi_i^{(0)} I_2 + \psi_i^{(0)} T_{2i} \varphi_i^{(0)}, \tag{22.19}$$

where T_{2i} is a bounded operator from $\mathcal{H}^{(2)}_s(\mathbf{R}^n)$ into $\mathcal{H}^{(2)}_{s+1}(\mathbf{R}^n)$. Thus

$$\mathfrak{A} R \Phi = \Phi + T_3 \Phi, \qquad \forall \Phi \in \mathcal{H}^{(2)}_s(G), \tag{22.20}$$

where T_3 is a bounded operator from $\mathcal{H}^{(2)}_s(G)$ into $\mathcal{H}^{(2)}_{s+1}(G)$ and hence is compact in $\mathcal{H}^{(2)}_s(G)$ (see Lemma 21.1).

It follows from (22.17) and (22.20) that $\mathfrak{A}$ is a Fredholm operator. Theorem 22.1 is completely proved.

We note that (22.17) implies the a priori estimate

$$\|u_0\|_s + \sum_{k=1}^{m_-} [v_k]_{s - \operatorname{Re} \gamma_k + \alpha + 1/2} \leqslant C\Bigg(\overset{+}{\|f\|}_{s-\alpha} + \sum_{j=1}^{m_+} [g_j]_{s - \operatorname{Re} \beta_j - 1/2} + \|u_0\|_{s-1} + \sum_{k=1}^{m_-} [v_k]_{s - \operatorname{Re} \gamma_k + \alpha - 1/2} \Bigg) \tag{22.21}$$

of the solution of the boundary value problem (22.4), (22.5).

REMARK 22.1. Let us determine when there exist p.o.'s $B_j(x, D)$, $C_k(x, D)$, $E_{jk}(x', D')$, $1 \leq k \leq m_-$, $1 \leq j \leq m_+$, such that the boundary value problem (22.4), (22.5) is normally solvable.

To this end it is convenient to use a special local coordinate system (see §21.1) in each neighborhood U_i intersecting Γ. Let $A_j^{(0)}(x'_{(j)}, x_n^{(j)}, \xi'_{(j)}, \xi_n^{(j)})$ be the p.h.s. of the operator $A(x, D)$ in U_j written in the jth special local coordinates. In $U_i \cap U_j$ we have

$$A_j^{(0)}(x'_{(j)}, x_n^{(j)}, \xi'_{(j)}, \xi_n^{(j)}) = A_i^{(0)}\left(s'_{ij}(x'_{(j)}), x_n^{(j)}, \left[\left(\frac{Ds'_{ij}(x'_{(j)})}{Dx'_{(j)}}\right)'\right]^{-1} \xi'_{(j)}, \xi_n^{(j)} \right) \tag{22.22}$$

by virtue of (21.1). Consequently, there is uniquely defined on $T_0^*(\Gamma)$ an infinitely differentiable matrix $A^{(0)}(x', 0, \xi'_{x'}, \xi_n)$, that coincides with $A_j^{(0)}(x'_{(j)}, 0, \xi'_{(j)}, \xi_n^{(j)})$ in U_j in the jth s.l.c.s.

For brevity we let $A^{(0)}(y, \xi_n) = A^{(0)}(x', 0, \xi'_{x'}, \xi_n)$, where $y = (x', \xi'_{x'})$. If (22.3) is satisfied and $y \in T_0^*(\Gamma)$, then $pA^{(0)}(y, D_n)$ is a Fredholm operator acting from $H_s^+(\mathbf{R}^1)$ into $H_{s-\alpha}(\mathbf{R}^1_+)$. Since the symbol $A^{(0)}(y, \xi_n)$ is homogeneous with respect to $(\xi'_{x'}, \xi_n)$ it is possible, by introducing a metric in $T_0^*(\Gamma)$, to confine oneself to the compact submanifold $S(\Gamma)$ of $T_0^*(\Gamma)$ that consists of the cotangent vectors of unit length in each fiber.

Analogously, we can associate functions $B_j^{(0)}(y, \xi_n)$ and $C_k^{(0)}(y, \xi_n)$ on $S(\Gamma)$ with the symbols $B_j(x, \xi)$ and $C_k(x, \xi)$, $1 \leq j \leq m_+$ and $1 \leq k \leq m_-$. The p.h.s. of the p.o. $E_{jk}(x', D')$ is also a function on $S(\Gamma)$ (see §21).

Thus, with each point $y \in S(\Gamma)$ we can associate the following boundary value problem in $\mathbf{R}^1_+$ of form (16.5), (16.6):

$$pA^{(0)}(y, D_n)v_+(x_n) + \sum_{k=1}^{m_-} \rho_k p C_k^{(0)}(y, D_n)\delta(x_n) = g(x_n), \tag{22.23}$$

$$p'B_j^{(0)}(y, D_n)v_+(x_n) + \sum_{k=1}^{m_-} E_{jk}^{(0)}(y)\rho_k = h_j, \qquad 1 \leq j \leq m_+. \tag{22.24}$$

The following theorem can now be proved analogously to Theorem 16.2.

THEOREM 16.2′. *For sufficiently large m_- and m_+ there exist p.o.'s ensuring the normal solvability of the boundary value problem* (22.4), (22.5) *if and only if the index of the family of Fredholm operators $pA^{(0)}(y, D_n)$ on $S(\Gamma)$ is equal to*

$$k(pA^{(0)}(y, D_n)) = [\underline{m}_+] - [\underline{m}_-] = \operatorname{sgn} m[\underline{m}], \tag{22.25}$$

where m is the index of $pA^{(0)}(y, D_n)$ for a fixed $y \in S(\Gamma)$ (for simplicity it is assumed that $S(\Gamma)$ is connected).

The following theorem is obtained by repeating the proof of Theorem 16.3.

THEOREM 16.3′. *The index of the family of operators $pA^{(0)}(y, D_n)$ on $S(\Gamma)$ is equal to* sgn $m[\underline{m}]$ *when $\alpha + i\beta = s = 0$ if and only if for sufficiently large N the matrix*

$$\left\| \begin{matrix} A^{(0)}(x', x_n, \xi'_{x'}, \xi_n) & 0 \\ 0 & I_N \end{matrix} \right\|,$$

which is defined for $(x', \xi'_x) \in T_0^*(\Gamma)$, $x_n \in (-\delta, \delta)$, $\xi_n \in \mathbf{R}$, *is homotopic to the matrix*

$$\left\| \begin{matrix} \left(\dfrac{\xi_n + i|\xi'_{x'}|}{\xi_n - i|\xi'_{x'}|} \right)^m I_1 & 0 \\ 0 & I_{N+p-1} \end{matrix} \right\|$$

in the class of elliptic matrices belonging to O_0 *and satisfying* (22.3) *when* $s = 0$.

REMARK 22.2. Besides the boundary value problem (22.4), (22.5) one can consider in a bounded domain the more general boundary value problem obtained from problem (22.4), (22.5) by adding to the operator pAu_0 the operator

$$pKu_0 = \sum_{j=1}^{m_0} pG_{j1}(x, D)(p'G_{j2}(x, D)u_0 \times \delta(\Gamma)),$$

where $G_{j2}(x, \xi) \in \hat{O}^{\infty}_{\delta_j}$, $\operatorname{Re} \delta_j < s - 1/2$, and $G_{j1}(x, \xi) \in \hat{D}^{(0)}_{\alpha+i\beta+1-\delta_k}(G)$ (see §§10 and 23). As was noted in Remark 17.4, the investigation of such a boundary value problem is completely analogous to the investigation of problem (22.4), (22.5).

2. *Scalar case.* A special case of problem (22.4), (22.5) is the case when $A(x, D)$ is a scalar elliptic p.o., $u_0(x) \in \mathring{H}_s(G)$ and $f(x) \in H_{s-\alpha}(G)$ are scalar functions, and $C_k(x, D)$ and $B_j(x, D)$ are scalar p.o.'s.

As was noted in §21, with each scalar elliptic p.o. $A(x, D)$ in a domain G it is possible to associate a function $\kappa(x')$ on Γ, viz. the factorization index of $A(x, \xi)$ in G. We assume in this subsection that the oscillation of $\operatorname{Re} \kappa(x')$ on Γ is less than 1, i.e.

$$\max_{x' \in \Gamma} \operatorname{Re} \kappa(x') - \min_{x' \in \Gamma} \operatorname{Re} \kappa(x') < 1, \tag{22.25'}$$

in which case it is possible to choose s so that

$$\operatorname{Re} \kappa(x') - s = m + \delta(x'), \tag{22.26}$$

where m is an integer and $|\delta(x')| < 1/2$ for all $x' \in \Gamma$. We note that (22.26) is equivalent to (22.3) in the scalar case.

The following theorem (cf. Theorem 12.1) is valid by virtue of Theorem 22.1.

THEOREM 22.2. *Suppose* $A(x, D)$ *is a scalar elliptic p.o. of class* $\hat{O}^{\infty}_{\alpha+i\beta}$, *i.e.*

$$A_0(x, \xi) \neq 0, \qquad \xi \neq 0, \tag{22.27}$$

where $A_0(x, \xi)$ *is the p.h.s. of* $A(x, D)$. *Under the fulfilment of conditions* (22.26) *and* (22.6), *the requirement that* $m_+ - m_- = m$ *and the Šapiro-Lopatinskiĭ condition (which is equivalent in each l.c.s. to condition* (12.27) *for* $m \geqslant 0$ *and to* (12.16) *for* $m < 0$) *the boundary value problem* (22.4), (22.5) *with scalar p.o.'s* $A(x, D)$, $B_j(x, D)$ *and* $C_k(x, D)$ *is normally solvable, i.e. defines a Fredholm operator from* $\mathcal{H}_s^{(1)}(G)$ *into* $\mathcal{H}_s^{(2)}(G)$.

We note three additional theorems, which are special cases of Theorem 22.2 and extend Theorems 7.1, 8.1 and 8.3 to the case of an arbitrary domain with a smooth boundary.

THEOREM 22.3. *Suppose* $A(x, \xi) \in \hat{O}^{\infty}_{\alpha+i\beta}$ *is a scalar elliptic symbol and* $\kappa(x')$ *is the factorization index of* $A_0(x, \xi)$ *in* G. *Suppose* (22.25′) *is satisfied and* s *is such that* $|\mathrm{Re}\,\kappa(x') - s| < 1/2$ *for all* $x' \in \Gamma$ *(i.e.* $m = 0$ *in* (22.26)*). Then the equation*

$$pA(x, D)u_0(x) = f(x) \tag{22.28}$$

defines a Fredholm operator from $\mathring{H}_s(G)$ *into* $H_{s-\alpha}(G)$.

THEOREM 22.3′. *Suppose* $A(x, \xi) \in \hat{O}^{\infty}_{\alpha+i\beta}$, *the ellipticity condition* (22.27) *is satisfied, and* (22.26) *holds with* $m > 0$. *Suppose* $B_j(x, \xi) \in \hat{O}^{\infty}_{\beta_j}$ *and* $\mathrm{Re}\,\beta_j < s - 1/2$, *for* $1 \leqslant j \leqslant m$. *Then under the fulfilment of the Šapiro-Lopatinskiĭ condition (which is equivalent in each l.c.s. to* (8.8)*) the boundary value problem*

$$pA(x, D)u_0(x) = f(x), \tag{22.29}$$

$$p'B_j(x, D)u_0(x) = g_j(x'), \qquad 1 \leqslant j \leqslant m, \tag{22.30}$$

is normally solvable for $u_0 \in \mathring{H}_s(G)$, *and*

$$(f(x), g_1(x'), \ldots, g_m(x')) \in H_{s-\alpha}(G) \times \prod_{j=1}^{m} H_{s-\mathrm{Re}\,\beta_j - 1/2}(\Gamma).$$

THEOREM 22.3″. *Suppose* $A(x, D)$ *is a scalar elliptic p.o. of class* $\hat{O}^{\infty}_{\alpha+i\beta}$ *and* (22.26) *holds with* $m < 0$. *Suppose* $C_k(x, \xi) \in \hat{O}^{\infty}_{\gamma_k}$ *and* $\mathrm{Re}\,\gamma_k < \alpha - s - 1/2$, *for* $1 \leqslant k \leqslant |m|$. *Then under the fulfilment of the Šapiro-Lopatinskiĭ condition (which is equivalent in each l.c.s. to* (8.34)*) the equation*

$$p\left(A(x, D)u_0 + \sum_{k=1}^{|m|} C_k(x, D)(v_k(x') \times \delta(\Gamma))\right) = f(x) \tag{22.31}$$

is normally solvable for

$$(u_0(x), v_1(x'), \ldots, v_{|m|}(x')) \in \mathring{H}_s(G) \times \prod_{k=1}^{|m|} H_{s-\alpha+\mathrm{Re}\,\gamma_k+1/2}(\Gamma),$$

and $f(x) \in H_{s-\alpha}(G)$.

EXAMPLE 22.1. *Singular integral equation in the domain* G. Suppose $A_0(x, \xi) \in O^{\infty}_0$. The p.o. $A_0(x, D)$ differs from the p.o. $A(x, D)$ with symbol $A(x, \xi) = A_0(x, \xi)(1 - \chi(\xi)) \in \hat{O}^{\infty}_0$ by an integral operator T with an infinitely differentiable kernel (see §3). Since G is a bounded domain, T is a compact operator from $\mathring{H}_s(G)$ into $H_s(G)$ for any s. Consequently, Theorems 22.2–22.3″ remain in force under the replacement of $A(x, D)$ by $A_0(x, D)$.

Let

$$c_0(x) = \frac{\Gamma(n/2)}{2\pi^{n/2}} \int_{S^{n-1}} A_0(x, \eta)\, ds_\eta,$$

and

$$a_0(x, z) = F_\xi^{-1}(A_0(x, \xi) - c_0(x)).$$

From §3.3 we have $a_0(x, z) \in O_{-n}^\infty$ and $\int_{S^{n-1}} a_0(x, z) ds_z = 0$. Consequently, $a_0(x, z)$ has the form

$$a_0(x, z) = \frac{b_0(x, z/|z|)}{|z|^n},$$

where $b_0(x, \frac{z}{|z|}) \in O_0^\infty$ and $\int_{S^{n-1}} b_0(x, z) ds_z = 0$. Thus the results of §3.3 imply that the p.o. $pA_0(x, D)u_0$ can be expressed in the form of a singular integral operator in G:

$$pA_0(x, D)u_0 \equiv c_0(x)u_0(x) + \text{p.v.}\int_G \frac{b_0(x, (x-y)/|x-y|)}{|x-y|^n} u_0(y)dy,$$

$$\forall u_0 \in C_0^\infty(G), \quad x \in G. \tag{22.32}$$

The factorization index of a symbol $A_0(x, \xi) \in O_0^\infty$ can be an arbitrary function on Γ, so that the condition $\alpha + i\beta = 0$ only slightly simplifies the formulation of boundary value problems for elliptic p.o.'s (see Theorems 22.2–22.3″). We note, in particular, that, according to Theorem 22.3, a sufficient condition for the normal solvability of the singular integral equation

$$c_0(x)u_0(x) + \text{p.v.}\int_G \frac{b_0(x, (x-y)/|x-y|)}{|x-y|^n} u_0(y)dy = f(x), \qquad x \in G, \tag{22.33}$$

in $L_2(G)$ is

$$|\text{Re}\,\kappa(x')| < 1/2, \qquad \forall x' \in \Gamma. \tag{22.34}$$

It can be shown that (22.34), as well as the conditions of Theorems 22.1–22.3″, is also necessary (cf. Remarks 8.4 and 15.1).

3. *Conditions for the unique solvability of the boundary value problem* (22.4), (22.5). By §4.6, the dual space of $\mathcal{H}_s^{(1)}(G)$ is

$$H_{-s}(G) \times \prod_{k=1}^{m_-} H_{-s+\alpha-\text{Re}\,\gamma_k-1/2}(\Gamma).$$

The value of a functional $V^* = (u_0^*(x), v_1^*(x'), \ldots, v_{m_-}^*(x')) \in (\mathcal{H}_s^{(1)}(G))^*$ at an element $V = (u_0(x), v_1(x'), \ldots, v_{m_-}(x')) \in \mathcal{H}_s^{(1)}(G)$ has the form

$$\{V^*, V\} = (u_0^*, u_0) + \sum_{k=1}^{m_-} [v_k^*, v_k]. \tag{22.35}$$

Analogously, the dual space of $\mathcal{H}_s^{(2)}(G) = H_{s-\alpha}(G) \times \prod_{j=1}^{m_+} H_{s-\text{Re}\,\beta_j-1/2}(\Gamma)$ is

$$(\mathcal{H}_s^{(2)}(G))^* = \mathring{H}_{\alpha-s}(G) \times \prod_{j=1}^{m_+} H_{-s+\text{Re}\,\beta_j+1/2}(\Gamma),$$

while the value of a functional $\Phi^* = (f^*(x), g_1^*(x'), \ldots, g_{m_+}^*(x')) \in (\mathcal{H}_s^{(2)}(G))^*$ at an element $\Phi = (f(x), g_1(x'), \ldots, g_{m_+}(x')) \in \mathcal{H}_s^{(2)}$ has the form

$$\{\Phi^*, \Phi\} = (f^*, f) + \sum_{k=1}^{m_+} [g_k^*, g_k]. \tag{22.36}$$

The adjoint $\mathfrak{A}^*$ of $\mathfrak{A}$ is a bounded operator from $(\mathcal{H}_s^{(2)}(G))^*$ into $(\mathcal{H}_s^{(1)}(G))^*$. To find the form of $\mathfrak{A}^*$, we note from (14.41) that

$$\left(pA(x, D)u_0 + p\sum_{k=1}^{m_-} C_k(x, D)(v_k(x') \times \delta(\Gamma)), f^*\right)$$
$$= (u_0, pA^*(x, D)f^*) + \sum_{k=1}^{m_-} (v_k(x') \times \delta(\Gamma), C^*(x, D)f^*), \tag{22.37}$$

where

$$f^* \in \mathring{H}_{-s+\alpha}(G), \quad u_0 \in \mathring{H}_s(G), \quad v_k(x') \in H_{s-\alpha+\mathrm{Re}\,\gamma_k+1/2}(\Gamma).$$

According to Theorems 19.1 and 19.2, $A^*(x, D)$ and $C_k^*(x, D)$ are also p.o.'s of classes $\hat{O}^\infty_{\alpha+i\beta}$ and $\hat{O}^\infty_{\gamma_k}$. On the other hand, from the definition of $\delta(\Gamma)$ (see (21.14)) we have

$$(v_k(x') \times \delta(\Gamma), \quad C_k^*(x, D)f^*) = [v_k(x'), p'C_k^*(x, D)f^*]. \tag{22.38}$$

Analogously,

$$\left[p'B_j(x, D)u_0 + \sum_{k=1}^{m_-} E_{jk}(x', D')v_k(x'), g_j^*\right]$$
$$= (u_0, B_j^*(x, D)(g_j^* \times \delta(\Gamma))) + \sum_{k=1}^{m_-} [v_k(x'), E_{jk}^*(x', D')g_j^*]. \tag{22.39}$$

Thus from (22.37)–(22.39) we get

$$\{\mathfrak{A}\Phi, \Phi^*\} = \left(u_0, pA^*(x, D)f^* + \sum_{j=1}^{m_+} pB_j^*(x, D)(g_j^* \times \delta(\Gamma))\right)$$
$$+ \sum_{k=1}^{m_-}\left[v_k, p'C_k^*(x, D)f^* + \sum_{j=1}^{m_+} E_{jk}^*(x', D')g_j^*\right]. \tag{22.40}$$

Consequently,

$$\{\mathfrak{A}\Phi, \Phi^*\} = \{\Phi, \mathfrak{A}^*\Phi^*\}, \tag{22.41}$$

where $\mathfrak{A}^*$ is defined by

$$pA^*(x, D)f^* + \sum_{j=1}^{m_+} pB_j^*(x, D)(g_j^* \times \delta(\Gamma)) = u_0^*, \tag{22.42}$$

$$p'C_k^*(x, D)f^* + \sum_{j=1}^{m_+} E_{jk}^*(x', D')g_j^* = v_k^*, \qquad 1 \leqslant k \leqslant m_-. \tag{22.43}$$

We note that the form of (22.42) and (22.43) is analogous to that of (22.4) and (22.5). The operator $\mathfrak{A}^*$ is a Fredholm operator since $\mathfrak{A}$ is. In order for the system (22.4), (22.5) to have a solution it is necessary and sufficient that the following "orthogonality" conditions be satisfied:

$$(f, f_i^*) + \sum_{k=1}^{m_+} [g_k, g_{ki}^*] = 0, \tag{22.44}$$

where $\Phi_i^* = (f_i^*, g_{1i}^*, \ldots, g_{m+i}^*) \in (\mathcal{H}_s^{(2)}(G))^*$, $1 \leqslant i \leqslant N_1$, is a basis of the space of solutions of the homogeneous equation $\mathfrak{A}^*\Phi^* = 0$. If, in particular, the equations $\mathfrak{A}V = 0$ and $\mathfrak{A}^*\Phi^* = 0$ have only the trivial solutions, then system (22.4), (22.5) is uniquely solvable for any $\Phi \in \mathcal{H}_s^{(2)}(G)$.

EXAMPLE 22.2. Consider in G the p.e.

$$p(-\Delta + 1)^{\alpha/2}u_0(x) = f(x), \tag{22.45}$$

where $(-\Delta + 1)^{\alpha/2}$ is the p.o. with symbol $A(\xi) = (|\xi|^2 + 1)^{\alpha/2} \in \hat{O}_\alpha^\infty$. The factorization index of $A_0(\xi) = |\xi|^\alpha$ in G is equal to $\alpha/2$. Consequently, by Theorem 22.3, equation (22.45) is normally solvable for $u_0 \in \overset{\circ}{H}_s(G)$, $f(x) \in H_{s-\alpha}(G)$, where $|s - \alpha/2| < 1/2$.

Let us show that (22.45) is uniquely solvable in the indicated function spaces. The unique solvability of (22.45) for $s = \alpha/2$ follows from Example 17.1 (the vanishing of the kernel and cokernel of the operator $p(-\Delta + 1)^{\alpha/2}$ can be proved with the use of (17.26) and (17.28)). Since $\overset{\circ}{H}_s(G) \subset \overset{\circ}{H}_{\alpha/2}(G)$ for $s > \alpha/2$, the kernel of $p(-\Delta + 1)^{\alpha/2}$ in $\overset{\circ}{H}_s(G)$ also vanishes for $s \in [\alpha/2, \alpha/2 + 1/2)$. According to Remark 16.2 applied to the operator $(-\Delta + 1)^{\alpha/2}$, a left regularizer of $(-\Delta + 1)^{\alpha/2}$ is bounded and a relation of form (22.17) is valid for any $s \in (\alpha/2 - 1/2, \alpha/2 + 1/2)$. It consequently follows from (22.17) that any solution $u_0(x) \in \overset{\circ}{H}_s(G)$, $s \in (\alpha/2 - 1/2, \alpha/2)$, of the equation $p(-\Delta + 1)^{\alpha/2} u_0(x) = 0$ actually belongs to $H_{\alpha/2+1/2-\varepsilon}(G)$, $\forall \varepsilon > 0$, and hence vanishes. Thus the kernel of $p(-\Delta + 1)^{\alpha/2}$ vanishes for all $s \in (\alpha/2 - 1/2, \alpha/2 + 1/2)$. The symbol $(|\xi|^2 + 1)^{\alpha/2}$ is real, so that the cokernel of $p(-\Delta + 1)^{\alpha/2}$ consists of the solutions of the equation $p(-\Delta + 1)^{\alpha/2}f^* = 0$ that belong to $\overset{\circ}{H}_{\alpha-s}(G)$, $s \in ((\alpha - 1)/2, (\alpha + 1)/2)$. Consequently, the cokernel also vanishes since $\alpha - s \in ((\alpha - 1)/2, (\alpha + 1)/2)$ whenever $s \in ((\alpha - 1)/2, (\alpha + 1)/2)$.

EXAMPLE 22.3. Let $\Lambda^\alpha(D)$ be the p.o. with symbol $|\xi|^\alpha \in O_\alpha^\infty$, $\alpha > -n$. As was noted in Example 22.1, $\Lambda^\alpha(D)$ differs from the p.o. Λ_1^α with symbol $|\xi|^\alpha(1 - \chi(\xi)) \in \hat{O}_\alpha^\infty$ by an integral operator with an infinitely differentiable kernel and, consequently, is normally solvable in the same function spaces as $\Lambda_1^\alpha(D)$. By virtue of Examples 3.1 and 3.3, $\Lambda^\alpha(D)$ can be represented in the form of an integral operator with a weak singularity for $-n < \alpha < 0$ and in the form of an integrodifferential operator for positive nonintegral $\alpha/2$. When $\alpha/2$ is equal to a positive integer k, $\Lambda^\alpha = (-\Delta)^k$.

From Theorem 22.3 it follows that the p.e.

$$p\Lambda^\alpha(D)u_0 = f \tag{22.46}$$

is normally solvable for $u_0 \in \overset{\circ}{H}_s(G)$ and $f \in H_{s-\alpha}(G)$, where $s \in ((\alpha - 1)/2, (\alpha + 1)/2)$. As in Example 22.2, it can be proved that (22.46) is uniquely solvable in these function spaces.

EXAMPLE 22.4 (Crack problem from the theory of elasticity). Consider in $\mathbf{R}^n_+$ the mixed boundary value problem (cf. Example 13.1)

$$\Delta u(x', x_n) = 0, \qquad x_n > 0, \tag{22.47}$$

$$\left.\frac{\partial u(x', 0)}{\partial x_n}\right|_G = g_1(x'), \qquad u(x', 0)|_{CG} = 0. \tag{22.48}$$

We will seek a solution of (22.47) in the form of the double layer potential (see (13.32′))

$$u(x', x_n) = F^{-1}_{\xi'} e^{-x_n|\xi'|}\tilde{v}(\xi') \tag{22.49}$$

where the density $v(x')$ is unknown. Substituting (22.49) in (22.48) (cf. Example 13.1), we get

$$v(x') = 0, \qquad x' \in CG, \tag{22.50}$$

$$\Lambda'(D')v(x')|_G = g_1(x'), \qquad x' \in G, \tag{22.51}$$

where $\Lambda'(D')$ is the p.o. in $\mathbf{R}^{n-1}$ with symbol $|\xi'|$. By virtue of Example 22.3, equation (22.51) has a unique solution $v(x') \in \overset{\circ}{H}_{1/2+\delta}(G)$ for arbitrary $g_1 \in H_{-1/2+\delta}(G)$ and arbitrary $\delta \in (-1/2, 1/2)$.

Equation (22.51) for $n = 3$ coincides with the simplest (in which the elastic constants are neglected) equation in the theory of elasticity for the normal displacement $v(x')$ of the surface points of a crack occupying a region G of the plane $x_3 = 0$ in an infinite elastic body under the assumption that the crack is opened out by the application of a pressure $g_1(x')$ to its free surface.

EXAMPLE 22.5 (Contact problem from the theory of elasticity). We pose the following boundary conditions for equation (22.47):

$$u(x', 0)|_G = g_2(x'), \qquad \left.\frac{\partial u(x', 0)}{\partial x_n}\right|_{CG} = 0. \tag{22.52}$$

In this case it is convenient to seek a solution of (22.47) in the form of a single layer potential. For $n \geqslant 3$ we have

$$u(x', x_n) = F^{-1}_{\xi'} e^{-x_n|\xi'|}|\xi'|^{-1}\tilde{v}(\xi') = \frac{\Gamma((n-2)/2)}{4\pi^{n/2}}\int_{-\infty}^{\infty}\frac{v(y')\,dy'}{[x_n^2 + |x' - y'|^2]^{(n-2)/2}}. \tag{22.53}$$

Substituting (22.53) in (22.52), we get

$$v(x') = 0, \qquad x' \in CG, \tag{22.54}$$

$$\Lambda'^{-1}(D')v|_G = g_2(x'), \qquad x' \in G, \tag{22.55}$$

where $\Lambda'^{-1}(D')$ is the p.o. in $\mathbf{R}^{n-1}$ with symbol $|\xi'|^{-1}$. By virtue of Example 22.3, equation (22.55) has a unique solution $v \in \overset{\circ}{H}_{-1/2+\delta}(G)$ for arbitrary $g_2 \in H_{1/2+\delta}(G)$ and arbitrary $\delta \in (-1/2, 1/2)$.

Equation (22.55) for $n = 3$ coincides, to within a constant factor depending on the elastic constants, with the equation for the normal stress $v(x')$ under the base of a cylindrical punch indenting an elastic halfspace $x_3 < 0$, where $g_2(x')$ is the normal displacement under the punch (determined by the shape and vertical displacement of the base of the punch). As in the case of (22.51), the coincidence with a boundary value problem for Laplace's equation is a consequence of the assumption that the shear stresses vanish.

For an approximate solution of equations (22.51) and (22.55), see §26.

§23. Smooth pseudodifferential operators in a domain

1. *Weighted function spaces.* For a fixed integer N we denote by $H_{s,N} = H_{s,N}(\mathbf{R}^n)$ the space of generalized functions with finite norm

$$\|u\|_{s,N} = \sum_{k=0}^{N} \|x_n^k u\|_{s+k}, \tag{23.1}$$

where $\|\cdot\|_s$ is the norm in $H_s(\mathbf{R}^n)$ (see §4). Clearly, $H_{s,N}(\mathbf{R}^n) \subset H_s(\mathbf{R}^n)$. The spaces $H_{s,N}^{\pm}$ and $H_{s,N}(\mathbf{R}_+^n)$ are defined analogously to $H_s^{\pm}$ and $H_s(\mathbf{R}_+^n)$; i.e. $H_{s,N}^{\pm}$ denotes the space of generalized functions $u(x) \in H_{s,N}(\mathbf{R}^n)$ with support in $\overline{\mathbf{R}_{\pm}^n}$, while $H_{s,N}(\mathbf{R}_+^n)$ denotes the space of generalized functions in $\mathbf{R}_+^n$ admitting an extension onto $\mathbf{R}^n$ that belongs to $H_{s,N}(\mathbf{R}^n)$.

LEMMA 23.1. *Suppose $\partial^k A(x, \xi)/\partial\xi_n^k \in S_{\alpha-k}^0$ for $0 \leqslant k \leqslant N$. Then*

$$x_n^N A(x, D)u = \sum_{k=0}^{N} \frac{(-1)^k}{k!} A_k(x, D)(D_n^k x_n^N)u, \tag{23.2}$$

where $A_k(x, D)$ is the p.o. with symbol $\partial^k A(x, \xi)/\partial\xi_n^k$ for $0 \leqslant k \leqslant N$.

PROOF. We have (see (19.1))

$$x_n^N A(x, D)u = \frac{1}{(2\pi)^n} \int_{-\infty}^{\infty} x_n^N A(x, \xi) u(y) e^{-i(x-y,\xi)}\, dy\, d\xi, \tag{23.3}$$

where the integral is regarded as an iterated integral. On the other hand, x_n^N has the following Taylor expansion about $x_n = y_n$:

$$x_n^N = \sum_{k=0}^{N} \frac{1}{k!} \frac{\partial^k y_n^N}{\partial y_n^k} (x_n - y_n)^k. \tag{23.4}$$

Therefore, since

$$(x_n - y_n)^k e^{-i(x_n - y_n)\xi_n} = i^k \frac{\partial^k}{\partial \xi_n^k} e^{-i(x_n - y_n)\xi_n},$$

formula (23.2) follows upon substituting (23.4) in (23.3) and integrating by parts with respect to ξ_n.

We note that the integration by parts can be justified, as in §19, by inserting the factor $\chi(\varepsilon\xi)$ in the integrand, carrying out the integration by parts, and then letting ε tend to zero.

LEMMA 23.2. *Suppose* $\dfrac{\partial^k A(x, \xi)}{\partial \xi_n^k} \in S_{\alpha-k}^0$, *for $0 \leqslant k \leqslant N$. Then the p.o. $A(x, D)$ is a bounded operator from $H_{s,N}$ into $H_{s-\alpha,N}$ for any s.*

PROOF. We have $\|Au\|_{s-\alpha,N} = \sum_{k=0}^{N}\|x_n^kAu\|_{s-\alpha+k}$. Applying Lemma 23.1 to x_n^kA, we get

$$x_n^kAu = \sum_{r=0}^{k} \frac{(-1)^r}{r!} A_r(x, D)D_n^r x_n^k u. \tag{23.5}$$

Consequently,

$$\begin{aligned}\|x_n^kAu\|_{s-\alpha+k} &\leqslant C\sum_{r=0}^{k} \|A_r x_n^{k-r}u\|_{s-\alpha+k} \leqslant C\sum_{r=0}^{k} \|x_n^{k-r}u\|_{s+k-r}\\ &\leqslant C\|u\|_{s,N},\end{aligned} \tag{23.6}$$

and the lemma follows upon summing (23.6) over k from 0 to N.

LEMMA 23.3. *Suppose the symbol $A(x, \xi)$ satisfies the same condition as in Lemma 23.2 and $s < -1/2$. Then the operator of potential type $A(x, D)(v(x') \times \delta(x_n))$ is a bounded operator from $H_{s+1/2}(\mathbf{R}^{n-1})$ into $H_{s-\alpha,N}(\mathbf{R}^n)$.*

PROOF. By Lemma 23.1,

$$x_n^kA(v(x') \times \delta(x_n)) = \sum_{r=0}^{k} \frac{(-1)^r}{r!} A_r(x, D)\big(v(x') \times (D_n^r x_n^k)\delta(x_n)\big). \tag{23.7}$$

Since $(D_n^r x_n^k)\delta(x_n) = 0$ for $r \leqslant k-1$, it follows that $x_n^kA(v \times \delta) = (-i)^k A_k(v \times \delta)$. Consequently (see Lemma 21.7),

$$\|x_n^kA(v \times \delta)\|_{s-\alpha+k} = \|A_k(v \times \delta)\|_{s-\alpha+k} \leqslant C\|v \times \delta\|_s \leqslant C[v]_{s+1/2}, \tag{23.8}$$

and the lemma follows upon summing (23.8) over k from 0 to N.

LEMMA 23.4. *θ^+ is a bounded operator from $H_{\delta,N}$ into $H_{\delta,N}^+$ for $|\delta| < 1/2$ and any N.*

PROOF. Clearly, for $|p| \leqslant k$, $D^p(x_n^k\theta^+u) = \theta^+D^p(x_n^ku)$. Consequently, the boundedness of θ^+ in $H_\delta(\mathbf{R}^n)$ for $|\delta| < 1/2$ (see Theorem 5.1) implies

$$\begin{aligned}\|\theta^+u\|_{\delta,N} &= \sum_{k=0}^{N} \|x_n^k\theta^+u\|_{\delta+k} \leqslant \sum_{k=0}^{N} \sum_{|p|\leqslant k} \|D^p(x_n^k\theta^+u)\|_\delta\\ &= \sum_{k=0}^{N} \sum_{|p|\leqslant k} \|\theta^+D^p(x_n^ku)\|_\delta \leqslant C\sum_{k=0}^{N} \sum_{|p|\leqslant k} \|D^p(x_n^ku)\|_\delta \leqslant C\|u\|_{\delta,N}.\end{aligned} \tag{23.9}$$

The following lemma for weighted spaces is analogous to Theorem 19.4.

LEMMA 23.5. *Suppose* $\dfrac{\partial^k A(x,\xi)}{\partial \xi_n^k} \in S^t_{\alpha-k}$, *for* $t > 0$ *and* $0 \leqslant k \leqslant N$, *and suppose we are given an arbitrary point* $x_0 \in \mathbf{R}^n$ *and any* $\varepsilon > 0$. *Then there exists a neighborhood* $U_0 \subset \mathbf{R}^n$ *such that for any* $\varphi(x) \in C_0^\infty(U_0)$ *the operator* $\varphi(x)A(x, D)$ *admits the decomposition*

$$\varphi(x)A(x, D) = \varphi(x)(A(x_0, D) + K(x, D) + T(x, D)), \tag{23.10}$$

in which

$$\|K(x, D)u\|_{s-\alpha,N} \leqslant \varepsilon\|u\|_{s,N}, \qquad \|T(x, D)u\|_{s-\alpha,N} \leqslant C\|u\|_{s-1,N}. \tag{23.11}$$

PROOF. We define $K(x, \xi)$ and $T(x, \xi)$ by the same formulas as in the proof of Theorem 19.4:

$$K(x, \xi) = \psi(x)(A(x, \xi) - A(x_0, \xi))\left(1 - \chi\left(\frac{\xi}{N_1}\right)\right),$$

$$T(x, \xi) = \psi(x)(A(x, \xi) - A(x_0, \xi))\chi\left(\frac{\xi}{N_1}\right), \tag{23.12}$$

where $\psi(x) \in C_0^\infty(U_0)$, $|\psi(x)| \leqslant 1$, $\psi(x)\varphi(x) \equiv \varphi(x)$, $\chi(\xi) \in C_0^\infty$, $\chi(\xi) = 1$ for $|\xi| \leqslant 1$, and N_1 is sufficiently large. By Lemma 23.2, $\|Tu\|_{s-\alpha,N} \leqslant C\|u\|_{s-M,N}$ for any M. Further, by Lemma 23.1, $x_n^k Ku = \sum_{r=0}^k K_r((D_n^r x_n^k)u)$, where $K_r(x, \xi) = \dfrac{(-1)^r}{r!}\dfrac{\partial^r K(x, \xi)}{\partial \xi_n^r}$. Therefore, applying (19.34) to the p.o.'s K_r, we obtain (23.11).

2. We now define weighted spaces in a domain.

Let G be a bounded domain with a smooth boundary Γ and let $r_k(x)$ denote a function of class $C^\infty(\mathbf{R}^n)$ that (i) doesn't vanish for $x \notin \Gamma$, (ii) has the form $x_n^k r_0(x)$ in an l.c.s. in a neighborhood of Γ, where $r_0(x) \in C^\infty$ and $r_0(x) \neq 0$, and (iii) is equal to unity for large $|x|$. We denote by $\mathring{H}_{s,N}(G)$ the space of generalized functions in $\mathbf{R}^n$ with support in $\overline{G}$ and finite norm

$$\|u\|_{s,N} = \sum_{k=0}^{N} \|r_k u\|_{s+k}. \tag{23.12$'$}$$

Further, we denote by $H_{s,N}(G)$ the space of generalized functions in G admitting an extension onto $\mathbf{R}^n$ with finite norm (23.12$'$). The norm in $H_{s,N}(G)$ is defined by

$$\overset{+}{\|u\|}_{s,N} = \inf\|lu\|_{s,N},$$

where the infimum is taken over all extensions lu.

Clearly, if $\varphi(x) \in C_0^\infty(G)$ and u belongs to either $\mathring{H}_{s,N}(G)$ or $H_{s,N}(G)$, then $\varphi u \in \mathring{H}_{s+N}(G)$, i.e. a function belonging to either $\mathring{H}_{s,N}(G)$ or $H_{s,N}(G)$ is smooth interior to G if N is sufficiently large. In a neighborhood of Γ the norm $\|\varphi u\|_{s,N}$ defined by (23.12$'$) is equivalent to the norm (23.1).

The following lemma can be obtained from Lemmas 23.2 and 23.3 by means of a partition of unity and Theorem 19.6 on a change of variables.

LEMMA 23.6. *Suppose* $A(x, \xi) \in \hat{O}^{\infty}_{\alpha+i\beta}$. *Then*

$$\|pA(x, D)u_0\|^{+}_{s-\alpha,N} \leqslant C\|u_0\|_{s,N}, \qquad \forall u_0 \in \mathring{H}_{s,N}(G), \tag{23.13}$$

$$\|pA(x, D)(v(x') \times \delta(\Gamma))\|^{+}_{s-\alpha,N} \leqslant C[v]_{s+1/2},$$
$$s < -1/2, \quad v \in H_{s+1/2}(\Gamma). \tag{23.14}$$

We note that the compactness of an imbedding of $\mathring{H}_{s+\varepsilon,N}(G)$ $(H_{s+\varepsilon,N}(G))$ into $\mathring{H}_{s,N}(G)$ $(H_{s,N}(G))$ for $\varepsilon > 0$ readily follows from Lemma 21.1.

3. *Smooth operators in a halfspace with variable symbol. The class* $\hat{D}^{t}_{\alpha+i\beta}$. Suppose $\dfrac{\partial^k A(x, \xi)}{\partial \xi_n^k} \in S^{t}_{\alpha-k}$, $0 \leqslant k < \infty$, and let

$$z = \frac{\xi_n + i(|\xi'| + 1)}{\xi_n - i(|\xi'| + 1)}.$$

Then

$$|z| = 1, \quad z = e^{i\varphi}, \quad 0 < \varphi < 2\pi, \quad \xi_n = i(|\xi'| + 1)\frac{z+1}{z-1} = (|\xi'| + 1)\operatorname{ctg}\frac{\varphi}{2}.$$

The symbol $A(x, \xi)$ is said to be of class $\hat{D}^{t}_{\alpha+i\beta}$ if the function

$$\mathcal{Q}_k(x', \xi', \varphi) = \hat{\Lambda}_{-}^{-(\alpha+i\beta)}(\xi)\frac{\partial^k}{\partial x_n^k} A\left(x', 0, \xi', (1 + |\xi'|)\operatorname{ctg}\frac{\varphi}{2}\right)$$

is an infinitely differentiable function of $z = e^{i\varphi}$ on the circle $|z| = 1$ for almost all $\xi' \in \mathbf{R}^{n-1}$ that satisfies the condition

$$\frac{\partial^r \mathcal{Q}_k(x', \xi', \varphi)}{\partial \varphi^r} \in S^{t_0}_{0}(\mathbf{R}^{n-1})$$

uniformly in φ in a neighborhood of $\varphi = 0$ for all nonnegative integers k and r, where $t_0 = \min(t, 1)$.

EXAMPLE 23.1. If $A(\xi) \in D_{\alpha+i\beta}$ (see §10), then

$$\hat{A}(\xi) = A\left((1 + |\xi'|)\frac{\xi'}{|\xi'|}, \xi_n\right) \in \hat{D}^{0}_{\alpha+i\beta}$$

by virtue of Remark 10.4.

LEMMA 23.7. *For any* N_1 *and* N_2 *a symbol* $A(x, \xi) \in \hat{D}^{t}_{\alpha+i\beta}$ *admits the decomposition*

$$A(x, \xi) = A_{-}(x, \xi) + R_{\alpha-N_1}(x, \xi)\hat{\Lambda}'^{N_1}(\xi') + x_n^{N_2}B(x, \xi), \tag{23.15}$$

where $A_{-}(x, \xi) \in S^{t_0}_{\alpha}$ *is a "minus" function (see* §20), $R_{\alpha-N_1}(x, \xi) \in S^{t_0}_{\alpha-N_1}$, $t_0 = \min(t, 1)$, $\hat{\Lambda}'^{N_1} = (1 + |\xi'|)^{N_1}$, *and* $B(x, \xi) \in S^{t}_{\alpha}$.

PROOF. Let $\chi(x_n)$ be a function of class $C_0^\infty(\mathbf{R}^1)$ that is equal to unity for $|x_n| \leqslant 1$. Expanding $A(x, \xi)$ about $x_n = 0$ with the use of Taylor's theorem, we get

$$A(x, \xi) = \sum_{k=0}^{N_2-1} \frac{x_n^k \chi(x_n)}{k!} \frac{\partial^k A(x', 0, \xi)}{\partial x_n^k} + x_n^{N_2} B(x, \xi), \tag{23.16}$$

where $B(x, \xi) \in S_\alpha^t$. On the other hand, about $\varphi = 0$ the function

$$\hat{\Lambda}_-^{-(\alpha+i\beta)} \frac{\partial^k}{\partial x_n^k} A(x', 0, \xi) = \mathcal{Q}_k(x', \xi', \varphi)$$

has the Taylor expansion

$$\mathcal{Q}_k(x', \xi', \varphi) = \sum_{r=0}^{N_1-1} \frac{\partial^r \mathcal{Q}_k(x', \xi', 0)}{\partial \varphi^r} \frac{\varphi^r}{r!} + r_{N_1 k}(x', \xi', \varphi)\varphi^{N_1}, \tag{23.16$'$}$$

where $r_{N_1 k}(x', \xi', \varphi) \in S_0^{t_0}(\mathbf{R}^{n-1})$ for $|\varphi| < \delta$, and $\dfrac{\partial^r}{\partial \varphi^r} \mathcal{Q}_k(x', \xi', 0) \in S_0^{t_0}(\mathbf{R}^{n-1})$. But

$$\varphi = \frac{1}{i} \ln\left(1 + \frac{2i\hat{\Lambda}'}{\hat{\Lambda}_-}\right),$$

since $z = 1 + \dfrac{2i\hat{\Lambda}'}{\hat{\Lambda}_-}$, while the inequality $|\varphi| < \delta$ is equivalent to an inequality of the form $|\xi_n| \geqslant C_\delta(1 + |\xi'|)$. Substituting in (23.16′) the expansion in powers of $\hat{\Lambda}'/\hat{\Lambda}_-$ of the function

$$\varphi = \frac{1}{i} \ln\left(1 + \frac{2i\hat{\Lambda}'}{\hat{\Lambda}_-}\right)$$

and collecting the coefficients of like powers of $\hat{\Lambda}'/\hat{\Lambda}_-$, we get for $|\xi_n| \geqslant C_\delta(1 + |\xi'|)$

$$\mathcal{Q}_k(x', \xi', \varphi) = \sum_{r=0}^{N_1-1} a_{kr}(x', \xi') \frac{\hat{\Lambda}'^r}{\hat{\Lambda}_-^r} + r^{(1)}_{N_1,k}(x', \xi', \xi_n) \frac{\hat{\Lambda}'^{N_1}}{\hat{\Lambda}_-^{N_1}}, \tag{23.17}$$

where $a_{kr}(x', \xi') \in S_0^{t_0}(\mathbf{R}^{n-1})$, and $r^{(1)}_{N_1,k}(x', \xi', \xi_n) \in S_0^{t_0}$. It therefore follows from (23.16) and (23.17) that (23.15) will hold if we put

$$A_-(x, \xi) = \sum_{k=0}^{N_2-1} \sum_{r=0}^{N_1-1} \frac{x_n^k \chi(x_n)}{k!} a_{kr}(x', \xi') \hat{\Lambda}'^r \hat{\Lambda}_-^{\alpha+i\beta-r}, \tag{23.18}$$

$$R_{\alpha-N_1}(x, \xi) = \sum_{k=0}^{N_2-1} \frac{x_n^k \chi(x_n)}{k!} r^{(1)}_{N_1,k}(x', \xi', \xi_n) \hat{\Lambda}_-^{\alpha+i\beta-N_1}. \tag{23.19}$$

Clearly, $A_-(x, \xi) \in S_\alpha^{t_0}$ is a "minus" function. The fact that $R_{\alpha-N_1}(x, \xi) \in S_{\alpha-N_1}^{t_0}$ follows (i) from (23.19) for $|\xi_n| \geqslant C_\delta(1 + |\xi'|)$, and (ii) directly from

(23.15) for $|\xi_n| \leqslant C_8(1 + |\xi'|)$, since then $A(x, \xi) - A_-(x, \xi) - x_n^{N_2}B(x, \xi) \in S_\alpha^{t_0}$ and $\hat{\Lambda}'^{N_1} = (1 + |\xi'|)^{N_1} \geqslant C_1(1 + |\xi|)^{N_1}$.

LEMMA 23.8. *Suppose $A(x, \xi) \in \hat{D}^0_{\alpha+i\beta}$. Then for any $r \geqslant 0$*

$$\|pA(x, D)u_+\|^+_{r-\alpha} \leqslant C\|u\|^+_r, \qquad \forall u \in H_r(\mathbf{R}^n_+), \tag{23.20}$$

$$\|pA(x, D)(v(x') \times \delta(x_n))\|^+_{r-\alpha} \leqslant C[v]_{r+1/2}, \qquad \forall v \in H_{r+1/2}(\mathbf{R}^{n-1}), \tag{23.21}$$

where u_+ denotes the extension that vanishes for $x_n < 0$ of a function $u \in H_r(\mathbf{R}^n_+)$.

PROOF (cf. the proofs of Theorem 10.1 and Lemma 8.1). Note that $u_+ = \chi u_+ + (1 - \chi)u_+$ and that, since $(1 - \chi)u_+ \in H_r^+(\mathbf{R}^n)$, the estimates

$$\|pA(1 - \chi)u_+\|^+_{r-\alpha} \leqslant \|A(1 - \chi)u_+\|_{r-\alpha} \leqslant C\|(1 - \chi)u_+\|_r \leqslant C\|u\|^+_r$$

follow from Theorem 18.1. It therefore suffices to prove (23.20) for any function $u(x) \in H_r(\mathbf{R}^n_+)$ whose support is compact with respect to x_n.

From Lemma 23.7 we have

$$pA(x, D)u_+ = p\big(A_-(x, D)u_+ + R_{\alpha-N_1}(x, D)\hat{\Lambda}'^{N_1}u_+ + x_n^{N_2}B(x, D)u_+\big).$$

Let $lu \in H_r(\mathbf{R}^n)$ be an extension with compact support with respect to x_n of $u(x)$ from $\mathbf{R}^n_+$ onto $\mathbf{R}^n$ such that $\|lu\|_r \leqslant 2\|u\|^+_r$, and let $u_- = lu - u_+$. Since $A_-(x, D)$ is a "minus" operator, we have $pA_-u_- = 0$ (see Lemma 20.5). Consequently,

$$pA_-(x, D)u_+ = pA_-(x, D)lu$$

and Theorem 18.1 implies

$$\|pA_-(x, D)u_+\|^+_{r-\alpha} \leqslant \|A_-(x, D)lu\|_{r-\alpha} \leqslant C\|lu\|_r \leqslant 2C\|u\|^+_r.$$

Further, choosing $N_1 \geqslant r$, we get by virtue of Theorem 18.1 that

$$\big\|pR_{\alpha-N_1}\hat{\Lambda}'^{N_1}u_+\big\|^+_{r-\alpha} \leqslant C\|\hat{\Lambda}'^{N_1}u_+\|_{r-N_1} \leqslant C\|\hat{\Lambda}'^{r}u_+\|_0.$$

But, analogously to (10.28), $\|\hat{\Lambda}'^r u_+\|_0 \leqslant \|u\|^+_r$. Therefore

$$\big\|pR_{\alpha-N_1}\hat{\Lambda}'^{N_1}u_+\big\|^+_{r-\alpha} \leqslant C\|u\|^+_r.$$

It remains to show that $\|px_n^{N_2}Bu_+\|^+_{r-\alpha} \leqslant C\|u\|^+_r$. Suppose first that $r - 1/2$ is not an integer. From (23.1) we have $\|px_n^{N_2}Bu_+\|^+_{r-\alpha} \leqslant \|Bu_+\|_{r-\alpha-N_2,N_2}$, while from Lemma 23.2 we have

$$\|Bu_+\|_{r-\alpha-N_2,N_2} \leqslant C\|u_+\|_{r-N_2,N_2} = C\|\theta^+lu\|_{r-N_2,N_2}.$$

But if N_2 is chosen so that $|r - N_2| < 1/2$, then Lemma 23.4 implies $\|\theta^+lu\|_{r-N_2,N_2} \leqslant C\|lu\|_{r-N_2,N_2} \leqslant C_1\|lu\|_r \leqslant C_2\|u\|^+_r$, since the support of lu is compact with respect to x_n.

Thus (23.20) is proved for nonintegral $r - 1/2$. It is consequently also valid for integral $r - 1/2$ by virtue of the interpolation theorem (see [62]).

Let us now prove (23.21). We have

$$pA_-(x, D)(v(x') \times \delta) = 0,$$

since $A_-(x, D)$ is a "minus" operator. Choosing $N_1 > r + 1/2$, we get from Lemma 21.7 that

$$\|pR_{\alpha - N_1}\hat{\Lambda}'^{N_1}(v \times \delta)\|^{+}_{r-\alpha} \leqslant C[\hat{\Lambda}'^{N_1}v]_{r-N_1+1/2} \leqslant C[v]_{r+1/2};$$

while choosing $N_2 > r + 1/2$, we get from Lemma 23.3 that

$$\|px_n^{N_2}B(v \times \delta)\|^{+}_{r-\alpha} \leqslant \|B(v \times \delta)\|_{r-\alpha-N_2,N_2} \leqslant C[v]_{r-N_2+1/2} \leqslant C[v]_{r+1/2}.$$

LEMMA 23.9. *Suppose $A(x, \xi) \in \hat{D}^t_{\alpha+i\beta}$ for $t > 0$, and suppose we are given a point $x_0 = (x_0', 0) \in \mathbf{R}^{n-1}$ and a positive number ε. Then there exists a neighborhood $U_0 \ni x_0$ such that for any $\varphi(x) \in C_0^\infty(U_0)$ the operator $\varphi(x)A(x, D)$ admits the decomposition* (23.10), *with*

$$\|pK(x, D)u_+\|^{+}_{r-\alpha} < \varepsilon\|u\|^{+}_{r}, \qquad \|pT(x, D)u_+\|^{+}_{r-\alpha} \leqslant C\|u\|^{+}_{r-1}. \tag{23.22}$$

PROOF. Let $\chi_-(x_n) \in C_0^\infty(\mathbf{R}^1_-)$ and $\tilde{\chi}_-(\xi_n) = F\chi_-$, with $\tilde{\chi}_-(0) = 1$ (cf. §20.2). We define the symbols $K(x, \xi)$ and $T(x, \xi)$ by (23.12), replacing $\chi(\xi/N_1)$ by $\tilde{\chi}_-(\xi_n/N_1)\chi(\xi'/N_1)$. The estimate of $pK(x, D)(1 - \chi(x_n))u_+$ follows directly from Theorem 19.4, while the estimate of $pK(x, D)\chi(x_n)u_+$ is obtained via (23.15) by applying (i) the proof of Theorem 19.4 separately to the summands with factors $A_-(x, D)$ and $R_{\alpha-N_1}(x, D)$, and (ii) Lemma 23.5 to the summand with factor $x_n^{N_2}B(x, D)$.

The class $\hat{D}^{(0)}_{\alpha+i\beta}(G)$. A symbol $A(x, \xi) \in \hat{O}^\infty_{\alpha+i\beta}$ is said to be of class $\hat{D}^{(0)}_{\alpha+i\beta}(G)$ if, in any neighborhood U_j intersecting Γ, (i) the symbol $A_j(x^{(j)}, \xi^{(j)})$ of the p.o. $A(x, D)$ written in the jth local coordinates is of class $\hat{D}^\infty_{\alpha+i\beta}$, and (ii) the p.h.s. $A_j^{(0)}(x^{(j)}, \xi^{(j)})$ and all of its derivatives with respect to $x_n^{(j)}$ are of class $D^{(0)}_{\alpha+i\beta}$ (see §10) for $x_n^{(j)} = 0$. We note that, in view of (21.4) and Theorem 19.6 on a change of variables, the class $\hat{D}^{(0)}_{\alpha+i\beta}(G)$ is independent of the choice of l.c.s.'s.

LEMMA 23.10. *Suppose $A(x, \xi) \in \hat{D}^{(0)}_{\alpha+i\beta}(G)$. Then for any $r \geqslant 0$*

$$\|pA(x, D)u_0\|^{+}_{r-\alpha} \leqslant C\|u\|^{+}_{r}, \qquad \|pA(x, D)(v(x') \times \delta(\Gamma))\|^{+}_{r-\alpha} \leqslant C[v]_{r+1/2}, \tag{23.23}$$

where u_0 denotes the extension of a function $u \in H_r(G)$ that vanishes for $x \notin G$.

PROOF. Let $\psi_j(x)$ and $\varphi_j(x)$ be the same functions as in §22. We have

$$pA(x, D)u_0 = \sum_j p\psi_j A\varphi_j u_0 + pT_\infty u_0, \qquad \text{where} \quad T_\infty u_0 = \sum_j (1 - \psi_j)A\varphi_j u_0.$$

Since T_∞ is clearly a bounded operator from $H_0(\mathbf{R}^n)$ into $H_{r-\alpha}(\mathbf{R}^n)$ for any r,

$$\|pT_\infty u_0\|_{r-\alpha}^{+} \leqslant C\|u_0\|_0 \leqslant C\|u\|_r^{+}.$$

Transforming $\psi_j A\varphi_j$ to the local coordinates in U_j and applying Lemma 23.8, we get that

$$\|p\psi_j A\varphi_j u_0\|_{r-\alpha}^{+} \leqslant C\|\varphi_j u\|_r^{+},$$

which implies the first estimate of (23.23). The second is proved analogously.

4. *Normal solvability of boundary value problems for smooth operators.* Let $A(\xi)$ be an elliptic symbol of class $D_{\alpha+i\beta}$. Since $A(0, -1) = e^{-(\alpha+i\beta)\pi i}A(0, +1)$ (see (10.2)), it follows in the notation of §14 that $a_j^{(0)} = 0$, $1 \leqslant j \leqslant p$ (see (14.38)). Consequently, Theorem 14.2 implies that $pA(\omega, D_n)$ is a Fredholm operator from $H_s^+(\mathbf{R}^1)$ into $H_{s-\alpha}(\mathbf{R}_+^1)$ for any $\omega \in S^{n-2}$ and any nonintegral $s - 1/2$. In particular, $pA(\omega, D_n)$ is a Fredholm operator from $H_0^+(\mathbf{R}^1)$ into $H_{-\alpha}(\mathbf{R}_+^1)$.

Let us show that $pA(\omega, D_n)u_+$ is a Fredholm operator from $H_r(\mathbf{R}_+^1)$ into $H_{r-\alpha}(\mathbf{R}_+^1)$ for any $r \geqslant 0$, where u_+ denotes the extension that vanishes for $x_n < 0$ of a function $u \in H_r(\mathbf{R}_+^1)$. The boundedness of $pA(\omega, D_n)u_+$ as an operator from $H_r(\mathbf{R}_+^1)$ into $H_{r-\alpha}(\mathbf{R}_+^1)$ follows from the proof of Theorem 10.1. Since $pA(\omega, D_n)u_+$ is a Fredholm operator for $r = 0$, the fact that it is a Fredholm operator for $r > 0$ will be proved if the following lemma on smoothness is established.

LEMMA 23.11. *Suppose $u \in H_0(\mathbf{R}_+^1)$ is a solution of the equation $pA(\omega, D_n)u_+ = f$, and $A(\xi) \in D_{\alpha+i\beta}$,* $\det A(0, +1) \neq 0$. *Then $u \in H_r(\mathbf{R}_+^1)$ iff $f \in H_{r-\alpha}(\mathbf{R}_+^1)$ and $r > 0$.*

For suppose Lemma 23.11 has been proved. If the function $f \in H_{r-\alpha}(\mathbf{R}_+^1) \subset H_{-\alpha}(\mathbf{R}_+^1)$ is orthogonal to the finite-dimensional kernel of the operator $pA^*(\omega, D_n)$ in $H_\alpha^+(\mathbf{R}^1) = (H_{-\alpha}(\mathbf{R}_+^1))^* \subset (H_{r-\alpha}(\mathbf{R}_+^1))^*$, then the equation $pA(\omega, D_n)u_+ = f$ has a solution $u_+ \in H_0^+(\mathbf{R}^1)$, since $pA(\omega, D_n)$ is a Fredholm operator for $r = 0$. But then Lemma 23.11 implies $u \in H_r(\mathbf{R}_+^1)$. Thus the solvability of the equation $pA(\omega, D_n)u_+ = f$ under the fulfilment of a finite number of conditions on the right side is proved. But the finite dimensionality of the kernel in $H_r(\mathbf{R}_+^1)$ follows from the finite dimensionality of the kernel in $H_0^+(\mathbf{R}^1)$, since $H_r(\mathbf{R}_+^1) \subset H_0(\mathbf{R}_+^1)$.

PROOF OF LEMMA 23.11. Let $\xi' = \omega$ and $N > r$. From the assumption $A(\xi) \in D_{\alpha+i\beta}$ and (10.10) we have

$$A(\omega, \xi_n) = a_+\Lambda_-^{\alpha+i\beta}(\xi_n) + \sum_{k=1}^{N-1} a_k(\omega)\Lambda_-^{\alpha+i\beta-k}(\xi_n) + A_N(\omega, \xi_n), \quad (23.24)$$

where

$$a_+ = A(0, +1), \quad \Lambda_-(\xi_n) = \xi_n - i, \quad |A_N| \leqslant C(1 + |\xi_n|)^{\alpha - N}.$$

Therefore, since $\det a_+ \neq 0$ and $\Lambda_-^{\alpha+i\beta}(D_n)$ is a "minus" operator, the equation $pA(\omega, D_n)u_+ = f$ can be written in the form

$$u = p\Lambda_-^{-\alpha-i\beta}(D_n)a_+^{-1}lf - pA_-(\omega, D_n)u_+ - pA_{N_1}(\omega, D_n)u_+, \quad (23.25)$$

where

$$A_-(\omega, \xi_n) = \sum_{k=1}^{N-1} a_+^{-1}a_k(\omega)\Lambda_-^{-k}(\xi_n) \qquad \text{and} \qquad A_{N_1} = \Lambda_-^{-\alpha-i\beta}(\xi_n)a_+^{-1}A_N.$$

The assumption $f \in H_{r-\alpha}(\mathbf{R}_+^1)$ implies $p\Lambda_-^{-\alpha-i\beta}(D_n)a_+^{-1}lf \in H_r(\mathbf{R}_+^1)$. Also the obvious estimate

$$|A_{N_1}(\omega, \xi_n)| \leqslant C(1 + |\xi_n|)^{-N}$$

implies $pA_{N_1}u_+ \in H_r(\mathbf{R}_+^1)$, since $u_+ \in H_0^+(\mathbf{R}^1)$ and $N > r$. Further, $A_-(\xi) \in D_{-1}$, so that

$$v = pA_-(\omega, D_n)u_+ \in H_{l+1}(\mathbf{R}_+^1),$$

for any $l \geqslant 0$ for which $u \in H_l(\mathbf{R}_+^1)$. Thus the assumption $u \in H_0(\mathbf{R}_+^1)$ implies $v \in H_1(\mathbf{R}_+^1)$, and hence Lemma 23.11 is proved for $r \leqslant 1$. If $r > 1$, then, having already proved that $u \in H_1(\mathbf{R}_+^1)$, we get from (23.25) that $u \in H_r(\mathbf{R}_+^1)$ for $1 < r \leqslant 2$, and so on ad infinitum.

We note that it follows from Lemma 23.11 that the kernel of $pA(\omega, D_n)u_+$ is contained in $H_N(\mathbf{R}_+^1)$ for all N and hence consists of infinitely differentiable functions.

THEOREM 23.1. *Suppose in the boundary value problem* (22.4), (22.5) *that* $A(x, D)$ *is a p.o. with elliptic symbol* $A(x, \xi) \in \hat{D}_{\alpha+i\beta}^{(0)}(G)$, *that* $B_j(x, D)$ *and* $C_k(x, D)$ *are p.o.'s with symbols* $B_j(x, \xi) \in \hat{D}_{\beta_j}^{(0)}(G)$, $C_k(x, \xi) \in \hat{D}_{\gamma_k}^{(0)}(G)$, *and that* $E_{jk}(x', D')$ *is a p.o. on* Γ *of class* $\hat{O}_{\beta_{jk}}^{\infty}$, *with* $\beta_{jk} = \beta_j + \gamma_k - \alpha - i\beta + 1$. *Suppose* $r \geqslant \max_{1 \leqslant j \leqslant m_+} (\operatorname{Re} \beta_j + 1/2)$. *Suppose, further, that for any* $x_0 \in \Gamma$ *and for any fixed* $x'_{(i)} = (x_1^{(i)}, \dots, x_{n-1}^{(i)})$, $\xi'_{(i)} \neq 0$ *in a corresponding l.c.s. the boundary value problem in* $\mathbf{R}_+^1$ *of form* (22.7), (22.8) *has a unique solution* $(v(x_n^{(i)}), \rho_1, \dots, \rho_{m_-}) \in H_r(\mathbf{R}_+^1) \times \mathbf{C}^{m_-}$ *for arbitrary* $(g(x_n^{(i)}), h_1, \dots, h_{m_+}) \in H_{r-\alpha}(\mathbf{R}_+^1) \times \mathbf{C}^{m_+}$. *Then the operator* $\mathfrak{A}$ *defined by the left sides of* (22.4) *and*

(22.5) *is a Fredholm operator from*

$$\mathcal{H}_r^{(1)} = \mathring{H}_r(G) \times \prod_{k=1}^{m_-} H_{r-\alpha+\operatorname{Re}\gamma_k+1/2}(\Gamma)$$

into

$$\mathcal{H}_r^{(2)} = H_{r-\alpha}(G) \times \prod_{k=1}^{m_+} H_{r-\operatorname{Re}\beta_k-1/2}(\Gamma).$$

We note that the expression $p'B_j(x, D)u_0$ in (22.5) for $\operatorname{Re}\beta_j \geqslant -1/2$ is obtained by applying the operator p' to an arbitrary extension belonging to $H_{r-\operatorname{Re}\beta_j}(\mathbf{R}^n)$ of $pB_j(x, D)u_0$ (cf. §10.3).

The proof is actually a repetition of the proof of Theorem 22.1. The boundedness of the operator $\mathfrak{A}$ from $\mathcal{H}_r^{(1)}$ into $\mathcal{H}_r^{(2)}$ is a consequence of Lemmas 23.10, 21.3 and 21.6. The case of a boundary value problem in a halfspace with constant symbols is proved in exactly the same way as Theorem 16.1. The general case is then proved in the same way as Theorem 22.1 with the use of Lemma 23.9 in place of Theorem 19.4.

EXAMPLE 23.2. *Scalar equation.* A special case of the boundary value problem of form (22.4), (22.5) is the boundary value problem for a scalar elliptic p.o. $A(x, D)$ of class $\hat{D}^{(0)}_{\alpha+i\beta}(G)$. In this case $B_j(x, D)$ and $C_k(x, D)$ are also scalar p.o.'s and the factorization index κ of the symbol $A(x, \xi)$ is equal to the integer $m_+ - m_-$. The analogs of Theorems 22.2, 22.3, 22.3′ and 22.3″, which extend Theorems 12.2, 11.1, 11.2 and 11.4 to the case of a p.o. in a domain, hold. Instead of referring to §16 and Lemma 23.11, one can use Theorems 11.1–11.4 and 12.2 to prove the corresponding theorems for a bounded domain.

We note, in particular, the extensions of Example 11.1 and Theorem 11.3. Let $A(x, \xi) \in \hat{D}^{(0)}_{\alpha+i\beta}$ be a scalar elliptic symbol. Then the following Dirichlet problem is normally solvable for $\kappa > 0$:

$$pA(x, D)u_0 = f, \qquad \left.\frac{\partial^{j-1}u}{\partial n^{j-1}}\right|_\Gamma = g_j(x'), \qquad 1 \leqslant j \leqslant \kappa, \tag{23.26}$$

where $f \in H_{r-\alpha}(G)$, $g_j(x') \in H_{r-j+1/2}(\Gamma)$, $u \in H_r(G)$ and n is the normal to Γ. When $\kappa < 0$, the following problem with potentials is normally solvable:

$$pA(x, D)\left(u_0 + \sum_{k=1}^{|\kappa|} v_k(x') \times \delta^{(k-1)}(\Gamma)\right) = f(x), \tag{23.27}$$

where $u \in H_r(G)$, $v_k(x') \in H_{r+k+1/2}(\Gamma)$ and the generalized function $v_k(x') \times \delta^{(k)}(\Gamma)$ acts on a C^∞ function φ in accordance with the formula

$$(v_k(x') \times \delta^{(k)}(\Gamma), \varphi) = (-i)^k \int_\Gamma v_k(x') \left.\frac{\partial^k \varphi}{\partial n^k}\right|_\Gamma ds_\Gamma$$

(cf. (21.16)).

EXAMPLE 23.3. *Boundary value problems for elliptic differential equations.*

THEOREM 23.2. *Suppose given a boundary value problem of the form*

$$L(x, D)u = f, \qquad f \in H_{r-2m}(G), \quad u(x) \in H_r(G), \tag{23.28}$$

$$B_j(x, D)u|_\Gamma = g_j(x'), \qquad 1 \leqslant j \leqslant m, g_j(x') \in H_{r-m_j-1/2}(\Gamma), \tag{23.29}$$

in which $L(x, D) = \|L_{ij}(x, D)\|_{i,j=1}^p$ *is an elliptic system of differential operators of order* $2m$ *and* $B_j(x, D)$ $(1 \leqslant j \leqslant m)$ *is a differential operator of order* m_j. *Suppose that at each point* $x_0 \in \Gamma$ *the Šapiro-Lopatinskiĭ condition (the condition that the corresponding differential boundary value problem on a halfline is uniquely solvable) is satisfied for each* $\xi'_{(J)} \neq 0$ *in a corresponding l.c.s. Then the boundary value problem* (23.28), (23.29) *is normally solvable for any* $r > \max_j(m_j + 1/2)$.

Theorem 23.2 is a special case of Theorem 23.1. We note that the case of a halfspace and constant coefficients can be considered directly in the same way as was done in Example 11.2 for a scalar elliptic differential equation.

REMARK 23.1. The more general elliptic systems of Douglis-Nirenberg type are frequently considered. Here it is assumed that (i) there exist numbers $s_1, \ldots, s_p$ and $t_1, \ldots, t_p$ such that each $L_{ij}(\xi)$ is a homogeneous function of degree at most $s_i + t_j$, i.e. $\deg_\xi L_{ij}(\xi) \leqslant s_i + t_j$, and (ii) the matrix $L_0(x, \xi)$ obtained from $L(x, \xi)$ by replacing every term of degree of homogeneity less than $s_i + t_j$ by zero is nonsingular for $\xi \neq 0$. Thus the principal part $L_0(x, \xi)$ (in the sense just indicated) of $L(x, \xi)$ is representable in the form

$$L_0(x, \xi) = \Lambda_1(\xi) L_0^{(1)}(x, \xi) \Lambda_2(\xi),$$

where $\deg_\xi L_0^{(1)}(x, \xi) = 0$, $\det L_0^{(1)}(x, \xi) \neq 0$ for $\xi \neq 0$, and $\Lambda_1 = \| \, |\xi|^{s_i}\delta_{ij}\|_{i,j=1}^p$ and $\Lambda_2 = \| \, |\xi|^{t_j}\delta_{ij}\|_{i,j=1}^p$ are diagonal matrices. Clearly, the theory of elliptic equations, both differential and pseudodifferential, easily carries over to the case of ellipticity in the Douglis-Nirenberg sense.

REMARK 23.2. A rational function $P(x, \xi)/Q(x, \xi)$ is an example of a symbol of class $\hat{D}^{(0)}_{\alpha+i\beta}(G)$. A boundary value problem in a bounded domain for a system of elliptic p.o.'s with rational symbols can be investigated by reducing it to a system of simultaneous elliptic differential equations (see Example 12.2).

§24. Mixed boundary value problems in a domain

1. *Function spaces with weight vanishing on* $\mathbf{R}^{n-2}$. In this section the notation $\mathbf{R}^n, \mathbf{R}^n_\pm, \mathbf{R}^{n-1}, \mathbf{R}^{n-1}_\pm, \mathbf{R}^{n-2}$ has the same meaning as in §13.

We denote by $W_{s,N} = W_{s,N}(\mathbf{R}^n)$ the space of generalized functions in $\mathbf{R}^n$ with finite norm

$$|[u]|_{s,N_1} = \sum_{k_1+k_2=0}^{N} \|x_n^{k_1} x_{n-1}^{k_2} u(x)\|_{s+k_1+k_2}. \tag{24.1}$$

Let $\chi(x_{n-1}, x_n)$ be a function of class C_0^∞ such that $\chi(x_{n-1}, x_n) = 1$ for $|x_{n-1}| + |x_n| \leqslant 1$. Since $(1 - \chi)(x_n^N + ix_{n-1}^N)^{-1} \in C^\infty$, the functions of class $W_{s,N}$ belong to H_{s+N} off $\mathbf{R}^{n-2}$, i.e.

$$\left(1 - \chi\left(\frac{x_{n-1}}{\varepsilon}, \frac{x_n}{\varepsilon}\right)\right)u \in H_{s+N}(\mathbf{R}^n)$$

for all $\varepsilon > 0$ if $u \in W_{s,N}$.

We denote by $H'_{s,N} = H_{s,N}(\mathbf{R}^{n-1})$ the space of generalized functions in $\mathbf{R}^{n-1}$ with finite norm

$$[v]_{s,N} = \sum_{k=0}^{N} [x_{n-1}^k v]_{s+k}. \tag{24.2}$$

The spaces $W_{s,N}(\mathbf{R}^n_+)$, $H_{s,N}(\mathbf{R}^{n-1}_\pm)$ and $H'^{\pm}_{s,N}$ are defined analogously to $H_{s,N}(\mathbf{R}^n_\pm)$ and $H^{\pm}_{s,N}(\mathbf{R}^n)$. Finally, we denote the norm in the Sobolev-Slobodeckiĭ space $H''_s = H_s(\mathbf{R}^{n-2})$ in $\mathbf{R}^{n-2}$ by $[\cdot]''_s$.

LEMMA 24.1. *Let p' be the operator of restriction to $\mathbf{R}^{n-1}$ of a function defined in $\mathbf{R}^n$ and let p'' be the operator of restriction to $\mathbf{R}^{n-2}$ of a function defined in $\mathbf{R}^n$ or $\mathbf{R}^{n-1}$. Then*

$$[p'u]_{s-1/2,N} \leqslant C|[u]|_{s,N}, \quad [p''v]''_{s-1/2} \leqslant C[v]_{s,N}, \qquad s > 1/2, \tag{24.3}$$

$$[p''u]''_{s-1} \leqslant C|[u]|_{s,N}, \qquad s > 1. \tag{24.3'}$$

PROOF. By Theorem 4.2,

$$[p'u]_{s-1/2,N} = \sum_{k=0}^{N} [x_{n-1}^k p'u]_{s+k-1/2} \leqslant C \sum_{k=0}^{N} \|x_{n-1}^k u\|_{s+k} \leqslant C|[u]|_{s,N}.$$

Analogously, by Theorem 4.2, $[p''v]''_{s-1/2} \leqslant C[v]_s \leqslant C[v]_{s,N}$. Finally,

$$[p''u]''_{s-1} \leqslant C[p'u]_{s-1/2} \leqslant C|[u]|_{s,N} \quad \text{for } s > 1.$$

LEMMA 24.2. *Suppose*

$$\frac{\partial^{k_1+k_2}}{\partial \xi_{n-1}^{k_1} \partial \xi_n^{k_2}} A(x, \xi) \in S^0_{\alpha-k_1-k_2}$$

for $0 \leqslant k_1 + k_2 \leqslant N$. Then

$$|[A(x, D)u]|_{s-\alpha,N} \leqslant C|[u]|_{s,N}. \tag{24.4}$$

LEMMA 24.3. *Suppose $A(x, \xi) \in S^0_\alpha$ and*

$$(1 + |\xi|)^{k_1}(1 + |\xi'|)^{k_2} \frac{\partial^{k_1+k_2}}{\partial \xi_n^{k_1} \partial \xi_{n-1}^{k_2}} A(x, \xi) \in S^0_\alpha$$

for $0 < k_1 + k_2 \leqslant N$. Then

$$|[pA(x, D)(v(x') \times \delta(x_n))]|_{s-\alpha,N} \leqslant C[v]_{s+1/2,N}, \qquad s < -1/2. \tag{24.5}$$

LEMMA 24.4. *Suppose*

$$\frac{\partial^{k_1+k_2}}{\partial \xi_n^{k_1} \partial \xi_{n-1}^{k_2}} A(x, \xi) \in S^t_{\alpha-k_1-k_2}$$

for $0 \leqslant k_1 + k_2 \leqslant N$ *and* $t > 0$, *and suppose we are given an arbitrary point* $x_0 \in \mathbf{R}^n$ *and an arbitrary positive number* ε. *Then* (23.10) *holds, with*

$$\|[K(x, D)u]\|_{s-\alpha,N} < \varepsilon\|[u]\|_{s,N}, \qquad \|[T(x, D)]\|_{s-\alpha,N} \leqslant C\|[u]\|_{s-1,N}. \quad (24.6)$$

Lemmas 24.2–24.4 are proved by means of a formula of form (23.2) in the same way as Lemmas 23.2, 23.3 and 23.5.

LEMMA 24.5. *Let* p_1' (p_2') *be the operator of restriction to* $\mathbf{R}_+^{n-1}$ ($\mathbf{R}_-^{n-1}$) *of a function defined in* $\mathbf{R}^n$, *and suppose we are given a noninteger* $s \leqslant 1/2$ *and an integer* $N > 1/2 - s$. *Then*

$$[p_1'u]^+_{s-1/2,N} \leqslant C\|[u]\|_{s,N}, \qquad [p_2'u]^-_{s-1/2,N} \leqslant C\|[u]\|_{s,N}, \quad (24.7)$$

where $[\cdot]^\pm_{s-1/2,N}$ *are the norms in* $H_{s-1/2,N}(\mathbf{R}^{n-1}_\pm)$.

PROOF. Since $N > 1/2 - s$, we have

$$\begin{aligned}[p_1'(1 - \chi(x_{n-1}, x_n))u]^+_{s-1/2,N} &\leqslant C[p'x^N_{n-1}(1-\chi)u]_{s-1/2+N}\\ &\leqslant C\|[(1-\chi)u]\|_{s,N}.\end{aligned}$$

The first estimate of (24.7) will consequently follow if we establish it for any function $u(x'', x_{n-1}, x_n)$ whose support is compact with respect to x_n and x_{n-1}.

Let r be the nonnegative integer such that $-1/2 < s + r \leqslant 1/2$. Since

$$\|[u]\|_{s,N} = \sum_{k_1+k_2=0}^{N} \|x_n^{k_1}x_{n-1}^{k_2}u\|_{s+k_1+k_2} \geqslant \sum_{k=0}^{N} \|x^k_{n-1}u\|_{s+k}, \quad (24.8)$$

we have $x^k_{n-1}u \in H_{s+r+1}$ for $k \geqslant r + 1$, and hence the restriction of $x^k_{n-1}u$ to $\mathbf{R}^{n-1}$ exists. If, on the other hand, $0 \leqslant k \leqslant r$, then the restriction of x^k_{n-1} to $\mathbf{R}^{n-1}$ generally does not exist, although its restriction to $\mathbf{R}^{n-1} \setminus \mathbf{R}^{n-2}$ exists.

We denote by $v(x')$ a generalized function in $\mathbf{R}^{n-1}$ such that

$$p'x^{r+1}_{n-1}u(x', x_n) = x^{r+1}_{n-1}v(x'). \quad (24.9)$$

For example, v can be defined on the functions $\varphi(x') \in S(\mathbf{R}^{n-1})$ by the formula

$$(v, \bar{\varphi}) = \int p'x^{r+1}_{n-1}u\frac{(\varphi(x') - \sum^r_{k=0}(1/k!)(\partial^k\varphi(x'', 0)/\partial x^k_{n-1})x^k_{n-1})}{x^{r+1}_{n-1}}\,dx'. \quad (24.10)$$

The function v is not uniquely determined by (24.9) but only to within a generalized function of the form

$$\sum_{k=0}^{r} c_k(x'')D^k_{n-1}\delta(x_{n-1}),$$

i.e. $\tilde{v}(\xi'', \xi_{n-1})$ is determined to within a polynomial in ξ_{n-1} of the form $\Sigma_{k=0}^{r} \tilde{c}_k(\xi'')\xi_{n-1}^k$. (We note that the support of v, like the support of u, is compact with respect to x_{n-1}.)

Let

$$\tilde{v}_1(\xi') = \tilde{v}(\xi') - \sum_{k=0}^{r} \frac{1}{k!} \frac{\partial^k \tilde{v}(\xi'', 0)}{\partial \xi_{n-1}^k} \xi_{n-1}^k.$$

Then $v_1(x')$ satisfies (24.9), and $\partial^k \tilde{v}_1(\xi'', 0)/\partial \xi_{n-1}^k = 0$ for $0 \leqslant k \leqslant r$. Thus (24.8) implies

$$|[u]|_{s,N} \geqslant \sum_{k=0}^{N} \|x_{n-1}^k u\|_{s+k} \geqslant \sum_{k=r+1}^{N} \|x_{n-1}^k u\|_{s+k} \geqslant C \sum_{k=r+1}^{N} [x_{n-1}^k v_1]_{s+k-1/2}. \tag{24.11}$$

On the other hand, since $s + r + 1/2 > 0$, we have

$$\begin{aligned} [x_{n-1}^{r+1} v_1]_{s+r+1/2}^2 &= \int (1 + |\xi'|)^{2(s+r+1/2)} |D_{n-1}^{r+1} \tilde{v}_1(\xi')|^2 \, d\xi' \\ &\geqslant \int |\xi_{n-1}|^{2(s+r+1/2)} |D_{n-1}^{r+1} \tilde{v}_1(\xi')|^2 \, d\xi'' \, d\xi_{n-1}. \end{aligned} \tag{24.12}$$

Let $\hat{v}_1(\xi'', z)$ denote the Mellin transform of $\xi_{n-1}^{s+r+1/2} D_{n-1}^{r+1} \tilde{v}_1(\xi')$ with respect to ξ_{n-1}:

$$\hat{v}_1(\xi'', z) = \int_0^\infty \xi_{n-1}^{s+r+1/2} D_{n-1}^{r+1} \tilde{v}_1(\xi'', \xi_{n-1}) \xi_{n-1}^{z-1} \, d\xi_{n-1}, \tag{24.13}$$

and suppose $0 \leqslant k \leqslant r$. Integrating $r + 1 - k$ times by parts in (24.13) while taking into account the fact that $\tilde{v}_1(\xi'', \xi_{n-1})$ has a zero of order $r + 1$ at $\xi_{n-1} = 0$, we get

$$\hat{v}_1(\xi'', z) = (-i)^{r+1-k} \left(s + r + z - \tfrac{1}{2}\right) \cdots \left(s + k + z - \tfrac{1}{2}\right) \hat{v}_2(\xi'', z), \tag{24.14}$$

where $\hat{v}_2(\xi'', z)$ is the Mellin transform of $\xi_{n-1}^{s+k-1/2} D_{n-1}^k \tilde{v}_1(\xi'', \xi_{n-1})$. Since s is not an integer, the factor in front of $\hat{v}_2$ in (24.14) does not vanish on the straight line $\operatorname{Re} z = 1/2$. Consequently, Parseval's equality for the Mellin transform (see §2) implies

$$\begin{aligned} &\int_{-\infty}^{\infty} \int_0^\infty |\xi_{n-1}|^{2(s+r+1/2)} |D_{n-1}^{r+1} \tilde{v}_1(\xi')|^2 \, d\xi' \\ &\qquad \geqslant C \int_{-\infty}^{\infty} \int_0^\infty |\xi_{n-1}|^{2(s+k-1/2)} |D_{n-1}^k \tilde{v}_1(\xi')|^2 \, d\xi'. \end{aligned} \tag{24.15}$$

An analogous proof with the use of the Mellin transform exists for the estimate

$$\int_{-\infty}^{\infty}\int_{-\infty}^{0}|\xi_{n-1}|^{2(s+r+1/2)}\left|D_{n-1}^{r+1}\tilde{v}_1(\xi')\right|^2 d\xi'$$
$$\geqslant C\int_{-\infty}^{\infty}\int_{-\infty}^{0}|\xi_{n-1}|^{2(s+k-1/2)}\left|D_{n-1}^{k}\tilde{v}_1(\xi')\right|^2 d\xi'. \qquad (24.15')$$

Thus, since $s + k - 1/2 \leqslant 0$ for $0 \leqslant k \leqslant r$,

$$\begin{aligned}\left[x_{n-1}^{r+1}v_1\right]_{s+r+1/2}^2 &\geqslant C\int_{-\infty}^{\infty}|\xi_{n-1}|^{2(s+k-1/2)}\left|D_{n-1}^{k}\tilde{v}_1(\xi')\right|^2 d\xi' \\ &\geqslant C\int_{-\infty}^{\infty}(1+|\xi'|)^{2(s+k-1/2)}\left|D_{n-1}^{k}\tilde{v}_1\right|^2 d\xi' \\ &= C\left[x_{n-1}^{k}v_1\right]_{s+k-1/2}^2. \end{aligned} \qquad (24.16)$$

It follows by virtue of (24.11) and (24.16) that

$$\|[u]\|_{s,N} \geqslant C\sum_{k=0}^{N}\left[x_{n-1}^{k}v_1\right]_{s+k-1/2} = C[v_1]_{s-1/2,N},$$

i.e. $v_1 \in H'_{s-1/2,N}$. But the restriction of $v_1(x')$ to $\mathbf{R}_+^{n-1}$ clearly coincides with $p_1'u$, so that v_1 is one of the extensions of $p_1'u$ from $\mathbf{R}_+^{n-1}$ onto $\mathbf{R}^{n-1}$. Consequently, by definition of the norm in $H_{s-1/2,N}(\mathbf{R}_+^{n-1})$, we conclude that

$$[p_1'u]_{s-1/2,N}^{+} = \inf_{l}\left[l(p_1'u)\right]_{s-1/2,N} \leqslant [v_1]_{s-1/2,N} \leqslant C\|[u]\|_{s,N}.$$

The estimate $[p_2'u]_{s-1/2,N}^{-} \leqslant C\|[u]\|_{s,N}$ is proved analogously.

2. *Mixed boundary value problems in a halfspace.* Consider in $\mathbf{R}_+^n$ an elliptic (in the Douglis-Nirenberg sense) system of differential equations

$$\sum_{j=1}^{p} L_{ij}(D)u_j(x) = f_i(x), \quad 1 \leqslant i \leqslant p, \quad \deg_\xi L_{ij}(\xi) = s_i + t_j \qquad (24.17)$$

with mixed differential boundary conditions

$$\sum_{j=1}^{p} B_{1ij}(D)u_j\big|_{\mathbf{R}_+^{n-1}} = g_{1i}(x'), \qquad \sum_{j=1}^{p} B_{2ij}(D)u_j\big|_{\mathbf{R}_-^{n-1}} = g_{2i}(x'), \qquad 1 \leqslant i \leqslant m, \qquad (24.18)$$

where $\deg_\xi B_{kij}(\xi) = m_{ki} + t_j$, $k = 1, 2$, and $2m = \sum_{i=1}^{p} s_i + \sum_{j=1}^{p} t_j$ is the order of $\det\|L_{ij}(\xi)\|$.

As in §13, let

$$\hat{L}_{ij}(\xi) = L_{ij}\left(\frac{1+|\xi''|}{|\xi''|}\xi'', \xi_{n-1}, \xi_n\right),$$

and let the notation $\hat{\hat{B}}_{kij}(\xi)$ have the analogous meaning. It is more convenient to consider instead of problem (24.17), (24.18) the mixed boundary value problem obtained by replacing $L_{ij}(D)$, $B_{kij}(D)$ by $\hat{\hat{L}}_{ij}(D)$, $\hat{\hat{B}}_{kij}(D)$.

In the case of an elliptic differential equation of order greater than two as well as in the case of a system of equations, it is not always possible to choose the function space so that a mixed problem is uniquely solvable. It is therefore necessary to consider the following generalized formulation of a mixed boundary value problem:

$$\sum_{j=1}^{p} \hat{\hat{L}}_{ij}(D)u_j(x) = f_i(x), \quad 1 \leqslant i \leqslant p, \quad u_j \in W_{s+t_j,N}(\mathbf{R}^n_+), \quad f_i \in W_{s-s_i,N}(\mathbf{R}^n_+), \tag{24.19}$$

$$p_1'\left(\sum_{j=1}^{p} \hat{\hat{B}}_{1ij}(D)lu_j(x) + \sum_{j=1}^{m_-} \hat{G}_{1ij}(D')(v_j(x'') \times \delta(x_{n-1}))\right) = g_{1i}(x'), \quad 1 \leqslant i \leqslant m, \tag{24.20}$$

$$p_2'\left(\sum_{j=1}^{p} \hat{\hat{B}}_{2ij}(D)lu_j(x) + \sum_{j=1}^{m_-} \hat{G}_{2ij}(D')(v_j(x'') \times \delta(x_{n-1}))\right) = g_{2i}(x'), \quad 1 \leqslant i \leqslant m, \tag{24.21}$$

$$p'' \sum_{j=1}^{p} \hat{\hat{B}}_{ij}^-(D)lu_j = h_i(x''), \qquad 1 \leqslant i \leqslant m_+, \tag{24.22}$$

where

$$\deg_{\xi'} \frac{\partial^r}{\partial \xi_{n-1}^r} G_{kij}(\xi') = m_{ki} + \gamma_j - r, \qquad 0 \leqslant r \leqslant N, \quad k = 1, 2,$$

the $B_{ij}^-(\xi', \xi_n)$ are "minus" functions with respect to ξ_n, $B_{ij}^-(\xi) \in O'_{\beta_i + t_j}$, $g_{1i} \in H_{s-m_{1i}-1/2,N}(\mathbf{R}^{n-1}_+)$, $g_{2i} \in H_{s-m_{2i}-1/2,N}(\mathbf{R}^{n-1}_-)$, $v_j \in H_{s+\operatorname{Re}\gamma_j-1/2}(\mathbf{R}^{n-2})$ and $h_i \in H_{s-\operatorname{Re}\beta_i-1}(\mathbf{R}^{n-2})$. The functions $p_k'\hat{\hat{B}}_{kij}(D)lu_j$ and $p''\hat{\hat{B}}_{ij}^-(D)lu_j$ do not depend on the choice of the extension of u since $\hat{\hat{B}}_{kij}$ and $\hat{\hat{B}}_{ij}^-$ are "minus" operators. It is assumed that

$$s + N > \max_{i,k}(m_{ki} + 1/2), \quad \operatorname{Re}\beta_i < s - 1, \quad \operatorname{Re}\gamma_j < -s + 1/2, \tag{24.23}$$

with s nonintegral if $s \leqslant \max_{i,k}(m_{ki} + 1/2)$. We note that the last condition in (24.23) is not necessary if $G_{kij}(\xi') \in D_{m_{ki}+\gamma_j}$.

Let $W_{s,N}^{(1)}$ denote the space of the unknown functions of problem (24.19)–(24.22), and let $W_{s,N}^{(2)}$ denote the space of its right sides. We fix ω'' so that $|\omega''| = 1$ and consider the associated (with problem (24.19)–(24.22)) family of mixed boundary value problems in $\mathbf{R}^2_+$

$$\sum_{j=1}^{p} L_{ij}(\omega'', D_{n-1}, D_n)u_j(x_{n-1}, x_n) = f_i(x_{n-1}, x_n), \qquad 1 \leqslant i \leqslant p, \tag{24.24}$$

$$p'_k\left(\sum_{j=1}^{p} B_{kij}(\omega'', D_{n-1}, D_n)lu_j(x_{n-1}, x_n) + \sum_{j=1}^{m_-} v_j G_{kij}(\omega'', D_{n-1})\delta(x_{n-1})\right)$$
$$= g_{ki}(x_{n-1}), \qquad 1 \leqslant i \leqslant m, \quad k = 1, 2, \tag{24.25}$$

$$p'' \sum_{j=1}^{p} B_{ij}^{-}(\omega'', D_{n-1}, D_n)lu_j(x_{n-1}, x_n) = h_i, \qquad 1 \leqslant i \leqslant m_+, \tag{24.26}$$

where v_j and h_i are complex numbers,

$$u_j \in W_{s+t_j,N}(\mathbf{R}^2_+), \quad f_i \in W_{s-s_i,N}(\mathbf{R}^2_+), \quad g_{1i} \in H_{s-m_{1i}-1/2,N}(\mathbf{R}^1_+),$$
$$g_{2i} \in H_{s-m_{2i}-1/2,N}(\mathbf{R}^1_-).$$

It will be assumed that

The boundary value problem (24.24)– (24.26) *is uniquely solvable for a given s and any ω''.* (24.27)

Taking the Fourier transform with respect to x'' of (24.19)–(24.22) and making the change of variables $x_{n-i} = y_{n-i}/(1 + |\xi''|)$, $i = 0, 1$, we get (completely analogously to the proof of Theorem 16.1) that the operator $\tilde{\mathfrak{L}}$ defined by the left sides of the resultant system is a pseudodifferential operator in $\mathbf{R}^{n-2}$ with an operator valued symbol $\mathfrak{L}(\omega'')$ defined by the left sides of (24.24)–(24.26). It follows, as in §16, that the invertibility of $\mathfrak{L}(\omega'')$ for all $\omega'' \in S^{n-3}$ ensures the unique solvability of problem (24.19)–(24.22). Thus the following theorem is proved.

THEOREM 24.1. *Suppose $L(D) = \|L_{ij}(D)\|_{i,j=1}^{p}$ is an elliptic operator in the Douglis-Nirenberg sense and conditions* (24.23) *and* (24.27) *are satisfied. Then the operator $\mathfrak{L}$ defined by the left sides of* (24.19)–(24.22) *bijectively and continuously maps $W_{s,N}^{(1)}$ onto $W_{s,N}^{(2)}$.*

Let us take up in more detail the question of conditions for the unique solvability of problem (24.24)–(24.26).

Suppose that the boundary operators $B_{1ij}(D)$ and $B_{2ij}(D)$ satisfy for each $\xi' \neq 0$ the Šapiro-Lopatinskiĭ condition of unique solvability of a differential boundary value problem on a halfline. Then Lemma 16.2 implies that the space of solutions of the system

$$\sum_{j=1}^{p} L_{ij}(\xi', D_n)u_j(x_n) = 0$$

on the halfline $\mathbf{R}^1_+$ has a basis

$$e_r(\xi', x_n) = (e_{1r}(\xi', x_n), \ldots, e_{pr}(\xi', x_n)), \qquad 1 \leqslant r \leqslant m,$$

whose elements smoothly depend on ξ' ($\xi' \neq 0$) and decrease as $x_n \to +\infty$. We normalize $e_{kj}(\xi', x_n)$ so that

$$e_{kj}\left(\tau\xi', \frac{x_n}{\tau}\right) = \tau^{-t_j} e_{kj}(\xi', x_n)$$

for $\tau > 0$ and let $E_{kj}(\xi', \xi_n)$ denote the Fourier transform with respect to x_n of the extension of $e_{kj}(\xi', x_n)$ that vanishes for $x_n < 0$. We note that $E_{kj}(\xi', \xi_n)$ is a rational function in ξ_n, with $\deg_\xi E_{kj}(\xi) = -t_j - 1$.

On the basis of the above remarks we will seek the solution of problem (24.24)–(24.26) in the form (of the general solution of equation (24.24))

$$u_i(x_{n-1}, x_n) = p\left(u_i^{(0)} + \sum_{j=1}^{p} E_{ij}(\omega'', D_{n-1}, D_n)(w_k(x_{n-1}) \times \delta(x_n))\right), \quad 1 \leqslant i \leqslant p, \tag{24.28}$$

where

$$u_i^{(0)} = \sum_{j=1}^{p} L_{ij}^{(1)}(\omega'', D_{n-1}, D_n) lf_j, \qquad 1 \leqslant i \leqslant p, \tag{24.29}$$

is a particular solution of system (24.24), $\|L_{ij}^{(1)}(\xi)\|_{i,j=1}^{p}$ is the inverse of the matrix $\|L_{ij}(\xi)\|_{i,j=1}^{p}$ and the $w_k(x_{n-1})$ are arbitrary functions of class $H_{s-1/2,N}(\mathbf{R}^1)$. We note that Lemmas 24.2 and 24.3 imply

$$|[u_i]|_{s+t_i,N} \leqslant C\left(\sum_{j=1}^{p} |[f_j]|_{s-s_j,N}^{+} + \sum_{k=1}^{m} [w_k]_{s-1/2,N}\right). \tag{24.30}$$

Let us substitute (24.28) in the boundary conditions (24.25). Representing the product of the matrices $\Lambda_-(\xi)\|B_{kij}(\xi)\|_{j=1,\dots,p}^{i=1,\dots,m}$ and $\|E_{jr}(\xi)\|_{r=1,\dots,m}^{j=1,\dots,p}$ in the form

$$\|a_{kir}^{(0)}(\xi) + a_{kir}^{(1)}(\xi)\|,$$

where the $a_{kir}^{(0)}(\xi)$ are polynomials in ξ_n and the $a_{kir}^{(1)}(\xi)$ decrease as $\xi_n \to \infty$, and taking into account the fact that

$$p\Lambda_-^{-1} a_{kir}^{(0)}(\omega, D_{n-1}, D_n)(w_r \times \delta(x_n)) = 0,$$

we get

$$p_k'\left(\sum_{r=1}^{m} a_{ir}^{(k)}(\omega'', D_{n-1}) w_r(x_{n-1}) + \sum_{j=1}^{m_-} v_j G_{kij}(\omega'', D_{n-1})\delta(x_{n-1})\right) = g_{ki}^{(1)}(x_{n-1}), \quad 1 \leqslant i \leqslant m, \quad k = 1, 2, \tag{24.31}$$

where

$$a_{ir}^{(k)}(\xi') = \frac{1}{2\pi}\int_{-\infty}^{\infty} \Lambda_-^{-1} a_{kir}^{(1)}(\xi', \xi_n)\, d\xi_n, \quad \deg_{\xi'} a_{ir}^{(k)}(\xi') = m_{ki},$$

$$g_{ki}^{(1)}(x_{n-1}) = g_{ki}(x_{n-1}) - p_k' \sum_{j=1}^{p} B_{kij} u_j^{(0)}.$$

We note that $g_{1i}^{(1)} \in H_{s-m_{1i}-1/2,N}(\mathbf{R}_+^1)$ and $g_{2i}^{(1)} \in H_{s-m_{2i}-1/2,N}(\mathbf{R}_+^1)$ by virtue of Lemma 24.5.

Let us now substitute (24.28) in (24.26). The rational function $E_{jk}(\xi)$ admits the decomposition $E_{jk}(\xi) = E_{jk}^-(\xi) + R_{jk}(\xi)$, where $E_{jk}^-(\xi)$ is a "minus" function and

$$|R_{jk}(\xi)| \leqslant C(1+|\xi|)^{-t_j-1-N_1}(1+|\xi'|)^{N_1}.$$

Letting $\|b_{ik}^{(1)}(\xi)\|_{k=1,\dots,m}^{i=1,\dots,m_+}$ denote the product of the matrices $\|B_{ij}^-(\xi)\|$ and $\|R_{jk}(\xi)\|$ and taking into account the fact that $N_1 > \operatorname{Re}\beta_i$ and

$$pB_{ij}^-E_{jk}^-(w_k(x_{n-1}) \times \delta(x_n)) = 0,$$

we get

$$p'' \sum_{j=1}^{p} b_{ij}^{(2)}(\omega'', D_{n-1})w_j(x_{n-1}) = h_i^{(1)}, \qquad 1 \leqslant i \leqslant m_+, \tag{24.32}$$

where

$$b_{ij}^{(2)}(\xi') = \frac{1}{2\pi}\int_{-\infty}^{\infty} b_{ik}^{(1)}(\xi', \xi_n)\,d\xi_n, \quad \deg_{\xi'} b_{ij}^{(2)}(\xi') = \beta_j,$$

$$h_i^{(1)} = h_i - p'' \sum_{j=1}^{p} B_{ij}^- u_j^{(0)}.$$

The Šapiro-Lopatinskiĭ condition for the boundary operators $B_{1ij}(D)$ and $B_{2ij}(D)$ is equivalent to the condition

$$\det A_1(\xi') \neq 0, \quad \det A_2(\xi') \neq 0, \quad \xi' \neq 0, \tag{24.33}$$

where $A_k(\xi') = \|a_{ir}^{(k)}(\xi')\|_{i,r=1}^m$.

System (24.31) is an elliptic system of paired pseudodifferential equations for $w_1(x_{n-1}), \dots, w_m(x_{n-1})$ (cf. §13). Extending it from $\mathbf{R}_+^1$ to $\mathbf{R}^1$ for $k = 1$, and from $\mathbf{R}_-^1$ to $\mathbf{R}^1$ for $k = 2$, we get

$$\sum_{r=1}^{m} a_{ir}^{(1)}w_r + \sum_{j=1}^{m_-} v_j G_{1ij}\delta(x_{n-1}) = lg_{1i}^{(1)}(x_{n-1}) + \Lambda_-^{m_{1i}}(D_{n-1})w_i^-(x_{n-1}),$$
$$1 \leqslant i \leqslant m, \tag{24.34}$$

$$\sum_{r=1}^{m} a_{ir}^{(2)}w_r + \sum_{j=1}^{m_-} v_j G_{2ij}\delta(x_{n-1}) = lg_{2i}^{(1)}(x_{n-1}) + \Lambda_+^{m_{2i}}(D_{n-1})w_i^+(x_{n-1}),$$
$$1 \leqslant i \leqslant m, \tag{24.35}$$

where the functions $w_i^\pm \in H_{s-1/2,N}^\pm(\mathbf{R}^1)$ are unknown and the operators $\Lambda_\pm^k(D_{n-1}) = (D_{n-1} \pm i)^k$ have been introduced for convenience. Solving (24.35) for the w_k and substituting the resultant expressions in (24.34) and (24.32), we get, after some obvious transformations, the following system of equations for the $w_i^{\pm}$ and v_j:

$$\sum_{k=1}^{m} a_{ik}(\omega'', D_{n-1})w_k^+(x_{n-1}) + \sum_{k=1}^{m_-} v_k C_{ik}(\omega'', D_{n-1})\delta$$
$$= lg_i^{(2)}(x_{n-1}) + w_i^-(x_{n-1}), \qquad 1 \leqslant i \leqslant m, \tag{24.36}$$

$$p'' \sum_{k=1}^{m} b_{ij}(\omega'', D_{n-1})w_k^+(x_{n-1}) + \sum_{k=1}^{m_-} E_{ik}^{(1)}(\omega'')v_k = h_i^{(2)}, \qquad 1 \leqslant i \leqslant m_+, \tag{24.37}$$

where $A(\xi') = \|a_{ik}(\xi')\|$ is an elliptic matrix such that $\deg_{\xi'} a_{ik}(\xi') = 0$ and

$$A(\xi') = \Lambda_-^{(1)} A_1(\xi')A_2^{-1}(\xi')\Lambda_+^{(1)}, \quad \Lambda_+^{(1)} = \left\|(\xi_{n-1} + i)^{m_{1j}}\delta_{jk}\right\|,$$
$$\Lambda_-^{(1)} = \left\|(\xi_{n-1} - i)^{-m_{2j}}\delta_{jk}\right\|. \tag{24.38}$$

(24.36) is equivalent to the system of ψ d.e.'s on a halfline

$$p'\left(\sum_{k=1}^{m} a_{ik}w_k^+ + \sum_{k=1}^{m} v_k C_{ik}(\omega'', D_{n-1})\delta(x_{n-1})\right) = g_i^{(2)}(x_{n-1}), \qquad 1 \leqslant i \leqslant m, \tag{24.36$'$}$$

where $g_i^{(2)}(x_{n-1})$ is the restriction of $lg_i^{(2)}$ to $\mathbf{R}_+^1$. Thus the mixed boundary value problem (24.24)–(24.26) reduces to the system of ψ d.e.'s (24.36′), (24.37), which has the same form as the system of ψ d.e.'s studied in Sobolev-Slobodeckiĭ spaces in §16.

Let us now take up the question of the unique solvability of system (24.36′), (24.37) in weighted function spaces. We will use the notation

$$\mathcal{H}_{s-1/2,N}^{(1)} = \prod_{k=1}^{m} H_{s-1/2,N}^+(\mathbf{R}^1) \times \mathbf{C}^{m_-}, \quad \mathcal{H}_{s-1/2,N}^{(2)} = \prod_{k=1}^{m} H_{s-1/2,N}(\mathbf{R}_+^1) \times \mathbf{C}^{m_+}.$$

THEOREM 24.2. *Suppose s satisfies* (14.38) *and the operator* $\mathfrak{A}(\omega'')$ *defined by the left sides of* (24.36′) *and* (24.37) *bijectively and continuously maps* $\mathcal{H}_{s-1/2,N}^{(1)}$ *onto* $\mathcal{H}_{s-1/2,N}^{(2)}$ *for* $N = 0$ *(i.e. in the case of Sobolev-Slobodeckiĭ spaces without a weight). Then* $\mathfrak{A}(\omega'')$ *bijectively and continuously maps* $\mathcal{H}_{s-1/2,N}^{(1)}$ *onto* $\mathcal{H}_{s-1/2,N}^{(2)}$ *for any* $N > 0$.

PROOF. Since $\mathcal{H}_{s-1/2,N}^{(k)} \subset \mathcal{H}_{s-1/2,0}^{(k)}$ for $k = 1, 2$, it follows, as in §23.4, that Theorem 24.2 will be proved if we establish that the solution of system (24.36′), (24.37) belongs to $\mathcal{H}_{s-1/2,N}^{(1)}$, $N > 0$, whenever it belongs to $\mathcal{H}_{s-1/2,0}^{(1)}$ and the right side belongs to $\mathcal{H}_{s-1/2,N}^{(2)}$. We note that

$$p' C_{ik}\delta(x_{n-1}) \in H_{s-1/2,N}(\mathbf{R}^1)$$

(see the proof of Lemma 23.2). It consequently remains to establish the following lemma on smoothness (cf. Lemma 23.11).

LEMMA 24.6. *Suppose*

$$\left|\frac{\partial^k}{\partial \xi_{n-1}^k} A(\xi')\right| \leqslant C(1+|\xi_{n-1}|)^{-k}, \qquad 0 \leqslant k \leqslant N+1, \quad \det A(\xi') \neq 0,$$

where $\xi' = (\omega'', \xi_{n-1})$, *and* s *satisfies* (14.38). *Then a function* $u_+(x_{n-1})$ *is of class* $H^+_{s,N}(\mathbf{R}^1)$ *if it is of class* $H_s^+(\mathbf{R}^1)$ *and satisfies the equation*

$$pA(\omega'', D_{n-1})u_+ = f, \tag{24.39}$$

where $f \in H_{s,N}(\mathbf{R}^1_+)$.

PROOF. Let $b_\pm(\xi_{n-1})$, $a_\pm$ and $P_\pm$ be the same as in §14, and suppose for simplicity that $s = 0$. Taking the Fourier transform with respect to x_{n-1} of the extension of (24.39) that vanishes for $x_{n-1} < 0$ and applying the operator

$$\tilde{R} = P_+A^{-1}a_+P_-a_+^{-1} = P_+A^{-1}(a_+b_-^{-1}b_+)P_+(b_+^{-1}b_-a_+^{-1})$$

(see §14), we get (cf. (17.16))

$$\tilde{u}_+(\xi_{n-1}) = P_+A^{-1}(\omega'', \xi_{n-1})a_+P_-a_+^{-1}\Pi^+lf + P_+Cb_+^{-1}b_-(\xi_{n-1})a_+^{-1}A\tilde{u}_+, \tag{24.40}$$

where $C = P_+A^{-1}(a_+b_-^{-1}b_+) - A^{-1}(a_+b_-^{-1}b_+)P_+$ is the same operator as in (17.16).

The operators $A^{-1}(\omega, \xi_{n-1})$ and Π^+ are bounded in $\tilde{H}_{0,N}(\mathbf{R}^1)$ (see Lemmas 23.2 and 23.4). We note that the norm $\|\tilde{v}(\xi_{n-1})\|_{0,N}$ in $\tilde{H}_{0,N}(\mathbf{R}^1)$ is clearly equivalent to the norm $\sum_{k=0}^N \|(\xi_{n-1}+i)^k \tilde{v}^{(k)}(\xi_{n-1})\|_0$, where $\tilde{v}^{(k)}(\xi_{n-1})$ denotes the derivative $\partial^k\tilde{v}(\xi_{n-1})/\partial\xi_{n-1}^k$. By (5.36),

$$\Pi^+\tilde{g}^{(k)}(\xi_{n-1}) = (\xi_{n-1}+i)^{-k}\Pi^+(\xi_{n-1}+i)^k\tilde{g}^{(k)}(\xi_{n-1}) \qquad \text{for any } \tilde{g} \in S(\mathbf{R}^1),$$

since $\Pi'(\xi_{n-1}+i)^p\tilde{g}^{(k)}(\xi_{n-1}) = 0$ for $p < k$ (cf. (17.4) and (17.5)). Consequently, with the use of Lemma 14.3 we get that the operators $P_\pm$ are also bounded in $\tilde{H}_{0,N}$. Thus

$$P_+A^{-1}a_+P_-a_+^{-1}\Pi^+lf \in \tilde{H}_{0,N},$$

since $lf \in \tilde{H}_{0,N}$. Further, since $\tilde{u}_+ \in \tilde{H}_0^+$, we have

$$\tilde{v} = b_+^{-1}b_-a_+^{-1}A\tilde{u}_+ \in \tilde{H}_0.$$

But it then follows from (17.7) that $C\tilde{v} \in \tilde{H}_{0,N}$ and hence that $P_+C\tilde{v} \in \tilde{H}^+_{0,N}$. Lemma 24.6 is proved.

3. *Mixed problem in a bounded domain.* Let G be a bounded domain in $\mathbf{R}^n$ with a smooth $(n-1)$-dimensional boundary Γ. We assume that Γ is divided by a smooth $(n-2)$-dimensional submanifold γ into two parts Γ_1 and Γ_2, so that $\Gamma_1 \cap \Gamma_2 = \gamma$ and $\Gamma_1 \cup \Gamma_2 = \Gamma$. The local coordinate system in a neighborhood U_j intersecting γ will always be chosen so that $x_n^{(j)} = 0$ is the equation for Γ and $x_n^{(j)} = 0$, $x_{n-1}^{(j)} = 0$ are the equations for γ.

Let $\rho_k(x)$, $k = 0, 1, \ldots, N$, denote a complex valued function of class $C^\infty(\mathbf{R}^n)$ that (i) does not vanish for $x \notin \gamma$, (ii) is equal to unity for large $|x|$, and (iii) has the form $(x_n^k + ix_{n-1}^k)\rho_0(x)$, where $\rho_0(x) \neq 0$ and $\rho_0(x) \in C^\infty$, in the corresponding l.c.s. in a neighborhood of γ. Further, let $\rho_k'(x')$, $k = 0, 1, \ldots, N$, denote an infinitely differentiable function on Γ that (i) does not vanish for $x' \notin \gamma$, and (ii) has the form $x_{n-1}^k \rho_0'(x')$, where $\rho_0'(x') \neq 0$ and $\rho_0'(x') \in C^\infty$, in the l.c.s. in a neighborhood of γ. We denote by $W_{s,N}(G)$ the space of generalized functions in G admitting an extension onto $\mathbf{R}^n$ with finite norm

$$\|[lu]\|_{s,N} = \sum_{k=0}^{N} \|\rho_k lu\|_{s+k},$$

and by $H_{s,N}(\Gamma_i)$, $i = 1, 2$, the space of generalized functions on Γ_i admitting an extension onto Γ with finite norm

$$[lg]_{s,N} = \sum_{k=0}^{N} [\rho_k' lg]_{s+k}.$$

We note that the norms

$$\inf_l \|[lu]\|_{s,N}, \qquad \inf_l [lg]_{s,N}$$

in $W_{s,N}(G)$, $H_{s,N}(\Gamma_i)$, $i = 1, 2$, are equivalent in a neighborhood of γ to the norms of form (24.1), (24.2).

THEOREM 24.3. *Suppose $\|L_{ij}(x, D)\|$ is a system of differential operators that is elliptic in the Douglis-Nirenberg sense. Suppose we are given in the domain G a generalized mixed problem of the form*

$$\sum_{j=1}^{p} L_{ij}(x, D)u_j(x) = f_i(x), \qquad 1 \leqslant i \leqslant p, \tag{24.41}$$

$$p_k'\left(\sum_{j=1}^{p} B_{kij}(x, D)lu_j(x) + \sum_{j=1}^{m_-} G_{kij}(x', D')(v_j(x'') \times \delta(\gamma))\right) = g_{ki}(x'),$$

$$k = 1, 2, \quad 1 \leqslant i \leqslant m, \qquad x' \in \Gamma_k, \quad x'' \in \gamma, \tag{24.42}$$

$$p'' \sum_{j=1}^{p} B_{ij}^- lu_j(x) = h_i(x''), \qquad x'' \in \gamma, \quad 1 \leqslant i \leqslant m_+, \tag{24.43}$$

in which the $B_{kij}(x, D)$ are differential operators of order $m_{ki} + t_j$ satisfying the Šapiro-Lopatinskiĭ condition on $\overline{\Gamma}_k$, the $G_{kij}(x', D')$ are p.o.'s on Γ of class

$\hat{O}^{\infty}_{m_{ki}+\gamma_j}$, *and the* B^-_{ij} *have the form*

$$B^-_{ij}lu_j = \sum_k{}'' \psi_k(x) B^-_{ijk}\varphi_k lu_j,$$

where the summation is taken over those k *for which* $U_k \cap \gamma \neq \varnothing$ *and the operator* B^-_{ijk}, *being written in the kth local coordinates, is a p.o. with a "minus" symbol that is homogeneous in* ξ *of degree* $\beta_i + t_j$ *and differentiable for* $\xi \neq 0$. *Let* $W^{(1)}_{s,N}(G)$ *denote the space of unknown functions* $u_j \in W_{s+t_j,N}(G)$, $v_j \in H''_{s+\mathrm{Re}\,\gamma_j-1/2}$ *(which is a Sobolev-Slobodeckiĭ space on* γ*) of problem* (24.41)–(24.43), *and let* $W^{(2)}_{s,N}(G)$ *denote the space of right sides* $f_j \in W_{s-s_j,N}(G)$, $g_{ki} \in H_{s-m_{ki}-1/2,N}(\Gamma_k)$, $h_i \in H_{s-\mathrm{Re}\,\beta_i-1}(\gamma)$ *of problem* (24.41)–(24.43). *Suppose* (24.23) *is satisfied and* (24.27) *is satisfied in the corresponding l.c.s. for any* $x'' \in \gamma$. *Then the operator* $\mathfrak{L}$ *defined by the left sides of* (24.41)–(24.43) *is a Fredholm operator from* $W^{(1)}_{s,N}(G)$ *into* $W^{(2)}_{s,N}(G)$.

The proof is essentially the same as the proof of Theorem 22.1. The boundedness of $\mathfrak{L}$ from $W^{(1)}_{s,N}(G)$ into $W^{(2)}_{s,N}(G)$ follows from Lemmas 23.6, 24.3 and 24.5, while the normal solvability is proved on the basis of Theorem 24.1 and Lemma 24.4 instead of Theorems 16.1 and 19.4, which were used in the proof of Theorem 22.1.

4. *Special cases.* We note some cases in which an ordinary mixed boundary value problem without potentials and the additional boundary conditions (24.43) can be considered instead of a generalized mixed problem. Suppose there exists an s such that the ψ d.e.

$$p'A(\omega'', D_{n-1})u_+(x_{n-1}) = f(x_{n-1})$$

(see (24.36′)) is uniquely solvable for $u_+ \in H^+_{s-1/2}(\mathbf{R}^1)$, $f \in H_{s-1/2}(\mathbf{R}^1_+)$. If this assumption holds for all $x'' \in \gamma$ and all ω'', then an ordinary mixed problem will be normally solvable. Such will be the case, for example, if the matrix $A(\xi')$ is close to the matrix $a_+b_-^{-1}(\xi')b_+(\xi')$ or is strongly elliptic (see §17).

In general, if s is taken sufficiently large, the operator $pA(\omega'', D_{n-1})$ will not have a kernel in $H_s^+(\mathbf{R}^1)$ (see §14.6 and Remark 17.3) for any $x'' \in \gamma$ and ω''. For such s the mixed problem (24.41), (24.42) is normally solvable without the additional boundary conditions (24.43).

If, on the other hand, one greatly widens the function space $W_{s,N}(G)$ by taking s sufficiently large and negative, the operator $pA(\omega'', D_{n-1})$ will not have a cokernel. In this case the mixed problem (24.41)–(24.43) is normally solvable without potentials.

Let us dwell in more detail on the mixed boundary value problem for an elliptic equation of second order

$$\sum_{i,j=1}^{n} a_{ij}(x)\frac{\partial^2 u}{\partial x_i \partial x_j} + \sum_{j=1}^{n} b_j(x)\frac{\partial u}{\partial x_j} + c(x)u = f(x),$$
$$u \in W_{s,N}(G), \quad f \in W_{s-2,N}(G), \tag{24.44}$$

$$B_1(x, D)u|_{\Gamma_1} = g_1(x'), \quad B_2(x, D)u|_{\Gamma_2} = g_2(x'), \quad g_k(x') \in H_{s-m_k-1/2,N}(\Gamma_k). \tag{24.45}$$

Suppose the boundary operator $B_k(x, D)$, $k = 1, 2$, satisfies the Šapiro-Lopatinskiĭ condition on $\overline{\Gamma}_k$. For each fixed $x'' \in \gamma$ a number $\kappa_0(x'')$ defined from (13.20) and (13.22) can be associated with the mixed problem (24.44), (24.45) in the corresponding l.c.s. We assume that

$$\max_{x'' \in \gamma} \operatorname{Re} \kappa_0(x'') - \min_{x'' \in \gamma} \operatorname{Re} \kappa_0(x'') < 1. \tag{24.46}$$

THEOREM 24.4. *Suppose s and N satisfy the inequality $0 < s - \operatorname{Re} \kappa_0(x'') < 1$, for $x'' \in \gamma$ and the inequality $s + N > \max_k(m_k + 1/2)$, with s being nonintegral if $s \leqslant \max_k(m_k + 1/2)$, and suppose $B_k(x, D)$, $k = 1, 2$, satisfies the Šapiro-Lopatinskiĭ condition on $\overline{\Gamma}_k$. Then the mixed problem* (24.44), (24.45) *is normally solvable.*

Theorem 24.4 is a special case of Theorem 24.3. We note that Theorem 24.4 can be directly proved from the results of §13 without using Theorems 24.1 and 24.2. We also note that Examples 13.1–13.3 carry over in a natural way to the case of an arbitrary bounded domain.

REMARK 24.1. The proof of Theorem 24.3 includes a proof of the normal solvability of problem (22.4), (22.5) in weighted function spaces. Theorem 24.3 is a special case of a more general theorem on the normal solvability of mixed problems in domains with edges (see [40]). The proof of this theorem requires a somewhat different technique, which is based on the Mellin transform (cf. §15), and will not be considered here.

REMARK 24.2. We note two problems similar to the mixed problem (24.41)–(24.43) for which the investigation of normal solvability is completely analogous.

Let G_1 and G_2 be bounded domains in $\mathbf{R}^n$ with smooth boundaries Γ_1 and Γ_2 such that $\overline{G}_1 \cap \overline{G}_2 = \Gamma_1 \cap \Gamma_2 = \Gamma_0$, where Γ_0 is a smooth $(n-1)$-dimensional surface in $\mathbf{R}^n$ with a smooth $(n-2)$-dimensional boundary γ. By an *elliptic matching problem* is meant a boundary value problem in which an elliptic equation (or system of equations) is given in G_1 and G_2 with (i) boundary conditions on $\Gamma_1 \setminus \Gamma_0$ and $\Gamma_2 \setminus \Gamma_0$ satisfying the Šapiro-Lopatinskiĭ condition on $\overline{\Gamma_1 \setminus \Gamma_0}$ and $\overline{\Gamma_2 \setminus \Gamma_0}$, and (ii) a matching condition on Γ_0 of the form

$$B_{j1}(x, D)u_1|_{\Gamma_0} - B_{j2}(x, D)u_2|_{\Gamma_0} = g_j(x'),$$

where u_1 (u_2) is the solution in G_1 (G_2). As in the case of the mixed problem (24.41)–(24.43), it turns out that it is generally necessary to assign additional boundary conditions on γ of form (24.43) and to add operators of potential type (coboundary operators) with unknown densities on γ to the boundary conditions on $\Gamma_1 \setminus \Gamma_0$ and $\Gamma_2 \setminus \Gamma_0$ (cf. (24.42)) as well as to the matching condition on Γ_0.

The second problem is a boundary value problem for a domain with a "cut" (or several "cuts"). Let G be a bounded domain in $\mathbf{R}^n$ with a smooth boundary Γ and let Γ_0 be a smooth $(n-1)$-dimensional surface in G with a smooth $(n-2)$-dimensional boundary γ. An elliptic system of differential equations is given in $G \setminus \Gamma_0$ with a boundary condition on Γ and a "jump" condition on Γ_0 of the form

$$B_{j1}(x, D)u_+ - B_{j2}(x, D)u_- = g_j(x'),$$

where $u_+(x)$ ($u_-(x)$) denotes the limit of the solution $u(x)$ for $x \to \Gamma_0$ from one (the other) side of Γ_0. As above, it is generally necessary to assign additional boundary conditions on γ and to add coboundary operators with densities on γ to the "jump" condition.

It is easily seen that in both problems, by taking a partition of unity and "freezing" the coefficients for $x \in \gamma$ (cf. §22), we can obtain a problem with constant coefficients that reduces, after transformations analogous to the transformations of §24.2, to a system of ψ d.e.'s on a halfline analogous to system (24.36′), (24.37).

§25. Spaces of functions of piecewise constant order of smoothness

1. When condition (24.46) is not satisfied, the class of function spaces must be widened in order for the mixed problem (24.44), (24.45) to be normally solvable. To this end it is expedient to consider spaces of functions in G whose smoothness is not constant in G.

Let $\{\varphi_j(x), U_j\}$ be a finite partition of unity in $\mathbf{R}^n$, let $\{s_j\}$ be a collection of real numbers such that

$$|s_i - s_j| < 1 \qquad \text{if } U_i \cap U_j \neq \varnothing, \tag{25.1}$$

and let G be a bounded domain in $\mathbf{R}^n$ with a smooth boundary Γ. We denote by $\mathring{H}_{(s_j)}(G)$ the space of generalized functions $u_0(x)$ in $\mathbf{R}^n$ with support in $\overline{G}$ and finite norm

$$\|u_0\|_{(s_j)} = \sum_j \|\varphi_j u_0\|_{s_j}, \tag{25.2}$$

where $\|\cdot\|_{s_j}$ is the norm in $H_{s_j}(\mathbf{R}^n)$. Further, we denote by $H_{(s_j)}(G)$ the space of generalize d functions $u(x)$ in G admitting an extension lu onto $\mathbf{R}^n$ w ith finite norm

$$\|lu\|_{(s_j)} = \sum_j \|\varphi_j lu\|_{s_j}. \tag{25.3}$$

The norm in $H_{(s_j)}(G)$ is given by

$$\overset{+}{\|u\|}_{s_j} = \inf_l \|lu\|_{(s_j)}$$

(cf. (4.51)), where the infimum is taken over all extensions lu.

Thus the spaces $\mathring{H}_{(s_j)}(G)$ and $H_{(s_j)}(G)$ consist of functions whose smoothness, roughly speaking, is of ord er s_j in U_j. The norms in these spaces depend on the functions $\varphi_j(x)$ in the partition of unity $\{\varphi_j(x), U_j\}$; but they are considered fixed. When $s_j = s = \text{const}$, the spaces $\mathring{H}_{(s_j)}(G)$ and $H_{(s_j)}(G)$ coincide with the Sobolev-Slobodeckiĭ spaces $\mathring{H}_s(G)$ and $H_s(G)$ (see §21). Clearly,

$$H_{s_+}(G) \subset H_{(s_j)}(G) \subset H_{s_-}(G), \tag{25.4}$$

where $s_+ = \max s_j$ and $s_- = \min s_j$. The analogous inclusions hold for $\mathring{H}_{(s_j)}(G)$.

We note the following properties of $\mathring{H}_{(s_j)}(G)$ and $H_{(s_j)}(G)$.

LEMMA 25.1. *The spaces* $\mathring{H}_{(s_j)}(G)$ *and* $H_{(s_j)}(G)$ *are complete.*

LEMMA 25.2. *Suppose we are given collections of numbers* $\{s_j\}$ *and* $\{\sigma_j\}$ *such that* (25.1) *is satisfied by each of them and*

$$s_j < \sigma_j \qquad \textit{for all } j. \tag{25.5}$$

Then $\mathring{H}_{(\sigma_j)}(G) \subset \mathring{H}_{(s_j)}(G)$ *and* $H_{(\sigma_j)}(G) \subset H_{(s_j)}(G)$, *with the corresponding imbedding operators being compact.*

The proofs of Lemmas 25.1 and 25.2 readily follow from the definitions of the spaces involved and from the proofs of the analogous assertions for the Sobolev-Slobodeckiĭ spaces (see Lemmas 4.2 and 21.1).

LEMMA 25.3. *Suppose* $A(x, \xi) \in \hat{O}^{\infty}_{\alpha+i\beta}$. *Then* $pA(x, D)u_0$ *is a bounded operator from* $\mathring{H}_{(s_j)}(G)$ *into* $H_{(s_j)-\alpha}(G)$, *i.e.*

$$\|pAu_0\|_{(s_j)-\alpha} \leqslant C\|u_0\|_{(s_j)}. \tag{25.6}$$

PROOF. We have

$$\|pAu_0\|_{(s_j)-\alpha} \leqslant \|Au_0\|_{(s_j)-\alpha} = \sum_j \|\varphi_j A u_0\|_{s_j-\alpha}. \tag{25.7}$$

Let $\psi_j(x)$ be a function of class $C_0^\infty(U_j)$ that is equal to unity in a neighborhood of $\operatorname{supp} \varphi_j(x)$, and note that

$$\varphi_j A u_0 = \varphi_j A \psi_j u_0 + \varphi_j A(1-\psi_j)u_0. \tag{25.8}$$

Since $\varphi_j A(1-\psi_j)$ is an operator of order $-\infty$ (see Corollary 19.1), we have

$$\|\varphi_j A(1-\psi_j)u_0\|_{s_j-\alpha} \leqslant C\|u_0\|_{s_-} \leqslant C\|u_0\|_{(s_j)}. \tag{25.9}$$

Further, permuting φ_j and A and taking into account the fact that $\varphi_j\psi_j = \varphi_j$, we get

$$\varphi_j A \psi_j u_0 = A\varphi_j u_0 + T_j \psi_j u_0, \tag{25.10}$$

where $\operatorname{ord} T_j \leqslant \alpha - 1$. By Theorem 18.1,

$$\begin{aligned} \|T_j\psi_j u_0\|_{s_j-\alpha} &\leqslant C\|\psi_j u_0\|_{s_j-1} \leqslant C \sum_{k\in k(j)} \|\varphi_k \psi_j u_0\|_{s_j-1} \\ &\leqslant C \sum_{k\in k(j)} \|\varphi_k u_0\|_{s_k} \leqslant C\|u_0\|_{(s_j)}, \end{aligned} \tag{25.11}$$

where $\sum_{k\in k(j)}$ denotes a summation over those k for which $U_k \cap U_j \neq \varnothing$. Here we have taken into account condition (25.1), which implies that $s_j - 1 < s_k$ for $k \in k(j)$.

Thus, by virtue of (25.8), (25.11) and Theorem 18.1,

$$\sum_j \|\varphi_j A u_0\|_{s_j-\alpha} \leq \sum_j \|A\varphi_j u_0\|_{s_j-\alpha} + C\|u_0\|_{(s_j)} \leq C\|u_0\|_{(s_j)}. \quad (25.12)$$

2. *Spaces of piecewise constant order on* Γ. Let $U_j' = U_j \cap \Gamma$ for $U_j \cap \Gamma \neq \varnothing$, and let $\varphi_j' = p'\varphi_j$, where $\{\varphi_j, U_j\}$ is the partition of unity defining the norm in $H_{(s_j)}(\mathbf{R}^n)$ and p' is the operator of restriction of a function to Γ. Clearly, $\{\varphi_j', U_j'\}$ is a partition of unity on Γ. We denote by $H_{(s_j)}(\Gamma)$ the space of generalized functions on Γ with finite norm

$$[v(x')]_{(s_j)} = \sum_j{}' [\varphi_j'(x')v(x')]_{s_j}, \quad (25.13)$$

where $[\cdot]_{s_j}$ is the norm in $H_{s_j}(\mathbf{R}^{n-1})$, taken in the jth l.c.s. (cf. (21.8)), and Σ_j' denotes a summation over those j for which $U_j \cap \Gamma \neq \varnothing$.

LEMMA 25.4. *The operator p' of restriction of a function $u(x) \in S(\mathbf{R}^n)$ to Γ extends by continuity to a bounded operator from $H_{(s_j)}(\mathbf{R}^n)$ into $H_{(s_j)-1/2}(\Gamma)$ if $s_j > 1/2$ for those j for which $U_j \cap \Gamma \neq \varnothing$.*

The proof is analogous to that of Lemma 21.6.

LEMMA 25.5. *Suppose $\rho(x') \in C^\infty(\Gamma)$ and $\rho \times \delta(\Gamma) \in S'(\mathbf{R}^n)$ is the generalized function defined by* (21.14). *If $s_j < -1/2$ for those j for which $U_j \cap \Gamma \neq \varnothing$, then*

$$\|\rho(x') \times \delta(\Gamma)\|_{(s_j)} \leq C[\rho]_{(s_j)+1/2}. \quad (25.14)$$

PROOF. Analogously to (4.4), we have

$$|[\rho, p'\varphi]| \leq C[\rho]_{s+1/2}[p'\varphi]_{-s-1/2}. \quad (25.15)$$

If $s < -1/2$, then Lemma 21.6 implies $[p'\varphi]_{-s-1/2} \leq C\|\varphi\|_{-s}$. Consequently,

$$|(\rho \times \delta(\Gamma), \varphi)| = |[\rho, p'\varphi]| \leq C[\rho]_{s+1/2}\|\varphi\|_{-s}. \quad (25.16)$$

Thus $\rho \times \delta(\Gamma)$ is a continuous functional on $H_{-s}(\mathbf{R}^n)$ and hence belongs to the dual space $H_s(\mathbf{R}^n)$ of $H_{-s}(\mathbf{R}^n)$ (see §4.5). It follows from (25.16) that

$$\|\rho \times \delta(\Gamma)\|_s \leq C[\rho]_{s+1/2}. \quad (25.17)$$

(We note that another proof of (25.17) was given in §21.7; see (21.19).) Suppose now $\rho \in H_{(s_j)+1/2}(\Gamma)$. Then, by (25.17),

$$\|\rho \times \delta(\Gamma)\|_{(s_j)} = \sum_j \|\varphi_j(\rho \times \delta(\Gamma))\|_{s_j} = \sum_j{}' \|\varphi_j'\rho \times \delta(\Gamma)\|_{s_j}$$

$$\leq C\sum_j{}' [\varphi_j'\rho]_{s_j+1/2} = C[\rho]_{(s_j)+1/2}.$$

LEMMA 25.6. *Suppose* $A(x, \xi) \in \hat{O}^{\infty}_{\alpha+i\beta}$. *Then the operator of potential type (coboundary operator)* $pA(x, D)(v(x') \times \delta(\Gamma))$ *is a bounded operator from* $H_{(s_j)+\alpha+1/2}(\Gamma)$ *into* $H_{(s_j)}(G)$ *if* $s_j < -\alpha - 1/2$ *for those* j *for which* $U_j \cap \Gamma \neq \varnothing$.

The proof follows from Lemmas 25.3 and 25.5.

LEMMA 25.7. *Suppose* $E(x', D')$ *is a p.o. on* Γ *of class* $\hat{O}^{\infty}_{\alpha+i\beta}$ *(see* §21.4). *Then* E *is a bounded operator from* $H_{(s_j)}(\Gamma)$ *into* $H_{(s_j)-\alpha}(\Gamma)$.

The proof is analogous to that of Lemma 25.3.

3. *Boundary value problems for p.o.'s in spaces of piecewise constant order.* In general, condition (24.46) is not satisfied by the function $\kappa_0(x'')$ defined by (13.20) and (13.22) for the mixed problem (24.44), (24.45). Let $\{s_j\}$ be a collection of real numbers satisfying (25.1) and the condition

$$0 < s_j - \operatorname{Re} \kappa_0(x'') < 1 \qquad \text{for all } x'' \in \overline{U}_j \cap \gamma. \tag{25.18}$$

We define the weighted spaces of piecewise constant order $W_{(s_j),N}(G)$, $H_{(s_j),N}(\Gamma_1)$ and $H_{(s_j),N}(\Gamma_2)$ analogously to §24.3.

The following theorem holds.

THEOREM 25.1. *Suppose* $B_k(x, D)$, $k = 1, 2$, *satisfies the Šapiro-Lopatinskiĭ condition on* $\overline{\Gamma}_k$. *Suppose* $\{s_j\}$ *satisfies* (25.1), (25.18) *and the condition that* s_j *be nonintegral if* $U_j \cap \gamma \neq \varnothing$ *and* $s_j \leqslant \max_{k=1,2} m_k + 1/2$. *Let* N *be such that* $s_j + N > m_k + 1/2$ *for those* s_j *for which* $U_j \cap \overline{\Gamma}_k \neq \varnothing$. *Then the mixed problem* (24.44), (24.45) *defines a Fredholm operator from* $W_{(s_j),N}(G)$ *into the space*

$$W_{(s_j)-2,N}(G) \times H_{(s_j)-m_1-1/2,N}(\Gamma_1) \times H_{(s_j)-m_2-1/2,N}(\Gamma_2).$$

Suppose now $A(x, D)$ is a scalar elliptic p.o. and $\kappa(x')$ is the factorization index of $A_0(x, \xi)$ on Γ (see §21.2). In general, $\kappa(x')$ will not satisfy (22.25′).

THEOREM 25.2. *Suppose* $\{s_j\}$ *satisfies* (25.1) *and the following two conditions*:

$$\operatorname{Re} \beta_k \leqslant \hat{s}_- - 1/2, \qquad \operatorname{Re} \gamma_p \leqslant \alpha - \hat{s}_+ - 1/2,$$
$$1 \leqslant k \leqslant m_+, \quad 1 \leqslant p \leqslant m_-, \tag{25.19}$$

where $\hat{s}_- = \min_\Gamma \operatorname{Re} \kappa(x') - m$ *and* $\hat{s}_+ = \max_\Gamma \operatorname{Re} \kappa(x') - m$, *and*

$$|s_j - \operatorname{Re} \kappa(x') + m| < 1/2 \qquad \text{for all } x' \in \overline{U}_j \cap \Gamma. \tag{25.20}$$

Suppose, further, that the Šapiro-Lopatinskiĭ condition is satisfied, i.e. that for any fixed $(x', \xi_{x'}) \in T_0^*(\Gamma)$ *the boundary value problem of form* (22.7), (22.8) *is uniquely solvable in* $\mathcal{H}_s^{(1)}(\mathbf{R}^1)$ *for any* s *satisfying the inequality* $|s - \operatorname{Re} \kappa(x') + m| < 1/2$. *Then the boundary value problem* (22.4), (22.5) *with scalar p.o.'s*

$A(x, D)$, $B_j(x, D)$ and $C_k(x, D)$ defines a Fredholm operator from the space

$$\mathcal{H}^{(1)}_{(s_j),N}(G) = \mathring{H}_{(s_j),N}(G) \times \prod_{k=1}^{m_-} H_{(s_j)-\alpha+\mathrm{Re}\,\gamma_k+1/2}(\Gamma)$$

into the space

$$\mathcal{H}^{(2)}_{(s_j),N}(G) = H_{(s_j)-\alpha,N}(G) \times \prod_{k=1}^{m_+} H_{(s_j)-\mathrm{Re}\,\beta_j-1/2}(\Gamma).$$

Theorem 25.2 is a special case of Theorem 25.3 below on the normal solvability of a boundary value problem of form (22.4), (22.5) in spaces of piecewise constant order. Analogously, Theorem 25.1 is a special case of a theorem on the normal solvability of a mixed problem of form (24.41)–(24.43) in the same spaces; however, for the sake of brevity, we will confine ourselves to the formulation and proof of Theorem 25.3.

Whereas in the scalar case it is possible for an arbitrary elliptic symbol $A_0(x, \xi)$ to find function spaces in which the corresponding operator is a Fredholm operator, this is no longer so in the case of systems of p.o.'s. We will therefore introduce a subclass of elliptic systems of p.o.'s for which there exist spaces of piecewise constant order in which the boundary value problem (22.4), (22.5) is normally solvable.

It is assumed that there exists a continuous function $s(x')$ on Γ such that for all $x' \in \Gamma$

$$s(x') - \mathrm{Re}\,\gamma_j(x') - \tfrac{1}{2} \in \mathbf{R} \setminus \mathbf{Z}, \qquad 1 \leqslant j \leqslant p, \tag{25.21}$$

where the $\gamma_j(x')$ are the same as in (22.3). For each $x' \in \Gamma$ we denote by $(s_-(x'), s_+(x'))$ the maximal open interval containing $s(x')$ in which every point satisfies (25.21). Clearly, $s_-(x')$ and $s_+(x')$ are continuous functions on Γ such that $0 < s_+(x') - s_-(x') \leqslant 1$.

THEOREM 25.3. *Suppose $\{s_j\}$ satisfies* (25.1) *and the condition*

$$s_-(x') < s_j < s_+(x') \qquad \textit{for all } x' \in \overline{U}_j \cap \Gamma. \tag{25.22}$$

Suppose also that (25.19) *is satisfied for $\hat{s}_+ = \max_\Gamma s_+(x')$ and $\hat{s}_- = \min_\Gamma s_-(x')$, that* (22.6′) *is satisfied and that the boundary value problem* (22.7), (22.8) *is uniquely solvable for any $(x', \xi'_{x'}) \in T^*_0(\Gamma)$ and any $s \in (s_-(x'), s_+(x'))$. Then the boundary value problem* (22.4), (22.5) *defines a Fredholm operator $\mathfrak{A}$ from $\mathcal{H}^{(1)}_{(s_j),N}(G)$ into $\mathcal{H}^{(2)}_{(s_j),N}(G)$ for any $N > 0$.*

PROOF. For simplicity we will confine ourselves to the case $N = 0$.

Let x_0' be an arbitrary point of Γ. For a sufficiently small neighborhood $U_0 = U_0(x_0')$ of x_0' the following decomposition holds in a corresponding l.c.s. (see (22.10)):

$$\psi_0\mathfrak{A}_0\varphi_0 = \psi_0(\hat{\mathfrak{A}}_0 + K_0 + T_0)_{\varphi_0}, \tag{25.23}$$

where $\varphi_0, \psi_0 \in C_0^\infty(U_0)$, $\mathfrak{A}_0$ is the operator $\mathfrak{A}$ written in the local coordinates, $\hat{\mathfrak{A}}_0$ is the operator analogous to the operator defined by (16.3) and (16.4), K_0 is a bounded operator from $\mathcal{H}_s^{(1)}(\mathbf{R}^n)$ into $\mathcal{H}_s^{(2)}(\mathbf{R}^n)$ with a small norm, and T_0 is a bounded operator from $\mathcal{H}_s^{(1)}(\mathbf{R}^n)$ into $\mathcal{H}_{s+1}^{(2)}(\mathbf{R}^n)$ for any s.

By virtue of the Šapiro-Lopatinskiĭ condition, the operator $\hat{\mathfrak{A}}_0$ has a bounded inverse R_0 from $\mathcal{H}_s^{(2)}(\mathbf{R}^n)$ into $\mathcal{H}_s^{(1)}(\mathbf{R}^n)$ for any $s \in (s_-(x_0'), s_+(x_0'))$. As was noted in Remark 16.2, the norm of R_0 is bounded with respect to s on any closed interval $[s_-(x_0') + \varepsilon, s_+(x_0') - \varepsilon]$, $\varepsilon > 0$. Since $s_-(x')$ and $s_+(x')$ are continuous functions and the inequalities (25.22) are strict, there clearly exists an $\varepsilon_0 > 0$ such that

$$s_-(x') + \varepsilon_0 < s_j < s_+(x') - \varepsilon_0 \qquad \text{for all } x' \in \overline{U}_j \cap \Gamma. \tag{25.24}$$

We choose the neighborhood $U(x_0')$ so small that (i)

$$\|K_0\|_{(s)} \leqslant 1/2\|R_0\|_{(s)} \quad \text{for all } s \in \left[s_-(x_0') + \varepsilon_0/2, s_+(x_0') - \varepsilon_0/2\right], \tag{25.25}$$

(ii) $U(x_0') \cap U_j = \varnothing$ if $x_0' \notin \overline{U}_j$, and (iii) the inequalities (25.24) remain in force for $x' \in \overline{U(x_0')} \cap \Gamma$ if $x_0' \in \overline{U}_j$.

Let

$$R^{(0)} = (I + R_0K_0)^{-1}R_0. \tag{25.26}$$

Analogously to (22.13), we have

$$R^{(0)}\mathfrak{A}_0 = I_1 + T_1 \qquad \text{and} \qquad \mathfrak{A}_0R^{(0)} = I_2 + T_2, \tag{25.27}$$

where $T_1 = R^{(0)}T_0$ and $T_2 = T_0R^{(0)}$. Since $R^{(0)}$ is a bounded operator from $\mathcal{H}_s^{(2)}(\mathbf{R}^n)$ into $\mathcal{H}_s^{(1)}(\mathbf{R}^n)$ for any $s \in [s_-(x_0') + \varepsilon_0/2, s_+(x_0') - \varepsilon_0/2]$ and T_0 is a bounded operator from $\mathcal{H}_s^{(1)}(\mathbf{R}^n)$ into $\mathcal{H}_{s+1}^{(2)}(\mathbf{R}^n)$ for any s, it follows that T_1 is a bounded operator from $\mathcal{H}_{s-1}^{(1)}(\mathbf{R}^n)$ into $\mathcal{H}_s^{(1)}(\mathbf{R}^n)$ for any $s \in [s_-(x_0') + \varepsilon_0/2, s_+(x_0') - \varepsilon_0/2]$ and T_2 is a bounded operator from $\mathcal{H}_s^{(2)}(\mathbf{R}^n)$ into $\mathcal{H}_{s+1}^{(2)}(\mathbf{R}^n)$ for the same s. In particular, since $s_+(x_0') - s_-(x_0') \leqslant 1$, each operator T_k ($k = 1, 2$) is a bounded operator from $\mathcal{H}_{s_-(x_0')+\varepsilon_0/2}^{(k)}(\mathbf{R}^n)$ into $\mathcal{H}_{s_+(x_0')-\varepsilon_0/2}^{(k)}(\mathbf{R}^n)$.

At this point in the proof of Theorem 25.3 we need

LEMMA 25.8. *Suppose $\varphi(x) \in C_0^\infty(U_0)$. Then the commutator $T = R^{(0)}\varphi - \varphi R^{(0)}$ is a bounded operator from $\mathcal{H}_{s_-(x_0')+\varepsilon_0/2}^{(2)}(\mathbf{R}^n)$ into $\mathcal{H}_{s_+(x_0')-\varepsilon_0/2}^{(1)}(\mathbf{R}^n)$.*

PROOF. We have

$$\mathfrak{A}_0 T = \mathfrak{A}_0 R^{(0)}\varphi - \mathfrak{A}_0 \varphi R^{(0)} = \varphi + T_2\varphi - \mathfrak{A}_0\varphi R^{(0)}. \tag{25.28}$$

Permuting $\mathfrak{A}_0$ and φ, we get by virtue of Corollary 18.1 that $\mathfrak{A}_0\varphi = \varphi\mathfrak{A}_0 + T_3$, where T_3 is a bounded operator from $\mathcal{H}_s^{(1)}(\mathbf{R}^n)$ into $\mathcal{H}_{s+1}^{(2)}(\mathbf{R}^n)$ for any s. Consequently,

$$\mathfrak{A}_0 T = T_2\varphi - \varphi T_2 - T_3 R^{(0)}. \tag{25.29}$$

Applying $R^{(0)}$ to (25.29) from the left, we get

$$T = R^{(0)}T_2\varphi - R^{(0)}\varphi T_2 - R^{(0)}T_3R^{(0)} - T_1(R^{(0)}\varphi - \varphi R^{(0)}). \tag{25.30}$$

But the right side of this equality is clearly a bounded operator from $\mathcal{H}^{(2)}_{s_-(x_0')+\varepsilon_0/2}(\mathbf{R}^n)$ into $\mathcal{H}^{(1)}_{s_+(x_0')-\varepsilon_0/2}(\mathbf{R}^n)$.

Continuation of the proof of Theorem 25.3. From the covering of Γ by neighborhoods of the form $U(x_0')$ we choose a finite subcovering and extend it by arbitrary neighborhoods not intersecting Γ to a finite covering of $\mathbf{R}^n$. Let $\{\alpha_k(x'), V_k\}$ be a partition of unity subordinate to this covering and let $U_{jk} = U_j \cap V_k$, $\varphi_{jk} = \varphi_j\alpha_k$ and $s_{jk} = s_j$. Clearly, the norm $\|u\|_{s_{j,k}} = \Sigma_{j,k}\|\varphi_{jk}u\|_{s_j}$ is equivalent to the norm $\|u\|_{(s_j)} = \Sigma_j\|\varphi_j u\|_{s_j}$. To avoid cumbersome notation we will write, as before, $\{\varphi_j, U_j\}$ instead of $\{\varphi_{jk}, U_{jk}\}$, and when $U_j \cap \Gamma \neq \varnothing$ we will let x_j' denote the point of Γ at which the decomposition of form (25.23) holds.

When $U_j \cap \Gamma \neq \varnothing$, we denote by R_j the operator having in the jth l.c.s. the form $R_j^{(0)}$ analogous to $R^{(0)}$. We note that $R_j^{(0)}$ is a bounded operator from $\mathcal{H}_s^{(2)}(\mathbf{R}^n)$ into $\mathcal{H}_s^{(1)}(\mathbf{R}^n)$ for any $s \in [s_-(x_j') + \varepsilon_0/2,\ s_+(x_j') - \varepsilon_0/2]$. When $U_j \cap \Gamma = \varnothing$, we denote by R_j the p.o. with symbol $A_0^{-1}(x, \xi)(1 - \chi(\xi))$ (cf. §22). In this way, as in §22, we define the operator (cf. (22.14))

$$R = \sum_j \psi_j R_j \varphi_j, \tag{25.31}$$

which has the representation $R = R' + R''$ in which $R' = \Sigma_j'\psi_jR_j\varphi_j$ and $R'' = \Sigma_j''\psi_jR_j\varphi_j$, where Σ_j' denotes a summation over those j for which $U_j \cap \Gamma \neq \varnothing$ and Σ_j'' denotes a summation over the remaining j.

To complete the proof of Theorem 25.3 we need

LEMMA 25.9. *R is a bounded operator from $\mathcal{H}^{(2)}_{(s_j)}(G)$ into $\mathcal{H}^{(1)}_{(s_j)}(G)$.*

PROOF. Since R'' is a p.o. in $\mathbf{R}^n$, its boundedness follows from Lemma 25.3. It therefore remains to prove the boundedness of R'.

By Lemma 25.8,

$$\varphi_k R' = \sum_{j(k)}{}' \psi_j R_j \varphi_k \varphi_j + \sum_{j(k)}{}' \psi_j T_{jl}\varphi_j,$$

where $\Sigma'_{j(k)}$ denotes a summation over those j for which $U_j \cap U_k \neq \varnothing$ and $U_j \cap \Gamma \neq \varnothing$ and the T_{jk} are bounded operators from $\mathcal{H}^{(2)}_{s_-(x'_j)+\varepsilon_0/2}(\mathbf{R}^n)$ into $\mathcal{H}^{(1)}_{s_+(x'_j)-\varepsilon_0/2}(\mathbf{R}^n)$. It follows from the construction of the partition of unity that

$$s_-(x'_j) + \varepsilon_0 < s_k < s_+(x'_j) - \varepsilon_0 \tag{25.32}$$

for those k for which $U_k \cap U_j \neq \varnothing$. Consequently, by (25.32),

$$\begin{aligned}|||R'\Phi|||_{1,(s_j)} &= \sum_k |||\varphi_k R'\Phi|||_{1,s_k} \leqslant C \sum_k \sum_{j(k)}{}' |||R_j\varphi_k\varphi_j\Phi|||_{1,s_k} \\ &\qquad + C\sum_k \sum_{j(k)}{}' |||T_{jk}\varphi_j\Phi|||_{1,s_k} \\ &\leqslant C\sum_k |||\varphi_k\Phi|||_{2,s_k} \\ &\qquad + C\sum_k \sum_{j(k)}{}' |||\varphi_j\Phi|||_{2,s_-(x'_j)+\varepsilon_0/2} \\ &\leqslant C\sum_j |||\varphi_j\Phi|||_{2,s_j} = C|||\Phi|||_{2,(s_j)},\end{aligned}$$

where $|||\cdot|||_{k,s}$ $(k = 1, 2)$ denotes the norm in $\mathcal{H}^{(k)}_s$. Lemma 25.9 is proved.

Completion of the proof of Theorem 25.3. Let us now verify that R is a regularizer of $\mathfrak{A}$. We have

$$R\mathfrak{A}V = \sum_j \psi_j R_j \varphi_j \mathfrak{A}\psi_j V + T^{(0)}V, \tag{25.33}$$

where $T^{(0)} = \Sigma_j \psi_j R_j T_j^\infty$ and $T_j^\infty = \varphi_j\mathfrak{A}(1 - \psi_j)$. Writing $\varphi_j\mathfrak{A}\psi_j$ in the jth local coordinates and permuting φ_j and $\mathfrak{A}_j$, we get

$$\varphi_j\mathfrak{A}_j\psi_j = \mathfrak{A}_j\varphi_j + T_j\psi_j, \tag{25.34}$$

where T_j is a bounded operator from $\mathcal{H}^{(1)}_s$ into $\mathcal{H}^{(2)}_{s-1}$ for any s. Thus, by (25.33) and (25.34),

$$R\mathfrak{A}V = V + T^{(1)}V, \tag{25.35}$$

where $T^{(1)} = T^{(0)} + T' + T''$, with $T'\,(T'') = \Sigma_j'{}^{('')}(\psi_j R_j T_j \psi_j + \psi_j T_{j1}\varphi_j)$, and T_{j1} is a bounded operator from $\mathcal{H}^{(1)}_{s_-(x'_j)-\varepsilon_0/2}(\mathbf{R}^n)$ into $\mathcal{H}^{(1)}_{s_+(x'_j)+\varepsilon_0/2}(\mathbf{R}^n)$ (p.o. of order at most -1) when $U_j \cap \Gamma \neq \varnothing$ $(= \varnothing)$.

Let us show that $T^{(1)}$ is a bounded operator from $\mathcal{H}^{(1)}_{(s_j)-\varepsilon_0/2}(G)$ into $\mathcal{H}^{(1)}_{(s_j)}(G)$. Since the p.o.'s R_jT_j and T_{j1} are of order at most -1 when $U_j \cap \Gamma \neq \varnothing$, Lemma 25.3 implies

$$|||T''V|||_{1,(s_j)} \leqslant C|||V|||_{1,(s_j)-1}. \tag{25.36}$$

Further, since T_j^∞ is of order $-\infty$, $s_k < s_+(x_j') - \varepsilon_0/2$ when $U_k \cap U_j \neq \varnothing$, and $s_-(x_j') + \varepsilon_0/2 < s_j - \varepsilon_0/2$ (see (25.32)), we get

$$\begin{aligned} |||T^{(0)}V|||_{1,(s_j)} + |||T'V|||_{1,(s_j)} &\leqslant C\sum_k \sum_{j(k)} |||R_j T_j^\infty V|||_{1,s_k} \\ &\quad + C\sum_k \sum_{j(k)}{}' \left(|||R_j T_j \psi_j V|||_{1,s_k} + |||T_{j1}\varphi_j V|||_{1,s_k}\right) \\ &\leqslant C\sum_k \sum_{j(k)} |||T_j^\infty V|||_{2,s_k} \\ &\quad + C\sum_k \sum_{j(k)}{}' \left(|||\psi_j V|||_{1,s_k - 1} + |||\varphi_j V|||_{1,s_-(x_j') + \varepsilon_0/2}\right) \\ &\leqslant C|||V|||_{1,(s_j)-\varepsilon_0/2} + C\sum_k \sum_{j(k)}{}' |||\psi_j V|||_{1,s_k - 1}. \end{aligned} \tag{25.37}$$

But from (25.32) and the inequality $s_+(x_j') - s_-(x_j') \leqslant 1$ we have $s_k - 1 < s_-(x_j') + \varepsilon_0/2 < s_k - \varepsilon_0/2$ for those k for which $U_k \cap U_j \neq \varnothing$. Consequently,

$$\begin{aligned} |||\psi_j V|||_{1,s_k - 1} &\leqslant |||\psi_j V|||_{1,s_-(x_j')+\varepsilon_0/2} \leqslant \sum_{i(j)} |||\varphi_i \psi_j V|||_{1,s_-(x_j')+\varepsilon_0/2} \\ &\leqslant C\sum_{i(j)} |||\varphi_i V|||_{1,s_i - \varepsilon_0/2} \leqslant C|||V|||_{1,(s_j)-\varepsilon_0/2}. \end{aligned}$$

Thus $T^{(1)}$ is a bounded operator from $\mathcal{H}^{(1)}_{(s_j)-\varepsilon_0/2}(G)$ into $\mathcal{H}^{(1)}_{(s_j)}(G)$ and hence, by Lemma 25.2, is a compact operator in $H^{(1)}_{(s_j)}(G)$, so that R is a left regularizer of $\mathfrak{A}$. It can be proved analogously that $\mathfrak{A}R = I_2 + T^{(2)}$, where $T^{(2)}$ is a bounded operator from $\mathcal{H}^{(2)}_{(s_j)}(G)$ into $\mathcal{H}^{(2)}_{(s_j)+\varepsilon_0/2}(G)$, so that R is also a right regularizer of $\mathfrak{A}$. Therefore $\mathfrak{A}$ is a Fredholm operator.

REMARK 25.1. The case when condition (24.46) or (22.25′) is not satisfied can also be treated by considering spaces of a variable order of smoothness (see [103]). We have confined ourselves here to spaces of piecewise constant order since they are simpler, especially in dealing with systems of ψ d.e.'s or mixed problems for elliptic differential equations.

REMARK 25.2. Under the conditions of Theorem 25.3, if the operator $\mathfrak{A}(\omega)$ defined by (16.5) and (16.6) is invertible for some $s_0 \in (s_-(x_0'), s_+(x_0'))$, then it is invertible for all $s \in (s_-(x_0'), s_+(x_0'))$. This can be proved by making use of the explicit form of the inverse operator constructed in §17.3 and the estimates for the operators $P_\pm$ (Lemma 14.3).

A simpler proof is based on the same idea as the proof of Theorem 24.2 (see also Lemma 24.6). Suppose that the right side of (16.5) belongs to $H_{s-\alpha}(\mathbf{R}^1_+)$ for $s_0 < s < s_+(x_0')$. We note that $pC_k(\omega, D_n)\delta \in H_{s_+ - \alpha - \varepsilon}(\mathbf{R}^1_+)$ for all $\varepsilon > 0$, since $\operatorname{Re} \gamma_k < \alpha - \hat{s}_+ - 1/2$ (see (25.19)). By applying a regularizer of $pA(\omega, D_n)v_+$ (e.g. one of the regularizers constructed in §14, Remark 15.2 and §17) to (16.5), one can easily get, as in the proof of Lemma 24.6, that $v_+(x_n) \in H_s^+(\mathbf{R}^1)$. Thus $\mathfrak{A}(\omega)$ is invertible for $s \in [s_0, s_+(x_0'))$. The fact that it is invertible for $s \in (s_-(x_0'), s_0]$ can be proved by going over to the adjoint operator.

CHAPTER VII

APPLICATIONS

§26. A finite element method for strongly elliptic pseudodifferential equations

1. In this section we shall describe a finite element method for strongly elliptic pseudodifferential equations in a bounded plane domain. Such equations are encountered, for example, in the analysis of mixed problems for an elastic halfspace or layer when the original three-dimensional problem is reduced to a two-dimensional problem in a bounded plane domain. Examples of such problems include crack and punch problems in the theory of elasticity and others (see Examples 22.4, 22.5 and [46], [64] and [113]). For some papers using such an approach to obtain a numerical solution of a three-dimensional problem see [14], [24], [53] and [68]. The present section is based on [53], [39], and [14] (see also [40]).

2. *Existence and uniqueness of the solution of a strongly elliptic pseudodifferential equation.* Let G be a bounded domain in $\mathbf{R}^n$ with a smooth boundary Γ and consider in G the system of pseudodifferential equations

$$pA(x, D)u_0 = f, \tag{26.1}$$

where $A(x, \xi) = \|a_{ij}(x, \xi)\|_{i,j=1}^{\infty} \in \hat{O}_{\alpha}^{\infty}$. It is assumed that the principal homogeneous symbol $A_0(x, \xi)$ is *strongly elliptic*, i.e. the matrix

$$A_R(x, \xi) = \tfrac{1}{2}(A_0(x, \xi) + A_0^*(x, \xi))$$

is positive definite for $x \in \overline{G}$ and $\xi \neq 0$ (cf. Example 17.1). It is also required that the selfadjoint part of the operator $A(x, D)$ be positive definite:

$$\operatorname{Re}(A(x, D)u_0, u_0) \geqslant C\|u_0\|_{\alpha/2}^2, \qquad \forall u_0 \in C_0^{\infty}(G). \tag{26.2}$$

Let $b_0(x')$, $x' \in \Gamma$, denote the well-defined matrix on Γ that is equal to

$$\left[A_j^{(0)}(x'_{(j)}, 0, 0, +1)\right]^{-1} A_j^{(0)}(x'_{(j)}, 0, 0, -1)$$

in each jth local coordinate system (l.c.s.)(cf. §22.1). Since $A_R(x, \xi)$ is positive definite, the matrix $b_0(x')$ has no negative eigenvalues.

For suppose on the contrary that for some $x'_0 \in \Gamma$ there exists an eigenvector e such that $b_0(x'_0)e_0 = -\mu e_0$, $\mu > 0$. Then in some l.c.s. we would have $b_0(x'_0) = (a_j^+)^{-1}a_j^-$, where $a_j^{\pm} = A_j^{(0)}(x_{(j)}^{(0)}, 0, 0, \pm 1)$, so that, letting $g_j = (a_j^+)^* e_0$, we would get

$$a_j^- e_0 \cdot e_0 = \left(a_j^+\right)^{-1} a_j^- e_0 \cdot g = -\mu e_0 \cdot g = -\mu a_j^+ e_0 \cdot e_0,$$

which contradicts the fact that the selfadjoint parts of a_j^+ and a_j^- are positive definite.

Therefore, as in §14.5, we can choose a matrix $b(x') = (1/2\pi i)\ln b_0(x')$ in such a way that

$$-\tfrac{1}{2} < \operatorname{Re} b_j(x') < \tfrac{1}{2}, \qquad \forall x' \in \Gamma, \quad 1 \leqslant j \leqslant p, \tag{26.3}$$

where the $b_j(x')$ are the eigenvalues of $b(x')$.

Let

$$s_+ = \min_{1 \leqslant j \leqslant p} \min_{x' \in \Gamma} \operatorname{Re} b_j(x') + \frac{\alpha}{2} + \frac{1}{2}$$

and

$$s_- = \max_{1 \leqslant j \leqslant p} \max_{x' \in \Gamma} \operatorname{Re} b_j(x') + \frac{\alpha}{2} - \frac{1}{2}.$$

It follows from (26.3) that $s_- < \alpha/2 < s_+$. In the particular case when the matrix $b_0(x')$ is symmetric we have $\operatorname{Re} b_j(x') \equiv 0$, $1 \leqslant j \leqslant p$, so that $s_- = \alpha/2 - \frac{1}{2}$ and $s_+ = \alpha/2 + \frac{1}{2}$.

THEOREM 26.1. *Suppose* (26.2) *holds. Then* (26.1) *has a unique solution* $u_0 \in \mathring{H}_s(G)$ *when* $f \in H_{s-\alpha}(G)$ *and* $s \in (s_-, s_+)$.

PROOF. The proof of Theorem 26.1 is essentially the same as in Example 22.2.

The normal solvability of system (26.1) for arbitrary $s \in (s_-, s_+)$ follows from Theorem 22.1. It therefore suffices to prove that the kernel and cokernel of the operator pAu_0 are zero.

For $s = \alpha/2$ the uniqueness of the solution of the homogeneous system (26.1) and of the dual system $pA^*(x, D)v_0 = 0$, $v_0 \in \mathring{H}_{\alpha/2}(G)$, follows from Example 17.1 (we assume that (26.2) also holds for the operator $A^*(x, D)$). Suppose $u_0 \in \mathring{H}_s(G)$, $pAu_0 = 0$ and $s \in (s_-, s_+)$. Since the left regularizor of pAu_0 exists for all $s \in (s_-, s_+)$ (see §§22 and 25), it follows from (22.17) that $u_0 \in \mathring{H}_{s_+ - \varepsilon}(G)$, $\forall \varepsilon > 0$. Therefore $u_0 \in \mathring{H}_{\alpha/2}(G)$, which implies $u_0 = 0$. The same is true for the solution of $pA^*v_0 = 0$, and the theorem follows.

In the particular case when (26.1) is a single p.e. and $A_0(x, \xi)$ is a strongly elliptic symbol, i.e. $\operatorname{Re} A_0(x, \xi) > 0$ for $x \in \overline{G}$ and $\xi \neq 0$, we have (see Example 6.1)

$$|\operatorname{Re} \kappa(x') - \alpha/2| < \tfrac{1}{2}, \qquad \forall x' \in \Gamma, \tag{26.4}$$

where $\kappa(x')$ is the factorization index of $A_0(x, \xi)$ in G. Therefore in this case $s_+ = \kappa_- + \frac{1}{2}$, $s_- = \kappa_+ - \frac{1}{2}$ and $s_- < \alpha/2 < s_+$, where $\kappa_- = \min_{x' \in \Gamma} \operatorname{Re} \kappa(x')$ and $\kappa_+ = \max_{x' \in \Gamma} \operatorname{Re} \kappa(x')$.

3. *Asymptotic behavior of the solution of an elliptic p.e. near a boundary.* Let $A(x, D)$ be an arbitrary elliptic system of p.o.'s with symbol $A(x, \xi) \in \hat{O}^{\infty}_{\alpha + i\beta}$, and consider the boundary value problem (22.4), (22.5). If all the conditions of Theorem 22.1 are satisfied, then the operator $\mathfrak{A}$ defined by (22.4) and (22.5) is a Fredholm operator from $\mathcal{H}^{(1)}_{s,N}(G)$ into $\mathcal{H}^{(2)}_{s,N}(G)$, where N is arbitrary (see Remark 24.1).

We shall first prove that the solution of problem (22.4), (22.5) is smooth in each jth l.c.s. with respect to $x'_{(j)}$, provided that f is also such a function and the $g_j(x')$, $1 \leqslant j \leqslant m_+$, are smooth on the boundary. Let $\{\varphi_j, U_j\}$ be a sufficiently refined partition of unity in $\overline{G}$. In a neighborhood $U(\Gamma)$ of Γ we use the special local coordinate system (s.l.c.s.) (see §22.1) obtained by taking the coordinates of the direct product $\Gamma \times [0, \delta)$. Let $\sigma(x) \geqslant 0$ be a smooth function in $\overline{G}$ that identically vanishes outside of $U(\Gamma)$, depends only on x_n in the s.l.c.s. when $0 \leqslant x_n < \delta$ and is a constant when $0 \leqslant x_n < \delta/2$.

$$\Lambda'_{\sigma(x)} u_0 = \sum_j{}' \psi_j \Lambda'_j \varphi_j u_0 + \sum_j{}'' \varphi_j u_0,$$

where Σ'_j denotes a summation over those j for which U_j is contained in $U(\Gamma)$, Σ''_j denotes a summation over the remaining j, φ_j and ψ_j are the same functions as in §21 and Λ'_j is the p.o. with symbol $(1 + |\xi'_{(j)}|^2)^{\sigma(x_n)/2}$ written in the special local coordinates in U_j. We denote by $\overset{\circ}{H}_{s,N,\sigma(x)}(G)$ the space of generalized functions with support in $\overline{G}$ and finite norm

$$\|u_0\|_{s,N,\sigma(x)} = \|\Lambda'_{\sigma(x)} u_0\|_{s,N} + \|u_0\|_s. \tag{26.5}$$

The space $H_{s,N,\sigma(x)}(G)$ is defined analogously, and $H_{s,\sigma(x)}(\Gamma)$ means $H_{s+\sigma}(\Gamma)$ (we note that $\sigma(x) = \sigma = \text{const}$ on Γ).

Since the symbol of $\Lambda'_{\sigma(x)}$ does not depend on ξ_n near the boundary in each s.l.c.s., it follows that the operator $\Lambda'_{\sigma(x)}$, commutes modulo lower terms with the operator $\mathfrak{A}$ and that $\mathfrak{A}$ is a bounded operator from $\mathcal{H}^{(1)}_{s,N,\sigma(x)}(G)$ into $\mathcal{H}^{(2)}_{s,N,\sigma(x)}(G)$ for arbitrary $\sigma(x)$. The same is true for the operators R and $T^{(3)}$ (see (22.17)). It follows upon applying $\Lambda'_{\sigma_0(x)}$ with $0 \leqslant \sigma_0(x) < 1$ to (22.17) from the left that $\Lambda'_{\sigma_0(x)} u_0 \in \overset{\circ}{H}_{s,N}(G)$. Repeating such arguments k times, we get that $(\Lambda'_{\sigma_0(x)})^k u_0 \in \mathring{H}_{s,N}(G)$. Since $\sigma_0(x)$ and k are arbitrary and $0 \leqslant \sigma_0(x) < 1$, we get that the solution of problem (22.4), (22.5) belongs to $\mathcal{H}^{(1)}_{s,N,\sigma(x)}(G)$, provided that the right sides of (22.4) and (22.5) belong to $\mathcal{H}^{(2)}_{s,N,\sigma(x)}(G)$, where $\sigma(x)$ can be taken arbitrarily large near the boundary. We have thus proved, in particular, that if $u_0 \in \mathring{H}_s(G)$ is a solution of (26.1), $s \in (s_-, s_+)$ and $f \in C^\infty(\overline{G})$, then u_0 is a C^∞ function interior to G (since N in (26.5) can be chosen arbitrarily) and $\partial^k u_0/\partial x'^k_{(j)} \in \mathring{H}_{s_+-\varepsilon}(G)$ for all k and ε in each s.l.c.s. near the boundary.

Let us now find an asymptotic expansion of the solution of (26.1) near the boundary. It is assumed that for each $N > 0$ and $|\xi| \geqslant 1$

$$A(x, \xi) = \sum_{k=0}^{N} A_k(x, \xi) + R_{N+1}(x, \xi), \tag{26.6}$$

where $A_k(x, \xi) \in O^\infty_{\alpha+i\beta-k}$ and $R_{N+1}(x, \xi) \in S^\infty_{\alpha-N-1}$.

We first consider the case in which $A(x, D)$ is a scalar operator. Let $a_0(x', \xi_n)$ denote the symbol that in each jth s.l.c.s. has the form

$$a_0(x'_{(j)}, \xi_n) = a_+(\xi_n - i)^{\alpha+i\beta-\kappa(x'_{(j)})}(\xi_n + i)^{\kappa(x'_{(j)})},$$

where $a_+ = A_j^{(0)}(x'_{(j)}, 0, 0, +1)$. As in §6, for $|\xi_n| \to \infty$

$$A_j^{(0)}(x'_{(j)}, 0, \xi'_{(j)}, \xi_n) - a_0(x'_{(j)}, \xi_n) = O(|\xi_n|^{\alpha+i\beta-1}|\xi'_{(j)}|). \tag{26.7}$$

Therefore in $U(\Gamma)$

$$A(x, D) = a_0(x', D_n) + B, \tag{26.8}$$

where B has the following form in the jth s.l.c.s.:

$$B_1^{(j)}(x^{(j)}, \xi^{(j)}) + x_n B_2^{(j)}(x^{(j)}, \xi^{(j)}),$$

in which $B_1^{(j)}(x^{(j)}, \xi^{(j)})$ satisfies (26.7) and $\operatorname{ord} B_2^{(j)} \leqslant \alpha$. Since $u_0 \in \mathring{H}_{\kappa_-+1/2-\varepsilon,N}(G)$ for arbitrary N and ε and since u_0 is a C^∞ function with respect to $x'_{(j)}$ in the jth s.l.c.s., it follows that $Bu_0 \in H_{\kappa_--\alpha+3/2-\varepsilon}(G)$, $\forall\varepsilon > 0$.

Let $v_0 = \Sigma'_j \varphi_j u_0$, where Σ'_j denotes a summation over those j for which $U_j \subset U(\Gamma)$. We have

$$p_U a_0(x', D_n) v_0 = -p_U B v_0 + g, \tag{26.9}$$

where p_U is the operator of restriction of a function to $U(\Gamma)$, and $g \in C^\infty(\overline{G})$ since $f \in C^\infty(\overline{G})$ and $u_0 - v_0 \in C_0^\infty(G)$. Inverting equation (26.9) as in §7, we get

$$v_0 = F^{-1}_{\xi_n} \frac{1}{(\xi_n + i)^{\kappa(x')}} \Pi^+ \frac{1}{(\xi_n - i)^{\alpha+i\beta-\kappa(x')}} \tilde{g}_1(x', \xi_n), \tag{26.10}$$

where

$$\tilde{g}_1(x', \xi_n) = a_+^{-1} F_{\xi_n}(-Bv_0 + lg). \tag{26.11}$$

We note that $\tilde{g}_1(x', \xi_n)$ is a C^∞ function of x' with values in $\tilde{H}_{\kappa_- - \alpha + 3/2 - \varepsilon}(\mathbf{R}^1)$ with respect to ξ_n, $\forall \varepsilon > 0$. For brevity we shall say that $\tilde{g}_1(x', \xi_n)$ belongs to C^∞ in x' and to $\tilde{H}_{\kappa_- - \alpha + 3/2 - \varepsilon}(\mathbf{R}^1)$ in ξ_n. Since

$$(\xi_n - i)^{-\alpha - i\beta + \kappa(x')} \tilde{g}_1(x', \xi_n) \in \tilde{H}_{3/2 + \kappa_- - \kappa_+ - \varepsilon}(\mathbf{R}^1)$$

and $\frac{3}{2} + \kappa_- - \kappa_+ > \frac{1}{2}$, (5.36) implies

$$\Pi^+ (\xi_n - i)^{-\alpha - i\beta + \kappa(x')} \tilde{g}_1(x', \xi_n) = \frac{1}{\xi_n + i} c_1(x') + c_2(x', \xi_n), \quad (26.12)$$

where

$$c_1(x') = \frac{i}{2\pi} \int_{-\infty}^{\infty} (\xi_n - i)^{-\alpha - i\beta + \kappa(x')} \tilde{g}_1(x', \xi_n)\, d\xi_n \in C^\infty(\Gamma),$$

$$c_2(x', \xi_n) = (\xi_n + i)^{-1} \Pi^+ (\xi_n + i)(\xi_n - i)^{-\alpha - i\beta + \kappa(x')} \tilde{g}_1(x', \xi_n).$$

We note that $c_2(x', \xi_n)$ belongs to C^∞ in x' and to $\tilde{H}^+_{3/2 + \kappa_- - \kappa_+ - \varepsilon}(\mathbf{R}^1)$, $\forall \varepsilon > 0$. It follows from (26.10) and (26.12) that v_0 can be represented near the boundary in the form

$$v_0 = x_{n,+}^{\kappa(x')} c_0(x') + v_1(x', x_n), \quad (26.13)$$

where $c_0(x') \in C^\infty(\Gamma)$ and

$$x_{n,+}^{\kappa(x')} = e^{i\pi(\kappa(x') + 1)/2} \Gamma(\kappa(x') + 1) F_{\xi_n}^{-1} (\xi_n + i0)^{-\kappa(x') - 1}$$

(see Example 2.3) and

$$v_1(x', x_n) = F_{\xi_n}^{-1} (\xi_n + i)^{-\kappa(x')} c_2(x', \xi_n) + x_{n,+}^{\kappa(x')} (e^{-x_n} - 1) c_0(x').$$

We note that $x_{n,+}^{\kappa(x')}(e^{-x_n} - 1)$ belongs to $H^+_{\kappa_- + 3/2 - \varepsilon}$ for $x_n < \delta$. Also, taking a sufficiently refined partition of unity in $U(\Gamma)$ that $\max \operatorname{Re} \kappa(x') - \min \operatorname{Re} \kappa(x') < \varepsilon$ in each neighborhood, we get that $F_{\xi_n}^{-1}(\xi_n + i)^{-\kappa(x')} c_2(x', \xi_n)$ belongs to $H^+_{3/2 + \kappa_- - \varepsilon}(\mathbf{R}^1)$ in ξ_n, and to C^∞ in x'.

To obtain additional terms in the asymptotic expansion of v_0 near the boundary, we let $A_k(x, \xi)$ be the same as in (26.6) and note from Taylor's formula that in each s.l.c.s.

$$A_k^{(j)}(x, \xi) = \sum_{r=0}^{N} \frac{\partial^r A_k^{(j)}(x', 0, \xi', \xi_n)}{\partial x_n^r} \frac{x_n^r}{r!} + x_n^{N+1} R_{N+1,k}^{(j)}(x', x_n, \xi^i, \xi_n),$$

(26.14)

where $R_{N+1,k}^{(j)} \in S^\infty_{\alpha - k}$ for $|\xi| \geqslant 1$. For simplicity in the notation we write (x', x_n, ξ', ξ_n) instead of $(x'_{(j)}, x_n, \xi'_{(j)}, \xi_n)$ in the jth s.l.c.s. As in (10.5) and (10.7), again using Taylor's formula, we get

$$\frac{\partial^r A_k^{(j)}(x', 0, \xi', \xi_n)}{\partial x_n^r} = \sum_{p=0}^{N-k} a_{kpr}^{(j)}(x', \xi', \operatorname{sgn} \xi_n) |\xi_n|^{\alpha + i\beta - k - p} + R_{N+1,k,r}^{(j)}(x, \xi),$$

(26.15)

where

$$\deg_{\xi'} a^{(j)}_{kpr}(x', \xi', \operatorname{sgn} \xi_n) = p$$

and

$$\left|\frac{\partial^m}{\partial x'^m} R^{(j)}_{N+1,k,r}(x, \xi)\right| \leqslant C_{mkr} \frac{|\xi'|^{N+1-k}}{|\xi_n|^{N+1-\alpha}}$$

for $|\xi'| \leqslant |\xi_n|$ and arbitrary m.Thus (26.6), (26.14) and (26.15) imply

$$Bv_0 = B_{N1}v_0 + B_{N2}v_0, \tag{26.16}$$

where B_{N1} has in the jth s.l.c.s. the form

$$B^{(j)}_{N1} = \sum_{k=1}^{N} \sum_{r=0}^{N-k-p} \sum_{p=0}^{N-k} \frac{x_n^r}{r!} a^{(j)}_{kpr}(x', \xi', \operatorname{sgn} \xi_n)|\xi_n|^{\alpha+i\beta-k-p} \tag{26.17}$$

for $|\xi'| \leqslant |\xi_n|$ and $B_{N2}v_0 \in H_{\kappa-\alpha+N+3/2-\varepsilon}(G)$, $\forall \varepsilon > 0$, $a^{(j)}_{000} = 0$.

Let $\tilde{g}_2(x', \xi_n) = a_+^{-1}(x')F_{\xi_n}(-B_{N2}v_0 + lg)$. Since $F_{\xi_n}(-B_{N2}v_0 + lg) \in \tilde{H}_{\kappa_- -\alpha+N+3/2-\varepsilon}(\mathbf{R}^1)$ in ξ_n and belongs to C^∞ in x', it follows from (5.36) that

$$\Pi^+(\xi_n + i)^{-\alpha-i\beta+\kappa(x')}\tilde{g}_2(x', \xi_n) = \sum_{k=1}^{N} (\xi_n + i)^{-k} g_{2,k}(x') + \tilde{g}_{2,N+1}(x', \xi_n), \tag{26.18}$$

where

$$g_{2,k}(x') = i\Pi'(\xi_n + i)^{k-1}(\xi_n - i)^{-\alpha-i\beta+\kappa(x')}\tilde{g}_2(x', \xi_n) \in C^\infty(\Gamma)$$

and $\tilde{g}_{2,N+1}(x', \xi_n) \in C^\infty$ in x' and belongs to $\tilde{H}^+_{3/2+N+\kappa_- -\kappa_+ -\varepsilon}(\mathbf{R}^1)$ in ξ_n. Further, for $|\xi_n| \geqslant 1$

$$\begin{aligned}(\xi_n - i)^{-\alpha-i\beta+\kappa(x')}a_+^{-1}(x')F_{\xi_n}\big(-B_{N1}\big(x_{n,+}^{\kappa(x')}c_0(x') + v_1(x', x_n)\big)\big) \\ = \sum_{k=2}^{N} b_k(x', \operatorname{sgn} \xi_n)\frac{1}{|\xi_n|^k} + \tilde{v}_2(x', \xi_n),\end{aligned} \tag{26.19}$$

where $b_k(x', \pm 1) \in C^\infty(\Gamma)$ and $\tilde{v}_2(x', \xi_n)$ belongs to C^∞ in x' and to $\tilde{H}_{5/2+\kappa_- -\kappa_+ -\varepsilon}(\mathbf{R}^1)$ in ξ_n. From (5.36) it follows as before that

$$\begin{aligned}(\xi_n + i)^{-\kappa(x')}\Pi^+\tilde{v}_2 &= (\xi_n + i)^{-\kappa(x')-1}i\Pi'\tilde{v}_2 \\ &+ (\xi_n + i)^{-\kappa(x')-2}i\Pi'(\xi_n + i)\tilde{v}_2 + \tilde{v}_3(x', \xi_n),\end{aligned} \tag{26.20}$$

where $\tilde{v}_3(x', \xi_n) \in \tilde{H}^+_{5/2+\kappa_- -\varepsilon}(\mathbf{R}^1)$ in ξ_n.

As was shown in Remark 10.3 (see (10.35)), we have

$$\frac{1}{|\xi_n|^k} b_k(x', \operatorname{sgn} \xi_n) = b_k(x', +1)\frac{1}{(\xi_n - i0)^k} + c_{k1}(x')\frac{\ln(\xi_n - i0)}{(\xi_n - i0)^k} + c_{k2}(x')\frac{\ln(\xi_n + i0)}{(\xi_n + i0)^k}. \tag{26.21}$$

Therefore

$$\Pi^+ \frac{1}{|\xi_n|^k} b_k(x', \operatorname{sgn} \xi_n) = c_{k2}(x')\frac{\ln(\xi_n + i0)}{(\xi_n + i0)^k}. \tag{26.22}$$

Here $|\xi_n|^{-k} b_k(x', \operatorname{sgn} \xi_n)$ is the generalized function defined by the right side of (26.21) and the application of Π^+ means taking the Fourier transform with respect to x_n of the function that is equal to the inverse Fourier transform for $x_n > 0$ and vanishes for $x_n < 0$. Consequently, combining (26.18)–(26.20), (26.22) and computing

$$F^{-1}_{\xi_n} \frac{\ln(\xi_n + i0)}{(\xi_n + i0)^{\kappa(x') + k}}$$

(see Example 2.3), we get

$$v_0(x) = c_0(x')x_{n,+}^{\kappa(x')} + c_{10}(x')x_{n,+}^{\kappa(x')+1} + c_{11}(x')x_{n,+}^{\kappa(x')+1}\ln x_n + v_4(x', x_n) \tag{26.23}$$

where $v_4(x', x_n)$ belongs to $H^+_{5/2+\kappa_- -\varepsilon}(\mathbf{R}^1)$ in x_n for $x_n < \delta$ and to C^∞ in x'.

Analogously, substituting (26.23) in (26.10), (26.11) and proceeding as before, we obtain after N steps the following asymptotic expansion of the solution $u_0 \in \mathring{H}_{s_+ - \varepsilon}(G)$ in an s.l.c.s. near the boundary:

$$u_0(x) = c_0(x')x_{n,+}^{\kappa(x')} + \sum_{k=1}^{N} \sum_{p=0}^{k} c_{kp}(x')x_{N,+}^{\kappa(x')+k}\ln^p x_n + u_{N+1}(x', x_n), \tag{26.24}$$

where $c_0(x') \in C^\infty(\Gamma)$, $c_{kp}(x') \in C^\infty(\Gamma)$ and $u_{N+1}(x', x_n) \in H^+_{\kappa_- +3/2+N-\varepsilon}$ in x_n for $x_n < \delta$. Since N is arbitrary, we can prove that

$$u_{N+1}(x', x_n) = O\left(x_n^{\kappa(x')+N+1-\varepsilon}\right)$$

by taking

$$u_{N+1}(x', x_n) = \sum_{k=N+1}^{N+2} \sum_{p=0}^{k} c_{kp}(x')x_{n,+}^{\kappa(x')+k}\ln^p x_n + u_{N+3}(x', x_n)$$

and estimating $u_{N+3}(x', x_n)$ as in §9.

Let us now consider the case in which (26.1) is a system of pseudodifferential equations. It is assumed that there exists a smooth nonsingular matrix $N(x')$ on Γ such that $N^{-1}b_0(x')N$ is a diagonal matrix. The matrix $N^{-1}bN$, where $b(x') = (1/2\pi i)\ln b_0(x')$, is then also a diagonal matrix. We note that the jth column of $N(x')$ is an eigenvector of $b(x')$ corresponding to the eigenvalue $b_j(x')$. Let $N(x', x_n) = N(x')$ for all $(x', x_n) \in U(\Gamma) = \Gamma \times [0, \delta)$, and let $w_0 = N^{-1}\Sigma_j'\varphi_j u_0$. Applying N^{-1} to the equation $pAu_0 = f$, we get

$$p_U N^{-1}A(x, D)Nw_0 = w_\infty, \tag{26.25}$$

where $w_\infty \in C^\infty(\overline{G})$. Since the matrix $b_0(x')$ corresponding to the operator $N^{-1}A(x, D)N$ is diagonal, it follows as in (26.8) that $N^{-1}AN$ has the representation

$$N^{-1}AN = a_0(x', D_n) + B, \tag{26.26}$$

where $a_0(x', \xi_n) = \|\delta_{ik}a_k(x', \xi_n)\|_{i,k=1}^p$ is a diagonal matrix in which

$$a_k(x', \xi_n) = a_k^+(x')(\xi_n - i)^{\alpha + i\beta - \alpha/2 - b_k(x')}(\xi_n + i)^{\alpha/2 + b_k(x')}, \qquad 1 \leqslant k \leqslant p,$$

Let $w_k(x)$, $1 \leqslant k \leqslant p$, be the components of the vector $w_0(x)$. As in (26.10), for $1 \leqslant k \leqslant p$ we have

$$w_k(x) = F_{\xi_n}^{-1}(\xi_n + i)^{-\alpha/2 - b_k(x')}\Pi^+(\xi_n - i)^{-\alpha/2 - i\beta + b_k(x')}\widetilde{g_{1k}}(x', \xi_n), \tag{26.27}$$

where $\widetilde{g_{1k}}(x', \xi_n)$ belongs to C^∞ in x' and to $\tilde{H}_{s_+ - \alpha + 1 - \varepsilon}(\mathbf{R}^1)$ in ξ_n, $\forall \varepsilon > 0$. Further, as in (26.13),

$$w_k(x) = c_{k0}(x')x_{n,+}^{\alpha/2 + b_k(x')} + w_{k1}(x), \qquad 1 \leqslant k \leqslant p, \tag{26.28}$$

where $c_{k0}(x') \in C^\infty(\Gamma)$ and $w_{k1}(x', x_n)$ belongs to C^∞ in x' and to $H^+_{s_+ + 1 - \varepsilon}$ for $x_n < \delta$, $\forall \varepsilon > 0$. Substituting (26.27) in (26.26) and proceeding as in the scalar case, we get for arbitrary N that

$$w_k(x) = c_{k0}(x')x_{n,+}^{\alpha/2 + b_k(x')} + \sum_{r=1}^{N}\sum_{m=1}^{p}\sum_{i=0}^{r} c_{kmri}(x')x_{n,+}^{\alpha/2 + b_m(x') + r}\ln^i x_n + w_{k,N+1}(x), \tag{26.29}$$

where $1 \leqslant k \leqslant p$, $c_{k0}(x') \in C^\infty(\Gamma)$, $c_{kmri}(x') \in C^\infty(\Gamma)$ and $w_{k,N+1}(x) = O(x_n^{s_+ + 1/2 + N - \varepsilon})$, $\forall \varepsilon > 0$.

Since $u_0 = Nw_0$ in $U(\Gamma)$,

$$u_0 = \sum_{k=1}^{p} n_k(x')w_k(x) = \sum_{k=1}^{p} x_{n,+}^{\alpha/2 + b_k(x')}n_k(x')w_k^{(1)}(x) \tag{26.30}$$

for $x_n < \delta$, where $n_k(x')$ is an eigenvector of $b(x')$ corresponding to the eigenvalue $b_k(x')$ and $w_k^{(1)}(x', x_n) = x_{n,+}^{-\alpha/2 - b_k(x')}w_k(x)$, $1 \leqslant k \leqslant p$. It follows from (26.29) that

$$w_k^{(1)}(x', x_n) = c_{k0}(x') + w_k^{(2)}(x', x_n),$$

where $w_k^{(2)}(x', x_n) = O(x_n^{s_+ - s_- - \varepsilon})$ and belongs to C^∞ in x' and to $H^+_{s_+ - s_- + 1/2 - \varepsilon}$ in x_n for $x_n < \delta$.

4. *The class* $\hat{D}^{(0)}_{\alpha,\gamma}(G)$. As was proved in §23, the solutions of the elliptic equation (26.1) belong to $C^\infty(\overline{G})$ if $A(x, \xi) \in \hat{D}^{(0)}_\alpha(G)$, $\alpha \geqslant 0$. We now introduce a class of symbols for which the solutions of (26.1) have particularly simple singularities. A symbol $A(x, \xi)$ is said to be of class $\hat{D}^{(0)}_{\alpha,\gamma}(G)$ if $\alpha - \gamma$ is an even integer, $|\gamma| \leqslant 1$ and $A(x, \xi) = A^{(1)}(x, \xi)|\xi|^\gamma$, where $A^{(1)}(x, \xi) \in \hat{D}^{(0)}_{\alpha-\gamma}$.

We will show that if $u_0 \in \mathring{H}_{s_+ - \varepsilon}(G)$ is a solution of (26.1), where $A(x, \xi) \in \hat{D}^{(0)}_{\alpha,\gamma}(G)$ and $f \in C^\infty(\overline{G})$, then there exists a function $u_1(x) \in C^\infty(\overline{G})$ such that

$$u_0(x) = x_{n,+}^{\alpha/2} u_1(x) \tag{26.31}$$

near the boundary Γ of G. When $x_{n,+}^{\alpha/2}$ is not a regular functional (i.e. $\alpha \leqslant -2$), we extend $u_1(x)$ as a C^∞ function to $\mathbf{R}^n$ and then take the product of $u_1(x)$ and $x_{n,+}^{\alpha/2}$. Clearly, if (26.31) holds, then for arbitrary N

$$u_0(x) = \sum_{k=0}^{N} c_k(x') x_{n,+}^{\alpha/2+k} + x_{n,+}^{\alpha/2+N+1} u_{N+1}(x), \tag{26.32}$$

where $c_k(x') \in C^\infty(\Gamma)$ and $u_{N+1}(x) \in C^\infty(\overline{G})$.

To prove (26.31) we note that in the jth s.l.c.s. in $U(\Gamma)$ it is possible to represent $A(x, D)$ in the form

$$A^{(j)}(x^{(j)}, \xi^{(j)}) = A_j^{(2)}(x^{(j)}, \xi^{(j)})(\xi_n + i)^{\alpha/2},$$

where $A_j^{(2)}(x^{(j)}, D^{(j)})$ is a smooth operator (see §23). Hence, if $v_0 = (D_n + i)^{\alpha/2} u_0$ near the boundary, the results of §23 imply that $v_0 \in C^\infty(\overline{G})$. Suppose $\alpha > 0$. Since

$$F^{-1}_{\xi_n}(\xi_n + i)^{\alpha/2} = C_\alpha x_{n,+}^{\alpha/2-1} e^{-x_n}$$

(see Example 2.3), for $0 < x_n < \delta$ we have

$$u_0 = F^{-1}_{\xi_n}(\xi_n + i)^{-\alpha/2} \tilde{v}_0(x', \xi_n) = C_\alpha \int_0^{x_n} (x_n - y_n)^{\alpha/2-1} e^{-(x_n - y_n)} v_0(x', y_n)\, dy_n.$$

Making the change of variables $y_n = x_n t$, we get $u_0 = C_\alpha x_{n,+}^{\alpha/2} w(x', x_n)$, where the function

$$w(x', x_n) = \int_0^1 (1 - t)^{\alpha/2-1} e^{-x_n(1-t)} v_0(x', t x_n)\, dt$$

clearly belongs to $C^\infty(\overline{G})$. Therefore (26.31) is proved for $\alpha > 0$. The proof for $\alpha < 0$ follows from the representation of $x_{n,+}^{\alpha/2-1}$ as a derivative of $x_{n,+}^{\gamma-1}$ with $\gamma > 0$.

5. *Description of the finite element method.* Consider a partition of the plane $\mathbf{R}^2$ into squares with sides of length h parallel to the coordinate axes. Let $G(h)$ be a polygonal domain consisting of such squares that approximates the domain G, and let $N(h)$ be the set of all pairs (p_1, p_2) such that p_1 and p_2 are integers and (p_1h, p_2h) are interior points of $G(h)$. One defines a *trial function* $\psi_{p_1p_2}(x)$ for each $(p_1, p_2) \in N(h)$ by the formula

$$\psi_{p_1p_2}(x) = \varphi_0(x_1/h - p_1, x_2/h - p_2), \tag{26.33}$$

where $\varphi_0(x_1, x_2) = (1 - |x_1|)(1 - |x_2|)$ for $|x_1|, |x_2| \leqslant 1$, and $\varphi_0(x) = 0$ outside of the square $|x_1|, |x_2| \leqslant 1$. An approximate solution of equation (26.1) is then sought in the form

$$u_h = \sum_{(p_1, p_2) \in N(h)} c_{p_1p_2}\psi_{p_1p_2}(x), \tag{26.34}$$

where the $c_{p_1p_2}$ are unknown numbers (vectors) when (26.1) is a single equation (system of equations). The coefficients $c_{p_1p_2}$ are determined by the Ritz-Galerkin method, i.e. by the equations

$$\big(A(x, D)u_h, \psi_{q_1q_2}\big) = (lf, \psi_{q_1q_2}), \qquad \forall (q_1, q_2) \in N(h), \tag{26.35}$$

which reduces to the algebraic system

$$\sum_{(p_1, p_2) \in N(h)} a_{q_1q_2,p_1p_2}c_{p_1p_2} = f_{q_1q_2}, \qquad (q_1, q_2) \in N(h), \tag{26.36}$$

in which $a_{q_1q_2,p_1p_2} = (A\psi_{p_1p_2}, \psi_{q_1q_2})$ and $f_{q_1q_2} = (lf, \psi_{q_1q_2})$. The function lf in (26.35) and (26.36) is a fixed extension of $f(x)$ onto $\mathbf{R}^2$ such that $\|lf\|_{s_+ - \alpha - \varepsilon} \leqslant 2\|f\|^+_{s_+ - \alpha - \varepsilon}$.

REMARK 26.1. Because $\alpha = \pm 1$ in most of the applications to elasticity problems, we consider here only piecewise linear trial functions. A more general application of the finite element method to p.e.'s is considered in [39].

REMARK 26.2. The finite element method (f.e.m.) described above is not always satisfactory, since the square elements generally do not yield a good approximation of the boundary. In subsection 8 below, we attempt to remedy this deficiency by equipping the trial functions with weight factors that vanish on the boundary.

6. *Calculation of the matrix* $\|a_{q_1q_2,p_1p_2}\|$. The problem of calculating the matrix of system (26.36) is clearly more complicated for pseudodifferential operators than for differential operators. Here we give some simple asymptotic formulas that permit one to readily calculate most of the elements of $\|a_{q_1q_2,p_1p_2}\|$.

Consider for simplicity the case when $A(x, \xi) = A_0(\xi) \in O_\alpha^\infty$ and $|\alpha| < 1$, i.e. when $A(x, \xi)$ does not depend on x and is homogeneous in ξ of degree α. From Parseval's equality we have

$$a_{q_1q_2p_1p_2} = \frac{1}{(2\pi)^2}\int_{-\infty}^{\infty} A_0(\xi)\tilde{\psi}_{p_1p_2}(\xi)\,\overline{\tilde{\psi}_{q_1q_2}(\xi)}\;d\xi_1 d\xi_2. \tag{26.37}$$

On the other hand, from a simple calculation we have

$$\widetilde{\psi_{q_1q_2}}(\xi) = \int_{-\infty}^{\infty} \varphi_0(x_1/h - q_1, x_2/h - q_2)e^{i(x,\xi)}dx_1dx_2 = h^2 e^{i(q_1\xi_1 + q_2\xi_2)h}\tilde{\varphi}_0(h\xi)$$

and

$$\tilde{\varphi}_1(h\xi) = \frac{\sin^2(h\xi_1/2)}{(h\xi_1/2)^2}\,\frac{\sin^2(h\xi_2/2)}{(h\xi_2/2)^2}.$$

Therefore, making the change of variables $\zeta = h\xi$ in (26.37), we get

$$a_{q_1q_2p_1p_2} = \frac{h^{2-\alpha}}{(2\pi)^2}\int_{-\infty}^{\infty} A_0(\zeta)e^{i(p-q,\zeta)}(\tilde{\varphi}_0(\zeta))^2\;d\zeta_1 d\zeta_2. \tag{26.38}$$

Since the $a_{q_1q_2p_1p_2}$ depend only on $r_1 = p_1 - q_1$ and $r_2 = p_2 - q_2$, we will write a_{r_1,r_2} below instead of $a_{q_1q_2p_1p_2}$. It is important to note that only those integrals of (26.38) for which $|p - q|$ is small need be calculated numerically. The rest of the integrals (26.38) can be calculated by means of the following asymptotic formula.

LEMMA 26.1. *The following asymptotic expansion holds for* $|r_1| + |r_2| \to \infty$:

$$a_{r_1r_2} = \frac{h^{2-\alpha}}{(2\pi)^2}\sum_{k_1+k_2=0}^{N}(-1)^{k_1+k_2}\varphi_{k_1k_2}\frac{\partial^{2k_1+2k_2}}{\partial r_1^{2k_1}\partial r_2^{2k_2}}a_0(r) + O\left(\frac{1}{|r|^{2N+2}}\right), \tag{26.39}$$

where $|r| = \sqrt{r_1^2 + r_2^2}$, *$N$ is an arbitrary nonnegative integer,* $a_0(r) = F_\xi A_0(\xi)$ *and the* $\varphi_{k_1k_2}$ *are the Taylor coefficients of the function*

$$(\tilde{\varphi}_0(\xi))^2 = \frac{\sin^4(\xi_1/2)\sin^4(\xi_2/2)}{(\xi_1/2)^4(\xi_2/2)^4},$$

i.e.

$$(\tilde{\varphi}_0(\xi))^2 = \sum_{k_1+k_2=0}^{N}\varphi_{k_1k_2}\xi_1^{2k_1}\xi_2^{2k_2} + r_{N+2}. \tag{26.40}$$

REMARK 26.3. For example, if $A_0(\xi) = |\xi|^\alpha$, then

$$a_0(r) = \frac{2^{\alpha+2}\pi\Gamma((\alpha+2)/2)}{\Gamma(-\alpha/2)}\,\frac{1}{|r|^{\alpha+2}}$$

(see Example 3.1).

PROOF OF LEMMA 26.1. It follows from (26.38) and (26.40) that

$$a_{r_1r_2} = \frac{h^{2-\alpha}}{(2\pi)^2}\left(\sum_{k_1+k_2=0}^{N} \varphi_{k_1k_2} F(\xi_1^{2k_1}\xi_2^{2k_2}A_0(\xi)) + F(A_0(\xi)r_{2N+2}(\xi))\right),$$

where the operator F is applied according to the definition of the Fourier transform of a generalized function (see §2). But the properties of the Fourier transform imply

$$F(\xi_1^{2k_1}\xi_2^{2k_2}A_0(\xi)) = (-1)^{k_1+k_2}\frac{\partial^{2k_1+2k_2}}{\partial r_1^{2k_1}\partial r_2^{2k_2}} a_0(r),$$

where $a_0(r) = FA_0(\xi) \in O^{\infty}_{-\alpha-2}$. It therefore remains to prove that

$$|F(A_0(\xi)r_{2N+2}(\xi))| \leqslant c|r|^{-2N-2}.$$

Let $\chi(\xi) \in C_0^\infty(\mathbf{R}^2)$, $\chi(\xi) = 1$ for $|\xi| \leqslant 1$, $\chi(\xi) = 0$ for $|\xi| \geqslant 2$ and $0 \leqslant \chi(\xi) \leqslant 1$ for all $\xi \in \mathbf{R}^2$. Since $r_{2N+2}(\xi) = o(|\xi|^{2N+2})$ for $|\xi| \leqslant 1$, the function

$$\left(\frac{\partial^2}{\partial \xi_1^2} + \frac{\partial^2}{\partial \xi_2^2}\right)^{N+1}(\chi A_0 r_{2N+2})$$

is absolutely integrable. Hence its Fourier transform is bounded:

$$\left|F\left(\frac{\partial^2}{\partial \xi_1^2} + \frac{\partial^2}{\partial \xi_2^2}\right)^{N+1}(\chi A_0 r_{2N+2})\right| = |r|^{2N+2}|F(\chi A_0 r_{2N+2})| \leqslant C.$$

Analogously, the fact that $r_{2N+2} = o(|\xi|^{2N})$ for $|\xi| \to \infty$ implies

$$\left|F\left(\frac{\partial^2}{\partial \xi_1^2} + \frac{\partial^2}{\partial \xi_2^2}\right)^{N+2}((1-\chi)A_0 r_{2N+2})\right| = |r|^{2N+4}|F(1-\chi)A_0 r_{2N+2}| \leqslant C.$$

Lemma 26.1 is proved.

7. *Convergence of the finite element method.* In this subsection we prove that the solution u_h of (26.35) converges at a certain rate to the solution u_0 of (26.1) as $h \to 0$.

THEOREM 26.2. *Suppose* $f \in H_{s_+-\alpha-\varepsilon}(G)$, $\forall \varepsilon > 0$. *Then, as* $h \to 0$, *the f.e.m. solution* u_h *converges in* $H_{\alpha/2}(\mathbf{R}^2)$ *to the exact solution* u_0 *at the rate indicated by the inequality*

$$\|u_0 - u_h\|_{\alpha/2} \leqslant C_\varepsilon h^{s_+-\alpha/2-\varepsilon}\|f\|^+_{s_+-\alpha-\varepsilon}, \qquad \forall \varepsilon > 0, \tag{26.41}$$

where s_+ *is the same as in Theorem 26.1 (in particular,* $s_+ = \min_{x'\in\Gamma} \operatorname{Re} \kappa(x') + \frac{1}{2} = \kappa_- + \frac{1}{2}$ *when* $A(x, D)$ *is a scalar p.o., and* $s_+ = \alpha/2 + \frac{1}{2}$ *when* $b_0(x')$ *is symmetric).*

The proof of Theorem 26.2 requires some lemmas, of which the following is a presentation of some results in [9] and [93].

LEMMA 26.2. *Let* $\varphi_0(x)$ *be the same function as in* (26.33), *let* $\varphi_1(x) = \varphi_0(x) * \varphi_0(x)$, *let* $\psi_p^{(k)}(x) = \varphi_k(x/h - p)$ *for* $k = 0, 1$ (*i.e.* $\psi_p^{(0)}(x) = \psi_p(x) = \varphi_0(x/h - p)$ *as in* (26.33), *and* $\psi_p^{(1)}(x) = \varphi_1(x/h - p)$) *and let*

$$P_k u = \frac{1}{h^2} \sum_p \left(u, \psi_p^{(k)}\right) \psi_p(x) \qquad \text{for } k = 0, 1.$$

Then for arbitrary s *and* γ *such that* $0 \leqslant \gamma \leqslant 2$, $s + \gamma \leqslant 2$ *and* $-\frac{3}{2} - 2k < s < \frac{3}{2}$

$$\|u - P_k u\|_s \leqslant C h^\gamma \|u\|_{s+\gamma}, \qquad \forall u \in H_{s+\gamma}(\mathbf{R}^2), \tag{26.42}$$

where C *does not depend on* $u(x)$.

PROOF. We have

$$\widetilde{\psi_p^{(k)}}(\xi) = h^2 e^{i(p,\xi)h} \tilde{\varphi}_k(h\xi) = h^2 e^{i(p,\xi)h} (\tilde{\varphi}_0(h\xi))^{1+k},$$

$k = 0, 1$, where

$$\tilde{\varphi}_0(h\xi) = \frac{\sin^2(h\xi/2)}{(h\xi_1/2)^2} \frac{\sin^2(h\xi_2/2)}{(h\xi_2/2)^2}.$$

Therefore Parseval's equality implies

$$\begin{aligned} \widetilde{P_k u}(\xi) &= \sum_p \frac{1}{h^2(2\pi)^2} \int_{-\infty}^{\infty} \tilde{u}(\zeta)\, \overline{\widetilde{\psi_p^{(k)}}(\zeta)}\; d\zeta \tilde{\psi}_p(\xi) \\ &= \sum_p \frac{h^2}{(2\pi)^2} \int_{-\infty}^{\infty} \tilde{u}(\zeta) e^{i(p,\xi-\zeta)h} \tilde{\varphi}_0^{1+k}(h\zeta)\, d\zeta \tilde{\varphi}_0(h\xi). \end{aligned} \tag{26.43}$$

On the other hand, Poisson's summation formula (see, for example, [49], Vol. I) implies

$$\begin{aligned} \sum_p \int_{-\infty}^{\infty} \tilde{f}(\zeta) e^{i(p,h\xi - h\zeta)}\, d\zeta &= (2\pi)^2 \sum_q \int_{-\infty}^{\infty} \tilde{f}(\zeta) \delta(h\xi - h\zeta - 2\pi q)\, d\zeta \\ &= \left(\frac{2\pi}{h}\right)^2 \sum_q \tilde{f}\left(\xi - \frac{2\pi q}{h}\right). \end{aligned}$$

Therefore

$$\widetilde{P_k u}(\xi) = \sum_p \tilde{u}(\xi - 2\pi p/h) \tilde{\varphi}_0^{1+k}(h\xi - 2\pi p) \tilde{\varphi}_0(h\xi). \tag{26.44}$$

For $0 \leqslant \gamma \leqslant 2$

$$|\tilde{\varphi}_0^{2+k}(h\xi) - 1| \leqslant C h^\gamma |\xi|^\gamma. \tag{26.45}$$

Therefore, taking into account only the term corresponding to $p = 0$ in (26.44),

$$\|\tilde{u}(\xi) - \tilde{u}(\xi)\widetilde{\varphi_0}^{2+k}(h\xi)\|_s^2 = \int_{-\infty}^{\infty} (1 + |\xi|)^{2s}|1 - \widetilde{\varphi_0}^{2+k}(h\xi)|^2|\tilde{u}(\xi)|^2\, d\xi$$
$$\leqslant Ch^{2\gamma}\|u\|_{s+\gamma}^2. \tag{26.46}$$

Choosing $\varepsilon > 0$ arbitrarily small, we get with the use of the Cauchy-Schwarz inequality that

$$\left|\sum_{p\neq 0} \tilde{u}\left(\xi - \frac{2\pi p}{h}\right)\widetilde{\varphi_0}^k(h\xi - 2\pi p)\widetilde{\varphi_0}^{3/4-\varepsilon/2}(h\xi - 2\pi p)\widetilde{\varphi_0}^{1/4+\varepsilon/2}(h\xi - 2\pi p)\right|^2$$
$$\leqslant \sum_{p\neq 0}\left|\tilde{u}\left(\xi - \frac{2\pi p}{h}\right)\right|^2 \widetilde{\varphi_0}^{2k+3/2-\varepsilon}(h\xi - 2\pi p)\cdot \sum_{p\neq 0}\widetilde{\varphi_0}^{1/2+\varepsilon}(h\xi - 2\pi p). \tag{26.47}$$

We note that $\sum_{p\neq 0}\widetilde{\varphi_0}^{1/2+\varepsilon}(h\xi - 2\pi p) \leqslant C$. Therefore, making the change of variables $\xi - 2\pi p/h = \zeta$, we get

$$\left\|\sum_{p\neq 0} \tilde{u}\left(\xi - \frac{2\pi p}{h}\right)\widetilde{\varphi_0}^{1+k}(h\xi - 2\pi p)\widetilde{\varphi_0}(h\xi)\right\|_s^2$$
$$\leqslant C\sum_{p\neq 0}\int_{-\infty}^{\infty}\left|\tilde{u}\left(\xi - \frac{2\pi p}{h}\right)\right|^2\widetilde{\varphi_0}^{2k+3/2-\varepsilon}(h\xi - 2\pi p)\widetilde{\varphi_0}^2(h\xi)(1 + |\xi|)^{2s}\, d\xi$$
$$= C\int_{-\infty}^{\infty}|\tilde{u}(\zeta)|^2(1 + |\zeta|)^{2s+2\gamma}Q(\zeta, h)\, d\zeta, \tag{26.48}$$

where

$$Q(\zeta, h) = \sum_{p\neq 0}\widetilde{\varphi_0}^{2k+3/2-\varepsilon}(h\zeta)\widetilde{\varphi_0}^2(h\zeta + 2\pi p)\frac{(1 + |\zeta + 2\pi p/h|)^{2s}}{(1 + |\zeta|)^{2s+2\gamma}}.$$

Lemma 26.2 will now follow if we show that $Q(\zeta, h) \leqslant Ch^{2\gamma}$. Suppose $s \geqslant 0$. Then

$$Q(\zeta, h) \leqslant \sum_{p\neq 0}\widetilde{\varphi_0}^{2k+3/2-\varepsilon}(h\zeta)\frac{\widetilde{\varphi_0}^2(h\zeta + 2\pi p)}{|h\zeta|^{2s+2\gamma}}(h^{2s} + |h\zeta + 2\pi p|^{2s})h^{2\gamma} \leqslant Ch^{2\gamma}$$

for $0 \leqslant s + \gamma \leqslant 2$ and $0 \leqslant s \leqslant \frac{3}{2} - \varepsilon$, since $\widetilde{\varphi_0}^{2k+3/2-\varepsilon}(h\zeta) \leqslant C$ and

$$\sum_{p\neq 0}\frac{\widetilde{\varphi_0}^2(h\zeta + 2\pi p)}{|h\zeta|^{2s+2\gamma}}|h\zeta + 2\pi p|^{2s} \leqslant C.$$

Suppose now $s < 0$. Then

$$Q(\zeta, h) \leqslant \sum_{p\neq 0}\frac{\widetilde{\varphi_0}^{2k+3/2-\varepsilon}(h\zeta)|h|\zeta||^{2|s|}}{|h\zeta + 2\pi p|^{2|s|}}\,\frac{\widetilde{\varphi_0}^2(h\zeta + 2\pi p)}{|h\zeta|^{2\gamma}}h^{2\gamma} \leqslant Ch^{2\gamma}$$

for $|s| \leqslant 2k + \frac{3}{2} - \varepsilon$ and $0 \leqslant \gamma \leqslant 2$, since

$$\frac{\widetilde{\varphi_0}^{2k+3/2-\varepsilon}(h\zeta)|h|\zeta||^{2|s|}}{|h\zeta + 2\pi p|^{2|s|}} \leqslant C \quad \text{and} \quad \sum_{p \neq 0} \frac{\widetilde{\varphi_0}^2(h\zeta + 2\pi p)}{|h\zeta|^{2\gamma}} \leqslant C.$$

Lemma 26.2 is proved.

The following well-known lemma characterizes the extremal property of the Ritz-Galerkin solution.

LEMMA 26.3. *Let u_h be the solution of* (26.35), *let v_h be an arbitrary function of form* (26.34) *and suppose $G(h) \subset G$. Then*

$$\|u_0 - u_h\|_{\alpha/2} \leqslant C\|u_0 - v_h\|_{\alpha/2}, \tag{26.49}$$

where u_0 is the solution of (26.1) *and C does not depend on v_h.*

PROOF. Since $\operatorname{supp} v_h \subset \overline{G(h)} \subset \overline{G}$, it follows from (26.1) that

$$(Au_0, v_h) = (lf, v_h), \tag{26.50}$$

and hence from (26.35) that

$$(A(u_0 - u_h), v_h) = (lf, v_h) - (lf, v_h) = 0 \tag{26.51}$$

for all v_h of form (26.34). Therefore

$$(A(u_0 - u_h), u_0 - u_h) = (A(u_0 - u_h), u_0 - v_h). \tag{26.52}$$

On the other hand, from (26.2) we have

$$C\|u_0 - u_h\|^2_{\alpha/2} \leqslant \operatorname{Re}(A(u_0 - u_h), u_0 - u_h), \tag{26.53}$$

while estimating the bilinear form $(A(u_0 - u_h), u_0 - v_h)$ from above, we get

$$\begin{aligned} |(A(u_0 - u_h), u_0 - v_h)| &\leqslant \|A(u_0 - u_h)\|^+_{-\alpha/2}\|u_0 - v_h\|_{\alpha/2} \\ &\leqslant C\|u_0 - u_h\|_{\alpha/2}\|u_0 - v_h\|_{\alpha/2}. \end{aligned} \tag{26.54}$$

The lemma now follows from (26.53), (26.52) and (26.54).

LEMMA 26.4. *Suppose $u_0 \in \mathring{H}_s(G)$. Then there exists a function $u_0^h \in \mathring{H}_s(G)$ such that the distance from* $\operatorname{supp} u_0^h$ *to $\Gamma = \partial G$ is at least $\sqrt{2}\, h$ and*

$$\|u_0 - u_0^h\|_{s-\delta} \leqslant Ch^\delta\|u_0\|_s, \qquad \forall\, 0 \leqslant \delta \leqslant 1. \tag{26.55}$$

PROOF. By taking a partition of unity and making a change of variables, we can reduce the proof to the case when $G = \mathbf{R}^n_+$, Γ is the plane $x_n = 0$ and $u_0 \in H_s^+(\mathbf{R}^n)$, where $n = 2$. If $u_0^h = u_0(x', x_n - \sqrt{2}\, h)$, then $u_0^h = 0$ for $x_n < \sqrt{2}\, h$ and $\widetilde{u_0^h}(\xi) = e^{i\sqrt{2}h\xi_n}\tilde{u}_0(\xi)$. Therefore (26.55) holds inasmuch as $|e^{i\sqrt{2}h\xi_n} - 1| \leqslant Ch^\delta|\xi_n|^\delta$ for $0 \leqslant \delta \leqslant 1$.

PROOF OF THEOREM 26.2. Suppose first that $G(h) \subset G$. Let u_0 be the solution of (26.1) and let u_0^h be a function with the properties indicated in Lemma 26.4. We have $\operatorname{supp} P_0u_0^h \subset \overline{G(h)}$, so that $P_0u_0^h$ has the form (26.34). Since $u_0 \in \mathring{H}_{s_+ - \varepsilon}(G)$ (see Theorem 26.1), it follows from Lemma 26.2 and

Lemma 26.4 with $\delta = 0$ that

$$\|u_0^h - P_0 u_0^h\|_{\alpha/2} \leqslant C_\varepsilon h^{s_+ - \alpha/2 - \varepsilon}\|u_0^h\|_{s_+ - \varepsilon} \leqslant C_\varepsilon h^{s_+ - \alpha/2 - \varepsilon}\|u_0\|_{s_+ - \varepsilon}.$$

On the other hand, it follows from Lemma 26.4 with $\delta = s_+ - \alpha/2 - \varepsilon$ that

$$\|u_0 - u_0^h\|_{\alpha/2} \leqslant C_\varepsilon h^{s_+ - \alpha/2 - \varepsilon}\|u_0\|_{s_+ - \varepsilon}.$$

Therefore

$$\|u_0 - P_0 u_0^h\|_{\alpha/2} \leqslant C_\varepsilon h^{s_+ - \alpha/2 - \varepsilon}\|u_0\|_{s_+ - \varepsilon} \leqslant C_\varepsilon h^{s_+ - \alpha/2 - \varepsilon}\|f\|^+_{s_+ - \alpha - \varepsilon}. \tag{26.56}$$

But Lemma 26.3 implies

$$\|u_0 - u_h\|_{\alpha/2} \leqslant C\|u_0 - P_0 u_0^h\|_{\alpha/2}, \tag{26.57}$$

and hence (26.41) holds when $G(h) \subset G$.

Suppose now that $G(h) \not\subset G$, and let G_0 be a fixed domain containing $G(h)$ and G. It is assumed that the distance between the boundaries of G and G_0 is less than Ch. Let $u^{(0)}$ be the solution of the equation $p_0 A u^{(0)} = p_0 lf$, where p_0 is the operator of restriction of a function to G_0 and lf is the same as in (26.35), with $\|lf\|_{s_+ - \alpha - \varepsilon} \leqslant 2\|f\|^+_{s_+ - \alpha - \varepsilon}$. We have

$$A u_0 = u_- + lf, \tag{26.58}$$

where u_0 is the solution of (26.1) and $u_- = 0$ in G. Applying p_0 to (26.58), we get $p_0 A u_0 = p_0 u_- + p_0 lf$. Therefore

$$p_0 A\big(u_0 - u^{(0)}\big) = p_0 u_-$$

and hence, according to Theorem 26.1,

$$\|u_0 - u^{(0)}\|_{\alpha/2} \leqslant C\|p_0 u_-\|^+_{-\alpha/2}. \tag{26.59}$$

Let us prove that

$$\|p_0 u_-\|^+_{-\alpha/2} \leqslant C_\varepsilon h^{s_+ - \alpha/2 - \varepsilon}\|u_-\|_{s_+ - \alpha - \varepsilon} \leqslant C_\varepsilon h^{s_+ - \alpha/2 - \varepsilon}\|f\|^+_{s_+ - \alpha - \varepsilon}. \tag{26.60}$$

Since $u_- = 0$ in G and the distance between ∂G and ∂G_0 is less than Ch, there exists, according to Lemma 26.4, a function u_-^h such that $u_-^h = 0$ in G_0 and

$$\|u_- - u_-^h\|_{-\alpha/2} \leqslant C_\varepsilon h^{s_+ - \alpha/2 - \varepsilon}\|u_-\|_{s_+ - \alpha - \varepsilon}.$$

Therefore

$$\|p_0\big(u_- - u_-^h\big)\|_{-\alpha/2} \leqslant C_\varepsilon h^{s_+ - \alpha/2 - \varepsilon}\|u_-\|_{s_+ - \alpha - \varepsilon},$$

and hence, since $p_0 u_-^h = 0$, (26.60) is valid.

On the other hand, since $G(h) \subset G_0$ and since Theorem 26.2, has already been proved for the case $G(h) \subset G$, we have

$$\|u^{(0)} - u_h\|_{\alpha/2} \leqslant C_\varepsilon h^{s_+ - \alpha/2 - \varepsilon} \|f\|^+_{s_+ - \alpha - \varepsilon}, \tag{26.61}$$

and the theorem follows from (26.59)–(26.61).

8. *The singular finite element method.* The rate of convergence of the finite element method (f.e.m.) described in subsections 5–7 is small and, indeed, as the numerical results in [53] show, is satisfactory only in the interior of the domain G and deteriorates near the boundary. In this subsection we introduce a weight factor that takes into account the asymptotic behaviour of the exact solution near the boundary.

We first consider the case in which (26.1) is a single equation. Let r_κ be a function of class C^∞ in $\mathbf{R}^2$ that is equal to $x_{n,+}^{\kappa(x')}$ in an s.l.c.s. near the boundary, differs from zero in G and vanishes outside G. We can always choose r_κ so that $r_\kappa = 1$ in $G \setminus U(\Gamma)$. Let $G(h)$ be a polygonal domain consisting of all squares Q_h such that $\mu(Q_h \cap G) \geqslant Ch^2$, where $\mu(Q_h \cap G)$ is the area of $Q_h \cap G$ and C is a given constant that does not depend on h, and let $N(h)$ be the same as in subsection 5. We will look for an approximate solution of (26.1) in the form

$$w_h = \sum_{(p_1, p_2) \in N(h)} c_{p_1 p_2} r_\kappa \psi_{p_1 p_2}(x), \tag{26.62}$$

where the $\psi_{p_1 p_2}(x)$ are the same functions as in (26.33). Analogously to subsection 5, the singular finite element method (s.f.e.m.) approximation w_h is the solution of the system of equations (cf. (22.35))

$$\big(Aw_h, r_\kappa \psi_{q_1 q_2}(x)\big) = \big(lf, r_\kappa \psi_{q_1 q_2}(x)\big), \qquad \forall (q_1, q_2) \in N(h). \tag{26.63}$$

In order to estimate the rate of convergence of the s.f.e.m. we need some lemmas.

LEMMA 26.5. *Suppose* $\kappa_- = \min_{x' \in \Gamma} \operatorname{Re} \kappa(x') > 0$ *and* $-\kappa_- - \frac{1}{2} < \beta < \kappa_- + \frac{1}{2}$. *Then*

$$\|r_\kappa u\|_\beta \leqslant C \|u\|_\beta \tag{26.64}$$

for all $u \in H_\beta(\mathbf{R}^n)$.

PROOF. By making use of a partition of unity, it can be assumed without loss of generality that supp r_κ is contained in a small neighborhood U_0 such that $U_0 \cap \Gamma \neq \varnothing$. We introduce local coordinates in U_0 such that $r_\kappa = x_{n,+}^{\kappa(x')} a(x)$ in these coordinates, where $a(x) \in C_0^\infty(\mathbf{R}^n)$. The Fourier transform $\tilde{r}_\kappa(\xi) = Fr_\kappa$ then satisfies the estimate

$$|\tilde{r}_\kappa(\xi)| \leqslant C_{\varepsilon\mu}(1 + |\xi'|)^{-\mu}(1 + |\xi_n|)^{-1-\kappa_- + \varepsilon}, \qquad \forall \varepsilon > 0, \quad \forall \mu > 0, \tag{26.65}$$

in which ε can be replaced by zero if $\kappa(x')$ is a constant.

We have

$$\|r_\kappa u\|_\beta^2 \leqslant \int (1 + |\xi|)^{2\beta} \left| \int |\tilde{r}_\kappa(\xi - \zeta)| \, |\tilde{u}(\zeta)| \, d\zeta \right|^2 d\xi.$$

By the Cauchy-Schwarz inequality we obtain for arbitrary γ

$$\left| \int_{-\infty}^{\infty} \tilde{r}_\kappa(\xi - \zeta) \tilde{u}(\zeta) \, d\zeta \right|^2 \leqslant \int_{-\infty}^{\infty} |\tilde{r}_\kappa(\xi - \zeta)| (1 + |\zeta|)^{-\gamma} \, d\zeta \cdot \int_{-\infty}^{\infty} |\tilde{r}_\kappa(\xi - \zeta)| (1 + |\zeta|)^{\gamma} |\tilde{u}(\zeta)|^2 \, d\zeta.$$

Therefore

$$\|r_\kappa u\|_\beta^2 \leqslant \int_{-\infty}^{\infty} \left(\int_{-\infty}^{\infty} I(\xi) |\tilde{r}_\kappa(\xi - \zeta)| (1 + |\xi|)^{2\beta} \, d\xi \right) (1 + |\zeta|)^{\gamma} |\tilde{u}(\zeta)|^2 \, d\zeta, \tag{26.66}$$

where

$$I(\xi) = \int_{-\infty}^{\infty} |\tilde{r}_\kappa(\xi - \zeta)| (1 + |\zeta|)^{-\gamma} \, d\zeta. \tag{26.67}$$

If

$$-\kappa_- < \gamma < 1 + \kappa_-, \tag{26.68}$$

then

$$I(\xi) \leqslant C(1 + |\xi|)^{-\gamma}. \tag{26.69}$$

In fact, if $\gamma > 0$, then

$$\begin{aligned} I(\xi) &= \int_{|\zeta| > |\xi|/2} + \int_{|\zeta| < |\xi|/2} \\ &\leqslant \int_{|\zeta| > |\xi|/2} \frac{C}{(1 + |\xi|)^{\gamma}} |\tilde{r}_\kappa(\xi - \zeta)| \, d\zeta + \frac{C_{\varepsilon\mu}}{(1 + |\xi|)^{1 + \kappa - \varepsilon}} \\ &\quad \cdot \int_{|\zeta| < |\xi|/2} \frac{d\zeta}{(1 + |\zeta|)^{\gamma} (1 + |\xi' - \zeta'|)^{\mu}} \\ &\leqslant \frac{C}{(1 + |\xi|)^{\gamma}} + \frac{C_{\varepsilon\mu}}{(1 + |\xi|)^{1 + \kappa_- - \varepsilon}} \int_{|\zeta_n| < |\xi|/2} \frac{d\zeta_n}{(1 + |\zeta_n|)^{\gamma}} \leqslant \frac{C}{(1 + |\xi|)^{\gamma}}. \end{aligned} \tag{26.70}$$

While, if $\gamma < 0$, then, clearly,

$$\begin{aligned} I(\xi) &= \int |\tilde{r}_\kappa(\zeta)| (1 + |\xi - \zeta|)^{|\gamma|} \, d\zeta \\ &\leqslant C \int_{-\infty}^{\infty} |\tilde{r}_\kappa(\zeta)| ((1 + |\xi|)^{|\gamma|} + (1 + |\zeta|)^{|\gamma|}) \, d\zeta \leqslant C(1 + |\xi|)^{|\gamma|}. \end{aligned} \tag{26.71}$$

Analogously, if

$$-\kappa_- < \gamma - 2\beta < 1 + \kappa_-, \tag{26.72}$$

then a repetition of (26.69) gives

$$\int_{-\infty}^{\infty} |\tilde{r}_\kappa(\xi - \zeta)| I(\xi)(1 + |\xi|)^{2\beta}\, d\xi \leqslant C \int_{-\infty}^{\infty} |\tilde{r}_\kappa(\xi - \zeta)|(1 + |\xi|)^{-(\gamma - 2\beta)}\, d\xi$$
$$\leqslant C(1 + |\zeta|)^{-(\gamma - 2\beta)}. \tag{26.73}$$

We note that there always exists a γ satisfying both (26.68) and (26.72). In fact, since $-\kappa_- - \frac{1}{2} < \beta < \kappa_- + \frac{1}{2}$, we have $-\kappa_- + 2\beta < 1 + \kappa_-$ and $-\kappa_- < 1 + \kappa_- + 2\beta$. Consequently, for $\beta \geqslant 0$ we can choose γ so that $-\kappa_- + 2\beta < \gamma < 1 + \kappa_-$ and for $\beta < 0$ we can choose γ so that $-\kappa_- < \gamma < 1 + \kappa_- + 2\beta$. It therefore follows from (26.66) and (26.73) that

$$\|r_\kappa u\|_\beta^2 \leqslant C \int (1 + |\zeta|)^{2\beta} |\tilde{u}(\zeta)|^2\, d\zeta = C\|u\|_\beta^2.$$

LEMMA 26.6. *Let* $\kappa_- = \min_{x' \in \Gamma} \operatorname{Re} \kappa(x')$. *If* $-1 < \kappa_- \leqslant 0$ *and* $-\frac{1}{2} < \beta < \kappa_- + \frac{1}{2}$, *then*

$$\|r_\kappa u\|_\beta \leqslant C_\varepsilon \|u\|_{\beta + |\kappa_-| + \varepsilon}, \qquad \forall \varepsilon > 0. \tag{26.74}$$

If $\kappa_- \leqslant -1$ *and* $\beta < \kappa_- + \frac{1}{2}$, *then*

$$\|r_\kappa u\|_\beta \leqslant C_\varepsilon \|u\|_{|\kappa_-| - 1/2 + \varepsilon}, \qquad \forall \varepsilon > 0. \tag{26.75}$$

PROOF. As in the proof of Lemma 26.5, we need to estimate the integral (26.66). If

$$|\kappa_-| < \gamma < 1, \tag{26.76}$$

then, analogously to (26.70), we get

$$I(\xi) = \int_{-\infty}^{\infty} |\tilde{r}_\kappa(\xi - \zeta)|(1 + |\zeta|)^{-\gamma}\, d\zeta \leqslant C_\varepsilon (1 + |\xi|)^{-\gamma + |\kappa_-| + \varepsilon}, \qquad \forall \varepsilon > 0, \tag{26.77}$$

the only difference from (26.70) being that there we made use of the fact that $(1 + |\zeta|)^{-\gamma} \leqslant C(1 + |\xi|)^{-\gamma}$ for $|\zeta| > |\xi|/2$, whereas here we have made use of the fact that

$$(1 + |\zeta|)^{-\gamma} \leqslant C_\varepsilon (1 + |\xi|)^{-\gamma + |\kappa_-| + 2\varepsilon} (1 + |\xi - \zeta|)^{-|\kappa_-| - 2\varepsilon}$$

for $|\eta| > |\xi|/2$. It follows that if

$$|\kappa_-| < \gamma - |\kappa_-| - 2\beta < 1, \tag{26.78}$$

then a repetition of (26.77) gives

$$\int_{-\infty}^{\infty} |\tilde{r}_\kappa(\xi - \zeta)|(1 + |\xi|)^{-(\gamma - |\kappa_-| - 2\beta + \varepsilon)}\, d\xi \leqslant C_\varepsilon (1 + |\zeta|)^{-(\gamma - |\kappa_-| - 2\beta + \varepsilon) + |\kappa_-| + \varepsilon}, \tag{26.79}$$

where $\varepsilon > 0$ is arbitrary. We note that, since $-\frac{1}{2} < \beta < \kappa_- + \frac{1}{2}$, we can always find a γ that satisfies both (26.76) and (26.78). In fact, for $\beta \geqslant 0$ we can take $2|\kappa_-| + 2\beta < \gamma < 1$ and for $\beta < 0$ we can take $|\kappa_-| < \gamma < 1 + 2\beta + |\kappa_-|$. Thus (26.66) and (26.79) imply

$$\|r_\kappa u\|_\beta^2 \leqslant C_\varepsilon \int_{-\infty}^{\infty} (1 + |\zeta|)^{2\beta + 2|\kappa_-| + 2\varepsilon} |\tilde{u}(\zeta)|^2 \, d\zeta = C_\varepsilon \|u\|^2_{\beta + |\kappa_-| + \varepsilon}, \qquad \forall \varepsilon > 0.$$

Let us now consider the case $\kappa_- \leqslant -1$. Here $x_{n,+}^{\kappa(x')}$ is defined as a generalized function (see Example 1.6). As in (26.71), for $\gamma > |\kappa_-|$

$$\int_{-\infty}^{\infty} |\tilde{r}_\kappa(\xi - \zeta)|(1 + |\zeta|)^{-\gamma} \, d\zeta \leqslant C_\varepsilon (1 + |\xi|)^{|\kappa_-| - 1 + \varepsilon}, \qquad \forall \varepsilon > 0, \quad (26.80)$$

inasmuch as $|\tilde{r}_\kappa(\xi)| \leqslant C_{\varepsilon\mu}(1 + |\xi_n|)^{|\kappa_-| - 1 + \varepsilon}(1 + |\xi'|)^{-\mu}$. Since $\beta < \kappa_- + \frac{1}{2}$, we have $-(2\beta + |\kappa_-| - 1) > |\kappa_-|$, and hence the same estimate as in (26.80) gives

$$\int_{-\infty}^{\infty} |\tilde{r}_\kappa(\xi - \zeta)|(1 + |\xi|)^{2\beta + |\kappa_-| - 1 + \varepsilon} \, d\xi \leqslant C_\varepsilon (1 + |\zeta|)^{|\kappa_-| - 1 + 2\varepsilon}, \qquad \forall \varepsilon > 0,$$

from which it follows that

$$\|r_\kappa u\|_\beta^2 \leqslant C_\varepsilon \int_{-\infty}^{\infty} (1 + |\zeta|)^{\gamma + |\kappa_-| - 1 + \varepsilon} |\tilde{u}(\zeta)|^2 \, d\zeta \leqslant C_\varepsilon \|u\|^2_{|\kappa_-| - 1/2 + \varepsilon} \quad (26.81)$$

since γ can be chosen arbitrarily close to $|\kappa_-|$.

We note that (26.81) implies $\|r_\kappa u\|_\beta \leqslant C\|u\|_{|\beta|}$ for $\kappa_- \leqslant -1$ and $\beta < \kappa_- + \frac{1}{2}$, since then $|\kappa_-| - \frac{1}{2} < |\beta|$.

REMARK 26.4. If $\kappa(x') = \text{const}$, then the following more precise estimate holds for $-1 < \operatorname{Re}\kappa < 0$ and $-\frac{1}{2} < \beta < \frac{1}{2} + \operatorname{Re}\kappa$:

$$\|r_\kappa u\|_\beta \leqslant C\|u\|_{\beta - \operatorname{Re}\kappa}. \quad (26.82)$$

In fact, for all $\mu > 0$ we have the estimate

$$|\tilde{r}_\kappa(\xi)| \leqslant C_\mu (1 + |\xi'|)^{-\mu}(1 + |\xi_n|)^{-1 + |\operatorname{Re}\kappa|},$$

which is independent of ε. Further, (26.77) holds for $\varepsilon = 0$. In fact,

$$I(\xi) \leqslant \int_{-\infty}^{\infty} C_\mu (1 + |\xi' - \zeta'|)^{-\mu} |\xi_n - \zeta_n|^{-1 + |\operatorname{Re}\kappa|}(1 + |\zeta'| + |\zeta_n|)^{-\gamma} \, d\zeta' d\zeta_n.$$

Consequently, if $|\xi_n| > 1 + |\xi'|$, then, replacing $(1 + |\zeta'| + |\zeta_n|)^{-\gamma}$ by $|\zeta_n|^{-\gamma}$ and making the change of variables $\zeta_n = |\xi_n|t$, we get $I(\xi) \leqslant C|\xi_n|^{-\gamma + |\operatorname{Re}\kappa|}$. If, on the other hand, $|\xi_n| < 1 + |\xi'|$, then, taking $\zeta_n = (1 + |\zeta'|)t$, we get

$$I(\xi) \leqslant \int_{-\infty}^{\infty} C_\mu (1 + |\xi' - \zeta'|)^{-\mu}(1 + |\zeta'|)^{-\gamma + |\operatorname{Re}\kappa|} \, d\zeta' \leqslant C(1 + |\xi'|)^{-\gamma + |\operatorname{Re}\kappa|}.$$

Therefore, in general,

$$I(\xi) \leqslant C(1 + |\xi|)^{-\gamma + |\operatorname{Re}\kappa|}.$$

We can now repeat verbatim the proof of Lemma 26.6.

THEOREM 26.3. *Suppose* (26.1) *is a single equation and* $f \in C^\infty(\overline{G})$. *Let* u_0 *be the solution of* (26.1), *let* u_h *be the solution of* (26.63) *and let* $\alpha \in [-1, 1]$. *If* $\kappa_- = \min_{x' \in \Gamma} \operatorname{Re} \kappa(x') \geqslant 0$, *then*

$$\|u_0 - u_h\|_{\alpha/2} \leqslant C_\varepsilon h^{3/2-\alpha/2-\varepsilon}, \qquad \forall \varepsilon > 0. \tag{26.83}$$

If $\kappa_- \leqslant 0$, *then*

$$\|u_0 - u_h\|_{\alpha/2} \leqslant C_\varepsilon h^{3/2-\alpha/2+\kappa_- -\varepsilon}, \qquad \forall \varepsilon > 0. \tag{26.84}$$

PROOF. An asymptotic expansion of u_0 near the boundary Γ gives

$$u_0 = r_\kappa(C_0(x') + r_1 C_{10}(x') + r_1 \ln r_1 C_{11}(x') + \cdots) = r_\kappa w_0(x),$$

where $w_0 \in H_{3/2-\varepsilon}(\mathbf{R}^n)$, $\forall \varepsilon > 0$. It therefore follows from Lemma 26.2 that there exists a function $w_h = \Sigma_p c_p \psi_p(x)$ such that

$$\|w_0 - w_h\|_t \leqslant C_\varepsilon h^{3/2-\varepsilon-t}, \qquad \forall \varepsilon > 0, \tag{26.85}$$

where $-\frac{1}{2} - \varepsilon \leqslant t \leqslant \frac{3}{2} - \varepsilon$. Consequently, letting $v_h = r_\kappa w_h$ and taking into account Lemmas 26.5 and 26.6, we get that if $\kappa_- > 0$, then

$$\|u_0 - v_h\|_{\alpha/2} = \|r_\kappa(w_0 - w_h)\|_{\alpha/2} \leqslant C_\varepsilon h^{3/2-\alpha/2-\varepsilon}, \qquad \forall \varepsilon > 0,$$

and if $\kappa_- \leqslant 0$, then

$$\|u_0 - v_h\|_{\alpha/2} = \|r_\kappa(w_0 - w_h)\|_{\alpha/2} \leqslant C_\varepsilon h^{3/2-\alpha/2+\kappa_- -\varepsilon}, \qquad \forall \varepsilon > 0.$$

The rest of the proof of Theorem 26.3 is the same as the proof of Theorem 26.2.

THEOREM 26.4. *Suppose* $A(x, \xi)$ *is a scalar symbol of class* $\hat{D}^{(0)}_{\alpha,\gamma}(G)$ *and* $f \in C^\infty(\overline{G})$. *Let* $\alpha \in [-1, 1]$, *let* u_0 *be the solution of* (26.1) *and let* u_h *be the solution of form* (26.62) *of the system* (26.63) *with* $\kappa(x') \equiv \alpha/2$. *Then*

$$\|u_0 - u_h\|_{\alpha/2} \leqslant Ch^{2-\alpha/2} \quad \text{for } \alpha > 0,$$

$$\|u_0 - u_h\|_{\alpha/2} \leqslant Ch^{2} \quad \text{for } -1 < \alpha \leqslant 0 \tag{26.87}$$

and

$$\|u_0 - u_h\|_{-1/2} \leqslant C_\varepsilon h^{2-\varepsilon}. \tag{26.88}$$

PROOF. Since $f \in C^\infty(\overline{G})$, we have $u_0 = r_{\alpha/2} w_0$, where $w_0 \in C^\infty(\overline{G})$. It follows from Lemma 26.2 that there exists a function $w_h = \Sigma_p c_p \psi_p(x)$ such that $\|w_0 - w_h\|_t \leqslant Ch^{2-t}\|w_0\|_2$ for $0 \leqslant t < \frac{3}{2} - \varepsilon$. Therefore, when $\alpha > 0$, we have

$$\|u_0 - r_{\alpha/2} w_h\|_{\alpha/2} \leqslant Ch^{2-\alpha/2}.$$

Analogously, using Remark 26.5 for $-1 < \alpha < 0$ and Lemma 26.6 for $\alpha = -1$, we get

$$\|u_0 - r_{\alpha/2} w_h\|_{\alpha/2} \leqslant C\|w_0 - w_h\|_0 \leqslant Ch^2$$

for $-1 < \alpha \leqslant 0$, since $\kappa = \alpha/2$ and $\|u_0 - r_{-1/2}w_h\|_{-1/2} \leqslant C_\varepsilon\|w_0 - w_h\|_\varepsilon \leqslant C_\varepsilon h^{2-\varepsilon}$ for $\alpha = -1$. We note that the estimate

$$\|u_0 - r_0 w_h\|_0 \leqslant C\|w_0 - w_h\|_0 \leqslant Ch^2 \qquad \text{for } \alpha = 0$$

follows from Lemma 5.3 since $x^0_{n,+} = \theta(x_n)$. The rest of the proof is the same as for Theorem 26.2.

REMARK 26.5. Consider the case in which (26.1) is a system of p.e.'s. Suppose, as at the end of subsection 3, there exists on Γ a smooth basis whose elements $n_k(x')$, $1 \leqslant k \leqslant p$, are eigenvectors of the matrix $b(x')$. Suppose there exists a smooth extension $n_k(x)$ of each $n_k(x')$, $1 \leqslant k \leqslant p$, onto $\overline{G}$ such that the $n_k(x)$ are linearly independent for all $x \in \overline{G}$. We also assume that there exists an s.l.c.s. in $U(\Gamma)$ in which $n_k(x', x_n) = n_k(x', 0)$ for $0 < x_n < \delta/2$, and that the $n_k(x)$ are constant vectors outside $U(\Gamma)$. Let $\psi_{p_1p_2}(x)$ and $r_{\alpha/2+b_k(x')}$ be the same functions as before. The s.f.e.m. solution of (26.1) is sought in the form

$$u_h = \sum_{(p_1,p_2)\in N(h)} \sum_{k=1}^{p} c_{p_1p_2k} r_{\alpha/2+b_k(x')} n_k(x)\psi_{p_1p_2}(x), \tag{26.89}$$

where the $c_{p_1p_2k}$ are unknown numbers. We note that in $G \setminus U(\Gamma)$ (26.89) is equivalent to (26.34) since $r_{\alpha/2+b_k(x')} = 1$ and the $n_k(x)$ are constants. By using the asymptotic expansion (26.30) and Lemmas 26.5 and 26.6, we can show analogously to the case of a single equation that u_h converges to u_0 in $\mathring{H}_{\alpha/2}(G)$, with

$$\|u_0 - u_h\|_{\alpha/2} \leqslant C_\varepsilon h^{1/2+s_+-s_--\alpha/2-\varepsilon}$$

when $\kappa_- = \min_{x',\gamma}(\alpha/2 + \operatorname{Re} b_\gamma(x')) > 0$, and

$$\|u_0 - u_h\|_{\alpha/2} \leqslant C_\varepsilon h^{1/2+s_+-s_--\alpha/2-|\kappa_-|-\varepsilon}$$

when $\kappa_- \leqslant 0$, $\forall\varepsilon > 0$. If $A(x, \xi) \in \hat{D}^{(0)}_{\alpha,\gamma}(G)$, then the solution u_h is sought in the form

$$u_h = \sum_{p_1,p_2} c_{p_1p_2} r_{\alpha/2}(x)\psi_{p_1p_2}(x),$$

where the $c_{p_1p_2}$ are now vectors. Theorem 26.4 remains valid for the case in which (26.1) is a system of p.e.'s.

REMARK 26.6. Numerical results obtained by Ezra Levi of the Hebrew Univsity of Jerusalem show that the s.f.e.m. gives better results than the f.e.m. used in [53]. The matrix

$$a_{q_1q_2,p_1p_2} = (Ar_{1/2}\psi_{p_1p_2}, r_{1/2}\psi_{q_1q_2})$$

was calculated for $A(x, \xi) = |\xi|$ ($\alpha = 1$) with the use of a piecewise linear approximation of $r_{1/2}\psi_{p_1p_2}$ by functions of form (26.34) after dividing each original square of size h into 4 or 16 equal squares. The matrix

$$(Ar_{1/2}\psi_{p_1p_2}, r_{1/2}\psi_{q_1q_2})$$

is then easily calculated by making use of the results of the calculation of the matrix $(A\psi_{p_1p_2}, \psi_{q_1q_2})$ for $h_1 = h/4$. We note that the latter matrix does not depend on the domain G.

9. *A punch problem with an unknown domain of contact.* Previously in this section the finite element method was considered for pseudodifferential equations. We will now consider the f.e.m. for a pseudodifferential variational inequality (see [61] for the theory of variational inequalities). Instead of studying the general case of a variational inequality for a strongly elliptic system of p.o.'s, we will confine ourselves to a particular punch problem in the theory of elasticity.

Let $z = \varphi(x)$, $x = (x_1, x_2)$, be the surface of rigid body (punch) indenting the elastic halfspace $z < 0$. It is assumed that $\varphi(x) \geqslant 0$, $\varphi(0) = 0$ and $\varphi(x) \to +\infty$ as $|x| \to \infty$. Let G be the unknown domain of contact and let δ be the imposed displacement of the punch parallel to the z-axis. Then for $z = 0$ we have

$$w(x) = \varphi(x) - \delta, \qquad \sigma_z(x) \leqslant 0 \text{ in } G, \tag{26.90}$$

$$w(x) \leqslant \varphi(x) - \delta, \qquad \sigma_z(x) = 0 \text{ outside } G, \tag{26.91}$$

where $w(x)$ is the normal displacement of the plane $z = 0$ and $\sigma_z(x)$ is the normal stress acting on that plane. Under the assumption that the shearing strains vanish for $z = 0$ it can be shown that $\sigma_z(x)$ and $w(x)$ are connected by the equation (see, for example, [64])

$$w(x) = C\iint_{-\infty}^{\infty} \frac{\sigma_z(y)}{|x - y|}\, dy_1 dy_2, \tag{26.92}$$

where C is a constant depending on the elastic properties of the halfspace $z < 0$. Thus our punch problem is to find the domain of contact G and the pressure $u(x) = -\sigma_z(x)$ under the punch such that relations (26.90)–(26.92) are satisfied.

Let us show that our problem reduces to a variational inequality. Relations (26.90)–(26.92) can be written in the form

$$\Lambda^{-1}u \geqslant \delta - \varphi(x), \qquad \forall x \in \mathbf{R}^2, \tag{26.93}$$

$$(\Lambda^{-1}u)u(x) = (\delta - \varphi(x))u(x), \qquad \forall x \in \mathbf{R}^2, \tag{26.94}$$

where $u(x) = -\sigma_z(x) \geqslant 0$ and Λ^{-1} is the operator defined by (26.92) (i.e. $w = \Lambda^{-1}\sigma_z$). The domain of contact G is the region in which $u(x) > 0$ or in which $\Lambda^{-1}u = \delta - \varphi(x)$.

Let D be a bounded domain containing the unknown domain G. An example of such a domain D is the region in which $\varphi(x) < \delta + \delta_0$, $\delta_0 > 0$. In fact, $\Lambda^{-1}u \geqslant 0$ for $u \geqslant 0$ since Λ^{-1} is an integral operator with a positive kernel. But then $\Lambda^{-1}u \geqslant 0 > \delta - \varphi(x)$ in $\mathbf{R}^2 \setminus D$ inasmuch as $\delta - \varphi(x) \leqslant -\delta_0 < 0$ for $x \in \mathbf{R}^2 \setminus D$, which implies $G \subset D$.

Suppose $u(x) \in L_2(D)$. Multiplying (26.93) by an arbitrary nonnegative function $v(x) \in L_2(D)$ and then integrating (26.93) and (26.94), we get

$$(\Lambda^{-1}u, v) \geqslant (\delta - \varphi, v), \qquad \forall v(x) \geqslant 0, \qquad v(x) \in L_2(D), \tag{26.93$'$}$$

$$(\Lambda^{-1}u, u) = (\delta - \varphi, u), \qquad u(x) \geqslant 0. \tag{26.94$'$}$$

It is easily seen that (26.93′), (26.94′) is equivalent to (26.93), (26.94).

The operator Λ^{-1} can be written in the form (see Example 22.5)

$$\Lambda^{-1}u = \frac{C}{(2\pi)^2} \iint_{-\infty}^{\infty} \frac{1}{|\xi|} \tilde{u}(\xi) e^{-i(x,\xi)}\, d\xi. \tag{26.95}$$

We note that Λ^{-1} is strongly elliptic and

$$C_1 \|v\|^2_{-1/2} \leqslant (\Lambda^{-1}v, v) \leqslant C_2 \|v\|^2_{-1/2}, \tag{26.96}$$

for all $v \in C_0^\infty(D)$.

Since inequalities (26.96) involve the norm of $\mathring{H}_{-1/2}(D)$, it is easier to study problem (26.93′), (26.94′) in $\mathring{H}_{-1/2}(D)$ instead of $L_2(D)$. It will be proved below that $u \in \mathring{H}_{1/2}(D) \subset L_2(D)$.

Let H^+ be the closed convex set in $\mathring{H}_{-1/2}(D)$ consisting of all nonnegative generalized functions, i.e. $u \in H^+$ if and only if $(u, \varphi) \geqslant 0$ for all $\varphi \in C_0^\infty(D)$, $\varphi \geqslant 0$. We wish to find $u \in H^+$ such that

$$(\Lambda^{-1}u, v) \geqslant (\delta - \varphi, v), \qquad \forall v \in H^+, \tag{26.97}$$

$$(\Lambda^{-1}u, u) = (\delta - \varphi, u). \tag{26.98}$$

When $u \in L_2(D)$, relations (26.97) and (26.98) coincide with (26.93′) and (26.94′).

It is easily seen that (26.97) and (26.98) are equivalent to the single inequality

$$(\Lambda^{-1}u, v - u) \geqslant (\delta - \varphi, v - u), \qquad \forall v \in H^+, \tag{26.99}$$

where $u \in H^+$. In fact, (26.99) is obtained by taking the difference of (26.97) and (26.98). On the other hand, replacing v in (26.99) by $v + u$ leads to (26.97), and replacing $v - u$ by $\pm u$ leads to $\pm(\Lambda^{-1}u, u) \geqslant \pm(\delta - \varphi, u)$, which implies (26.98).

The following proposition is established with the same arguments as in [61].

LEMMA 26.7. *For each $\varphi(x) \in H_{1/2}(D)$ there exists a unique solution $u \in H^+$ of (26.99). The solution u minimizes the following functional on H^+:*

$$F(v) = \tfrac{1}{2}(\Lambda^{-1}v, v) - (\delta - \varphi, v), \tag{26.100}$$

i.e. $F(u) = \min_{v \in H^+} F(v)$.

PROOF. Since

$$|(\delta - \varphi, v)| \leqslant \|\delta - \varphi\|^{+}_{1/2}\|v\|_{-1/2} \leqslant \varepsilon\|v\|^2_{-1/2} + C_\varepsilon(\|\delta - \varphi\|^{+}_{1/2})^2,$$

it follows from (26.96) that

$$F(v) \geqslant (c_1/2 - \varepsilon)\|v\|^2_{-1/2} - C_\varepsilon(\|\delta - \varphi\|^{+}_{1/2})^2 \geqslant -C_\varepsilon(\|\delta - \varphi\|^{+}_{1/2})^2,$$

and hence the functional $F(v)$ is bounded from below. Let $\mu = \inf_{v \in H^+} F(v)$, and let $v_n \in H^+$ be a minimizing sequence, i.e. $\lim_{n \to \infty} F(v_n) = \mu$. Since $F(v)$ is a quadratic functional, we have for arbitrary v_m and v_n

$$\left(\Lambda^{-1}\left(\frac{v_m - v_n}{2}\right), \frac{v_m - v_n}{2}\right) = F(v_m) + F(v_n) - 2F\left(\frac{v_m + v_n}{2}\right). \tag{26.101}$$

We note that $(v_m + v_n)/2 \in H^+$ inasmuch as H^+ is a convex set. Since $F((v_m + v_n)/2) \geqslant \mu$, it follows from (26.96) and (26.101) that

$$C_1\|v_m - v_n\|^2_{-1/2} \leqslant 4\left[(F(v_m) - \mu) + (F(v_m) - \mu)\right]. \tag{26.102}$$

Therefore $\{v_n\}$ is a Cauchy sequence in H^+ and hence there exists a $u \in H^+$ such that $v_n \to u$ in $\mathring{H}_{-1/2}(D)$. But then $F(v_n) \to F(u)$ since F is a continuous functional in $\mathring{H}_{-1/2}(D)$. Consequently, $F(u) = \mu = \inf_{v \in H^+} F(v)$.

We note that the following useful estimate is obtained by letting $v_m \to u$ in (26.102):

$$\|v_n - u\|^2_{-1/2} \leqslant (4/C_1)(F(v_n) - F(u)). \tag{26.103}$$

The fact that u is the only function minimizing $F(v)$ follows from the fact that the quadratic functional $F(v)$ is strictly convex, i.e.

$$F((v_1 + v_2)/2) \leqslant \tfrac{1}{2}(F(v_1) + F(v_2))$$

and the equality holds if and only if $v_1 = v_2$.

Let us now prove that u is a solution of (26.99). For every $v \in H^+$ and $\theta \in (0, 1)$ we have $u + \theta(v - u) \in H^+$. Therefore

$$\frac{F(u + \theta(v - u)) - F(u)}{\theta} \geqslant 0,$$

from which, passing to the limit for $\theta \to 0$, we get

$$(\Lambda^{-1}u, v - u) \geqslant (\delta - \varphi, v - u), \qquad \forall v \in H^+.$$

It is also easily seen that a solution of (26.99) minimizes $F(v)$ on H^+. Lemma 26.7 is proved.

We now consider the application of the f.e.m. to (26.99). Let $\psi_{p_1p_2}(x)$ be the same trial function as in (26.33) and let $G(h)$ be a polygonal domain such that $G \subset G(h)$. It is assumed that $G(h)$ and D have been chosen so that $G \subset G(h) \subset D$ and the distance from $\overline{G}$ to $\partial G(h)$ is greater than $2\sqrt{2}\,h$. We wish to approximately minimize the functional (26.100) by means of a function of the form

$$v_h = \sum_{(p_1,p_2)\in N(h)} c_{p_1p_2}\psi_{p_1p_2}(x), \tag{26.104}$$

where $N(h)$ is the same as in (26.34).

Since $\psi_{p_1p_2}(p_1h, p_2h) = 1$ and $\psi_{p_1p_2}(q_1h, q_2h) = 0$ when $(p_1, p_2) \neq (q_1, q_2)$, the condition $v_h \geqslant 0$, i.e. $v_h \in H^+$, implies

$$c_{p_1p_2} \geqslant 0, \qquad \forall (p_1, p_2) \in N(h). \tag{26.105}$$

Substituting (26.104) in (26.100), we get

$$F(v_h) = \frac{1}{2}\sum_{p,\,q\in N(h)} a_{qp}c_pc_q - \sum_{p\in N(h)} (\delta h^2 - \varphi_p)c_p, \tag{26.106}$$

where $p = (p_1, p_2)$, $q = (q_1, q_2)$, $a_{qp} = (\Lambda^{-1}\psi_p, \psi_q)$ and $\varphi_p = (l\varphi, \psi_p)$. Thus, our punch problem reduces to the quadratic programming problem of minimizing the quadratic function (26.106) subject to the constraints (26.105).

We note that the matrix a_{qp} can be obtained from the matrix of subsection 6 by setting $a_0(r) = 2\pi/|r|$.

10. *Convergence of the finite element method for the punch problem.* Let

$$u_h = \sum_{p\in N(h)} C_p^{(0)}\psi_p(x)$$

be the f.e.m. solution of the punch problem, viz. the solution of the quadratic programming problem.

THEOREM 26.5. *Suppose $\varphi \in H_{3/2}(D)$. Then the f.e.m. solution u_h converges in $\mathring{H}_{-1/2}(D)$ to the solution u of inequality* (26.99). *More precisely, there exists a constant C not depending on h such that*

$$\|u - u_h\|_{-1/2} \leqslant Ch. \tag{26.107}$$

The proof of Theorem 26.5 is based on Lemma 26.7 and

LEMMA 26.8. *Suppose $\varphi(x) \in H_{3/2}(D)$. Then the solution of the variational inequality* (26.99) *belongs to $\mathring{H}_{1/2}(D)$.*

PROOF. We will establish Lemma 26.8 by means of the Brezis-Stampaccia method (see [61]).

Since $\delta - \varphi(x) < 0$ in a neighborhood of ∂D (see subsection 9), there exists an extension $f_0(x)$ of $\delta - \varphi(x)$ from D to $\mathbf{R}^2$ that has a compact support, is nonpositive in $\mathbf{R}^2 \setminus D$, belongs to $H_{3/2}(\mathbf{R}^2)$ and satisfies the inequality $\|f_0\|_{3/2} \leqslant 2\|\delta - \varphi\|^+_{3/2}$. Let $D(\Lambda^{-1/2})$ denote the set of all $v \in H_{-1/2}(\mathbf{R}^2)$ such that

$$\int\!\!\int_{-\infty}^{\infty} \frac{|\tilde{v}(\xi)|^2}{|\xi|}\, d\xi_1 d\xi_2 < +\infty.$$

If $v \in H_{-1/2}(\mathbf{R}^2)$ and has a compact support, then $v \in D(\Lambda^{-1/2})$ since $\tilde{v}(\xi)$ is analytic and hence bounded in a neighborhood of $\xi = 0$. It is obvious that the bilinear form $(\Lambda^{-1}v_1, v_2)$ is well defined for $v_1, v_2 \in D(\Lambda^{-1/2})$.

We have

$$(\Lambda^{-1}u - f_0, v - u) \geqslant 0 \tag{26.108}$$

for all $v \in D(\Lambda^{-1/2})$, $v \geqslant 0$, where u is the solution of (26.99). In fact $\Lambda^{-1}u > 0$ for all $x \in \mathbf{R}^2$, inasmuch as $u \geqslant 0$ and $u \not\equiv 0$. Therefore $\Lambda^{-1}u - f_0 \geqslant 0$ for all $x \in \mathbf{R}^2$, which implies

$$(\Lambda^{-1}u, v) \geqslant (f_0, v), \qquad \forall v \in D(\Lambda^{-1/2}). \tag{26.109}$$

On the other hand, (26.98) implies

$$(\Lambda^{-1}u, u) = (f_0, u). \tag{26.110}$$

Consequently, taking the difference of (26.109) and (26.110), we get (26.108).

It follows from (26.108) that

$$(\Lambda^{-1}v - f_0, v - u) \geqslant 0 \tag{26.111}$$

for all $v \geqslant 0$, $v \in D(\Lambda^{-1/2})$. In fact,

$$(\Lambda^{-1}v - f_0, v - u) = (\Lambda^{-1}u - f_0, v - u) + (\Lambda^{-1}(v - u), v - u) \geqslant 0.$$

Let v_ε be the solution in $\mathbf{R}^2$ of the equation

$$u = v_\varepsilon - \varepsilon\Delta v_\varepsilon,$$

so that

$$v_\varepsilon = F_\xi^{-1}\frac{\tilde{u}(\xi)}{1 + \varepsilon|\xi|^2} = G_\varepsilon * u, \qquad \text{where } G_\varepsilon = F_\xi^{-1}/(1 + \varepsilon|\xi|^2).$$

We note that $v_\varepsilon \in D(\Lambda^{-1/2})$. Since $G_\varepsilon(x)$ is a positive function (see for example [91]), we have $v_\varepsilon \geqslant 0$. But then (26.111) implies

$$(\Lambda^{-1}v_\varepsilon - f_0, v_\varepsilon - u) \geqslant 0, \tag{26.112}$$

which means, since $v_\varepsilon - u = \varepsilon\Delta v_\varepsilon$, that $(\Lambda^{-1}v_\varepsilon - f_0, \varepsilon\Delta v_\varepsilon) \geqslant 0$ or, equivalently, that

$$(\Lambda v_\varepsilon, v_\varepsilon) \leqslant -(f_0, \Delta v_\varepsilon),$$

where $\Lambda = -\Delta\Lambda^{-1}$ is the p.o. with symbol $|\xi|$. Therefore

$$\|\Lambda^{1/2}v_\varepsilon\|_0^2 \leqslant C\|\Lambda^{3/2}f_0\|_0 \, \|\Lambda^{1/2}v_\varepsilon\|_0$$

and hence $\|\Lambda^{1/2}v_\varepsilon\|_0 \leqslant C\|\Lambda^{3/2}f_0\|_0$, i.e.

$$\int_{-\infty}^{\infty} \frac{|\xi| \, |\tilde{u}(\xi)|^2}{1 + \varepsilon|\xi|^2} \, d\xi \leqslant C\|f_0\|_{3/2}^2.$$

Consequently, letting $\varepsilon \to 0$, we get $\|\Lambda^{1/2}u\|_0 \leqslant C\|f_0\|_{3/2}$, from which it follows that $u \in \mathring{H}_{1/2}(D)$.

PROOF OF THEOREM 26.5. Let $v_h = P_1 u$, where P_1 is the same operator as in Lemma 26.2. The fact that $u \geqslant 0$ implies $v_h \geqslant 0$, while the fact that supp $u \subset G$ and the distance from $\bar{G}$ to $\partial G(h)$ is greater than $2\sqrt{2}\, h$ implies supp $v_h \subset G(h)$. It follows from Lemmas 26.2 and 26.8 that

$$\|u - v_h\|_s \leqslant C_s h^{1/2-s}\|u\|_{1/2}, \qquad -\tfrac{3}{2} \leqslant s \leqslant \tfrac{1}{2}. \tag{26.113}$$

Let u_h be the f.e.m. solution. Then $F(u_h) \leqslant F(v_h)$, since u_h minimizes the functional $F(v)$ on the subspace of functions of form (26.34) subject to the constraints $v_h \geqslant 0$. Therefore, by (26.103),

$$\|u - u_h\|_{-1/2}^2 \leqslant \frac{4}{C_1}(F(u_h) - F(u)) \leqslant \frac{4}{C_1}(F(v_h) - F(u)).$$

On the other hand, we have

$$F(v_h) - F(u) = \tfrac{1}{2}(\Lambda^{-1}(v_h - u), v_h - u) + (\Lambda^{-1}u - (\delta - \varphi), v_h - u)$$

and hence

$$|F(v_h) - F(u)| \leqslant C\|v_h - u\|_{-1/2}^2 + \|\Lambda^{-1}u - (\delta - \varphi)\|_{3/2}^{+} \, \|v_h - u\|_{-3/2},$$

in which $\Lambda^{-1}u - (\delta - \varphi) \in H_{3/2}(D)$ since $u \in H_{1/2}(D)$ and $\varphi \in H_{3/2}(D)$. Thus, taking (26.113) into account,

$$|F(v_h) - F(u)| \leqslant Ch^2\|u\|_{1/2}^2 + C\|\Lambda^{-1}u - (\delta - \varphi)\|_{3/2}^{+} h^2\|u\|_{1/2} \leqslant Ch^2,$$

which implies that $\|u - u_h\|_{-1/2} \leqslant Ch$.

REMARK 26.7. Let $Q = \iint_G u(x)\,dx$ be the force exerted by the punch and let

$$Q_h = \iint_{G_h} u_h \, dx = \sum_{(p_1, p_2) \in N(h)} c_{p_1 p_2}^{(0)} h^2$$

be its approximate value. It follows from the Cauchy-Schwarz inequality and Theorem 26.5 that

$$|Q - Q_h| = |(1, u - u_h)| \leqslant \|1\|_{1/2}^{+}\|u - u_h\|_{-1/2} \leqslant C_1 h.$$

Therefore $Q_h \to Q$ as $h \to 0$.

REMARK 26.8. In books on the theory of elasticity (see [46] and [64]) the domain of contact G is determined by the condition that $u(x)$ be a bounded solution of the equation $\Lambda^{-1}u = \delta - \varphi$ in G. The solution of the variational inequality satisfies this condition under the assumption that $\varphi(x)$ is smooth and G is a domain with a smooth boundary Γ. In fact, if G is an arbitrary domain with a smooth boundary Γ, $\varphi(x) \in C^\infty(\overline{G})$ and $u(x) \in \mathring{H}_{-1/2}(G)$ satisfies the equation $\Lambda^{-1}u = \delta - \varphi$ in G, then $u(x)$ has the following behavior near the boundary in an s.l.c.s.:

$$u(x) = x_{n,+}^{-1/2}c_0(x') + u_1(x),$$

where $c_0(x') \in C^\infty(\Gamma)$ and $u_1(x)$ is bounded in G. But the fact that $u \in \mathring{H}_{1/2}(D) \subset L_2(D)$ (see Lemma 26.8) implies $c_0(x') \equiv 0$, and hence $u(x)$ is bounded.

11. *Punch problem with a given force*. Suppose that in the preceding punch problem the displacement δ is not known but the force $Q = \iint_{\mathbf{R}^2} u(x)\,dx$ is given. In this case $u(x)$ must satisfy not only the variational inequality (26.99), in which δ is now not known, but the condition

$$(1, u) = Q. \qquad (26.114)$$

LEMMA 26.9. *Let $F_0(v)$ be the functional*

$$F_0(v) = \tfrac{1}{2}(\Lambda^{-1}v, v) + (\varphi, v), \qquad (26.115)$$

and let $H_1^+ = \{v\colon v \in H^+$ and $(1, v) = Q\}$. There exists a unique solution $u \in H_1^+$ of the variational inequality (26.99) *for some $\delta > 0$. This solution minimizes $F_0(v)$ on H_1^+, i.e.*

$$F_0(u) = \min_{v \in H_1^+} F_0(v).$$

PROOF. As in Lemma 26.7, it is easily proved that the functional F_0 is minimized by a unique function $v \in H_1^+$ and that the problem of minimizing it is equivalent to the problem of solving the inequality

$$(\Lambda^{-1}u + \varphi, v - u) \geqslant 0, \qquad \forall v \in H_1^+. \qquad (26.116)$$

We determine δ by means of the relation

$$\delta = (\Lambda^{-1}u + \varphi, u)/Q. \qquad (26.117)$$

Since $(1, v) = Q$, it follows from (26.117) that

$$(\Lambda^{-1}u + \varphi, u) - (\delta, v) = 0. \qquad (26.118)$$

Taking the sum of (26.116) and (26.118), we obtain the inequality

$$(\Lambda^{-1}u, v) \geqslant (\delta - \varphi, v), \qquad \forall v \in H_1^+, \qquad (26.119)$$

which holds for all $v \in H^+$ without the constraint $(1, v) = Q$ inasmuch as it is homogeneous with respect to v. Lemma 26.9 is proved.

We note that the displacement δ is a Lagrange multiplier for the problem of minimizing $F_0(v)$ on H_1^+.

12. *Punch problem for an elastic layer.* Suppose that in the punch problem of subsection 9 the elastic halfspace $z < 0$ is replaced by an elastic layer of thickness a resting on a rigid foundation $z \leqslant -a$. Then for $z = 0$ we have (26.90) and (26.91). Under the assumption that the shearing strains vanish for $z = 0$ the functions $u(x) = -\sigma_z(x)$ and $w(x)$ for $z = 0$ are connected by the equation (see, for example, [113])

$$w(x) = Au, \tag{26.120}$$

where

$$Au = \frac{C}{(2\pi)^2} \iint_{\mathbf{R}^2} \frac{2\sinh^2 a|\xi|\tilde{u}(\xi_1, \xi_2)e^{-ix_1\xi_1 - ix_2\xi_2}}{|\xi|(2a|\xi| + \sinh 2a|\xi|)}\, d\xi_1 d\xi_2.$$

Since

$$C_1(1 + |\xi|)^{-1} \leqslant \frac{2\sinh^2 a|\xi|}{|\xi|(2a|\xi| + \sinh 2a|\xi|)} \leqslant C_2(1 + |\xi|)^{-1},$$

it follows that A is strongly elliptic and

$$C_1\|v\|^2_{-1/2} \leqslant (Av, v) \leqslant C_2\|v\|^2_{-1/2}, \qquad \forall v \in \mathring{H}_{-1/2}(D).$$

Therefore the results of subsections 9 and 10 remain valid for an elastic layer if we replace the operator Λ^{-1} by the operator $-A$.

§27. Asymptotic solution of elliptic pseudodifferential equations with a small parameter

1. In this section we will consider elliptic pseudodifferential equations depending on a small parameter. Differential equations of this kind have been considered by Višik and Ljusternik (see [110]). Many pseudodifferential equations depending on a small parameter arise in various problems of physics and mechanics (see Examples 27.2 and 27.3 below).

A. S. Demidov has considered pseudodifferential operators of class $D_\alpha^{(0)}$ with a small parameter (see [25]–[27]). Here we consider the general case by a new method that was briefly described in [41]. we note that the methods of Višik and Ljusternik and Demidov do not work in the general case, while our method, when applied to pseudodifferential operators of class $D_\alpha^{(0)}$, is simpler and yields an expression for the leading term of an asymptotic expansion that is simpler and more concise even in the case of differential equations. In addition, our method permits one to prove that a given equation is uniquely solvable if its limiting form for $\varepsilon \to 0$ is uniquely solvable.

2. *A symbol class.* Suppose $A_0(x, \xi) \in C^\infty(\mathbf{R}^n \times \mathbf{R}^n)$ and

$$A_0(x, \xi) = \sum_{k=0}^{N} a_{k0}(x, \omega) r^{m_0 - k} + A_{N0}(x, \xi), \qquad \forall N, \quad \forall r \geqslant 1, \quad (27.1)$$

where $r = |\xi|$, $\omega = \xi/|\xi|$, $a_{k0}(x, \omega) \in C^\infty$ for $x \in \mathbf{R}^n$, $|\omega| = 1$,

$$a_{00}(x, \omega) \neq 0, \qquad \forall x \in \mathbf{R}^n, \qquad |\omega| = 1, \quad (27.2)$$

and

$$|D_x^p D_\omega^q D_r^k A_{N0}(x, \xi)| \leqslant C_{pqk} r^{m_0 - N - k - 1} \quad \text{for } r \geqslant 1, \quad \forall p, q, k. \quad (27.3)$$

Suppose $A_1(x, \omega, t) \in C^\infty$ for $x \in \mathbf{R}^n$, $|\omega| = 1$ and $t \geqslant 0$ and

$$A_1(x, \omega, t) = \sum_{k=0}^{N} a_{k1}(x, \omega) t^{m_1 - k} + A_{N1}(x, \omega, t), \qquad t \geqslant 0, \quad \forall N, \quad (27.4)$$

where

$$a_{01}(x, \omega) \neq 0, \qquad \forall x \in \mathbf{R}^n, \quad |\omega| = 1, \quad (27.5)$$

and $A_{N1}(x, \omega, r)$ satisfies the estimates of form (27.3) with m_0 replaced by m_1. It is also assumed that

$$A_1(x, \omega, t) \neq 0 \qquad \forall x \in \mathbf{R}^n, \quad |\omega| = 1, \quad t \geqslant 0, \quad (27.6)$$

and

$$A_1(x, \omega, 0) \equiv 1. \quad (27.7)$$

We will consider the pseudodifferential operators with symbol of the form

$$A(x, \xi, \varepsilon) = (A_0(x, \xi) + \varepsilon A_2(x, \omega, r, \varepsilon r, \varepsilon)) A_1(x, \omega, \varepsilon|\xi|) \quad (27.8)$$

where $A_2(x, \omega, r, \varepsilon r, \varepsilon) \in C^\infty$ for $x \in \mathbf{R}^n$, $|\omega| = 1$, $r \geqslant 0$, $t \geqslant 0$ and $\varepsilon \geqslant 0$, and

$$|D_x^p D_\omega^q D_r^{k_1} D_t^{k_2} D_\varepsilon^{k_3} A_2(x, \omega, r, t, \varepsilon)| \leqslant C_{pqk_1k_2k_3}(1 + r)^{m_0 - k_1}. \quad (27.9)$$

We note that the class of symbols of form (27.8) is invariant under an arbitrary change of variables for the p.o. $A(x, D, \varepsilon)$.

We need some lemmas on the expansions of such symbols.

LEMMA 27.1. *Suppose $B(x, \omega, r, \varepsilon r, \varepsilon)$ has the same properties as $A_2(x, \omega, r, \varepsilon r, \varepsilon)$ in (27.9). Then for arbitrary N*

$$B(x, \omega, r, \varepsilon r, \varepsilon) = \sum_{k=0}^{N} \varepsilon^k B_k(x, \omega, r) + \varepsilon^{N+1} B_{N+1}(x, \omega, r, \varepsilon r, \varepsilon), \quad (27.10)$$

where

$$|D_x^p D_\omega^q D_r^{k_1} B_k(x, \omega, r)| \leqslant C_{pqk_1} r^{m_0 + k - k_1}, \qquad r \geqslant 1, \quad (27.11)$$

and $B_{N+1}(x, \omega, r, \varepsilon r, \varepsilon)$ satisfies the estimates of form (27.9) with m_0 replaced by $m_0 + N + 1$.

Lemma 27.1 readily follows from the Taylor expansion of $B(x, \omega, r, \varepsilon r, \varepsilon)$ in powers of εr and ε.

LEMMA 27.2. *Suppose $A_1(x, \omega', \omega_n, \varepsilon r)$ is a symbol satisfying* (27.4)–(27.7). *Then for arbitrary N*

$$A_1(x, \omega', \omega_n, \varepsilon r)A_1^{-1}(x, 0, \operatorname{sgn} \xi_n, \varepsilon|\xi_n|)$$
$$= 1 + \sum_{k=1}^{N} \varepsilon^k B_{1k}(x, \operatorname{sgn} \xi_n, \varepsilon|\xi_n|)b_{1k}(\xi') + \sum_{k=1}^{N} \varepsilon^k B_{2k}(x, \xi)$$
$$+ \varepsilon^{N+1}R_{N+1}(x, \omega, r, \varepsilon r, \varepsilon), \tag{27.12}$$

where

$$\omega' = \frac{\xi'}{\sqrt{|\xi'|^2 + \xi_n^2}}, \qquad \omega_n = \frac{\xi_n}{\sqrt{|\xi'|^2 + \xi_n^2}}$$

and

$$|B_{1k}(x, \operatorname{sgn} \xi_n, t)| \leq C_k(1 + t)^{-1}, \qquad |b_{1k}(\xi')| \leq (1 + |\xi'|)^{k+N},$$
$$|B_{2k}(x, \xi)| \leq C_k(1 + |\xi'|)^{2N+2}, \qquad |R_{N+1}| \leq C_N(1 + |\xi'|)^{2N+2}.$$

PROOF. Let $\chi(\xi_n) \in C_0^\infty(\mathbf{R}^1)$, $\chi(\xi_n) = 1$ for $|\xi_n| \leq 1$, and $\chi(\xi_n) = 0$ for $|\xi_n| \geq 2$. The Taylor expansion of the function

$$A_1(x, \omega, \varepsilon r)A_1^{-1}(x, 0, \operatorname{sgn} \xi_n, \varepsilon|\xi_n|)\chi(\xi_n)$$

in powers of ε is

$$A_1(x, \omega, \varepsilon r)A_1^{-1}(x, 0, \operatorname{sgn} \xi_n, \varepsilon|\xi_n|)\chi(\xi_n) = \sum_{k=0}^{N} \varepsilon^k B_{2k}^{(1)}(x, \xi) + \varepsilon^{N+1}R_{N+1}^{(1)}, \tag{27.13}$$

where $|B_{2k}^{(1)}(x, \xi)| \leq C_k(1 + |\xi'|)^k$ and $|R_{N+1}^{(1)}| \leq C(1 + |\xi'|)^{N+1}$. For $|\xi_n| > 1$ we have

$$A_1(x, \omega', \omega_n, \varepsilon r) = A_1\left[x, \frac{\xi'}{\sqrt{|\xi'|^2 + \xi_n^2}}, \frac{\xi_n}{\sqrt{|\xi'|^2 + \xi_n^2}}, \varepsilon\sqrt{|\xi'|^2 + \xi_n^2}\right]$$
$$= A_1\left(x, \frac{\xi'}{|\xi|}, \operatorname{sgn} \xi_n + r_1, \varepsilon|\xi_n| + \varepsilon r_2\right), \tag{27.14}$$

where

$$r_1 = \frac{\xi_n}{|\xi|} - \frac{\xi_n}{|\xi_n|} = -\operatorname{sgn} \xi_n \frac{|\xi'|^2}{(|\xi| + |\xi_n|)|\xi|}, \qquad r_2 = |\xi| - |\xi_n| = \frac{|\xi'|^2}{|\xi| + |\xi_n|},$$

and $|\xi| = \sqrt{|\xi'|^2 + \xi_n^2}$. Therefore Taylor's formula implies

$$A_1(x, \omega', \omega_n, \varepsilon|\xi|) = \sum_{|k_1|+k_2+k_3=0}^{N} \frac{(-\operatorname{sgn}\xi_n)^{k_2}\varepsilon^{k_3}}{k_1!k_2!k_3!} \frac{\partial^{k_1+k_2+k_3}A_1(x, 0, \operatorname{sgn}\xi_n, \varepsilon|\xi_n|)}{\partial\omega'^{k_1}\partial\omega_n^{k_2}\partial t^{k_3}}$$
$$\times \frac{\xi'^{k_1}}{|\xi|^{|k_1|}} \frac{|\xi'|^{2(k_2+k_3)}}{|\xi|^{k_2}(|\xi| + |\xi_n|)^{k_2+k_3}} + R_{N+1}^{(2)}(x, \xi, \varepsilon), \tag{27.15}$$

where

$$|R_{N+1}^{(2)}| \leqslant C_N \frac{|\xi'|^{N+1}}{|\xi|^{N+1}}\left[(1 + \varepsilon|\xi_n|)^{m_1} + \frac{\varepsilon^{N+1}|\xi'|^{N+1}}{(1 + \varepsilon|\xi_n|)^{N+1-m_1}} + (\varepsilon|\xi'|)^{m_1^-}\right], \tag{27.16}$$

$m_1^- = m_1$ if $m_1 > 0$ and $m_1^- = 0$ if $m_1 \leqslant 0$.

To complete the proof of Lemma 27.2 we need

LEMMA 27.3. *Suppose* $B(x, \operatorname{sgn}\xi_n, t) \in C^\infty$ *with respect to* $x \in \mathbf{R}^n$ *and* $t \geqslant 0$, *and suppose for all* $N \geqslant 0$

$$B(x, \operatorname{sgn}\xi_n, t) = \sum_{k=0}^{N} B_k(x, \operatorname{sgn}\xi_n)t^{-t_1-k} + B_N(x, \operatorname{sgn}\xi_n, t), \tag{27.17}$$

where $t_1 \geqslant 0$ *and* $|B_N(x, \operatorname{sgn}\xi_n, t)| \leqslant C_N(1 + t)^{-t_1-N-1}$. *Further, suppose for all* $N \geqslant 0$

$$b(\xi', \xi_n) = \sum_{k=0}^{N} b_k(\xi')|\xi_n|^{-t_2-k} + b_N(\xi', \xi_n) \qquad \text{for } |\xi_n| \geqslant 1, \tag{27.18}$$

where t_2 *is a positive integer,* $|b_k(\xi')| \leqslant C_k(1 + |\xi'|)^{t_3+k}$ and

$$|b_N(\xi', \xi_n)| \leqslant C_N(1 + |\xi'|)^{t_3+N+1}|\xi_n|^{-t_2-N-1}$$

for some real t_3. *Then for arbitrary* N

$$B(x, \operatorname{sgn}\xi_n, \varepsilon|\xi_n|)b(\xi', \xi_n) = \sum_{k=0}^{N} \varepsilon^{t_2+k}C_{k1}(x, \operatorname{sgn}\xi_n, \varepsilon|\xi_n|)C_{k2}(\xi')$$
$$+ \sum_{k=0}^{N} \varepsilon^k C_{k3}(x, \xi', \xi_n) + \varepsilon^{N+1}R_{N+1}^{(3)}(x, \xi', \xi_n, \varepsilon), \tag{27.19}$$

where

$$|C_{k1}(x, \operatorname{sgn}\xi_n, \varepsilon|\xi_n|)| \leqslant C_k(1 + \varepsilon|\xi_n|)^{-1},$$
$$|C_{k2}(\xi')| \leqslant C_k(1 + |\xi'|)^{t_3+k},$$
$$|C_{k3}(x, \xi', \xi_n)| \leqslant C_k(1 + |\xi_n|)^{-1}(1 + |\xi'|)^{t_3+N+1},$$
$$|R_{N+1}^{(3)}| \leqslant C_N(1 + \varepsilon|\xi_n|)^{-1}(1 + |\xi'|)^{t_3+N+1}.$$

PROOF. From the Taylor expansion of $B(x, \operatorname{sgn} \xi_n, t)$ in powers of t we get

$$B(x, \operatorname{sgn} \xi_n, \varepsilon|\xi_n|)b(\xi', \xi_n)\chi(\xi_n) = \sum_{k=0}^{N} \varepsilon^k C_{k4}(x, \xi', \xi_n) + \varepsilon^{N+1}R_{N+1}^{(4)}(x, \xi', \xi_n, \varepsilon), \quad (27.20)$$

where

$$C_{k4}(x, \xi', \xi_n)| \leqslant C_{k\mu}\frac{(1 + |\xi'|)^{t_3}}{(1 + |\xi_n|)^{\mu}}, \quad |R_{N+1}^{(4)}(x, \xi, \varepsilon)| \leqslant C_{N\mu}(1 + |\xi'|)^{t_3}(1 + |\xi'|)^{-\mu}$$

and μ is arbitrary. From the same Taylor expansion of $B(x, \operatorname{sgn} \xi_n, t)$ in powers of t we also get

$$\begin{aligned} &B(x, \operatorname{sgn} \xi_n, \varepsilon|\xi_n|)|\xi_n|^{-t_2-k}(1 - \chi(\xi_n)) \\ &= \sum_{p=0}^{t_2+k-1} \frac{1}{p!}\frac{\partial^p B(x, \operatorname{sgn} \xi_n, 0)}{\partial t^p}\varepsilon^p|\xi_n|^p|\xi_n|^{-t_2-k}(1 - \chi(\xi_n)) \\ &+ B_{t_2+k}(x, \operatorname{sgn} \xi_n, \varepsilon|\xi_n|)\varepsilon^{t_2+k}|\xi_n|^{t_2+k}|\xi_n|^{-t_2-k}(1 - \chi(\xi_n)), \end{aligned} \quad (27.21)$$

where $|B_{t_2+k}(x, \operatorname{sgn} \xi_n, \varepsilon|\xi_n|)| \leqslant C(1 + \varepsilon|\xi_n|)^{-1}$. Analogously,

$$\begin{aligned} &(1 - \chi(\xi_n))B(x, \operatorname{sgn} \xi_n, \varepsilon|\xi_n|)b_N(\xi', \xi_n) \\ &= \sum_{p=0}^{t_2+N} \frac{1}{p!}\frac{\partial^p B(x, \operatorname{sgn} \xi_n, 0)}{\partial t^p}\varepsilon^p|\xi_n|^p b_N(1 - \chi(\xi_n)) \\ &+ B_{N+t_2}(x, \operatorname{sgn} \xi_n, \varepsilon|\xi_n|)\varepsilon^{N+t_2}|\xi_n|^{N+t_2}b_N(\xi', \xi_n)(1 - \chi(\xi_n)), \end{aligned} \quad (27.22)$$

and the lemma follows from (27.17), (27.20), (27.21) and (27.22).

Completion of the proof of Lemma 27.2. we now apply Lemma 27.3 for the case when

$$B(x, \operatorname{sgn} \xi_n, \varepsilon|\xi_n|) = \frac{\partial^{k_1+k_2+k_3}A_1(x, 0, \operatorname{sgn} \xi_n, \varepsilon|\xi_n|)}{\partial\omega'^{k_1}\partial\omega_n^{k_2}\partial t^{k_3}}A_1^{-1}(x, 0, \operatorname{sgn} \xi_n, \varepsilon|\xi_n|)$$

and

$$b(\xi') = \frac{\xi'^{k_1}|\xi'|^{2(k_2+k_3)}}{|\xi|^{|k_1|+k_2}(|\xi| + |\xi_n|)^{k_2+k_3}}(1 - \chi(\xi_n)).$$

Since

$$|B(x, \operatorname{sgn} \xi_n, \varepsilon|\xi_n|)| \leqslant C(1 + \varepsilon|\xi_n|)^{-k_3}$$

and

$$|b(\xi')| \leqslant C(1 + |\xi'|)^{|k_1|+2k_2+2k_3}(1 + |\xi_n|)^{-|k_1|-2k_2-k_3},$$

we have $t_1 = k_3$, $t_2 = |k_1| + 2k_2 + k_3$ and $t_3 = |k_1| + 2k_2 + 2k_3$. Thus from (27.15) and the Taylor expansion of $R^{(2)}_{N+1}A_1^{-1}(x, 0, \operatorname{sgn} \xi_n, \varepsilon|\xi_n|)$ in powers of ε we obtain (27.12). Lemma 27.2 is proved.

REMARK 27.1. we will use the following refinement of Lemma 27.3.

LEMMA 27.3′. *Suppose* $B(x, \operatorname{sgn} \xi_n, t)$ *has a logarithmic behavior for* $t \to 0$, *i.e.*

$$B(x, \operatorname{sgn} \xi_n, t) = B_0(x) + \sum_{k=1}^{N} \sum_{p=0}^{k} B_{kp}(x, \operatorname{sgn} \xi_n) t^k \ln^p t + B^{(1)}_{N+1}(k, \operatorname{sgn} \xi_n, t) \tag{27.23}$$

where $|B^{(1)}_{N+1}(x, \operatorname{sgn} \xi_n, t)| \leqslant C_{N\delta} t^{N+1-\delta}$, $\forall \delta > 0$, *and suppose* $b(\xi)$ *has a logarithmic behavior for* $|\xi_n| \to \infty$, *i.e.*

$$b(\xi) = \sum_{k=0}^{N} \sum_{p=0}^{k} b_{kp}(\xi')|\xi_n|^{-t_2-k} \ln^p|\xi_n| + b^{(1)}_N(\xi', \xi_n), \tag{27.24}$$

where

$$|b^{(1)}_N(\xi', \xi_n)| \leqslant C_{N\delta}|\xi_n|^{-t_2-N-1+\delta}(1 + |\xi'|)^{t_3+N+1}$$

and t_2, t_3 *are the same as in* (27.18). *Then*

$$\begin{aligned} B(x, \operatorname{sgn} \xi_n, \varepsilon|\xi_n|) b(\xi') &= \sum_{k=0}^{N} \sum_{p=0}^{k} \varepsilon^{t_2+k} \ln^p \varepsilon C_{kp1}(x, \operatorname{sgn} \xi_n, \varepsilon|\xi_n|) C_{kp2}(\xi') \\ &+ \sum_{k=0}^{N} \sum_{p=0}^{k} \varepsilon^k \ln^p \varepsilon C_{kp3}(x, \xi', \xi_n) \\ &+ \varepsilon^{N+1-\delta} R^{(\delta)}_{N+1}(x, \xi', \xi_n, \varepsilon|\xi_n|, \varepsilon), \quad \forall \delta > 0, \end{aligned} \tag{27.25}$$

where C_{kp1}, C_{pk2}, $R^{(\delta)}_{N+1}$ *admit the same estimates as* C_{k1}, C_{k2}, $R^{(3)}_{N+1}$ *in* (27.19), *and*

$$|C_{kp3}(x, \xi', \xi_n)| \leqslant C_{kp\delta}(1 + |\xi_n|)^{-1+\delta}(1 + |\xi'|)^{t_3+N+1}, \qquad \forall \delta > 0.$$

PROOF. From (27.23) we have

$$\begin{aligned} B(x, \operatorname{sgn} \xi_n, t) &= B(x, \operatorname{sgn} \xi_n, t)(1 - \chi(t)) + B_0(x)\chi(t) \\ &+ \sum_{k=1}^{N} \sum_{p=0}^{k} B_{kp}(x, \operatorname{sgn} \xi_n) t^k \chi(t) \ln^p t + B^{(1)}_{N+1}(x, \operatorname{sgn} \xi_n, t)\chi(t). \end{aligned}$$

Consider the product

$$B^{(0)}(x, \operatorname{sgn} \xi_n, \varepsilon|\xi_n|) \ln^q \varepsilon|\xi_n| \cdot |\xi_n|^{-t_2-k} \ln^p|\xi_n|(1 - \chi(\xi_n)),$$

where $B^{(0)}(x, \operatorname{sgn} \xi_n, t)$ is a smooth function in t for $t = 0$ and has an expansion of form (27.17) for $t \to +\infty$. From the proof of Lemma 27.3 it follows that the function

$$B^{(0)}(x, \operatorname{sgn} \xi_n, \varepsilon|\xi_n|)|\xi_n|^{-t_2-k}(1 - \chi(\xi_n))$$

has an expansion of form (27.21). Multiplying this expansion by $\ln^q \varepsilon|\xi_n| \ln^p |\xi_n|$ and using the identity

$$\ln^q \varepsilon|\xi_n|\ln^p|\xi_n| = \ln^q \varepsilon|\xi_n|(\ln \varepsilon|\xi_n| - \ln \varepsilon)^p = \sum_{i=0}^{p} \frac{(-1)^i p!}{i!\,(p-i)!} \ln^i \varepsilon \ln^{q+p-i} \varepsilon|\xi_n|$$

for the last summand in (27.21) and the identity

$$\ln^q \varepsilon|\xi_n|\ln^p|\xi_n| = (\ln \varepsilon + \ln|\xi_n|)^q \ln^p|\xi_n| = \sum_{i=0}^{q} \frac{q!}{i!\,(q-i)!} \ln^i \varepsilon \ln^{p+q-i}|\xi_n|$$

for the sum in (27.21), we obtain (27.25).

3. *Factorization of the symbol* $A_1(x, \omega, \varepsilon|\xi|)$. Consider the p.o. A_1 with symbol $A_1(x, \xi/|\xi|, \varepsilon|\xi|)$ and special local coordinate systems (s.l.c.s.'s; see §21) near the boundary. Let $A_1^{(j)}(x'_{(j)}, x_n, \xi^{(j)}/|\xi^{(j)}|, \varepsilon|\xi^{(j)}|)$ be the principal symbol of the operator A_1 expressed in the jth s.l.c.s. To simplify the notation we will write (x', x_n, ξ', ξ_n) in the jth s.l.c.s. instead of $(x'_{(j)}, x_n, \xi'_{(j)}, \xi_n)$.

We factor $A_1^{(j)}(x', x_n, \xi/|\xi|, \varepsilon|\xi|)$ with respect to ξ_n:

$$A_1^{(j)}(x', x_n, \xi/|\xi|, \varepsilon|\xi|) = A_-^{(j)}(x', x_n, \varepsilon\xi', \varepsilon\xi_n)A_+^{(j)}(x', x_n, \varepsilon\xi', \varepsilon\xi_n), \qquad (27.26)$$

where the functions $A_\pm^{(j)}$ are defined by the formulas (cf. §6)

$$A_+^{(j)} = \exp\left(-i\kappa_1^{(j)}\frac{\pi}{2}\right)\hat{\Lambda}^{\kappa_1^{(j)}}{}_+(\varepsilon\xi', \varepsilon\xi_n)\exp B_+^{(j)}, \qquad (27.27)$$

$$A_-^{(j)} = \exp\left(i\frac{\pi}{2}(m_1 - \kappa_1^{(j)})\right)\hat{\Lambda}^{m_1-\kappa_1^{(j)}}{}_-(\varepsilon\xi', \varepsilon\xi_n)\exp B_-^{(j)}, \qquad (27.28)$$

in which $\kappa_1^{(j)} = \kappa_1^{(j)}(x', x_n)$ is the factorization index of A_1 and $\hat{\Lambda}_\pm(\varepsilon\xi', \varepsilon\xi_n) = \varepsilon\xi_n \pm i\sqrt{\varepsilon^2|\xi'|^2 + 1}$,

$$B_\pm^{(j)}(x, \varepsilon\xi', \varepsilon\xi_n) = \frac{i}{2\pi}\int_{-\infty}^{\infty} \frac{1}{\xi_n \pm i0 - \zeta_n} \cdot \ln \frac{A_1^{(j)}\left(x, \dfrac{\xi'}{\sqrt{|\xi'|^2 + \zeta_n^2}}, \dfrac{\zeta_n}{\sqrt{|\xi'|^2 + \zeta_n^2}}, \varepsilon\sqrt{|\xi'|^2 + \zeta_n^2}\right)}{\exp\left(i\frac{\pi}{2}(m_1 - 2\kappa_1^{(j)})\right)(\hat{\Lambda}_+(\varepsilon\xi', \varepsilon\zeta_n))^{\kappa_1^{(j)}}\hat{\Lambda}^{m_1-\kappa_1^{(j)}}(\varepsilon\xi', \varepsilon\zeta_n)}\, d\zeta_n.$$

It is assumed that the $A_\pm^{(j)}$ are chosen so that $A_\pm^{(j)}(x, 0, 0) = 1$ (cf. condition (6.6)), which means that $A_1^{(j)}(x', x_n, \xi/|\xi|, 0) = 1$ and hence that $\varepsilon^{-m_1}A_1^{(j)}(x', x_n, \xi/|\xi|, \varepsilon|\xi|)$ is a homogeneous function of $(\xi, 1/\varepsilon)$ of degree m_1

that is differentiable on the sphere $|\xi|^2 + 1/\varepsilon^2 = 1$. It therefore follows from the proofs of Lemmas 6.1 and 17.1 that $\varepsilon^{-\kappa_1^{(j)}} A_+^{(j)}$ and $\varepsilon^{-m_1+\kappa_1^{(j)}} A_-^{(j)}$ are also homogeneous functions of $(\xi, 1/\varepsilon)$ and that the derivatives of $A_\pm^{(j)}$ satisfy the estimates

$$\left|\frac{\partial}{\partial \xi_i} A_+^{(j)}\right| \leqslant C_\delta \varepsilon^{1-\delta}(1 + \varepsilon|\xi|)^{\operatorname{Re}\kappa_1^{(j)}-1+\delta}, \qquad 1 \leqslant i \leqslant n, \quad \forall \delta > 0, \tag{27.30}$$

$$\left|\frac{\partial}{\partial \xi_i} A_-^{(j)}\right| \leqslant C_\delta \varepsilon^{1-\delta}(1 + \varepsilon|\xi|)^{m_1-\operatorname{Re}\kappa_1^{(j)}+\delta}, \qquad 1 \leqslant i \leqslant n, \quad \forall \delta > 0. \tag{27.31}$$

We will establish expansions of the $A_\pm^{(j)}$ analogous to (27.12).

An application of Lemma 27.2 to the function

$$C(x', \varepsilon\xi', \varepsilon\zeta_n) = A_1^{(j)}\left(x, \frac{\xi'}{\sqrt{|\xi'|^2 + \zeta_n^2}}, \frac{\zeta_n}{\sqrt{|\xi'|^2 + \zeta_n^2}}, \varepsilon\sqrt{|\xi'|^2 + \zeta_n^2}\right)$$
$$\cdot \hat{\Lambda}_+^{-\kappa_1^{(j)}}(\varepsilon\xi', \varepsilon\zeta_n)\hat{\Lambda}_-^{\kappa_1^{(j)}-m_1}(\varepsilon\xi', \varepsilon\zeta_n)$$

yields the representation

$$C(x, \varepsilon\xi', \varepsilon\zeta_n) = A_1^{(j)}(x, 0, \operatorname{sgn}\zeta_n, \varepsilon|\zeta_n|)\hat{\Lambda}_+^{-\kappa_1^{(j)}}(0, \varepsilon\zeta_n)$$
$$\cdot \hat{\Lambda}_-^{\kappa_1^{(j)}-m_1}(0, \varepsilon\zeta_n)(1 + \varepsilon C_0), \tag{27.32}$$

$$C_0 = \sum_{k=1}^{N} \varepsilon^{k-1} C_{1k}^{(0)}(x, \operatorname{sgn}\zeta_n, \varepsilon|\zeta_n|) C_{2k}^{(0)}(\xi') + \sum_{k=1}^{N} \varepsilon^{k-1} C_{3k}^{(0)}(x, \xi', \zeta_n)$$
$$+ \varepsilon^N R_{N+1}(x, \xi', \zeta_n, \varepsilon|\zeta_n|, \varepsilon), \tag{27.33}$$

where $C_{1k}^{(0)}$, $C_{2k}^{(0)}$, $C_{3k}^{(0)}$ and R_{N+1} have the same properties as B_{1k}, b_{1k}, B_{2k} and R_{N+1} in (27.12). Taking the Taylor expansion of $\ln(1 + \varepsilon C_0)$ in powers of εC_0 and applying Lemma 27.3, we obtain an expansion of form (27.12) for $\ln(1 + \varepsilon C_0)$. Then, applying the Cauchy operators $\Pi^\pm$ in the right side of (27.29), we obtain the following expansion of form (27.12) for the $B_\pm^{(j)}$ (we note that the result of applying $\Pi^\pm$ to a function depending on $\varepsilon\xi_n$ is also a function depending on $\varepsilon\xi_n$, as can be verified by making the change of variables $\varepsilon\xi_n = \hat{\xi}_n$, $\varepsilon\zeta_n = \hat{\zeta}_n$):

$$B_\pm^{(j)}(x, \varepsilon\xi)\left(B_\pm^{(j)}(x, 0, \varepsilon\xi_n)\right)^{-1} = \sum_{k=1}^{N} \varepsilon^k B_{k1}^\pm(x, \operatorname{sgn}\xi_n, \varepsilon|\xi_n|) b_k^\pm(\xi')$$
$$+ \sum_{k=1}^{N} \varepsilon^k B_{k2}^\pm(x, \xi) + \varepsilon^{N+1} R_{N+1}^\pm(x, \xi, \varepsilon). \tag{27.34}$$

Since we do not assume that $A_1^{(j)}(x, \xi/|\xi|, \varepsilon|\xi_n|)$ is of class $\hat{D}_{m_1}^{\infty}$, it turns out, as in §9, that the asymptotic expansion of $B_{k1}^{\pm}(x, \operatorname{sgn} \xi_n, \varepsilon|\xi_n|)$ for $\varepsilon|\xi_n| \to \infty$ and the asymptotic expansion of $B_{k2}^{\pm}(x, \xi', \xi_n)$ for $|\xi_n| \to \infty$ can contain logarithmic terms, as in (9.29). Analogously, since the derivative

$$\frac{\partial^{k_1+k_2+k_3} A_1(x, 0, \operatorname{sgn} \xi_n, \varepsilon|\xi_n|)}{\partial \omega'^{k_1} \partial \omega_n^{k_2} \partial t^{k_3}}$$

can be discontinuous at the point $\varepsilon|\xi_n| = 0$, i.e. it can have different limits as $\varepsilon|\xi_n| \to 0$, for $\xi_n > 0$ and $\xi_n < 0$, the expansions for $\varepsilon|\xi_n| \to 0$ of the Cauchy integrals $B_{\pm}^{(j)}$ can contain terms with the logarithmic factors $\ln(\varepsilon\xi_n \pm i0)$. It follows upon taking the Taylor expansion of $\exp B_{\pm}^{(j)}$ in powers of $B_{\pm}^{(j)}$ and applying Lemma 27.3′ that the $A_{\pm}^{(j)}$ have the expansions

$$\begin{aligned} &A_{\pm}^{(j)}(x, \varepsilon\xi', \varepsilon\xi_n)\left(A_{\pm}^{(j)}(x, 0, \varepsilon\xi_n)\right)^{-1} \\ &\quad = 1 + \sum_{k=1}^{N} \sum_{p=0}^{k} \varepsilon^k \ln^p \varepsilon A_{1kp}^{\pm}(x, \varepsilon\xi_n \pm i0) a_{kp}^{\pm}(\xi') \\ &\qquad + \sum_{k=1}^{N} \sum_{p=0}^{k} \varepsilon^k \ln^p \varepsilon A_{2kp}^{\pm}(x, \xi', \xi_n) + \varepsilon^{N+1-\delta} R_{\delta,N+1}^{\pm}(x, \xi', \xi_n, \varepsilon), \\ &\qquad\qquad \forall \delta > 0, \ \forall N, \end{aligned} \tag{27.35}$$

in which for $\varepsilon\xi_n \to \infty$

$$A_{1kp}^{\pm}(x, \varepsilon\xi_n \pm i0) = \sum_{r=0}^{N} \sum_{m=0}^{r+1} a_{kp2m}^{\pm}(x)(\varepsilon\xi_n \pm i0)^{-1-r} \ln^m(\varepsilon\xi_n \pm i0) + R_{kpN}^{(1)}, \quad \forall N, \tag{27.36}$$

and for $\varepsilon\xi_n \to 0$

$$\begin{aligned} &A_{1kp}^{\pm}(x, \varepsilon\xi_n \pm i0) \\ &\quad = b_{kp1}^{\pm}(x) + \sum_{r=1}^{N} \sum_{m=0}^{r} b_{kprm}^{\pm}(x)(\varepsilon\xi_n \pm i0)^r \ln^m(\varepsilon\xi_n \pm i0) + R_{kpN}^{(2)}, \quad \forall N, \end{aligned} \tag{27.37}$$

the $A_{2kp}^{\pm}(x, \xi', \xi_n)$ have analogous asymptotic expansions for $\xi_n \to \infty$ and $\xi_n \to 0$, and

$$|A_{2kp}^{\pm}(x, \xi', \xi_n)| \leqslant C_N(1 + |\xi'|)^{2N+2}, \qquad |a_{kp}^{\pm}(\xi')| \leqslant C_k(1 + |\xi'|)^{k+N},$$

$$|R_{\delta,N+1}^{\pm}| \leqslant C_\delta(1 + |\xi'|)^{2N+2}, \qquad \forall \delta > 0,$$

In order to simplify the notation we omitted the index j in the right side of (27.35). we note that expansions of form (27.35) also hold for the derivatives of the $A_{\pm}^{(j)}$ with respect to x and $\xi = (\xi', \xi_n)$ and can be obtained by differentiating (27.35).

REMARK 27.2. If $A_1(x, 0, \operatorname{sgn} \xi_n, \varepsilon|\xi_n|)$ and all of the derivatives of $A_1(x, \omega', \omega_n, \varepsilon|\xi_n|)$ with respect to ω', ω_n and $t = \varepsilon|\xi_n|$ are continuous at the point $\xi_n = 0$ when $\omega' = 0$ and $\omega_n = \operatorname{sgn} \xi_n$, then there will not exist any logarithmic terms in ε in the expansion (27.35). We note that such a condition on $A_1(x, \xi/|\xi|, \varepsilon|\xi|)$ is similar to conditions (10.2) and (10.2′) (see the definitions of the classes $D_{\alpha+i\beta}$ and $\hat{D}_{\alpha+i\beta}^{\infty}$ in §§10 and 23).

4. ***Differential equation with a small parameter in a bounded domain.*** Let G be a bounded domain in $\mathbf{R}^n$ with a smooth boundary Γ, and consider in G the p.e.

$$pA(x, D, \varepsilon)u_\varepsilon = f, \tag{27.38}$$

where $A(x, D, \varepsilon)$ is a p.o. with symbol $A(x, \xi, \varepsilon)$ of form (27.8) and p is the operator of restriction to G.

We introduce some function spaces depending on a parameter ε_0 that are appropriate for the structure (27.8) of $A(x, \xi, \varepsilon)$. Let $H^{s,r}(\mathbf{R}^n)$ be the space of generalized functions $u_\varepsilon(x)$ in $\mathbf{R}^n$ with finite norm

$$\langle u \rangle_{s,r}^2 = \sup_{0<\varepsilon<\varepsilon_0} \int_{-\infty}^{\infty} (1 + \varepsilon|\xi|)^{2s}(1 + |\xi|)^{2r}|\tilde{u}_\varepsilon(\xi)|^2 \, d\xi \tag{27.39}$$

for some $\varepsilon > 0$. Let $s(x)$ be a smooth function in $\mathbf{R}^n$ and $\Lambda_0^{s(x)}(D)$ be the p.o. with the symbol $(1 + |\xi|^2)^{s(x)/2}$. A function space depending on a parameter ε whose members have a variable order of smoothness is obtained by letting $H^{s(x),r}(\mathbf{R}^n)$ denote the space of generalized functions with finite norm

$$\langle u \rangle_{s(x),r} = \sup_{0<\varepsilon<\varepsilon_0} \|\Lambda_0^{s(x)}(\varepsilon D)\Lambda_0^r(D)u\|_0. \tag{27.40}$$

The spaces $\mathring{H}^{s(x),r}(G)$ and $H^{s(x),r}(G)$ are defined analogously to $\mathring{H}_r(G)$ and $H_r(G)$ (see §21).

Let $\kappa_0(x')$ be the factorization index of the principal symbol $A^{(0)}(x, \xi)$ (see §21) of $A_0(x, \xi)$: $A^{(0)}(x, \xi) = a_{00}(x, \omega)|\xi|^{m_0}$ (see (27.1)). we assume for simplicity that the oscillation of $\operatorname{Re} \kappa_0(x')$ on Γ is less than 1. There then exists an $r \in \mathbf{R}^1$ such that

$$|r - \operatorname{Re} \kappa_0(x')| < \tfrac{1}{2}, \qquad x' \in \Gamma. \tag{27.41}$$

Let $\kappa_1^{(j)}(x'_{(j)}, x_n)$ be the factorization index of $A_1^{(j)}$ in the jth s.l.c.s. It follows from the results of §21.2 that there exists a C^∞ function $\kappa_1(x', x_n)$ in $U_\delta = U(\Gamma) \approx \Gamma \times [0, \delta)$ that coincides with $\kappa_1^{(j)}(x'_{(j)}, x_n)$ in each jth s.l.c.s. we take an arbitrary C^∞ extension of $\kappa_1(x)$ onto G. If $\kappa_1(x) = \text{const}$ in U_δ, we extend $\kappa_1(x)$ so that $\kappa_1(x) = \text{const}$ in G.

We note that the limiting equation

$$pA_0(x, D)u_0 = f \tag{27.42}$$

is normally solvable for $u_0 \in \mathring{H}_r(G)$ when $f \in H_{r-m_0}(G)$ and r satisfies (27.41) (see Theorem 22.3).

The following theorem holds, the proof of which contains an algorithm for the approximate solution of equation (27.38).

THEOREM 27.1. *Suppose* $A(x, \xi, \varepsilon)$ *is of form* (27.8), *where* A_0, A_1 *and* A_2 *satisfy conditions* (27.1)–(27.9), *and suppose* (27.42) *has a unique solution* $u_0 \in \mathring{H}_r(G)$ *for all* $f \in H_{r-m_0}(G)$. *Then there exists an* $\varepsilon_0 > 0$ *such that* (27.38) *has a unique solution* $u_\varepsilon \in \mathring{H}^{\kappa_1(x),r}(G)$ *for all* $f \in H^{\kappa_1(x)-m_1, r-m_0}(G)$ *and* $\epsilon < \epsilon_0$.

PROOF. We begin by constructing operators $A_\pm^{(-1)}$ that transform the singular perturbation problem (27.38) into a regular perturbation problem.

Let $\tau(x_n) \in C_0^\infty(\mathbf{R}^1)$, $\tau(x_n) = 1$ for $x_n < \delta$ and $\tau(x_n) = 0$ for $x_n \geqslant 2\delta$, where δ is small. In each jth s.l.c.s. we have for $0 \leqslant x_n < \delta$ the factorization of $A_1^{(j)}$ given by (27.26)–(27.28). Since this factorization is unique, we have in $T^*(U_\delta) \approx T^*(\Gamma) \times [0, \delta) \times \mathbf{R}^1$ two well defined functions $A_+(x', x_n, \xi', \xi_n)$ and $A_-(x', x_n, \xi', \xi_n)$ that are equal to $A_+^{(j)}$ and $A_-^{(j)}$ in each jth s.l.c.s. For simplicity we denote a cotangent vector at a point $x' \in \Gamma$ by ξ' instead of $\xi'_{x'}$ (see subsections 2 and 5 of §21). we define functions $\mathring{A}_\pm(x', x_n, \xi', \xi_n)$ in $T^*(U_{3\delta}) \approx T^*(\Gamma) \times [0, 3\delta) \times \mathbf{R}^1$ by the formulas

$$\mathring{A}_+(x', x_n, \varepsilon\xi', \varepsilon\xi_n) = A_+^{\tau(x_n)}(x', x_n, \varepsilon\xi', \varepsilon\xi_n)(1 + \varepsilon^2|\xi|^2)^{((1-\tau(x_n))/2)\kappa_1(x)} \tag{27.43}$$

$$\mathring{A}_-(x, \varepsilon\xi', \varepsilon\xi_n) = A_-^{\tau(x_n)}(x, \varepsilon\xi', \varepsilon\xi_n)(1 + \varepsilon^2|\xi|^2)^{((1-\tau(x_n))/2)(m_1-\kappa_1)}. \tag{27.44}$$

Thus, for $2\delta \leqslant x_n < 3\delta$ we have

$$\mathring{A}_+(x', x_n, \xi', \xi_n) = (1 + \varepsilon^2|\xi|^2)^{\kappa_1(x', x_n)/2},$$

$$\mathring{A}_-(x', x_n, \xi', \xi_n) = (1 + \varepsilon^2|\xi|^2)^{(m_1-\kappa_1(x', x_n))/2}.$$

We note that $\mathring{A}_\pm = A_\pm$ for $0 \leqslant x_n < \delta$.

When we return from the jth s.l.c.s. to the original (global) coordinate system in $G \subset \mathbf{R}^n$, we find that the p.o. with symbol $(1 + \varepsilon^2|\xi|^2)^{\kappa_1/2}$ in the jth s.l.c.s. is transformed into a p.o. with principal symbol $(1 + \varepsilon^2|t_j(x)\xi|^2)^{\kappa_1/2}$, $t_j(x) = ((Ds_j(x)/Dx)^*)^{-1}$ and $y = s_j(x)$ is the transformation from the original coordinate system to the jth s.l.c.s. (see Theorem 19.6). Thus, in the original coordinate system we have well defined symbols

$$(1+\varepsilon^2|t(x)\xi|^2)^{\kappa_1/2} \qquad \text{and} \qquad (1+\varepsilon^2|t(x)\xi|^2)^{(m_1-\kappa_1)/2}$$

for $x \in U_{3\delta} - U_{2\delta}$ that will be denoted by $\mathring{A}_+(x, \varepsilon\xi)$ and $\mathring{A}_-(x, \varepsilon\xi)$. we next smoothly extend $\mathring{A}_\pm(x, \varepsilon\xi)$ from $U_{3\delta} \setminus U_\delta$ to $G \setminus U_{3\delta}$.

We now define the operators

$$A_\pm^{(-1)} v = \sum_j{}' \psi_j R_j^\pm \varphi_j v + \sum_j{}'' \psi_j A_\pm^{(-1)} \varphi_j v, \qquad v \in C_0^\infty(\mathbf{R}^n), \quad (27.45)$$

where $\varphi_j(x)$ and $\psi_j(x)$ are the same functions as in §22, Σ' denotes a summation over those j for which $0 \leqslant x_n \leqslant 3\delta$ in U_j, $R_j^\pm$ are the p.o.'s with symbols $R_j^\pm = (\mathring{A}_\pm^{(j)})^{-1}$ in the jth s.l.c.s., Σ_j'' denotes a summation over those j for which $x_n > 2\delta$ in U_j, and $A_\pm^{(-1)}$ are the p.o.'s with symbols $\mathring{A}_\pm^{-1}(x, \varepsilon\xi)$ in the original (global) coordinate system.

At this point in the proof of Theorem 27.1 we need

LEMMA 27.4. *For $\varepsilon > 0$ sufficiently small and for arbitrary $s(x)$ and r the operator $A_+^{(-1)}$ isomorphically maps $\mathring{H}^{s(x),r}(G)$ onto $\mathring{H}^{s(x)+\kappa_1(x),r}(G)$ and the operator $pA_-^{(-1)}lf$ isomorphically maps $H^{s(x),r}(G)$ onto $H^{s(x)+m_1-\kappa_1(x),r}(G)$, where lf is an arbitrary extension of $f(x)$.*

PROOF. Since $(A_+^{(j)}(x', x_n, \varepsilon\xi', \varepsilon\xi_n))^{-1}$ is analytic in $\zeta_n = \xi_n + i\tau$ for $\tau > 0$ and $x_n < \delta$, the operator $A_+^{(-1)}$ has the following property (cf. Theorem 4.4 and Lemma 20.2): if supp $u \subset \overline{G}$, then supp $A_+^{(-1)}u \subset \overline{G}$. we call such operators "*plus*" *operators* (see §20). It follows from Theorem 18.1 (when $\kappa_1(x)$ and $s(x)$ are constants) or from Theorem 18.3 (when $\kappa_1(x)$ and $s(x)$ are functions) that $A_+^{(-1)}$ is a bounded operator from $\mathring{H}^{s(x),r}(G)$ into $\mathring{H}^{s(x)+\kappa_1(x),r}(G)$.

Let A_+ denote the operator defined by the formula analogous to (27.45) in which $\mathring{A}_+^{(j)}$ is taken instead of $(\mathring{A}_+^{(j)})^{-1}$ for j such that $x_n < 3\delta$ in U_j and $\mathring{A}_+(x, \varepsilon\xi)$ is taken instead of $\mathring{A}_+^{-1}(x, \varepsilon\xi)$ for j such that $x_n > 2\delta$ in U_j. Clearly, A_+ is also a "plus" operator.

From (27.30) and Theorem 18.3 it follows that

$$A_+ A_+^{(-1)} v = v + C_+ v, \quad (27.46)$$

where C_+ is a "plus" operator and

$$\langle C_+ v\rangle_{\kappa_1(x)+s(x),r} \leqslant C_\delta \varepsilon^{1-\delta} \langle v\rangle_{s(x),r}, \qquad \forall\delta > 0. \quad (27.47)$$

Hence the operator $I + C_+$ is invertible for ε sufficiently small, and $(I + C_+)^{-1}A_+$ is a left inverse of $A_+^{(-1)}$. Analogously, one can prove that there exists a right inverse of $A_+^{(-1)}$. Therefore $A_+^{(-1)}$ maps $\mathring{H}^{s(x),r(x)}(G)$ isomorphically onto $\mathring{H}^{s(x)+\kappa_1(x),r}(G)$.

On the other hand, $A_-^{(-1)}$ is a "*minus*" *operator*, i.e. $A_-^{(-1)}v = 0$ in G if $v = 0$ in G. Therefore $pA_-^{(-1)}lf$ does not depend on the extension lf of f (cf. Lemma 4.6). In the same way as for $A_+^{(-1)}$, we can prove that $pA_-^{(-1)}lf$ is invertible for small ε as an operator from $H^{s(x),r}(G)$ into $H^{s(x)+m_1-\kappa_1(x),r}(G)$. Lemma 27.4 is proved.

Completion of the proof of Theorem 27.1. Let lf be an arbitrary extension of f onto $\mathbf{R}^n$. From (27.38) we have

$$A(x, D, \varepsilon)u_\varepsilon = lf + f_\varepsilon^-, \tag{27.48}$$

where $f_\varepsilon^- = 0$ for $x \in G$. we apply the operator $pA_-^{(-1)}$ to (27.48) and express the solution of (27.38) in the form

$$u_\varepsilon = A_+^{(-1)} v_\varepsilon, \tag{27.49}$$

where $v_\varepsilon \in \mathring{H}_r(G)$. Then, since $pA_-^{(-1)}f_\varepsilon^- = 0$,

$$pA_-^{(-1)}AA_+^{(-1)} v_\varepsilon = pA_-^{(-1)}lf, \tag{27.50}$$

where $pA_-^{(-1)}lf \in H_{r-m_0}(G)$.

Since $A_1(x, \omega, 0) = 1$ and $A_\pm^{(j)}(x, 0, 0) = 1$, it follows from (27.27)–(27.29) that

$$\left|\frac{\partial}{\partial x_k} A_+^{(j)}(x', x_n, \varepsilon\xi', \varepsilon\xi_n)\right| \leqslant C_\delta \varepsilon^{1-\delta}|\xi|(1 + \varepsilon|\xi|)^{\operatorname{Re}\kappa_1(x)-1+\delta}, \qquad \forall\delta > 0, \tag{27.51}$$

$$\left|\frac{\partial}{\partial x_k} A_-^{(j)}(x', x_n, \varepsilon\xi', \varepsilon\xi_n)\right| \leqslant C_\delta \varepsilon^{1-\delta}|\xi|(1 + \varepsilon|\xi|)^{m_1-\operatorname{Re}\kappa_1(x)-1+\delta}, \qquad \forall\delta > 0, \tag{27.52}$$

where $1 \leqslant k \leqslant n$. These estimates are a consequence of the fact that the $A_\pm^{(j)}$ are Hölder continuous of order $1 - \delta$ on the sphere $|\xi|^2 + 1/\varepsilon^2 = 1$, $\forall\delta > 0$. Using (27.30), (27.31), (27.51) and (27.52) and applying Theorem 18.3, we get

$$p(A_0 + A_3)v_\varepsilon = pA_-^{(-1)}lf, \tag{27.53}$$

where

$$\|A_3 v_\varepsilon\|_{s-m_0} \leqslant C_\delta \varepsilon^{1-\delta}\|v_\varepsilon\|_s, \qquad \forall s, \quad \forall\delta > 0. \tag{27.54}$$

Thus the original singular perturbation problem of solving equation (27.38) has been transformed by means of the operators $A_+^{(-1)}$ and $A_-^{(-1)}$ into the problem of solving (27.53), which is a regular perturbation problem since $\operatorname{ord} A_3 \leqslant \operatorname{ord} A_0$.

Since (27.42) has a unique solution, there exists a bounded operator R_0 from $H_{r-m_0}(G)$ into $\mathring{H}_r(G)$ that is the inverse of pA_0. Therefore

$$v_\varepsilon + R_0 pA_3 v_\varepsilon = R_0 pA_-^{(-1)}lf. \tag{27.55}$$

But R_0pA_3 is bounded and has a small norm as an operator in $\mathring{H}_r(G)$ when ε is small. Consequently, $(I + R_0pA_3)^{-1}$ exists for small ε, and hence

$$v_\varepsilon = (I + R_0pA_3)^{-1}R_0pA_-^{(-1)}lf = \sum_{k=0}^{\infty}(-1)^k(R_0pA_3)^kR_0pA_-^{(-1)}lf \tag{27.56}$$

is the unique solution of (27.53) in $\mathring{H}_r(G)$, while

$$u_\varepsilon = A_+^{(-1)}\,v_\varepsilon = A_+^{(-1)}\,(I + R_0pA_3)^{-1}R_0pA_-^{(-1)}lf \tag{27.57}$$

is the unique solution of (27.38) in $\mathring{H}^{\kappa_1(x),r}(G)$.

Since the series (27.56) converges in the operator norm, formulas (27.56) and (27.57) provide a means of approximately calculating the solution u_ε. In subsections 5 and 7 below, under the assumption that $f \in C^\infty(\overline{G})$, we will give a simpler and more explicit way of calculating the asymptotic behavior of u_ε.

REMARK 27.3. The existence and uniqueness theorems for pseudodifferential operators with a small parameter have much in common with the existence and uniqueness theorems for parabolic equations (see [21]). The presence of a parameter allows one to prove existence and uniqueness theorems instead of normal solvability theorems (cf. Lemma 27.4).

5. *An explicit expression for the principal asymptotic term.* Suppose $f \in H^{\kappa_1(x)-m_1,r_0-m_0}(G)$. Then (27.57) implies

$$u_\varepsilon = A_+^{(-1)}\,v_\varepsilon^{(0)} + u_\varepsilon^{(1)}. \tag{27.58}$$

Here $v_\varepsilon^{(0)}$ is the unique solution in $\mathring{H}_r(G)$ of the equation

$$pA_0v_\varepsilon^{(0)} = pA_-^{(-1)}lf \tag{27.59}$$

and

$$\langle u_\varepsilon^{(1)}\rangle_{\kappa_1(x),r} \leqslant C_\delta\varepsilon^{1-\delta}, \qquad \forall\delta > 0, \tag{27.60}$$

where C_δ clearly depends on $f(x)$.

Suppose now that $f \in C^\infty(\overline{G})$, and take $lf \in C^\infty(\mathbf{R}^n)$. The symbols

$$R_j^{\pm}(x', x_n, \varepsilon\xi', \varepsilon\xi_n) = \left(A_\pm^{(j)}(x', x_n, \varepsilon\xi', \varepsilon\xi_n)\right)^{-1}$$

for $x_n < \delta$ (see (27.45)) have expansions of form (27.35) in which the $A_{1kp}^{\pm}(x, \varepsilon\xi_n \pm i0)$ satisfy (27.37) for $\varepsilon\xi_n \to 0$. Hence

$$pA_-^{(-1)}lf = f + \sum_{k=1}^{N}\sum_{p=0}^{k}\varepsilon^k\ln^p\varepsilon f_{kp}(x) + \varepsilon^{N+1-\delta}f_{\delta,N}(x,\varepsilon), \tag{27.61}$$

where $\|f_{\delta,N}(x,\varepsilon)\|_s^+ \leqslant C_s$, $\forall s$, and $f_{kp}(x) \in C^\infty(\overline{G})$. It follows from Remarks 16.2 and 25.2 that if (27.42) is uniquely solvable for some $r \in (\operatorname{Re}\kappa_0(x') - \frac{1}{2}, \operatorname{Re}\kappa_0(x') + \frac{1}{2})$, then it is uniquely solvable for any $r' \in (\operatorname{Re}\kappa_0(x') - \frac{1}{2}, \operatorname{Re}\kappa_0(x') + \frac{1}{2})$, provided that $f \in H_{r'-m_0}(G)$. Therefore, since $pA_-^{(-1)}lf \in C^\infty(\overline{G})$, we can take $r' = r_+ - \delta$, where $r_+ = \min_{x'\in\Gamma}\operatorname{Re}\kappa_0(x') + \frac{1}{2}$ and

$\delta > 0$ is arbitrarily small. Thus (27.57) and (27.61) imply

$$u_\varepsilon = A_+^{(-1)} u_0 + u_\varepsilon^{(2)}, \tag{27.62}$$

where u_0 is the solution of (27.42) in $\mathring{H}_{r_+-\delta}(G)$, $\forall\delta > 0$, and $\langle u_\varepsilon^{(2)}\rangle_{\kappa_1(x),r_+-\delta} < C_\delta\varepsilon^{1-\delta}$, $\forall\delta > 0$.

Since u_0 is a C^∞ function with respect to x' in each s.l.c.s. (see §26), we can represent $R_j^+(x, \varepsilon\xi', \varepsilon\xi_n) = (A_+^{(j)}(x, \varepsilon\xi', \varepsilon\xi_n))^{-1}$ by the expansion (27.35) for $x_n < \delta$, and therefore

$$u_\varepsilon = A_+^{(-1)}(x', x_n, 0, \varepsilon D_n)u_0 + u_\varepsilon^{(3)}, \tag{27.63}$$

where $A_+^{(-1)}(x', x_n, 0, \varepsilon D_n)$ is the p.o. with symbol $A_+^{-1}(x', x_n, 0, \varepsilon\xi_n)$ and $u_\varepsilon^{(3)}$ has the same estimate as $u_\varepsilon^{(2)}$. we note that the symbol $A_+(x', x_n, 0, \varepsilon\xi_n)$ that equals $A_+^{(j)}(x', x_n, 0, \varepsilon\xi_n)$ in each jth s.l.c.s. is a well defined function for $x' \in \Gamma$, $x_n \in [0, \delta)$, $\xi_n \in \mathbf{R}^1$.

The fact that $A_+^{-1}(x', x_n, 0, 0) = 1$ and $A_+^{-1}(x', x_n, 0, \varepsilon\xi_n)$ has an expansion of form (27.37) for $\varepsilon\xi_n \to 0$ implies that

$$A_+^{-1}(x', x_n, 0, \varepsilon\xi_n) = A_+^{-1}(x', 0, 0, \varepsilon\xi_n) + B_+(x', x_n, \varepsilon\xi_n),$$

where

$$B_+(x', x_n, \varepsilon\xi_n) = x_n(\varepsilon\xi_n B_1^+(x', x_n, \varepsilon\xi_n) + \varepsilon\xi_n \ln(\varepsilon\xi_n + i0)B_2^+(x', x_n, \varepsilon\xi_n))$$

and

$$\operatorname{ord}_t B_k^+(x', x_n, t) < \operatorname{ord}_t A_+^{-1}(x', 0, t) - 1 + \delta, \qquad \forall\delta > 0,\quad k = 1, 2.$$

Moreover, since $x_n D_n u_0$ and $x_n D_n \ln(D_n + i0)u_0$ belong to $\mathring{H}_{r_+-\delta}(G)$, it follows that

$$\langle B_+(x', x_n, \varepsilon D_n)u_0\rangle_{\kappa_1(x),r_+-\delta} \leqslant C_\delta\varepsilon^{1-\delta}, \qquad \forall\delta > 0.$$

Thus

$$u_\varepsilon = A_+^{(-1)}(x', 0, 0, \varepsilon D_n)u_0 + u_\varepsilon^{(4)}, \tag{27.64}$$

where

$$\langle u_\varepsilon^4\rangle_{\kappa_1(x),r_+-\delta} \leqslant C_\delta\varepsilon^{1-\delta}, \qquad \forall\delta > 0. \tag{27.65}$$

Suppose $\operatorname{Re}\kappa_1(x') > -1$ and $\operatorname{Re}\kappa_0(x') > -1$. In this case it is possible to represent the principal asymptotic term of the solution u_ε in a simpler form (as a product of two functions). Let $\kappa_0(x') = k_0 + \gamma(x')$, where k_0 is a nonnegative integer and $1 > \operatorname{Re}\gamma(x') > -1$. The asymptotic behavior of $u_0(x)$ near the boundary takes the following form in an s.l.c.s. (see §26):

$$u_0(x) = c_0(x')x_{n,+}^{k_0+\gamma(x')} + \sum_{k=1}^{N}\sum_{p=0}^{k} c_{pk}(x')x_{n,+}^{k_0+k+\gamma(x')}\ln^p x_n$$

$$+ u_{0N}(x) = x_{n,+}^{\gamma(x')}\sum_{k=0}^{N} w_k(x)\ln^k x_n + u_{0N}(x), \tag{27.66}$$

where $c_0(x') \in C^\infty(\Gamma)$, $c_{pk}(x') \in C^\infty(\Gamma)$, $u_{0N}(x) \in C^N(\overline{G})$, $x_{n,+}^\gamma = x_n^\gamma$ for $x_n > 0$ and $x_{n,+}^\gamma = 0$ for $x_n < 0$, $w_k(x) \in C^\infty(\overline{G})$ and $w_k = O(x_n^k)$.

Let

$$a(x', x_n) = \frac{1}{2\pi}\int_{-\infty}^{\infty} A_+^{-1}(x', 0, 0, \xi_n)e^{-ix_n\xi_n}\,d\xi_n, \tag{27.67}$$

where the inverse Fourier transform is taken according to definition (2.28). Then

$$\frac{1}{2\pi}\int_{-\infty}^{\infty} A_+^{-1}(x', 0, 0, \varepsilon\xi_n)e^{-ix_n\xi_n}\,d\xi_n = \frac{1}{\varepsilon}a\left(x', \frac{x_n}{\varepsilon}\right). \tag{27.68}$$

Since $A_+(x', 0, 0, \varepsilon\xi_n) \in C^\infty$ in x' and $\varepsilon\xi_n$, $A_+(x', 0, 0, 0) = 1$ and $A_+(x', 0, 0, \varepsilon\xi_n)$ admits an expansion of form (27.36) for $\varepsilon\xi_n \to \infty$ with principal term $(\varepsilon\xi_n + i0)^{\kappa_1(x')}$, we get that $a(x', x_n)$ has the following properties: $a(x', x_n) \in C^\infty$ for $x' \in \Gamma$, $x_n > 0$, $|x_n^k a(x', x_n)| \leqslant C_k$, $\forall k$, $a(x', x_n) = 0$ for $x_n < 0$ and

$$\frac{1}{\varepsilon}a\left(x', \frac{x_n}{\varepsilon}\right) \to \delta(x_n) \qquad \text{as } \varepsilon \to 0, \tag{27.69}$$

$$a\left(x', \frac{x_n}{\varepsilon}\right) = O\left(\left(\frac{x_n}{\varepsilon}\right)^{\kappa_1(x')-1}\right) \qquad \text{as } \frac{x_n}{\varepsilon} \to 0, \quad x_n > 0. \tag{27.70}$$

In addition, (27.68) implies

$$\frac{1}{\varepsilon}\int_0^{x_n} a\left(x', \frac{y_n}{\varepsilon}\right) dy_n = \int_0^{x_n/\varepsilon} a(x', t)\,dt \to \int_0^\infty a(x', t)\,dt = 1 \quad \text{as } \frac{x_n}{\varepsilon} \to +\infty, \tag{27.71}$$

since

$$\int_0^\infty a(x', t)\,dt = \int_{-\infty}^\infty a(x', t)e^{it\xi_n}\,dt\Big|_{\xi_n=0} = A_+(x', 0, 0, \varepsilon\xi_n)\big|_{\xi_n=0} = 1.$$

While the representation of a p.o. as a convolution (cf. §3) implies

$$\begin{aligned} A_+^{(-1)}(x', 0, 0, \varepsilon D_n)u_0(x) &= \int_0^{x_n} \frac{1}{\varepsilon}a\left(x', \frac{y_n}{\varepsilon}\right)u_0(x', x_n - y_n)\,dy_n \\ &= \int_0^{x_n} \frac{1}{\varepsilon}a\left(x', \frac{y_n}{\varepsilon}\right)\left(\sum_{k=0}^{N}(x_n - y_n)^{\gamma(x')}\ln^k(x_n - y_n)w_k(x', x_n - y_n)\right. \\ &\qquad \left. + u_{0N}(x', x_n - y_n)\right) dy_n. \end{aligned} \tag{27.72}$$

Let

$$a_1\left(x', x_n, \frac{x_n}{\varepsilon}\right) = \frac{1}{\varepsilon}\int_0^{x_n} a\left(x', \frac{y_n}{\varepsilon}\right)(x_n - y_n)^{\gamma(x')}\,dy_n. \tag{27.73}$$

We note (see (2.34)) that

$$a_1\left(x', x_n, \frac{x_n}{\varepsilon}\right) = \frac{1}{\varepsilon} a\left(x', \frac{x_n}{\varepsilon}\right) * x_{n,+}^{\gamma(x')}$$
$$= \frac{1}{2\pi}\int_{-\infty}^{\infty} \frac{\left[\exp i\frac{\pi}{2}(\gamma(x')+1)\right]\Gamma(\gamma(x')+1)e^{-ix_n\xi_n}}{A_+(x', 0, 0, \varepsilon\xi_n)(\xi_n + i0)^{\gamma(x')+1}}\, d\xi_n. \tag{27.73'}$$

Putting first $y_n = \varepsilon\tau$ and then $y_n = x_n t$ in (27.73), we obtain the representations

$$a_1\left(x', x_n, \frac{x_n}{\varepsilon}\right) = \int_0^{x_n/\varepsilon} a(x', \tau)(x_n - \varepsilon\tau)^{\gamma(x')}\, d\tau$$
$$= \frac{x_n^{\gamma(x')+1}}{\varepsilon}\int_0^1 a\left(x', \frac{x_n}{\varepsilon}t\right)(1-t)^{\gamma(x')}\, dt, \tag{27.74}$$

which together with (27.70) and (27.71) imply

$$a_1\left(x', x_n, \frac{x_n}{\varepsilon}\right) \to x_n^{\gamma(x')} \qquad \text{as } x_n/\varepsilon \to \infty,\quad x_n > 0, \tag{27.75}$$

and

$$a_1\left(x', x_n, \frac{x_n}{\varepsilon}\right) = O\left(x_n^{\gamma(x')}\cdot\left(\frac{x_n}{\varepsilon}\right)^{\kappa_1(x')}\right) \qquad \text{for } x_n/\varepsilon \to 0,\quad x_n > 0. \tag{27.76}$$

Taking now the Taylor expansions of $w_k(x', x_n - y_n)$ and $u_{0N}(x', x_n - y_n)$ in powers of $-y_n$, we get from (27.72) that

$$u_\varepsilon = a_1\left(x', x_n, \frac{x_n}{\varepsilon}\right)\cdot\frac{u_0(x', x_n)}{x_n^{\gamma(x')}} + u_\varepsilon^{(5)}, \tag{27.77}$$

where (as can be proved analogously to (27.75) and (27.76))

$$u_\varepsilon^{(5)}(x', x_n, \varepsilon) = O(\varepsilon^{1-\delta}x_n^{\kappa_0(x')}) \qquad \text{for } x_n/\varepsilon \to \infty,\quad x_n > 0,\quad \forall\delta > 0, \tag{27.78}$$

$$u_\varepsilon^{(5)}(x', x_n, \varepsilon) = O\left(\left(\frac{x_n}{\varepsilon}\right)^{\kappa_1(x')} x_n^{\kappa_0(x')+1-\delta}\right) \quad \text{for } x_n/\varepsilon \to 0,\quad x_n > 0,\quad \forall\delta > 0, \tag{27.79}$$

which implies that $u_\varepsilon^{(5)}$ is a lower order term and

$$\langle u_\varepsilon^{(5)}\rangle_{\kappa_1(x'), r_+ - \delta} < C_\delta\varepsilon^{1-\delta}, \qquad \forall\delta > 0. \tag{27.80}$$

Thus the solution of (27.38) can be represented in the form (27.77), where $a_1(x', x_n, x_n/\varepsilon)$ is given by (27.73) and $u_\varepsilon^{(5)}$ satisfies (27.78)–(27.80). we note that $a_1(x', x_n, x_n/\varepsilon) \to x_n^{\gamma(x')}$ as $x_n/\varepsilon \to +\infty$. Therefore, in the interior of G, $u_\varepsilon(x) \to u_0(x)$ as $\varepsilon \to 0$; but in the boundary layer where x_n/ε is small, the behavior of u_ε and u_0 are different. The function $a_1(x', x_n, x_n/\varepsilon)$ describes (modulo the lower order asymptotic terms) the transition from u_0 to u_ε near the boundary.

When $\kappa_0(x')$ is an integer, i.e. $\gamma(x') \equiv 0$, (27.77) takes the simpler form

$$u_\varepsilon = a_1(x', x_n/\varepsilon) \cdot u_0(x', x_n) + u_\varepsilon^{(5)}, \tag{27.77'}$$

where

$$a_1(x', x_n/\varepsilon) = \int_0^{x_n/\varepsilon} a(x', \tau)\, d\tau$$

and $u_\varepsilon^{(5)}$ is the same as in (27.78)–(27.80). In this case $a_1(x', x_n/\varepsilon) \to 1$ as $x_n/\varepsilon \to +\infty$ and

$$a_1(x', x_n/\varepsilon) = O\big((x_n/\varepsilon)^{\kappa_1(x')}\big) \qquad \text{for } x_n/\varepsilon \to 0.$$

6. *Examples of pseudodifferential equations with a small parameter.*

EXAMPLE 27.1. *Elliptic differential equations with a small parameter.* Consider the differential equation

$$(\varepsilon^2\Delta^2 - \Delta)u_\varepsilon = f, \qquad x \in G, \tag{27.81}$$

with boundary conditions

$$u_\varepsilon|_\Gamma = 0, \qquad \frac{\partial u_\varepsilon}{\partial n}\Big|_\Gamma = 0, \tag{27.82}$$

where $\partial/\partial n$ is the normal derivative.

In the notation of subsection 1 we have

$$A_0(x, \xi) = |\xi|^2, \qquad A_1(x, \varepsilon\xi) = \varepsilon^2|\xi|^2 + 1, \qquad A(x, \xi, \varepsilon) = A_0(x, \xi)A_1(x, \xi)$$

and $\kappa_0 = 1$, $\kappa_1 = 1$. Since $A_1(\varepsilon\xi) = \varepsilon^2|\xi|^2 + 1$ is invariant under an orthogonal transformation, the symbol $A_+^{(-1)}(x', 0, 0, \varepsilon\xi_n)$ is equal to $i(\varepsilon\xi_n + i)^{-1}$, where

$$i^{-1}(\varepsilon\xi_n + i) \cdot i(\varepsilon\xi_n - i) = \varepsilon^2\xi_n^2 + 1$$

is the factorization of $A_1(0, \varepsilon\xi_n)$ in the global coordinate system. Therefore

$$a(x', x_n) = \frac{1}{2\pi}\int_{-\infty}^{\infty} \frac{i}{\xi_n + i} e^{-ix_n\xi_n}\, d\xi_n = e^{-x_n}$$

and

$$a_1(x', x_n/\varepsilon) = \int_0^{x_n/\varepsilon} a(x', r)\, dr = \int_0^{x_n/\varepsilon} e^{-r}\, dr = 1 - e^{-x_n/\varepsilon}.$$

By Theorem 27.1, for each $f \in H^{-1,-1}(G)$ there exists a unique solution of (27.81) in $\mathring{H}^{1,1}(G)$. If $f \in C^\infty(\overline{G})$, then

$$u_\varepsilon \in \mathring{H}^{3/2-\delta,3/2-\delta}(G) \cap H^{s;3/2-\delta}(G), \qquad \forall\delta > 0, \quad \forall s \geqslant 3/2,$$

and u_ε has the following asymptotic representation for $\varepsilon \to 0$:

$$u_\varepsilon(x', x_n) = (1 - e^{-x_n/\varepsilon}) \cdot u_0(x) + w_\varepsilon(x), \tag{27.83}$$

where $u_0(x)$ is the solution of the Dirichlet problem

$$-\Delta u_0 = f, \qquad u_0|_\Gamma = 0 \tag{27.84}$$

and $\langle \omega_\varepsilon \rangle_{1,1} < C\varepsilon$.

Problem (27.81), (27.82) is a particular case of the following problem considered by Višik and Ljusternik (see [110]):

$$\left(L_{2r_0}(x, D) + \sum_{k=1}^{r} \varepsilon^{2k} L_{2r_0+2k}(x, D)\right) u_\varepsilon = f, \tag{27.85}$$

$$u_\varepsilon|_\Gamma = \frac{\partial u_\varepsilon}{\partial n}\bigg|_\Gamma = \dots = \frac{\partial^{r_0+r-1} u_\varepsilon}{\partial n^{r_0+r-1}}\bigg|_\Gamma = 0, \tag{27.86}$$

where $L_{2r_0+2k}(x, D)$ is a differential operator of order $2r_0 + 2k$. Let $L^{(0)}_{2r_0+2k}(x, \xi)$ denote the principal part of $L_{2r_0+2k}(x, \xi)$. It is assumed (as in subsection 1) that

$$L^{(0)}_{2r_0}(x, \xi) \neq 0 \qquad \text{for } x \in \overline{G}, \quad \xi \neq 0 \tag{27.87}$$

(i.e. $L_{2r_0}(x, D)$ is an elliptic operator), that

$$L^{(0)}_{2r_0+2r}(x, \xi) \neq 0 \qquad \text{for } x \in \overline{G}, \quad \xi \neq 0, \tag{27.88}$$

and that

$$A_1(x, \varepsilon\xi) = 1 + \sum_{k=1}^{r} \varepsilon^{2k} \frac{L^{(0)}_{2r_0+2k}(x, \xi)}{L^{(0)}_{2r}(x, \xi)} \neq 0 \tag{27.89}$$

for all $x \in \overline{G}$, $\varepsilon \geqslant 0$ and $\xi \neq 0$.

For this example $m_0 = 2r_1$, $m_1 = 2r_0 + 2r$ and $\kappa_0 = r_0$. Since $A_1(x, 0, \varepsilon\xi_n)$ is a polynomial in $\varepsilon\xi_n$ of degree $2r$, it follows that the factorization index of A_1 in G is equal to r, i.e. $\kappa_1 = r$, and $A_+(x', 0, 0, \varepsilon\xi_n)$ is a polynomial in $\varepsilon\xi_n$ of degree r.

The following theorem is a consequence of Theorem 27.1 and the asymptotic representation (27.77′).

THEOREM 27.2. *Suppose conditions* (27.87)–(27.89) *are satisfied and for each* $f \in H_{-r_0}(G)$ *the limiting problem*

$$L_{2r_0}(x, D)u_0 = f, \tag{27.90}$$

$$u_0|_\Gamma = 0, \dots, \frac{\partial^{r_0-1} u_0}{\partial n^{r_0-1}}\bigg|_\Gamma = 0 \tag{27.91}$$

has a unique solution $u_0 \in \mathring{H}_{r_0}(G)$. *Then for sufficiently small* ε *problem* (27.85), (27.86) *has a unique solution* $u_\varepsilon \in \mathring{H}^{r,r_0}(G)$. *If* $f \in C^\infty(\overline{G})$, *then*

$$u_\varepsilon \in \mathring{H}^{r+1/2-\delta, r_0+1/2-\delta}(G) \cap H^{s, r_0+1/2-\delta}(G), \qquad \forall \delta > 0, \quad \forall s \geqslant r + \tfrac{1}{2},$$

and u_ε has the representation

$$u_\varepsilon(x', x_n) = a_1(x', x_n/\varepsilon) \cdot u_0(x) + w_\varepsilon, \tag{27.92}$$

where

$$a_1\left(x', \frac{x_n}{\varepsilon}\right) = \int_0^{x_n/\varepsilon} a(x', y_n)\, dy_n, \qquad a(x', x_n) = F_{\xi_n}^{-1} A_+^{-1}(x', 0, 0, \xi_n)$$

and $\langle w_\varepsilon \rangle_{r+s, r_0} \leqslant C_s \varepsilon$, $\forall s \geqslant 0$.

The fact that $A_+(x', 0, 0, \varepsilon\xi_n)$ is a polynomial in $\varepsilon\xi_n$ means that one can easily calculate $a(x', x_n)$ and $a_1(x', x_n/\varepsilon)$. In particular, we have $a_1(x', x_n/\varepsilon) = 1 + O(e^{-cx_n/\varepsilon})$ for some $c > 0$ since $a_1(x', x_n/\varepsilon)$ is a sum of exponential polynomials in x_n/ε.

EXAMPLE 27.2. Consider the following equation which appears in the inverse problems of classical field theory (see [58] and [26]):

$$\varepsilon^2 u_\varepsilon(x) + \frac{1}{4\pi} \int_G \frac{1}{|x - y|} u_\varepsilon(y)\, dy = f(x), \tag{27.93}$$

where $G \subset \mathbf{R}^3$, $(1/4\pi)\int_G |x - y|^{-1} u(y)\, dy$ is a Newtonian potential (see (3.15″)) and ε^2 is the dielectric constant.

Here

$$A(x, \xi, \varepsilon) = \varepsilon^2 + \frac{1}{|\xi|^2} = \frac{\varepsilon^2|\xi|^2 + 1}{|\xi|^2},$$

so that $A_0 = 1/|\xi|^2$, $A_1 = \varepsilon^2|\xi|^2 + 1$, $A = A_0 A_1$, $\kappa_1 = 1$ and $\kappa_0 = -1$. we note that A_1 is the same as for (27.81). Therefore

$$A_+^{-1}(x', 0, 0, \varepsilon\xi_n) = i/(\varepsilon\xi_n + i).$$

The limiting equation is

$$\frac{1}{4\pi} \int_G \frac{1}{|x - y|} u_0(y)\, dy = f(x). \tag{27.94}$$

It follows from Example 22.3 for $\alpha = -2$ and $n = 3$ that (27.94) has a unique solution $u_0 \in \mathring{H}_{-1}(G)$ for each $f \in H_1(G)$. Suppose $f \in H_{2+r}(G)$ and $r \geqslant 0$. Then Examples 23.2 (see (23.27)) and 22.3 show that the solution $u_0 \in \mathring{H}_{-1}(G)$ has the following representation in an s.l.c.s.:

$$u_0(x', x_n) = u_0^{(1)}(x) + v_0^{(1)}(x')\delta(x_n), \tag{27.95}$$

where $u^{(1)}(x) \in H_r(G)$, $v_0^{(1)}(x') \in H_{r+3/2}(\Gamma)$, $u_0^{(1)} = u^{(1)}$ for $x \in G$, $u_0^{(1)} = 0$ for $x \notin G$ and $\delta(x_n)$ is the delta function.

By repeating the proof of Theorem 27.1 and using the fact that A_0, A_1, $A_+^{(-1)}$ and $A_-^{(-1)}$ are of class $\hat{D}^\infty$ (see §23), it can be shown that (27.93) has a unique solution

$$u_\varepsilon \in \mathring{H}^{3/2-\delta,-1/2-\delta}(G) \cap H^{r+5/2+\delta,-1/2-\delta}(G), \qquad \forall\delta > 0.$$

Suppose for simplicity that $f \in C^\infty(\overline{G})$. Then (27.64) and (27.95) imply

$$\begin{aligned} u_\varepsilon &= A_+^{(-1)}(x', 0, 0, \varepsilon D_n)u_0(x) + w_\varepsilon(x) \\ &= \varepsilon^{-1}e^{-x_n/\varepsilon} * \left(u_0^{(1)} + v_0^{(1)}\delta(x_n)\right) + w_\varepsilon(x) \\ &= (1 - e^{-x_n/\varepsilon}) \cdot u_0^{(1)} + \varepsilon^{-1}e^{-x_n/\varepsilon}v_0^{(1)}(x') + w_\varepsilon^{(1)}(x), \end{aligned} \tag{27.96}$$

where

$$\langle w_\varepsilon^{(1)}\rangle_{1,-1} \leqslant C\varepsilon. \tag{27.97}$$

Consider now the more general case

$$pA(x, D, \varepsilon)u_\varepsilon = f, \tag{27.98}$$

in which

$$A(x, \xi, \varepsilon) = A_1(x, \xi/|\xi|, \varepsilon|\xi|)A_0(x, \xi),$$

where A_0 and A_1 satisfy all of the conditions of subsection 1 and, in addition,

$$A_0(x, \xi) \in \hat{D}^\infty_{m_0}, \qquad A_1(x, \xi/|\xi|, |\xi|) \in \hat{D}^\infty_{m_1}$$

and $\kappa_0 < 0$, $\kappa_1 \geqslant 0$. we note that κ_0 and κ_1 are integers inasmuch as the factorization indices of the symbols of class $\hat{D}^\infty$ are integers. Since $A_1 \in \hat{D}^\infty_{m_1}$ and $\kappa_1 \geqslant 0$, we find that $A_+(x', 0, 0, \varepsilon\xi_n)$ is a polynomial in $\varepsilon\xi_n$ of degree κ_1, and hence that $a(x', x_n)$ and $a_1(x', x_n/\varepsilon)$ have the same properties as $a(x', x_n)$ and $a_1(x', x_n/\varepsilon)$ in Example 27.1.

Suppose the limiting equation

$$pA_0(x, D)u_0 = f \tag{27.99}$$

has a unique solution $u_0 \in \mathring{H}_{\kappa_0}(G)$ for $f \in H_{\kappa_0 - m_0}(G)$, and suppose $f \in H_{r-m_0}(G)$. It follows from Example 23.2 and from the uniqueness of the solution of (27.99) in $\mathring{H}_{\kappa_0}(G)$ that, in an s.l.c.s.,

$$u_0 = u_0^{(1)} + \sum_{k=1}^{|\kappa_0|} v_k(x')\delta^{(k-1)}(x_n), \tag{27.100}$$

where $u^{(1)} \in H_r(G)$, $v_k(x') \in H_{r+k+1/2}(\Gamma)$, $u_0^{(1)} = u^{(1)}$ for $x \in G$ and $u_0^{(1)} = 0$ for $x \notin G$.

As before, we find that for small $\varepsilon > 0$ there exists a unique solution of (27.98) that belongs to

$$\mathring{H}^{\kappa_1+1/2-\delta,\kappa_0+1/2-\delta}(G) \cap H^{r+m_1-\kappa_0-1/2+\delta,\kappa_0+1/2-\delta}(G), \qquad \forall\delta > 0,$$

and has the following asymptotic expansion in an s.l.c.s.:

$$\begin{aligned}u_\varepsilon(x) &= A_+^{(-1)}(x', 0, 0, \varepsilon D_n)u_0(x) + w_\varepsilon \\ &= \frac{1}{\varepsilon}a\left(x', \frac{x_n}{\varepsilon}\right) * \left(u_0^{(1)} + \sum_{k=1}^{|\kappa_0|} v_k(x')\delta^{(k-1)}(x_n)\right) + w_\varepsilon \\ &= a_1\left(x', \frac{x_n}{\varepsilon}\right)\cdot u_0^{(1)}(x) + \sum_{k=1}^{|\kappa_0|}\frac{i^{k-1}}{\varepsilon^k}a^{(k-1)}\left(x', \frac{x_n}{\varepsilon}\right)v_k(x') + w_\varepsilon^{(1)},\end{aligned} \tag{27.101}$$

where

$$a^{(k)}(x', t) = \frac{\partial^k a(x', t)}{\partial t^k} \qquad \text{and} \qquad \langle w_\varepsilon^{(1)}\rangle_{\kappa_1+1/2-\delta, \kappa_0+1/2-\delta} < C_\delta \varepsilon.$$

We note that w_ε and $w_\varepsilon^{(1)}$ are of order ε in Examples 27.1 and 27.2 (and not of order $\varepsilon^{1-\delta}$, $\forall\delta$, as in Theorem 27.1), since the symbols in these examples satisfy the condition of Remark 27.2. Therefore $\ln\varepsilon$ does not appear in the asymptotic expansions.

EXAMPLE 27.3. *Punch indenting a thin layer*. As in §26.11, consider a punch indenting a layer of thickness 2ε resting on a rigid foundation. Suppose that the domain of contact is given and that the shearing strains are zero. Then the pressure under the base of the punch satisfies the following equation (see [113], [11] and (26.120)):

$$pA(\varepsilon D)u_\varepsilon = (2/\varepsilon)f(x), \tag{27.102}$$

where

$$A(\varepsilon\xi) = \frac{4\sinh^2\varepsilon|\xi|}{\varepsilon|\xi|(2\varepsilon|\xi| + \sinh 2\varepsilon|\xi|)}$$

(we include the elasticity constants in f).

Here $A_0(\xi) = 1$, $A_1(\varepsilon\xi) = A(\varepsilon\xi)$, $\kappa_0 = 0$ and $\kappa_1 = -\frac{1}{2}$. Since $A_1(\varepsilon|\xi|)$ depends only on $|\xi|$, the function $A_+(x', 0, 0, \varepsilon\xi_n) = A_+(\varepsilon\xi_n)$ does not depend on x' and can be calculated by factoring the function

$$A_1(0, \varepsilon|\xi_n|) = \frac{4\sinh^2\varepsilon|\xi_n|}{\varepsilon|\xi_n|(2\varepsilon|\xi_n| + \sinh 2\varepsilon|\xi_n|)}.$$

In this example we cannot find $A_+(\varepsilon\xi_n)$ in explicit form and must therefore use an approximate method (cf. Example 17.2). Suppose for simplicity that the domain of contact does not depend on ε and that $f \in C^\infty(\overline{G})$. Since $A_0(\xi) = 1$, the limiting equation is $pu_0 = f$, while the unique solution $u_\varepsilon \in \mathring{H}^{-1/2,0}(G)$ of (27.102) has the form

$$u_\varepsilon = 2\varepsilon^{-1}(a_1(x_n/\varepsilon)\cdot f_0(x', x_n) + w_\varepsilon), \tag{27.103}$$

where $f_0 = f$ for $x \in G$, $f_0 = 0$ outside G,

$$a_1(x_n/\varepsilon) = \int_0^{x_n/\varepsilon} a(t)\,dt, \qquad a(t) = \frac{1}{2\pi}\int_{-\infty}^{\infty} \frac{e^{-it\xi_n}}{A_+(\xi_n)}\,d\xi_n$$

and $\langle w_\varepsilon \rangle_{-1/2,0} \leqslant C_\delta \varepsilon^{1-\delta}$. The function $a_1(x_n/\varepsilon)$ can also be written in the form

$$a_1\left(\frac{x_n}{\varepsilon}\right) = \frac{1}{2\pi}\exp i\frac{\pi}{4}\cdot\sqrt{\pi}\int_{-\infty}^{\infty} \frac{e^{-ix_n\xi_n}\,d\xi_n}{A_+(\xi_n)(\xi_n + i0)^{1/2}}$$

(see (27.73)).

We note that there are other mixed problems in the theory of elasticity that reduce to p.e. with a small parameter (e.g. a crack in a thin layer).

7. *Asymptotic expansion of the solution of a pseudodifferential equation with a small parameter*. For brevity we will call a summand of the form $\varepsilon^k B_k(x, \xi)$ a *regular summand* and a summand of the form $\varepsilon^k B_k(x, \varepsilon\xi_n) b_k(\xi')$ a *singular summand*. As was shown in Lemma 27.3, a product of a singular and a regular summand can be represented as a sum of regular and singular summands.

It was noted in subsection 4 that

$$A_-^{(-1)} A(x, D, \varepsilon) A_+^{(-1)} = A_0 + A_3$$

(see (27.53)), where $\|A_3 v_\varepsilon\|_r \leqslant C_\delta \varepsilon^{1-\delta}\|v_\varepsilon\|_r$, $\forall r$, $\forall \delta > 0$. Applying Lemma 27.2 to $A_+^{(-1)}$ and $A_-^{(-1)}$, Theorem 18.3 to the product $A_-^{(-1)} A A_+^{(-1)}$, and Lemma 27.3 to A_3, we get

$$\begin{aligned} &A_3(x, \xi, \varepsilon) \\ &= A^{(0)}(x, \xi)\left(\sum_{k=1}^{N}\sum_{p=0}^{k} \varepsilon^k \ln^p \varepsilon A_{3kp}(x, \varepsilon\xi_n) a_{3k}(\xi') + \sum_{k=1}^{N}\sum_{p=0}^{k} \varepsilon^k \ln^p \varepsilon A_{4kp}(x, \xi)\right) \\ &\quad + \varepsilon^{N+1-\delta} R_{\delta,N}(x, \xi, \varepsilon). \end{aligned} \tag{27.104}$$

where the order of $R_{\delta,N}$ with respect to ξ_n is less than or equal to m_0, $\operatorname{ord}_{\xi_n} A_{4kp} \leqslant 0$, $\operatorname{ord}_t A_{3kp}(x, t) \leqslant -1 + \delta$, $\forall \delta > 0$, and $A^{(0)}(x, \xi)$ is the principal part of $A_0(x, \xi)$. we wish to add lower order terms to $A_+^{(-1)}$ and $A_-^{(-1)}$ so as to eliminate the singular summands in (27.104), i.e. we wish to find $A_{+,1}, \ldots, A_{+,N}$ and $A_{-,1}, \ldots, A_{-,N}$ such that

$$\begin{aligned} &(A_-^{(-1)} + \varepsilon A_{-,1} + \ldots + \varepsilon^N A_{-,N}) A(x, D, \varepsilon)(A_+^{(-1)} + \varepsilon A_{+,1} + \ldots + \varepsilon^N A_{+,N}) \\ &\qquad = A_0(x, D) + \sum_{k=1}^{N}\sum_{p=0}^{k} \varepsilon^k \ln^p \varepsilon^k A_{5kp}(x, D) + \varepsilon^{N+1-\delta} R_{N1\delta}(x, D, \varepsilon), \end{aligned} \tag{27.105}$$

where $\operatorname{ord}_{\xi_n} A_{5k}(x, \xi) \leqslant m_0$, $\operatorname{ord}_{\xi_n} R_{N1\delta}(x, \xi, \varepsilon) \leqslant m_0$ and $\operatorname{ord}_t A_{\pm,k} \leqslant \operatorname{ord}_t A_{\pm} - 1 + \delta$.

Let

$$S(x, \xi', \varepsilon\xi_n, \varepsilon) = \sum_{k=1}^{N} \sum_{p=0}^{k} \varepsilon^k \ln^p \varepsilon A_{3kp}(x, \varepsilon\xi_n) a_{3k}(\xi'). \tag{27.106}$$

By making the change of variables $\varepsilon\xi_n = \zeta_n$, it is easily seen that $S_+(x, \xi', \varepsilon\xi_n, \varepsilon) = \Pi^+ S$ and $S_-(x, \xi, \varepsilon) = \Pi^- S$ are also sums of singular summands of form (27.106):

$$S_{\pm}(x, \xi', \varepsilon\xi_n, \varepsilon) = \sum_{k=1}^{N} \sum_{p=0}^{k} \varepsilon^k \ln^p \varepsilon A^{\pm}_{3kp}(x, \varepsilon\xi_n) a_{3k}(\xi'), \tag{27.107}$$

where $\operatorname{ord}_t A^{\pm}_{3kp}(x, t) \leqslant -1 + \delta$, $\forall \delta > 0$, and the $A^{\pm}_{3kp}(x, t)$ have expansions of form (27.36) for $t \to \infty$ and expansions of form (27.37) for $t \to 0$. Now let

$$\varepsilon A_{-,1}(x, \xi', \varepsilon\xi_n, \varepsilon) = S_-(x, \xi', \varepsilon\xi_n, \varepsilon) A_-^{-1}(x', 0, 0, \varepsilon\xi_n), \tag{27.108}$$

$$\varepsilon A_{+,1}(x, \xi', \varepsilon\xi_n, \varepsilon) = S_+(x, \xi', \varepsilon\xi_n, \varepsilon) A_+^{-1}(x', 0, 0, \varepsilon\xi_n). \tag{27.109}$$

Then $\operatorname{ord}_t A_{\pm,1} \leqslant \operatorname{ord}_t A_{\pm} - 1 + \delta$, $\forall \delta > 0$, while from Theorem 18.3 and Lemma 27.2 we obtain the relation

$$\left(A_-^{(-1)} + \varepsilon A_{-,1}\right) A(x, D, \varepsilon) \left(A_+^{(-1)} + \varepsilon A_{+,1}\right) = A_0 + A_4, \tag{27.110}$$

in which

$$A_4(x, \xi, \varepsilon) = A_0(x, \xi)(S_1(x, \xi', \varepsilon\xi_n, \varepsilon) + r_1(x, \xi, \varepsilon)) + \varepsilon^{N+1-\delta} R_{N2\delta}(x, \xi, \varepsilon), \tag{27.111}$$

where $R_{N2\delta}$ has the same properties as $R_{N1\delta}(x, \xi, \varepsilon)$, $r_1(x, \xi, \varepsilon)$ is a sum of regular summands analogous to the second sum in (27.104) and $S_1(x, \xi', \varepsilon\xi_n, \varepsilon)$ is a sum of singular summands of the form

$$S_1(x, \xi', \varepsilon\xi_n, \varepsilon) = \sum_{k=2}^{N} \sum_{p=0}^{k} \varepsilon^k \ln^p \varepsilon A_{6pk}(x, \varepsilon\xi_n)\, a_{6k}(\xi'). \tag{27.112}$$

We note that if we have a product $B(x, \varepsilon\xi_n) b(\xi)$, where $\operatorname{ord}_{\xi_n} b \leqslant -k_0$, and $B(x, t)$ has the asymptotic behavior (27.36), (27.37) with respect to t, then Lemmas 27.3 and 27.3′ imply that the singular part of $B(x, \varepsilon\xi_n) b(\xi)$ in the expansion of form (27.19) will be of order k_0 in ε, i.e. $O(\varepsilon^{k_0})$. Therefore the product of a regular lower order term and a singular term increases the exponent of ε in the expansion of form (27.19). For other types of products the increase in the exponent of ε is obvious.

Thus, continuing the procedure by next defining

$$\varepsilon^2 A_{\pm,2}(x, \xi, \varepsilon\xi_n, \varepsilon) = A_{\pm}^{-1}(x', 0, 0, \varepsilon\xi_n) \Pi^{\pm} S_1,$$

and so on, we arrive at (27.105).

An asymptotic expansion of the solution of (27.38) can now be obtained by looking for the solution in the form

$$u_\varepsilon = \left(A_+^{(-1)} + \varepsilon A_{+,1} + \dots + \varepsilon^N A_{+,N}\right) v_\varepsilon. \tag{27.113}$$

Extending (27.38) onto $\mathbf{R}^n$ (see (27.48)) and applying the operator $p(A_-^{(-1)} + \varepsilon A_{-,1} + \dots + \varepsilon^N A_{-,N})$, we get as in subsection 4,

$$p\left(A_0 + \sum_{k=1}^{N} \sum_{p=0}^{k} \varepsilon^k \ln^p \varepsilon A_{5kp} + \varepsilon^N R_{N1}\right) v_\varepsilon = p\left(A_-^{(-1)} + \dots + \varepsilon^N A_{-,N}\right) lf. \tag{27.114}$$

Further, expanding $A_-^{(-1)}, \dots, A_{-,N}$ in ε by means of Lemmas 27.1 and 27.3′, we get

$$p\left(A_-^{(-1)} + \dots + \varepsilon^N A_{-,N}\right) lf = f + \sum_{k=1}^{N} \sum_{p=0}^{k} \varepsilon^k \ln^p \varepsilon f_{kp0}(x) + \varepsilon^{N+1-\delta} f_{N\delta}(x, \varepsilon), \tag{27.115}$$

where $f_{kp0}(x) \in C^\infty(\overline{G})$, $\forall \delta > 0$, and $\|f_{N\delta}(x, \varepsilon)\|_s^+ \leqslant C_s$, $\forall s$ (cf. (27.61)).

The problem of solving (27.114) is a simple regular perturbation problem. Since (27.42) is uniquely solvable, we have (cf. (27.56))

$$v_\varepsilon = u_0 + \sum_{k=1}^{N} \sum_{p=0}^{k} \varepsilon^k \ln^p \varepsilon v_{kp}(x) + v_\varepsilon^N, \tag{27.116}$$

where $\|v_\varepsilon^N\|_{r_+ - \delta} \leqslant C_\delta \varepsilon^{N+1-\delta}$, $\forall \delta > 0$, and the $v_{kp}(x)$ are solutions of the equations

$$pA_0 v_{kp} = f_{kp} - \sum_{r+r'=k} \sum_{q+q'=p} A_{5rq} v_{r'q'}, \tag{27.117}$$

in which $v_{00} = u_0$, $0 \leqslant q \leqslant r$ and $0 \leqslant q' \leqslant r'$.

We note that the functions u_0 and $v_{kp}(x)$ have asymptotic expansions of form (27.66) near the boundary. In §26.2 such an expansion was proved for the solution of (27.117) when the right side $g(x)$ is a C^∞ function in $\overline{G}$. But the same proof is also valid for the case when $g(x)$ has an asymptotic expansion of form (27.66). Therefore, the representation of form (27.35) using for $A_+^{(-1)}, A_{+,k}$, we get

$$\begin{aligned} u_\varepsilon &= \left(A_+^{(-1)} + \dots + \varepsilon^N A_{+,N}\right) v_\varepsilon \\ &= A_+^{(-1)}(x', x_n, 0, \varepsilon D_n)\left(u_0 + \sum_{k=1}^{N} \sum_{p=0}^{k} \varepsilon^k \ln^p \varepsilon v_{kp}(x)\right) \\ &\quad + \sum_{k=1}^{N} \sum_{p=0}^{k} \varepsilon^k \ln^p \varepsilon A_{6pk}^{+}(x', x_n, \varepsilon D_n) v_{kp2}(x) + u_\varepsilon^N, \end{aligned} \tag{27.118}$$

where the $v_{kp1}(x)$ and $v_{kp2}(x)$ have asymptotic expansions of form (27.66),

$$\langle u_\varepsilon^N \rangle_{\kappa_1(x), r_+ - \delta} \leqslant C_\delta \varepsilon^{N+1-\delta}, \qquad \forall \delta > 0, \tag{27.119}$$

$$\operatorname{ord}_t A^+_{6pk}(x', x_n, t) A_+^{-1}(x', x_n, 0, t) \leqslant -1 + \delta, \qquad \forall \delta > 0,$$

and the $A_{6pk}(x', x_n, t)$ have asymptotic expansions of forms (27.36) and (27.37) for $t \to \infty$, $t \to 0$ respectively.

We now expand $A_+^{(-1)}(x', x_n, \varepsilon\xi_n)$ and $A^+_{6pk}(x', x_n, \varepsilon\xi_n)$ in powers of x_n by Taylor's formula.

Consider, for example, a term of the form

$$B^+(x', \varepsilon D_n)(x_n^k v_0), \tag{27.120}$$

where v_0 has an expansion of form (27.66), $k > 0$,

$$\operatorname{ord}_t B^+(x', t) A_+^{-1}(x', 0, 0, t) \leqslant 0$$

and $B^+(x', t)$ has the asymptotic behaviors (27.36) and (27.37) for $t \to \infty$ and $t \to 0$.

Taking the asymptotic expansion of $B^+(x, \varepsilon\xi_n) A_+^{-1}(x', 0, 0, \varepsilon\xi_n)$ for $\varepsilon\xi_n \to 0$ up to terms of order $k - 1$ in ε, we find (as in the proofs of Lemmas 27.3 and 27.3′) that

$$\begin{aligned} &B^+(x', \varepsilon D_n)(x_n^k v_0) \\ &\quad = A_+^{(-1)}(x', 0, 0, \varepsilon D_n)\left(b_+(x' x_n) + \sum_{l=0}^{k-1} \sum_{p=0}^{l} \varepsilon^l \ln^p \varepsilon b^+_{lp}(x', x_n)\right) \\ &\quad + \varepsilon^{k-1} B^+_{k-1}(x', \varepsilon D_n) w_0, \end{aligned} \tag{27.121}$$

where b_+, b^+_{lp} and w_0 have an expansion of form (27.66). we note that, since

$$A_+^{-1}(x', x_n, 0, 0) = A_+^{-1}(x', 0, 0, 0) = 1,$$

the Taylor expansion of $[A_+^{(-1)}(x', x_n, 0, \varepsilon D_n) - A_+^{(-1)}(x', 0, 0, \varepsilon D_n)]u_0$ in powers of x_n has $b_+(x', x_n) = 0$ in the expression of form (27.121) (cf. (27.64)). Therefore we finally get

$$\begin{aligned} u_\varepsilon &= A_+^{(-1)}(x', 0, 0, \varepsilon D_n)\left(u_0 + \sum_{k=1}^{N} \sum_{p=0}^{k} \varepsilon^k \ln^p \varepsilon w_{kp1}(x)\right) \\ &\quad + \sum_{k=1}^{N} \sum_{p=0}^{k} \varepsilon^k \ln^p \varepsilon A^+_{7pk}(x', \varepsilon D_n) w_{kp2}(x) + w_\varepsilon^N, \end{aligned} \tag{27.122}$$

where w_{kp1}, w_{kp2}, w_ε^N and $A^+_{7pk}(x', \varepsilon\xi_n)$ have the same properties as v_{kp1}, v_{kp2}, u_ε^N and $A^+_{6pk}(x', x_n, \varepsilon\xi_n)$ in (27.118).

When Re $\kappa_0(x) > -1$ and Re $\kappa_1(x) > -1$ as in subsection 5, we can represent u_ε in the form (see (27.72) and (27.77))

$$u_\varepsilon = a_1\left(x', x_n, \frac{x_n}{\varepsilon}\right) \cdot \frac{u_0(x', x_n)}{x_n^{\gamma(x)}}$$

$$+ \sum_{k=1}^{N} \sum_{p=0}^{k} \varepsilon^k \ln^p \varepsilon a_{pk}\left(x', x_n, \frac{x_n}{\varepsilon}\right) \frac{w_{kp}^{(0)}(x)}{x_n^{\gamma(x)}} + w_{\varepsilon,N}.$$

where the $a_{pk}(x', x_n, x_n/\varepsilon)$ behave analogously to $a_1(x', x_n, x_n/\varepsilon)$, the $\omega_{kp}^{(0)}(x)$ have the asymptotic expansion (27.66) and $\omega_{\varepsilon,N}$ has the estimate of form (27.119).

It follows from (27.122) that u_ε together with all of its derivatives with respect to x tends to u_0 uniformly on each compact set $K \subset G$ as $\varepsilon \to 0$.

BIBLIOGRAPHY

Items marked by an asterisk were added for the English translation

1. *Shmuel Agmon, *Lectures on elliptic boundary value problems*, Van Nostrand, Princeton, N. J., 1965.

2. S. Agmon, A. Douglis and L. Nirenberg, *Estimates near the boundary for solutions of elliptic partial differential equations satisfying general boundary conditions*. I, Comm. Pure Appl. Math. **12** (1959), 623–727.

3. ______, *Estimates near the boundary for solutions of elliptic partial differential equations satisfying general boundary conditions*. II, Comm. Pure Appl. Math. **17** (1964), 35–92.

4. M. S. Agranovič, *Elliptic singular integro-differential operators*, Uspehi Mat. Nauk **20** (1965), no. 5 (125), 3–120 = Russian Math. Surveys **20** (1965), no. 5, 1–121.

5. M. S. Agranovič and A. S. Dynin, *General boundary-value problems for elliptic systems in an n-dimensional domain*, Dokl. Akad. Nauk SSSR **146** (1962), 511–514 = Soviet Math. Dokl. **3** (1962), 1323–1327.

6. M. S. Agranovič and M. I. Višik, *Elliptic problems with a parameter and parabolic problems of general type*, Uspehi Mat. Nauk **19** (1964), no. 3(117), 53–161 = Russian Math. Surveys **19** (1964), no. 3, 53–157.

7. M. F. Atiyah, *K-theory*, Lecture Notes, Harvard Univ., Cambridge, Mass., 1964; reprint, Benjamin, New York, 1968.

8. M. F. Atiyah and R. Bott, *The index problem for manifolds with boundary*, Differential Analysis (Bombay Colloq., 1964), Oxford Univ. Press, London, 1964, pp. 175–186.

9. *Jean-Pierre Aubin, *Approximation of elliptic boundary-value problems*, Wiley, New York, 1972.

10. K. I. Babenko, *On conjugate functions*, Dokl. Akad. Nauk SSSR **62** (1948), 157–160. (Russian)

11. *V. A. Babeško, *Asymptotic solution of some two-dimensional convolutions, and their applications*, Dokl. Akad. Nauk SSSR **204** (1972), 572–574 = Soviet Phys. Dokl. **17** (1972/73), 504–506.

12. Ju. M. Berezanskiĭ, *Expansion in eigenfunctions of selfadjoint operators*, "Naukova Dumka", Kiev, 1965; English transl., Amer. Math. Soc., Providence, R. I., 1968.

13. A. V. Bicadze, *Boundary value problems for second order elliptic equations*, "Nauka", Moscow, 1966; English transl., North-Holland, Amsterdam; Interscience, New York, 1968.

14. *A. Bogomolnii [Bogomol'nyĭ], G. Èskin and S. Zuchowizkii [Zuhovickiĭ], *Numerical solution of the stamp problem*, Comput. Methods Appl. Mech. Engrg. **15** (1978), 241–258.

15. Louis Boutet de Monvel, *Comportement d'un opérateur pseudo-différentiel sur une variété à bord*. I, II, J. Analyse Math. **17** (1966), 241–253, 255–304.

16. ______, *Opérateurs pseudo-différentiels analytiques et problèmes aux limites elliptiques*, Ann. Inst. Fourier (Grenoble) **19** (1970), fasc. 2, 169–268.

17. ______, *Boundary problems for pseudo-differential operators*, Acta Math. **126** (1971), 11–51.

18. Felix E. Browder, *Estimates and existence theorems for elliptic boundary value problems*, Proc. Nat. Acad. Sci. USA **45** (1959), 365–372.

19. A. P. Calderón and A. Zygmund, *On the existence of certain singular integrals*, Acta Math. **88** (1952), 85–139.

20. ______, *Singular integral operators and differential equations*, Amer. J. Math. **79** (1957), 901–921.

21. Čan Zuĭ Ho [Chan Zui Kho] and G. I. Èskin, *Boundary value problems for parabolic systems of pseudodifferential equations*, Dokl. Akad. Nauk SSSR **198** (1971), 50–53 = Soviet Math. Dokl. **12** (1971), 739–743.

22. H. O. Cordes, *An algebra of singular integral operators with two symbol homomorphisms*, Bull. Amer. Math. Soc. **75** (1969), 37–42.

23. ______, *Pseudo-differential operators on a half-line*, J. Math. Mech. **18** (1968/69), 893–908.

24. *T. A. Cruse, *Application of the boundary-integral equation method to three dimensional stress analysis*, Computers and Structures **3** (1973), 509–527.

25. *A. S. Demidov, *Boundary effects and the asymptotic nature of the degeneration of certain elliptic pseudodifferential operators*, Uspehi Mat. Nauk **27** (1972), no. 1(163), 245–246. (Russian)

26. *______, *Asymptotic behavior of the solution of a boundary value problem for elliptic pseudodifferential equations with a small parameter multiplying the leading operator*, Trudy Moskov. Mat. Obšč. **32** (1975), 119–146 = Trans. Moscow Math. Soc. **32** (1975), 115–142.

27. *______, *Elliptic pseudodifferential boundary value problems with a small parameter in the coefficient of the leading operator*, Mat. Sb. **91(133)** (1973), 421–444 = Math. USSR Sb. **20** (1973), 439–463.

28. A. S. Dikanskiĭ, *Problems adjoint to elliptic pseudodifferential boundary value problems*, Dokl. Akad. Nauk SSSR **200** (1971), 1020–1023 = Soviet Math. Dokl. **12** (1971), 1520–1525.

29. *______, *Conjugate problems of elliptic differential and pseudodifferential boundary value problems in a bounded domain*, Mat. Sb. **91(133)** (1973), 62–77 = Math. USSR Sb. **20** (1973), 67–83.

30. *Nelson Dunford and Jacob T. Schwartz, *Linear operators*. Vol. I, Interscience, New York, 1958.

31. A. S. Dynin, *Singular operators of arbitrary order on a manifold*, Dokl. Akad. Nauk SSSR **141** (1961), 21–23 = Soviet Math. Dokl. **2** (1961), 1375–1377.

32. ______, *Multidimensional elliptic boundary value problems with a single unknown function*, Dokl. Akad. Nauk SSSR **141** (1961), 285–287 = Soviet Math. Dokl. **2** (1961), 1431–1433.

33. ______, *On the theory of pseudodifferential operators on a manifold with boundary*, Dokl. Akad. Nauk SSSR **186** (1969), 251–253 = Soviet Math. Dokl. **10** (1969), 575–578.

34. G. I. Èskin, *Boundary value problems and the parametrix for systems of elliptic pseudodifferential equations*, Trudy Moskov. Mat. Obšč. **28** (1973), 75–116 = Trans. Moscow Math. Soc. **28** (1973), 74–115. (See also Preprint no. 11, Institute for Problems of Mechanics, Acad. Sci. USSR, Moscow, 1972. (Russian))

35. ______, *General boundary value problems for equations of principal type in a plane region with corners*, Uspehi Mat. Nauk **18** (1963), no. 3(111), 241–242. (Russian)

36. ______, *Generalized functions and pseudodifferential operators*, Preprint no. 5, Institute for Problems of Mechanics, Acad. Sci. USSR, Moscow, 1971. (Russian)

37. ______, *Boundary value problems in a half space for elliptic pseudodifferential equations*, Preprint no. 7, Institute for Problems of Mechanics, Acad. Sci. USSR, Moscow, 1971. (Russian)

38. ______, *The conjugacy problem for equations of principal type with two independent variables*, Trudy Moskov. Mat. Obšč. **21** (1970), 245–292 = Trans. Moscow Math. Soc. **21** (1970), 263–316.

39. *______, *A variational-difference method for solving elliptic pseudodifferential equations*, Uspehi Mat. Nauk **28** (1973), no. 5(173), 255–256. (Russian)

40. *______, *The numerical solution of some pseudo-differential equations of elasticity by finite element methods*, Sém. Inst. Recherche Informat. Automat. (Paris, 1977) (to appear).

41. *______, *Asymptotics of solutions of elliptic pseudodifferential equations with a small parameter*, Dokl. Akad. Nauk SSSR **211** (1973), 547–550 = Soviet Math. Dokl. **14** (1973), 1080–1084.

42. *______, *Asymptotics near the boundary of spectral functions of elliptic self-adjoint boundary problems*, Israel J. Math. **22** (1975), 214–246.

43. *______, *General mixed problems for elliptic differential equations*, Centro Internaz. Mat. Estivo, II Ciclo (Bressanone, 1977), Napoli, 1977.

44. K. O. Friedrichs, *Pseudo-differential operators*, Lecture Notes, Courant Inst. Math. Sci., New York Univ., New York, 1968; rev. ed., 1970.

45. F. D. Gahov, *Boundary value problems*, Fizmatgiz, Moscow, 1958; English transl., Pergamon Press, Oxford; Addison-Wesley, Reading, Mass., 1966.

46. *L. A. Galin, *Contact problems in the theory of elasticity*, GITTL, Moscow, 1953; English transl., North Carolina State College, Raleigh, N. C., 1961.

47. F. R. Gantmaher, *The theory of matrices*, 2nd ed., "Nauka", Moscow, 1968; English transl. of 1st ed., vols. I, II, Chelsea, New York, 1959.

48. Lars Gårding, *Transformation de Fourier des distributions homogènes*, Bull. Soc. Math. France **89** (1961), 381–428.

49. I. M. Gel'fand and G. E. Šilov, *Generalized functions*. Vols. 1, 2, 3, Fizmatgiz, Moscow, 1958; English transl., Academic Press, New York, 1964, 1967, 1968.

50. I. C. Gohberg and I. A. Fel'dman, *Convolution equations and projection methods for their solution*, "Nauka", Moscow, 1971; English transl., Amer. Math. Soc., Providence, R. I., 1974.

51. I. C. Gohberg and M. G. Kreĭn, *Systems of integral equations on a half line with kernels depending on the difference of the arguments*, Uspehi Mat. Nauk **13** (1958), no. 2(80), 3–72; English transl., Amer. Math. Soc. Transl. (2) **14** (1960), 217–287.

52. I. C. Gohberg and N. Ja. Krupnik, *The algebra generated by the one-dimensional singular integral operators with piecewise continuous coefficients*, Funkcional. Anal. i Priložen. **4** (1970), no. 3, 26–36 = Functional Anal. Appl. **4** (1970), 193–201.

53. *R. V. Gol'dšteĭn, I. S. Kleĭn and G. I. Èskin, *A variational-difference method for solving certain integral and integro-differential equations of three-dimensional problems of elasticity theory*, Preprint no. 33, Institute for Problems of Mechanics, Acad. Sci. USSR, Moscow, 1973. (Russian) RŽ Mat. **1974** 5Б1073.

54. Lars Hörmander, *Linear partial differential operators*, Springer-Verlag, Berlin; Academic Press, New York, 1963.

55. ______, *Pseudo-differential operators*, Comm. Pure Appl. Math. **18** (1965), 501–517.

56. ______, *Pseudo-differential operators and hypoelliptic equations*, Proc. Sympos. Pure Math., vol. 10, Amer. Math. Soc., Providence, R. I., 1967, pp. 138–183.

57. ______, *Fourier integral operators*. I, Acta Math. **127** (1971), 79–183.

58. *D. Ivanenko and A. Sokolov, *The classical theory of fields* (*new problems*), 2nd ed., GITTL, Moscow, 1951; German transl. of 1st ed., Akademie-Verlag, Berlin, 1953.

59. J. J. Kohn and L. Nirenberg, *An algebra of pseudo-differential operators*, Comm. Pure Appl. Math. **18** (1965), 269–305.

60. M. G. Kreĭn, *Integral equations on a half line with kernels depending on the difference of the arguments*, Uspehi Mat. Nauk **13** (1958), no. 5 (83), 3–120; English transl., Amer. Math. Soc. Transl. (2)**22** (1962), 163–288.

61. *Jacques-Louis Lions, *Quelques méthodes de résolution des problèmes aux limites non linéaires*, Dunod; Gauthier-Villars, Paris, 1969.

62. J.-L. Lions and E. Magenes, *Problèmes aux limites non homogènes et applications*. Vols. 1, 2, 3, Dunod, Paris, 1968, 1970; English transl., Springer-Verlag, Berlin and New York, 1972, 1973.

63. Ja. B. Lopatinskiĭ, *A method of reducing boundary value problems for a system of differential equations of elliptic type to regular integral equations*, Ukrain. Mat. Ž. **5** (1953), no. 2, 123–151; English transl., Amer. Math. Soc. Transl. (2)**89** (1970), 149–183.

64. *A. I. Lur'e, *Elasticity theory*, "Nauka", Moscow, 1970. (Russian)

65. S. G. Mihlin, *Multidimensional singular integrals and integral equations*, Fizmatgiz, Moscow, 1962; English transl., Pergamon Press, New York, 1965.

66. ______, *Variational methods in mathematical physics*, 2nd rev. aug. ed., "Nauka", Moscow, 1970; English transl. of 1st ed., Macmillan, New York, 1964.

67. N. I. Mushelišvili, *Singular integral equations*, 3rd ed., "Nauka", Moscow, 1968; English transl. of 1st ed., Noordhoff, Groningen, 1953; reprinted, 1972.

68. *J. C. Nédélec, *Curved finite element methods for the solution of singular integral equations on surfaces in* R^3, Comput. Methods Appl. Mech. Engrg. **8** (1976), 61–80.

69. *L. Nirenberg, *Pseudo-differential operators*, Proc. Sympos. Pure Math., vol. 16, Amer. Math. Soc., Providence, R. I., 1970, pp. 149–167.

70. B. Noble, *Methods based on the Wiener-Hopf technique for the solution of partial differential equations*, Pergamon Press, New York, 1958.

71. O. A. Oleĭnik and E. V. Radkevič, *Equations of second order with nonnegative characteristic form*, Itogi Nauki: Mat. Anal., 1969, VINITI, Moscow, 1971; English transl., Plenum Press, New York, 1973.

72. R. E. A. C. Paley and Norbert Wiener, *Fourier transforms in the complex domain*, Amer. Math. Soc., Providence, R. I., 1934.

73. Jaak Peetre, *Mixed problems for higher order elliptic equations in two variables*. I, Ann. Scuola Norm. Sup. Pisa (3) **15** (1961), 337–353.

74. ______, *Another approach to elliptic boundary problems*, Comm. Pure Appl. Math. **14** (1961), 711–731.

75. Z. Ja. Šapiro, *On general boundary value problems for equations of elliptic type*, Izv. Akad. Nauk SSSR Ser. Mat. **17** (1953), 539–562. (Russian)

76. Martin Schechter, *General boundary value problems for elliptic partial differential equations*, Comm. Pure Appl. Math. **12** (1959), 457–486.

77. L. Schwartz, *Théorie des distributions*. Tomes 1, 2, Actualités Sci. Indust., nos. 1089, 1122, Hermann, Paris, 1950, 1951.

78. R. T. Seeley, *Integro-differential operators on vector bundles*, Trans. Amer. Math. Soc. **117** (1965), 167–204.

79. Eliahu Shamir, *Elliptic systems of singular integral operators*. I, Trans. Amer. Math. Soc. **127** (1967), 107–124.

80. ______, *Mixed boundary value problems for elliptic equations in the plane. The* L_p *theory*, Ann. Scuola Norm. Sup. Pisa (3) **17** (1963), 117–139.

81. *______, *Regularization of mixed second-order elliptic problems*, Israel J. Math. **6** (1968), 150–168.

82. *______, *Boundary value problems for elliptic convolution systems*, Centro Internaz. Mat. Estivo, II Ciclo (Stresa, 1968), Edizione Cremonese, Rome, 1969, pp. 308–331.

83. G. E. Šilov, *Mathematical analysis. Special course*, 2nd ed., Fizmatgiz, Moscow, 1961; English transl., Pergamon Press, New York, 1965.

84. ______, *Mathematical analysis. Second special course*, "Nauka", Moscow, 1965; English transl., *Generalized functions and partial differential equations*, Gordon & Breach, New York, 1968.

85. I. B. Simonenko, *A new general method for investigating linear operator equations of singular integral equation type*. I, II, Izv. Akad. Nauk SSSR Ser. Mat. **29** (1965), 567–586, 757–782. (Russian)

86. L. N. Slobodeckiĭ, *Generalized Sobolev spaces and their application to boundary value problems for partial differential equations*, Leningrad. Gos. Ped. Inst. Učen. Zap. **197**(1958), 54–112; English transl., Amer. Math. Soc. Transl. (2)**57** (1966), 207–276.

87. A. I. Šnirel'man, *Convolution equations in a half space*, Mat. Sb. **82(124)** (1970), 476–493 = Math. USSR Sb. **11** (1970), 441–458.

88. S. L. Sobolev, *Applications of functional analysis in mathematical physics*, Izdat. Leningrad. Gos. Univ., Leningrad, 1950; English transl., Amer. Math. Soc., Providence, R. I., 1963.

89. Norman Steenrod, *The topology of fibre bundles*, Princeton Univ. Press, Princeton, N. J., 1951.

90. Elias M. Stein, *Note on singular integrals*, Proc. Amer. Math. Soc. **8** (1957), 250–254.

91. *______, *Singular integrals and differentiability properties of functions*, Princeton Univ. Press, Princeton, N. J., 1970.

92. *Gilbert Strang and George J. Fix, *An analysis of the finite element method*, Prentice-Hall, Englewood Cliffs, N. J., 1973.

93. *______, *A Fourier analysis of the finite element variational method*, Centro Internaz. Mat. Estivo II Ciclo (Erice, 1971), Edizione Cremonese, Rome, 1973, pp. 793–840.

94. M. A. Šubin, *On the index of families of Wiener-Hopf operators*, Mat. Sb. **84(126)** (1971), 537–558 = Math. USSR Sb. **13** (1971), 529–551.

95. ______, *Factorization of matrices depending on a parameter, and elliptic equations in a half space*, Mat. Sb. **85(127)** (1971), 65–84 = Math. USSR Sb. **14** (1971), 65–84.

96. E. C. Titchmarsh, *Introduction to the theory of Fourier integrals*, Clarendon Press, Oxford, 1937.

97. B. R. Vaĭnberg and V. V. Grušin, *Uniformly nonelliptic problems*. I, II, Mat. Sb. **72(114)** (1967), 602–636; **73(115)** (1967), 126–154 = Math. USSR Sb. **1** (1967), 543–568; **2** (1967), 111–134.

98. I. N. Vekua, *New methods for solving elliptic equations*, OGIZ, Moscow, 1948; English transl., North-Holland, Amsterdam; Interscience, New York, 1967.

99. N. P. Vekua, *Systems of singular integral equations and some boundary value problems*, 2nd rev. ed., "Nauka", Moscow, 1970; English transl. of 1st ed., Noordhoff, Groningen, 1967.

100. M. I. Višik and G. I. Èskin, *Boundary value problems for general singular integral equations in a bounded domain*, Dokl. Akad. Nauk SSSR **155** (1964), 24–27 = Soviet Math. Dokl. **5** (1964), 325–329.

101. ______, *Convolution equations in a bounded domain*, Uspehi Mat. Nauk **20** (1965), no. 3(123), 89–152 = Russian Math. Surveys **20** (1965), no. 3, 85–151.

102. ______, *Elliptic convolution equations in a bounded domain and their applications*, Uspehi Mat. Nauk **22** (1967), no. 1(133), 15–76 = Russian Math. Surveys **22** (1967), no. 1, 13–75.

103. ______, *Convolution equations of variable order*, Trudy Moskov. Mat. Obšč. **16** (1967), 25–50 = Trans. Moscow Math. Soc. **1967**, 27–52.

104. ______, *Mixed boundary value problems for elliptic systems of differential equations*, Thbilis. Sahelmç Univ. Gamogeneb. Math. Inst. Šrom. **2** (1969), 31–48. (Russian)

105. ______, *Variable order Sobolev-Slobodeckiĭ spaces with weighted norms and their application to mixed boundary value problems*, Sibirsk. Mat. Ž. **9** (1968), 973–997 = Siberian Math. J. **9** (1968), 723–740.

106. ______, *Normally solvable problems for elliptic systems of convolution equations*, Mat. Sb. **74(116)** (1967), 326–356 = Math. USSR Sb. **3** (1967), 303–330.

107. *______, *Convolution equations in a bounded domain in spaces with weighted norms*, Mat. Sb. **69(111)** (1966), 65–110; English transl., Amer. Math. Soc. Transl. (2) **67** (1968), 33–82.

108. *______, *General boundary value problems with discontinuous boundary conditions*, Dokl. Akad. Nauk SSSR **158** (1964), 25–28 = Soviet. Math. Dokl. **5** (1964), 1154–1157.

109. *______, *Parabolic convolution equations in a bounded domain*, Mat. Sb. **71(113)** (1966), 162–190; English transl. in Amer. Math. Soc. Transl. (2) **95** (1970), 131–162.

110. *M. I. Višik and L. A. Ljusternik, *Regular degeneration and boundary layer for linear differential equations with a small parameter*, Uspehi Mat. Nauk **12** (1957), no. 5(77), 3–122; English transl., Amer. Math. Soc. Transl. (2) **20** (1962), 239–364.

111. V. S. Vladimirov, *Equations of mathematical physics*, 2nd rev. ed., "Nauka", Moscow, 1971; English transl. of 1st ed., Dekker, New York, 1971.

112. L. R. Volevič and B. P. Panejah, *Some spaces of generalized functions and imbedding theorems*, Uspehi Mat. Nauk **20** (1965), no. 1(121), 3–74 = Russian Math. Surveys **20** (1965), no. 1, 1–73.

113. I. I. Vorovič, V. M. Aleksandrov and V. A. Babeško, *Nonclassical mixed problems of elasticity theory*, "Nauka", Moscow, 1974. (Russian)

NOTATION INDEX

SUBJECT INDEX

ABCDEFGHIJ–CM–8987654321